Green Energy Choices:

THE BENEFITS, RISKS AND TRADE-OFFS OF LOW-CARBON TECHNOLOGIES FOR ELECTRICITY PRODUCTION

Foreword

Renewable energy is a cornerstone of a future of human prosperity without environmental sacrifice. The international community has recognized this. Through the Secretary-General's Sustainable Energy for All Initiative (SE4All), governments and stakeholders across the globe have demonstrated a commitment to ensuring universal access to affordable, reliable and modern energy services by 2030, while increasing the share of renewable energy in the global energy mix.

With this acknowledgment, the world community has a unique opportunity to steer investments over the next two decades towards energy systems that meet the demands of an increasing population while reducing greenhouse gas (GHG) emissions; water, air and soil pollution; and habitat loss.

This report from the International Resource Panel provides a comprehensive comparison of the GHG mitigation potential of various energy generation technologies, including hydro, solar, geothermal and wind. It also examines the environmental and human health impacts of these options, and their implications for resource use. Their impacts are compared with those of fossil fuels, including coal- and gas-fired power, with and without carbon capture and storage (CCS).

The report provides strong evidence that electricity generated from renewable sources causes substantially less pollution than that generated from fossil fuels. A business-as-usual expansion of fossil fuel-based generation would lead to increased pollution, with serious impacts on human health and the environment, and a doubling of GHG emissions by 2050. Meanwhile, renewable electricity generation produces only 5-6% of the GHG emissions of coal-fired power plants and 8-10% of those of gas-fired power plants.

The right mix of low-carbon electricity generation technologies will help to stabilise and potentially reduce pollution and impacts on the environment, including climate change and acidification. It is crucial to determine the optimal mix of these technologies, as well as policy objectives that will support these efforts. It is my hope that decision-makers will use the scientific evidence in this report to select the cleanest, safest and most sustainable mix of energy technologies for the coming decades.

Achim Steiner
UN Under-Secretary-General
UNEP Executive Director

Preface

Demand for energy is expected to double over the coming decades in order to meet the needs of a growing and developing global population. Responding to this demand will require significant investment over the next 20 years to develop and install new energy systems. With this challenge comes the opportunity to design systems and select technologies that will minimize adverse impacts on the environment, climate, and human health, as well as address the additional pressure on natural resources.

With this in mind, the International Resource Panel's experts have analysed nine key electricity generation technologies, including coal- and gas-fired power plants, technologies for solar power, hydropower, wind power, and geothermal. They examined their greenhouse gas (GHG) mitigation potential, and trade-offs in terms of environmental impacts, effects on human health, and the implications for natural resource use (including concrete, metals, energy, water and land). They also assessed the consequences of implementing the International Energy Agency (IEA) BLUE Map Scenario of a global energy mix consistent with limiting the average global temperature increase to 2°C.

The findings are crucial in terms of helping policy-makers choose appropriate mixes of energy technologies. The modelling carried out for the report found that during the life cycle of renewable energy technologies, GHG emissions are 5-6 per cent those of coal and 8-10 per cent those of natural gas fired power plants. Other damage to the environment from renewable energy technologies is 3-10 times lower than from fossil fuel based systems. Renewable energy systems also have considerable health benefits. Air pollution from renewable energy is around 10-30 per cent that of state-of-the-art fossil fuel power generation. Implementation of the BLUE Map scenario would therefore see electricity generation double while GHG emissions would fall by a factor of five, and human health, ecosystem and land use would all either stabilize or decline.

However, as the findings of the report demonstrate, there are potential trade-offs to the deployment of renewable sources, including in terms of land and water use, material use, and site-specific impacts which will need to be taken into account and minimized to the extent possible in the deployment of these technologies.

Green Energy Choices will be followed by a second report, following the same approach and methodology, but examining energy efficiency technologies, including for mobility, buildings and industry.

We would like to thank International Resource Panel Members Edgar Hertwich, lead author of this report, and Sangwon Suh for their vision and leadership in coordinating this extremely important body of work.

Janez Potočnik
Co-Chair, IRP

Ashok Khosla
Co-Chair, IRP

Contents

List of Acronyms and Abbreviations

1,4-DB	1,4-dichlorobenzene
2-BE	2-butoxyethanol
AC	alternating current
ACC	air-cooled condenser
AFC	alkaline fuel cell
AP	acidification potential
AR5	Fifth Assessment Report [IPCC]
a-Si	amorphous silicon
BFB	bubbling fluidized beds
bGHG	biogenic greenhouse gas
BLUE Map	scenarios [IEA ETP]
BOS	balance of system
bpd	barrels per day
C_3H_4	propane
C_4H_{10}	butane
CAES	compressed air energy storage
CBM	coal bed methane
CCS	carbon [dioxide] capture and storage
CDM	Clean Development Mechanism
CdS	cadmium sulfide
CdTe	cadmium telluride
CED	cumulative non-renewable energy demand
CFB	circulating fluidized beds
CH_4	methane
CHAT	cascaded humidified advanced turbine
CHOPS	cold heavy oil production with sand
CHP	combined heat and power
CHPDHC	combined heat and power district heating and cooling
CI	compression ignition
CIGS	copper indium gallium selenide
CIS	copper indium diselenide
CLFR	compact linear Fresnel reflectors
CLRTAP	Convention on Long-Range Transboundary Air Pollution
CO_2	carbon dioxide
COD	chemical oxygen demand
CSP	concentrating solar power
CSS	cyclic steam stimulation
DALY	Disability Adjusted Life Year
DC	direct current
DG	distributed generation of electricity
DHC	district heating and cooling
DIC	dissolved inorganic carbon

DMEA	dimethylethanolamine
DNI	direct normal irradiation
DPO	diphenyl oxide
EC	European Commission
ECBM	enhanced coalbed methane recovery
ecoinvent	[life cycle inventory database]
EEIOA	environmentally extended input-output analysis
EFLH	equivalent full load hours
EGS	engineered geothermal systems
EI	energy intensity
ELR	environmental loading ratio
EOL	end-of-life
EOR	enhanced oil recovery
EP	eutrophication potential
ET-DSP	electro-thermal dynamic stripping process
ETP	Energy Technology Perspectives [IEA]
EU	European Union
EUMENA	Europe, the Middle East and North Africa
EXIOBASE	[multiregional environmentally extended supply and use input-output database]
EXIOPOL	[environmental accounting framework project]
EXPC	sub-critical pulverized coal fired power plant
EXPC	existing pulverized coal
EYR	energy yield ratio
FAETP	freshwater aquatic ecotoxicity potential
FBC	fluidized bed combustion
FET	freshwater ecotoxicity
FEU	freshwater eutrophication
FOAK	first of a kind
GB	gravity-based foundation
GDP	Gross Domestic Product
GHD	gas hydrates deposit
GHG	greenhouse gas
GT	gas turbine
GWP	Global Warming Potential
GWP100	Global Warming Potential [100 years]
H_2CO_3	carbonic acid
H_2S	hydrogen sulfide
HCFC	hydrochlorofluorocarbon
HFC	hydrofluorocarbon
Hg^2+	oxidized mercury
HHV	higher heating value
HLCA	hybrid life cycle assessment
HPP	hydropower project
HPR	heat-to-power ratio
HRSG	heat recovery steam generator
HSTPT	hybrid solar thermal parabolic trough
HTF	heat transfer fluid
HTP	human toxicity potential
HVAC	high-voltage AC
HVDC	high-voltage direct current

I&P intermediate and peak

IAEA International Atomic Energy Agency

ICE internal combustion engine

IEA International Energy Agency

IEC International Electrotechnical Commission

IGA International Geothermal Association

IGCC integrated gasification combined cycle

IGIP initial gas in place

IO input-output

IPCC Intergovernmental Panel on Climate Change

IPCC SRREN Special Report on Renewable Energy

IRP International Resource Panel [UNEP]

ISCC integrated solar combined cycle

ISO independent system operator

ISO International Organization for Standardization

ITO independent transmission operator

KNO_3 potassium nitrate

LCA life cycle assessment

LCI life cycle inventory

LCOE levelised cost of electricity

LFR linear Fresnel reflector

LHTES latent heat thermal energy storage

LHV lower heating value

Li-ion lithium ion

LNG liquid natural gas

MAETP marine aquatic ecotoxicity potential

MCFC molten carbonate fuel cell

MD mineral depletion

MEA monoethanolamine

MG-silicon metallurgical grade silicon

MVA monitoring, verification and accounting

N_2O nitrous oxide

$NaNO_3$ sodium nitrate

NaOH sodium hydroxide

NCG non-condensable gases

NdFeB neodymium-iron-boron [alloy]

NEEDS New Energy Externalities Development for Sustainability [LCI database]

NGCC natural gas combined cycle

NH_3 ammonia

NiCd nickel cadmium

NiMH nickel metal hydride

NMVOC non-methane volatile organic compound

NOAK nth of a kind

NORM naturally occurring radioactive material

NO_x nitrogen oxides

NPP net primary production

O&M operation and maintenance

O_3 ozone

OECD Organization for Economic Cooperation and Development

PAFC phosphoric acid fuel cell

PAFS	potentially affected fraction of species
PbA	lead acid
PC	pulverized coal-fired
PCM	phase-change material
PEM	polymer electrolyte membrane
PEMFC	proton exchange membrane fuel cell
PFC	perfluorocarbon
PGIP	producible gas in place
PM	particulate matter
PO_4^{-3}	phosphate ion
POF	photochemical oxidant formation
Poly-Si	polycrystalline silicon
PSH	pumped storage hydro
PTC	parabolic trough collector
PV	photovoltaic
quantum dot	QD
ReCiPe	[model]
ribbon-Si	ribbon-silicon modules
SAGD	steam assisted gravity drainage
SCO	crude oil
SCPC	supercritical pulverized coal
SE4All	Sustainable Energy for All Initiative [UN Secretary General]
SF_6	sulfur hexafluoride
SHTES	sensible heat thermal energy storage
SI	spark ignition
SnO_2	tin oxide
SOFC	solid oxide fuel cell
SoG	solar grade silicon
SOR	steam-to-oil ratio
STE	solar thermal electricity
STLC	Soluble Threshold Limit Concentration
syngas	synthesis gas
TA	terrestrial acidification
TCLP	Toxicity Characteristic Leaching Procedure
TCO	transparent conducting oxide
TES	thermal energy storage
TETP	terrestrial ecotoxicity potential
TF	thin-film
TTLC	Total Threshold Limit Concentration
UCG	underground coal gasification
UNEP	United Nations Environment Programme
UNESCO	United Nations Educational, Scientific and Cultural Organization
UNFCCC	United Nations Framework Convention on Climate Change
VOC	volatile organic compound
WCD	World Commission on Dams
WHO	World Health Organization
WSG	water soluble gas
WT	wind turbine
ZnO	zinc oxide

Glossary

air pollution
The introduction into Earth's atmosphere of one or more substances (particulates, gases, biological molecules), or other harmful chemicals, materials or physical conditions (such as excess heat or noise) in high enough concentrations to cause harm to humans, other animals, vegetation or materials. Air pollution may come from anthropogenic or natural sources. (Wikipedia and UNFCCC)

albedo
The fraction of solar radiation reflected by a surface or object, often expressed as a percentage. Snow-covered surfaces have a high albedo, the surface albedo of soils ranges from high to low and vegetation-covered surfaces and oceans have a low albedo. The Earth's planetary albedo varies mainly through varying cloudiness, snow, ice, leaf area and land cover changes. (IPPC SYR Appendix)

anthropogenic emissions
Emissions of pollution associated with human activities, including the burning of fossil fuels, deforestation, land-use changes, livestock, fertilisation, etc. (IPPC SYR Appendix)

biomass
Renewable energy from living (or recently living) plants and animals, e.g. wood chippings, crops and manure. Plants store energy from the Sun while animals get their energy from the plants they eat. (IEA)

biomass
Organic material produced by living organisms. The quantity of biomass is expressed as a dry weight or as the energy, carbon or nitrogen content. (IPPC SYR Appendix)

carbon dioxide (CO_2)
A naturally occurring gas, also a by-product of burning fossil fuels from fossil carbon deposits, such as oil, gas and coal, of burning biomass and of land use changes and other industrial processes. It is the principal anthropogenic greenhouse gas (GHG) that affects the Earth's radiative balance. It is the reference gas against which other GHGs are measured and therefore has a Global Warming Potential (GWP) of 1. (IPPC SYR Appendix)

carbon [dioxide] capture and storage (CCS)
A process consisting of separation of carbon dioxide from industrial and energy-related sources, transport to a storage location and long-term isolation from the atmosphere. (IPPC SYR Appendix)

carbon dioxide equivalent
A metric measure used to compare the emissions of the different GHGs based upon their GWP. GHG emissions in the United States are most commonly expressed as "carbon equivalents," which are CO_2 equivalents measured in terms of the mass of carbon and not carbon dioxide. GWPs are used to convert GHGs to carbon dioxide equivalents. (UNFCCC)

Clean Development Mechanism (CDM)
Defined in Article 12 of the Kyoto Protocol, the CDM is intended to meet two objectives: (1) to assist parties not included in Annex I in achieving sustainable development and in contributing to the ultimate objective of the convention; and (2) to assist parties included in Annex I in achieving compliance with their quantified emission limitation and reduction commitments. Certified Emission Reduction Units from CDM projects undertaken in non-Annex I countries that limit or reduce GHG emissions, when certified by operational entities designated by Conference of the Parties/Meeting of the Parties, can be accrued to the investor (government

or industry) from parties in Annex B. A share of the proceeds from the certified project activities is used to cover administrative expenses as well as to assist developing country parties that are particularly vulnerable to the adverse effects of climate change to meet the costs of adaptation. (IPPC SYR Appendix)

climate change
Climate change refers to a change in the state of the climate that can be identified (e.g., by using statistical tests) by changes in the mean and/or the variability of its properties, and that persists for an extended period, typically decades or longer. Climate change may be due to natural internal processes or external forcings, or to persistent anthropogenic changes in the composition of the atmosphere or in land use. Note that the United Nations Framework Convention on Climate Change (UNFCCC), in its Article 1, defines climate change as: "a change of climate which is attributed directly or indirectly to human activity that alters the composition of the global atmosphere and which is in addition to natural climate variability observed over comparable time periods." The UNFCCC thus makes a distinction between climate change attributable to human activities altering the atmospheric composition, and climate variability attributable to natural causes. (IPPC SYR Appendix)

coal
Refers to a variety of solid, combustible, sedimentary, organic rocks that are composed mainly of carbon and varying amounts of other components such as hydrogen, oxygen, sulfur and moisture. Coal is formed from vegetation that has been consolidated between other rock strata and altered by the combined effects of pressure and heat over millions of years. Many different classifications of coal are used around the world, reflecting a broad range of ages, compositions and properties. (IEA)

co-generation (or combined heat and power, CHP)
The simultaneous generation of both electricity and heat from the same fuel, for useful purposes. The fuel varies greatly and can include coal, biomass, natural gas, nuclear material, the Sun or the heat stored in the Earth. (IEA)

concentrating solar power (CSP)
Devices that concentrate energy from the Sun's rays to heat a receiver to high temperatures. This heat is transformed first into mechanical energy (by turbines or other engines) and then into electricity. (IEA)

consumption
The use of products and services for (domestic) final demand, i.e. for households, government and investments. The consumption of resources can be calculated by attributing the life cycle-wide resource requirements to those products and services (e.g. by input-output calculation). (IRP)

ecosystem
A system of living organisms interacting with each other and their physical environment. The boundaries of what could be called an ecosystem are somewhat arbitrary, depending on the focus of interest or study. Thus, the extent of an ecosystem may range from very small spatial scales to, ultimately, the entire Earth. (IPPC SYR Appendix)

electricity generation
The total amount of electricity generated by power only or combined heat and power plants including generation required for own use. This is also referred to as gross generation. (IEA)

electricity production
The total amount of electricity generated by a power plant. It includes own-use electricity, as well as transmission and distribution losses. (IEA)

energy, geothermal
Heat transferred from the Earth's molten core to underground deposits of dry steam (steam with no water

droplets), wet steam (a mixture of steam and water droplets), hot water or rocks lying fairly close to the Earth's surface. (UNFCCC)

energy, heat
Heat is obtained from fuels combustion, nuclear reactors, geothermal reservoirs, capture of sunlight, exothermic chemical processes and heat pumps which can extract it from ambient air and liquids. It may be used for heating or cooling or converted into mechanical energy for transport vehicles or electricity generation. Commercial heat sold is reported under total final consumption with the fuel inputs allocated under power generation. (IEA)

energy, renewable
Energy that is derived from natural processes (e.g. sunlight and wind) that are replenished at a higher rate than they are consumed. Solar, wind, geothermal, hydro and biomass are common sources of renewable energy. (IEA)

energy, solar
Solar radiation exploited for hot water production and electricity generation by: flat plate collectors, mainly of the thermosyphon type, for domestic hot water or for the seasonal heating of swimming pools; photovoltaic cells; or, solar thermal-electric plants. (OECD)

energy
The amount of work or heat delivered. Energy is classified in a variety of types and becomes useful to human ends when it flows from one place to another or is converted from one type into another. Primary energy (also referred to as energy sources) is the energy embodied in natural resources (e.g., coal, crude oil, natural gas, uranium) that has not undergone any anthropogenic conversion. This primary energy needs to be converted and transported to become usable energy (e.g. light). Renewable energy is obtained from the continuing or repetitive currents of energy occurring in the natural environment, and includes non-carbon technologies such as solar energy, hydropower, wind, tide and waves, and geothermal heat, as well as carbon neutral technologies such as biomass. Embodied energy is the energy used to produce a material substance (such as processed metals, or building materials), taking into account energy used at the manufacturing facility (zero order), energy used in producing the materials that are used in the manufacturing facility (first order), and so on. (IPPC SYR Appendix)

eutrophication potential
An aggregate measure of the contribution of effluents to eutrophication. In this publication's impact assessment methods, phosphorus is treated as the limited nutrient for freshwater eutrophication and the freshwater eutrophication potential captures the contribution of different forms of phosphorus to freshwater eutrophication. Nitrogen is considered the limiting nutrient of marine ecosystems and the marine eutrophication potential captures the contribution of different forms of nitrogen to marine eutrophication.

fossil fuel combustion
Burning of coal, oil (including gasoline), or natural gas. The burning needed to generate energy release carbon dioxide by-products that can include unburned hydrocarbons, methane, and carbon monoxide. Carbon monoxide, methane, and many of the unburned hydrocarbons slowly oxidize into carbon dioxide in the atmosphere. Common sources of fossil fuel combustion include cars and electric utilities. (UNFCCC)

fossil fuels
Carbon-based fuels from fossil hydrocarbon deposits, including coal, peat, oil, and natural gas. (IPPC SYR Appendix)

fugitive emissions
Emissions as by-products or waste or loss in the process of fuel production, storage, or transport, such as methane given off during oil and gas drilling and refining, or leakage of natural gas from pipelines. (UNFCCC)

gas, natural
Underground deposits of gases consisting of 50—90% methane (CH_4) and small amounts of heavier gaseous hydrocarbon compounds such as propane (C_3H_4) and butane (C_4H_{10}). (UNFCCC)

gas, unconventional
Sources of gas trapped deep underground by impermeable rocks, such as coal, sandstone and shale. The three main types of "unconventional" gas are: shale gas (found in shale deposits); coalbed methane (extracted from coal beds), and tight gas (which is trapped underground in impermeable rock formations). While different techniques are applied, depending on the type of gas being extracted, one common method is known as hydraulic fracturing: large volumes of water (mixed with some sand and chemicals) are injected underground to create cracks in the rock, freeing the trapped gas so it can flow into the well bore created by the drill and be collected. Another key technology is horizontal drilling which enables the exposure of significantly more surface to the well. (IEA)

global warming potential (GWP)
An index, based upon radiative properties of well mixed GHGs, measuring the radiative forcing of a unit mass of a given well mixed GHG in today's atmosphere integrated over a chosen time horizon, relative to that of carbon dioxide. The GWP represents the combined effect of the differing times these gases remain in the atmosphere and their relative effectiveness in absorbing outgoing thermal infrared radiation. The Kyoto Protocol is based on GWPs from pulse emissions over a 100-year time frame. (IPPC SYR Appendix)

global warming
The observed increase of the global average temperature as a result of human and other activities, including through the increased concentration of GHGs such as CO_2 from energy. (IEA)

greenhouse effect
GHGs effectively absorb thermal infrared radiation, emitted by the Earth's surface, by the atmosphere itself due to the same gases, and by clouds. Atmospheric radiation is emitted to all sides, including downward to the Earth's surface. Thus GHGs trap heat within the surface-troposphere system. This is called the greenhouse effect. Thermal infrared radiation in the troposphere is strongly coupled to the temperature of the atmosphere at the altitude at which it is emitted. In the troposphere, the temperature generally decreases with height. Effectively, infrared radiation emitted to space originates from an altitude with a temperature of, on average, –19 °C, in balance with the net incoming solar radiation, whereas the Earth's surface is kept at a much higher temperature of, on average, +14 °C. An increase in the concentration of GHGs leads to an increased infrared opacity of the atmosphere, and therefore to an effective radiation into space from a higher altitude at a lower temperature. This causes a radiative forcing that leads to an enhancement of the greenhouse effect, the so-called enhanced greenhouse effect. (IPPC SYR Appendix)

greenhouse gas (GHG)
Any gas that absorbs infrared radiation in the atmosphere. Greenhouse gases include, but are not limited to, water vapor, carbon dioxide (CO_2), methane (CH_4), nitrous oxide (N_2O), hydrochlorofluorocarbons (HCFCs), ozone (O_3), hydrofluorocarbons (HFCs), perfluorocarbons (PFCs), and sulfur hexafluoride (SF_6). (UNFCCC)

heat
Form of kinetic energy that flows from one body to another when there is a temperature difference between the two bodies. Heat always flows spontaneously from a hot sample of matter to a colder sample of matter. This is one way to state the second law of thermodynamics. (UNFCCC)

hydropower
The electrical energy derived from turbines being spun by fresh flowing water. This can be from rivers or from man-made installations, where water flows from a high-level reservoir down through a tunnel and away from a dam. (IEA)

life cycle
Life cycle is a concept used to describe the environmental burden (resource requirements and environmental impacts) of products and services from the cradle to the grave, i.e. along the extraction-production-consumption-recycling-disposal chain. (IRP)

life cycle assessment (LCA)
Compilation and evaluation of the inputs, outputs and the potential environmental impacts of a product system throughout its life cycle. [IEC (ISO 14040:2006, definition 3.2)]

life cycle inventory (LCI)
The second step of LCA wherein extractions and emissions, the energy and raw materials used, and emissions to the atmosphere, water and land, are quantified for each process, then combined in the process flow chart and related to the functional basis. (UNEP)

low-carbon technologies oil
Technologies that produce low—or zero—GHG emissions while operating. In the power sector this includes fossil-fuel plants fitted with CCS, nuclear plants and renewable-based generation technologies. (IEA)

methane (CH4)
Methane is one of the six GHGs to be mitigated under the Kyoto Protocol. It is the major component of natural gas and associated with all hydrocarbon fuels. It is produced by anaerobic digestion of biomass and anthropogenic sources include animal husbandry, rice farming, and artificial water bodies in addition to fossil fuel systems. Coal-bed methane is the gas found in coal seams.

mitigation
In the context of climate change, a human intervention to reduce the sources or enhance the sinks of GHGs. Examples include using fossil fuels more efficiently for industrial processes or electricity generation, switching to solar energy or wind power, improving the insulation of buildings, and expanding forests and other “sinks” to remove greater amounts of CO_2 from the atmosphere. (UNFCCC)

nitrogen oxides (NO_x)
Gases consisting of one molecule of nitrogen and varying numbers of oxygen molecules. Nitrogen oxides are produced, for example, by the combustion of fossil fuels in vehicles and electric power plants. In the atmosphere, nitrogen oxides can contribute to formation of photochemical ozone (smog), impair visibility, and have health consequences; they are considered pollutants. (UNFCCC)

nitrous oxide (N_2O)
A powerful GHG with a GWP evaluated at 310. Major sources of nitrous oxide include soil cultivation practices, especially the use of commercial and organic fertilizers, fossil fuel combustion, nitric acid production and biomass burning. (UNFCCC)

oil
As defined by the IEA, includes crude oil, condensates, natural gas liquids, refinery feedstocks and additives, other hydrocarbons (including emulsified oils, synthetic crude oil, mineral oils extracted from bituminous minerals such as oil shale, bituminous sand and oils from coal-to-liquid and gas-to-liquid) and petroleum products (refinery gas, ethane, liquefied petroleum gas, aviation gasoline, motor gasoline, jet fuels, kerosene, gas/diesel oil, heavy fuel oil, naphtha, white spirit, lubricants, bitumen, paraffin waxes and petroleum coke). (IEA)

oil, shale oil
Underground formation of a fine-grained sedimentary rock containing varying amounts of kerogen, a solid, waxy mixture of hydrocarbon compounds. Heating the rock to high temperatures converts the kerogen to a vapor, which can be condensed to form a slow-flowing heavy oil called shale oil. (UNFCCC)

oil, unconventional oil
Includes oil shale, oil sands-based extra heavy oil and bitumen, derivatives such as synthetic crude products, and liquids derived from natural gas (gas-to-liquid or coal-to-liquid). (IEA)

photovoltaic (PV)
Directly convert solar energy into electricity using a photovoltaic cell; this is a semiconductor device. (IEA)

power
The rate of doing work, rate of electrical or mechanic energy flow.

power, electric
Electric energy produced by hydro-electric, geothermal, nuclear and conventional thermal power stations, excluding energy produced by pumping stations, measured by the calorific value of electricity (3.6 TJ/GWh). (OECD)

power, ocean
Energy available for recovery through different types of technologies that exploit the following phenomena: tidal rise and fall (barrages), tidal/ocean currents, waves, temperature gradients, and salinity gradients. (IEA)

radiative forcing
A change in the balance between incoming solar radiation and outgoing infrared (i.e., thermal) radiation. Without any radiative forcing, solar radiation coming to the Earth would continue to be approximately equal to the infrared radiation emitted from the Earth. The addition of GHGs to the atmosphere traps an increased fraction of the infrared radiation, reradiating it back toward the surface of the Earth and thereby creates a warming influence. Typically, radiative forcing is quantified at the tropopause in units of watts per square meter of the Earth's surface. (UNFCCC and Wikipedia)

TABLE 1

Resources for Definitions

Publisher	Publication	Link*
IEA	Glossary	http://www.iea.org/aboutus/glossary/
IEC	Electropedia	http://www.electropedia.org/
IPCC	"Glossary of Terms used in the IPCC Fourth Assessment Report"	https://www.ipcc.ch/publications_and_data/publications_and_data_glossary.shtml
IRP	Draft Glossary of Terms Used by the International Resource Panel	http://www.unep.org/resourcepanel/KnowledgeResources/GlossaryofTerms/tabid/133339/Default.aspx
OECD	Glossary of Statistical Terms	https://stats.oecd.org/glossary/search.asp
UNEP	Resource Efficiency, Consumption	http://www.unep.org/resourceefficiency/Consumption/StandardsandLabels/MeasuringSustainability/LifeCycleAssessment/tabid/101348/Default.aspx
UNFCCC	Glossary of climate change acronyms	http://unfccc.int/essential_background/glossary/items/3666.php
UNFCCC	Glossary for Greenhouse Gas Emissions Inventories	http://unfccc.int/resource/cd_roms/na1/ghg_inventories/english/8_glossary/Glossary.htm

**Definitions accessed in January 2016.*

©RogerWaleson/Freepik

Technical summary

Lead authors: by Edgar G. Hertwich, Thomas Gibon, Anders Arvesen, Peter Bayer, Evert Bouman, Joe Bergesen, Garvin Heath, Jacqueline Aloisi de Larderel, Claudia Peña, Andrea Ramirez, Sangwon Suh
Secretariat at UNEP: Lowri Rees, Shaoyi Li, Zura Nukusheva-Béguin

1. INTRODUCTION

Faced with an expected doubling in world demand for energy by 2050, massive investment will be needed to develop and install systems that can not only meet the energy needs of nine billion people but at the same time reduce greenhouse gas (GHG) emissions, air pollution, toxicity, the impacts on land, water and other eco-systems. This investment need presents the perfect opportunity to select the best electricity generation technologies to meet these aims (Chapter 1).

This report consists of this Technical Summary, and ten chapters constituting the full report. It identifies important environmental characteristics of low-carbon electricity generation technologies and provides decision makers with essential information on these characteristics. It assesses the impacts of building, operating and dismantling renewable power generation technologies such as hydropower, wind power, photovoltaics, and concentrated solar power on human health, ecosystems and natural resources. It also assesses the impacts of coal- and gas-fired power with carbon capture and storage (CCS). The impacts of these technologies are compared with those of modern coal and gas-fired power without CCS, but with state-of-the-art pollution control.

This report focuses specifically on mainstream commercial renewables and promising medium-term CCS options. Bioenergy is not included because it is covered in a recent IRP report (Bringezu et al., 2009). Nuclear power generation is not included because UNEP sees this technology as being under the responsibility of a different UN agency (IAEA); oil fired steam power plants were excluded because they are seen as less relevant for the future. Marine energy technologies such as wave, tidal, ocean-thermal or salt power, were not assessed because they are still relatively immature.

This report presents the first in-depth international comparative assessment of the environmental and resource impacts of different energy technologies, modelled over the whole life cycle of each technology, from cradle to grave. It is the work of an international scientific and technical expert team.

Over the coming twenty years, the world will invest around $2.5 trillion a year in new energy installations and energy conservation (IEA, 2014). Meeting the rising energy demands of a growing world population presents an ideal opportunity to make technology choices that also address the climate, environmental and health issues caused by fossil fuels. Technology change, efficiency and pollution concerns are expected to drive an increasing share of electricity in the world's total energy mix. Future energy scenarios suggest that a rising carbon price, as nations seek to avoid and mitigate the impacts of climate change, will progressively shift final energy demand away from gas, petrol, diesel, and coal and towards electricity generated from sources with low carbon emissions (Riahi et al., 2012; Bashmakov et al., 2014). Under these scenarios, the electricity supply is transformed by the large-scale deployment of renewable energy sources, nuclear power and fossil fuel power plants equipped with CCS. According to the IEA, massive investment in the development and deployment of low-carbon electricity supply technologies will be needed to limit global warming to 2°C, the goal set by the international community at the Cancun climate summit[1]. To choose the best technologies for a national or local energy system means that attributes other than costs and greenhouse emissions are equally important. There is a risk that shifting the burden of curbing emissions to other parts of the economic chain may simply cause new environmental and social problems, such as heavy metal pollution, habitat destruction, or resource depletion. The ideal solution/s will mitigate a range of problems at the same time as maximizing the energy benefits and minimizing economic costs.

Before asking the public and the private sector to invest trillions of dollars in the large-scale development and deployment of new energy technologies, we need to understand their wider potential repercussions, both positive and negative. Such an assessment of repercussions forms part of the "due diligence" required for such long-term investments, to avoid unintended consequences and help decision-makers select the cleanest, safest, most efficient mix – for a nation, a region or a local community.

2. ASSESSMENT APPROACH

This report presents an assessment of the impact of key power plant technologies on human health, ecosystem health, and resources, using a life cycle approach (Gibon et al., 2015). It models the life cycle of various kinds of power plants in nine different world regions for 2010, 2030 and 2050 based on technology performance characteristics and energy mixes of the Energy Technology Perspectives of the International Energy Agency (IEA) (IEA, 2010).

The assessment focuses on environmental impacts and resource requirements that lend themselves to quantitative comparison; it also contains a qualitative discussion of impacts which are considered important but for which mature assessment approaches are not yet available. The comparison is based on life cycle assessment (LCA), a well-established method to address not only the impacts that occur during power production, e.g. fossil carbon emissions, but also impacts resulting from fuel production and the production, construction, maintenance and disposal of the power equipment. Other social impacts – positive and negative, from employment creation to social acceptance - are recognized as important but were not included in this assessment.

The International Resource Panel (IRP) recruited teams of experts for each technology who then reviewed and provided a written assessment of the existing technical and scientific literature. For each technology, the expert teams collected quantitative resource use and emissions data for power plants over their entire life cycle, including construction, operations and fuel supply, in a consistent data format. The LCAs used an integrated model capable of describing impacts of the various energy technologies reflecting specific requirements in the IEA's nine regions (Gibon et al., 2015). They then evaluated the total emissions resulting from increasing use of low-carbon technologies following a mitigation scenario. We evaluated the life cycle inventories (LCIs) of low-carbon technologies with their likely deployment under the IEA's BLUE Map scenario (IEA, 2010), which is consistent with the goal of limiting global warming to 2°C (Arvesen and Hertwich, 2011). We compared the resulting global emissions rates and resource use with the deployment of energy technologies foreseen in the IEA's baseline scenario. Our assessment covers both the physical infrastructure of power generation and related processes such as materials production and transport, as well as manufacturing and installation using cost data and a global input-output model. Such a hybrid life cycle assessment (HLCA), combining physical and economic data, yields a fuller representation of life cycle impacts (Chapter 2). Furthermore, the research also covered site-specific effects of generation technology use, such as habitat change and wildlife impacts (Chapter 1).

1 http://cancun.unfccc.int/cancun-agreements/main-objectives-of-the-agreements/#c33

The environmental impacts of energy technologies will change as progressively cleaner technologies enter the global energy mix, driven by factors such as mitigation needs, economies of scale and accumulated experience. In our assessment, the manufacture of future energy equipment is modeled using the energy mixes expected in those future years when the technology is built, while we also take account of technology improvements leading to higher yields and efficiency. For fossil power plants, we assume the deployment of successive generations of technologies; for photovoltaics, we use foresight studies on critical parameters such as solar cell thickness.

Different indicators are available to evaluate the potential impacts of technologies, reflecting (1) resource use or emissions in basic physical units, (2) environmental impacts, such as climate change or the eutrophication of water bodies, or (3) measures of damages to human health and ecosystems. This report includes results for the use of land, non-renewable energy, and selected base materials (iron, copper, aluminium, cement) as direct resource use indicators, the ReCiPe (H) set of 16 midpoint indicators, and the ReCiPe (H) endpoint indicators of Disability Adjusted Life Years for human health damages and Potentially Affected Fraction of Species for ecosystem damages (Goedkoop et al., 2008).

3. TECHNOLOGY SUMMARIES

3.1 HYDROPOWER

Hydropower is currently the world's most important source of renewable electricity, providing 6.1% of total energy supply and growing at 3% per year. There is potential for a three- to five-fold increase in hydropower production (Chapter 4.1). Hydropower dams also serve other purposes, such as water storage, irrigation and transport. Their environmental and social impacts have received much attention (Asmal et al. 2000). Recognizing that these impacts depend on site and project characteristics, assessments need to be made on a case by case basis. Some impacts can be mitigated through appropriate flow management regimes or technical adaptations (e.g., fish ladders). In this report, we reviewed the literature on ecological impacts primarily associated with the disruption of the natural river flow regime and migration routes for aquatic life, and the generation of fugitive methane emissions from the decomposition of biomass in reservoirs. These impacts are not easily included in LCA.

Ecological impacts

Chapter 4.4.2 describes ecological impacts. The most significant ecological impacts of hydropower (Chapter 4.3) are connected to habitat change due to changes in the flow regime and flooding of reservoir area, habitat fragmentation and the obstruction of migration routes. Habitat and flow changes affect fish and other aquatic species and may threaten those adapted to river environments. Table 1 summarizes the main ecological effects, some of which can be mitigated through measures such as environmental flow control or sediment management.

Climate impacts from hydropower

Climate impacts are discussed in Chapter 4.4.3. Contrary to popular belief, hydropower plants can also release significant volumes of GHGs in the form of CO_2, CH_4 and N_2O as a result of bacteria digesting organic matter in the reservoir. The main concern from a climate perspective is methane, which is around 30 times more potent as a climate change driver than CO_2. The process begins when organic matter (leaves, soil, plants, etc.) is washed into the reservoir by feeder streams, or inundated when it fills, creating a rich feed source for the bacteria (Demarty and Bastien, 2011). Although CO_2 emissions are part of the natural biogenic carbon cycle, the impact that dams have on world carbon emissions is still being researched and is not yet clearly quantified.

TABLE 1

Potential ecological effects of hydropower plants

- Obstruction of fish and other migratory aquatic species.
- Habitat change and fragmentation in riverine and shallow water ecosystems.
- Water quality reduction in the reservoir due to the growth of phytoplankton and algae and development of thermal stratification.
- Reduction of freshwater storage capacity through sedimentation by 0.5-1% per year, lessening flood protection.
- Changes in flooding, sediment flow and associated nutrient deposition, affecting the extent and fertility of floodplains and deltas.
- Turbidity, sedimentation, stagnation and eutrophication of downstream waters
- Changes in the timing and volume of water flow, affecting species whose life cycle is adapted to seasonal water flow patterns but potentially beneficial to other species.
- Reduced temperature and increased gas content (supersaturation) of water released from dams, which affects fish.

FIGURE 1

LCA results for two different hydropower plants implemented in Latin America, normalized to the global average electricity mix. The investigated hydropower plants have significantly lower environmental impacts than the global electricity mix. Impacts of the two plants in the same area are quite different. Abbreviations for the impact indicators are:
CC-climate change;
FET-freshwater ecotoxicity;
FEU-freshwater eutrophication;
HT-human toxicity; MD-metal depletion; PM-particulate matter formation;
POF-photochemical oxidant formation;
TA-terrestrial acidification;
LO-land occupation.

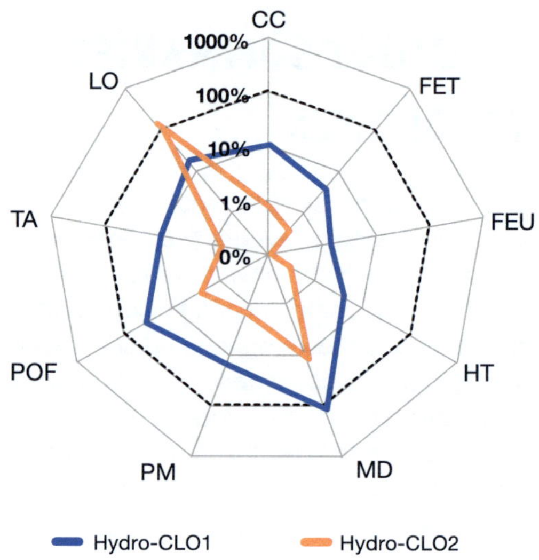

FIGURE 2

LCA results for OECD Europe onshore and offshore wind power systems normalized to global electricity mix. The results indicate that environmental impacts are between 1-10% of those of the global mix, but metal depletion is higher than the global mix. Abbreviations for the impact indicators are:
CC-climate change;
FET-freshwater ecotoxicity;
FEU-freshwater eutrophication;
HT-human toxicity; MD-metal depletion; PM-particulate matter formation;
POF-photochemical oxidant formation; TA-terrestrial acidification;
LO-land occupation.

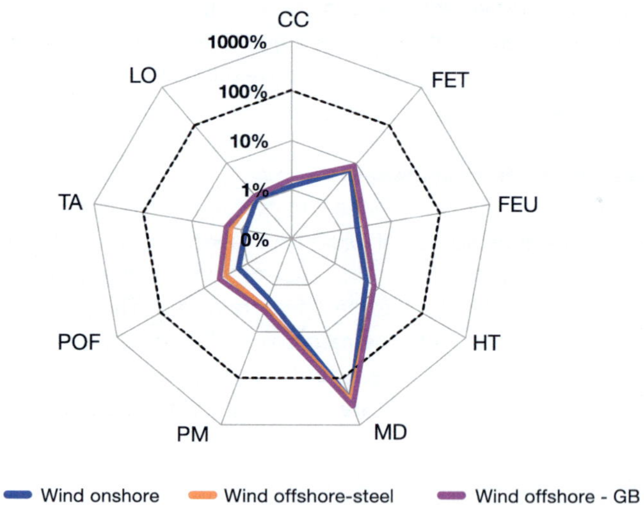

Most of the existing measurements of hydropower emissions come from completed reservoirs and do not quantify the changes in GHG emissions from before to after dam construction (Kumar et al. 2011). The amount of emissions from a reservoir area depend on the volume of biomass and nutrients entering it, in relation to its area, age and climate; emissions per unit of energy generated depend strongly on the reservoir area per unit energy generated. Studies to date show these emission factors can vary by several orders of magnitude across dams, which creates significant uncertainty when trying to assess the global contribution of hydropower to climate change. A small fraction of dams is responsible for the majority of emissions. Based on current data our estimate for the global methane emissions from hydropower plants is around 10 (-6/+10) million tons per year, which corresponds to 70 g CO_2e/kWh.

LCA results

The material and energy required to build a hydropower plant depends entirely on its site. Reservoir volume and head can vary enormously among hydro plants. The LCIs used in this study are based on two reservoir hydropower plants in Chile that have a lower land use, and therefore fewer GHG emissions than the global average. One of these plants, however, is located at a site so remote that the transport involved in its construction contributed substantially to the impact (Figure 1). The LCAs show that the environmental performance of hydropower plants can differ substantially. The investigated plants have lower pollution impacts than fossil fuel based power plants, especially for toxicity, eutrophication, and acidification. The land occupation and metal depletion impacts are of the same order of magnitude as those of fossil power plants.

3.2 WIND POWER

Wind energy is experiencing steady global growth. Over the past ten years, cumulative global installed capacity grew at an average annual rate of around 22%, mainly owing to markets in Asia, North America and Europe, reaching 318 GW by the end of 2013. Most of current installed capacity is onshore (98%), but the offshore segment is growing. Wind power technology is characterized by an increasing size of power plants and technical improvements resulting in increasing capacity factors (more energy harvested) and lower costs. Novel technologies are increasing generation reliability and further reducing costs. Wind power plays an important role in practically all mitigation scenarios. The LCIs for wind power in this study are adjusted from (Arvesen et al., 2013; Arvesen and Hertwich, 2011).

Land use

Some land or water area is used exclusively by wind turbines, their dedicated roads and other infrastructure, and this area cannot be combined with other human or wildlife uses. Wind power plants tend to affect a much larger area than other forms of power generation because of the scattered arrangement of the turbines; however part of this space can be conserved as natural habitat, used for agriculture or other purposes. A much larger area may be regarded as impacted, if indirect effects on wildlife or landscape visual quality are considered. The necessary spacing between power plants limits the overall capacity of wind power.

Wildlife mortality

The numbers of bird and bat fatalities recorded at wind farms vary widely and depend on the species, region and site characteristics, among other factors. The overall ecological significance of bird and bat mortality remains unclear and a topic of research and debate. There are concerns that wind power has become a significant mortality factor for bats in North America. Spatial planning, plant operation and other measures can potentially alleviate some mortality due to wind power.

Scarce materials

The direct-drive wind turbines used predominantly in offshore wind power plants employ permanent magnets containing rare earth elements such as neodymium and dysprosium, although the most common wind turbine

FIGURE 3

LCA results for OECD Europe onshore and offshore wind power systems by main components.

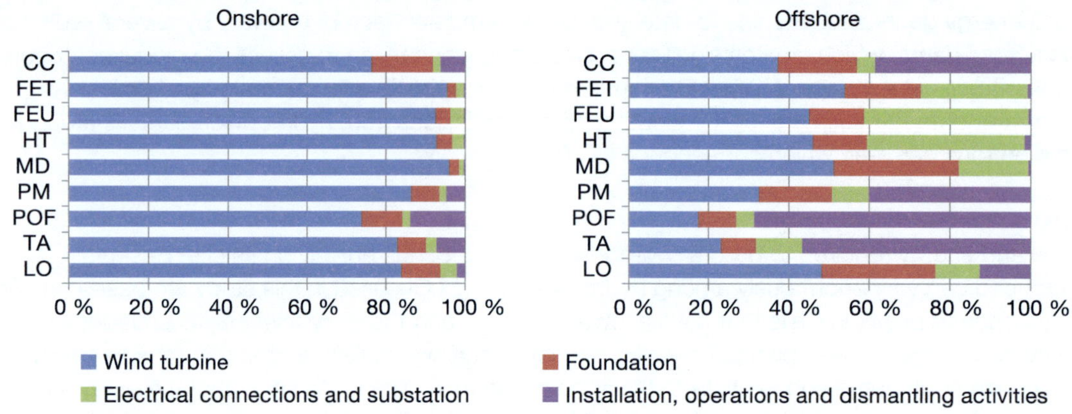

designs do not rely on such elements. In recent years, the constrained availability of rare earth elements and environmental damage caused by rare earth mining and processing have emerged as subjects of concern. A combination of limited resources, local environmental costs and geopolitical factors may limit the market uptake of large wind turbines containing rare earth elements. Wind power faces an increasing geopolitical risk from environmental and export restrictions by countries holding the largest strategic reserves of these materials, which is limiting access to them and could become an economic constraint (EC, 2010).

LCA results

Wind power scores one to two orders of magnitude better than fossil power generation technologies for all the assessed impact categories except metal depletion (Figure 2). It should be noted that the land use indicator results includes only the area occupied by wind farm infrastructure, not the spaces in between. If the total wind farm area was considered, land use would be about two orders of magnitude higher.

Offshore wind systems consume more materials and energy than onshore, but on the other hand, benefit from more favorable capacity factor and lifetime assumptions. Figure 2 shows that onshore and offshore wind facilities have similar life cycle impacts, though the offshore system exhibits worse performance in acidification, photochemical oxidants and particulate matter exposure. The relative contribution of components differs between onshore and offshore systems, however, as is evident from Figure 3. Production of wind turbine components contributes 70-90% to all impact indicators for the onshore system but only 20-50% for the offshore system. The installation, operations and decommissioning activities contribute significantly to the impact of offshore wind power. The contribution of the electrical connections is also larger than for the onshore system.

3.3 CONCENTRATED SOLAR POWER

Concentrated solar power (CSP) systems use sunlight falling on a surface kept perpendicular to the sun's rays to produce high-temperature steam for electricity generation. Areas particularly suitable for CSP are those with strong sunshine and clear skies. The global installed CSP capacity was 2500 MW at the end of 2012. In this study, two types of CSP plants were selected for LCA: parabolic trough, which is the most widely-applied technology to date, and power tower, also known as central receiver. The trough plant is assumed to be wet-cooled and the power tower dry-cooled. Other major CSP technology alternatives are linear Fresnel and dish/engine systems. Integration with low-cost thermal storage adds considerable value to CSP energy generation. The LCIs for CSP analyzed in this study are adjusted from (Burkhardt et al., 2011; Whitaker et al., 2013).

Water use

Unlike PV or wind, wet-cooled CSP plants require a considerable amount of water for cooling. The water use of wet-cooled CSP plants is similar to that of thermal power plants using fossil fuel or nuclear fission. Water is also needed for cleaning the mirrors. As good CSP sites also typically occur in dry climates, water use can be a critical constraint on large-scale deployment of wet-cooled CSP. Air-cooling is technologically feasible and can reduce operational water use by about 90%, but this also reduces efficiency and increases electricity production costs.

LCA results

With two exceptions, results shown in Figure 4 indicate that CSP has a far superior performance compared to the global electricity mix. The main exception is its high metal depletion burden, which appears greater than for other power generators. The other exception is land use, where CSP is generally comparable with other energies. The area occupied by CSP plants can seldom be combined with larger wildlife or other human uses, but CSP plants may provide valuable habitat for smaller animals and various plants and may be used for grazing.

Figure 5 shows the contribution of the main components for the tower and trough plants. The collector system, which includes the mirrored surfaces used to concentrate direct solar radiation, causes 40-50% of total impact for the tower and 30-40% for the trough for most impact categories. The trough plant uses a synthetic oil heat transfer fluid combined with molten salt storage while the tower plant uses salt as both as a heat transfer fluid and as a storage medium and hence does not have a separate heat transfer fluid system. Far less salt is used in the tower plant compared with trough, which in large part explains the lower relative contributions from thermal energy storage in the tower case. The results shown in Figure 5 depend on specific plant design, which may vary considerably depending on site features and project design.

3.4 PHOTOVOLTAIC POWER

Photovoltaic (PV) solar power is growing rapidly, with 41 GW of newly installed capacity in 2014 alone. This brings total global installed capacity to 177 GW, up fourfold in just five years (Chapter 7.1). This rapid and continued growth has been driven by renewable energy portfolio policies, feed-in tariffs and the decreasing cost of PV collectors and systems. Solar insolation is abundant on the earth's surface, and even cloudy countries like Germany - which is a leader in installed capacity - have sufficient areas of available land and roof space for generating the large quantities of PV electricity prescribed by climate change mitigation scenarios like the IEA BLUE Map.

Photovoltaic technologies

There are a number of viable, substitutable technologies that can provide PV solar power. This report analyzes a cross section of mature PV technologies: polycrystalline silicon (Poly-Si), cadmium telluride (CdTe), and copper indium gallium selenide (CIGS). Crystalline silicon technologies are the most mature, and account for most of the PV market. China currently dominates global production of Poly-Si PV, providing 73% of the world's production capacity of crystalline silicon modules in 2012. This report therefore uses Chinese production data in its LCA of silicon PV (EPIA, 2013). CdTe and CIGS are the most mature thin-film (TF) technologies and are steadily gaining market share. Unlike silicon PV, most CIGS and CdTe production is based in Europe, Japan, Malaysia and the United States. Thin film modules are thought to have a substantial potential for technological improvement, increasing in energy conversion efficiency and decreasing in their materials requirements by 2050 (Goodrich, 2011; Woodhouse et al., 2011). In addition to the technologies considered in this report, several emerging PV technologies (organic polymers, quantum dot, and dye sensitized PV) may play a significant role in the PV market by 2050.

Life cycle assessment results

LCA of PV technologies shows clear environmental benefits in terms of climate change, particulates, ecotoxicity, human health and eutrophication relative to fossil fuel technologies. However, PV electricity requires a greater amount of metals, especially copper, and, for roof-mounted PV, aluminium. The environmental and resource impacts of Poly-Si, CdTe and CIGS ground-mounted and roof-mounted systems have similar magnitude, despite differing technological composition. By 2030 and 2050, all three PV technologies will show major improvements in impacts and metal consumption due to expected increases in material efficiency of their modules, increased power generation efficiency and changes in the electricity grid. Figure 6 shows the environmental impacts in 2010 for PV technologies relative to the global mix.

Generally, thin film technologies show lower environmental impacts than crystalline silicon. Energy use during module manufacture contributes most to climate change, particulates, and toxicity results. Crystalline silicon requires a greater quantity of electricity and has higher direct emissions during the production of metallurgical grade silicon, polycrystalline silicon wafers and modules. In this report it has higher life cycle GHG emissions and energy consumption than in some other studies. This is mainly because of the lower material efficiencies in the production of Chinese silicon wafers, cells and modules, and the widespread use of coal-fired electricity generation in China.

The largest contributors to metal use in PV systems are the inverters, transformers, wiring, mounting and construction. Although metal use for these applications is significant, these system components can be recycled or reused, allowing the recovery of many of the metals. Silicon is the second most abundant element in the earth's crust but PV uses substantial amounts of silver as a conductor. Thin film technologies rely on semiconductor layers composed of byproduct metals, namely cadmium, tellurium, gallium, indium and selenium. As thin film technologies using these elements capture larger market shares, they may encounter shortages if the recovery of these metals from primary copper and zinc production is not increased (Woodhouse et al., 2013; Woodhouse et al., 2011). Metal supply shortage is a particular concern for tellurium in CdTe technology. Due to the toxicity of the involved metals, a proper recovery and recycling is important.

3.5 GEOTHERMAL POWER

Geothermal energy is thermal energy generated by and stored in the Earth's crust. Ninety-nine per cent of the earth's volume has temperatures up to 1000°C, while only 0.1% of it is less than 100 °C. The total heat content of the Earth is immense, and is estimated at about 1013 EJ. The main sources of geothermal energy come from the residual energy left over from planet formation and the energy continuously generated by radionuclide decay. The Earth transfers about 40,000 GW of this heat to the atmosphere. Thus, the geothermal resource base is large and ubiquitous. This resource, however, is widely distributed and the power density is low (0.1W/m^2 compared to 300-500 W/m^2 for solar radiation). Heat of useful temperature is not always easily accessible from the surface, except in a few geologically active regions. Geothermal resources consist of thermal energy stored within the earth in both rock and trapped steam or liquid water. This energy source can be used both indirectly for electricity generation and directly for heating buildings, baths, greenhouses, food processing etc.

Geothermal power plants are typically 20-60 MW in size and require several wells. Plant designs include direct steam plants, flash steam plants, double flash plants, and binary systems (Chapter 8.2). The design is driven by local resource characteristics such as whether a well is dry or has geofluids present, the temperature of those fluids, and gas content. Plant efficiency typically varies between 10-23% and depends on the temperature of the reservoir as well as the cooling system.

Technology-specific impacts

Just as the geological circumstances vary from site to site, so do the environmental impacts of geothermal energy (Chapter 8.3). These are the main issues to consider.

FIGURE 4

LCA results for Africa and Middle East CSP trough and tower systems normalized to global electricity mix. The results show low pollution-related impacts but high metal depletion and land occupation comparable to the global mix. The tower system has higher impacts, partly related to its lower efficiency as it is modelled as an air-cooled system. Abbreviations for the impact indicators are:
CC-climate change;
FET-freshwater ecotoxicity;
FEU-freshwater eutrophication;
HT-human toxicity;
MD-metal depletion;
PM-particulate matter formation;
POF-photochemical oxidant formation; TA-terrestrial acidification;
LO-land occupation.

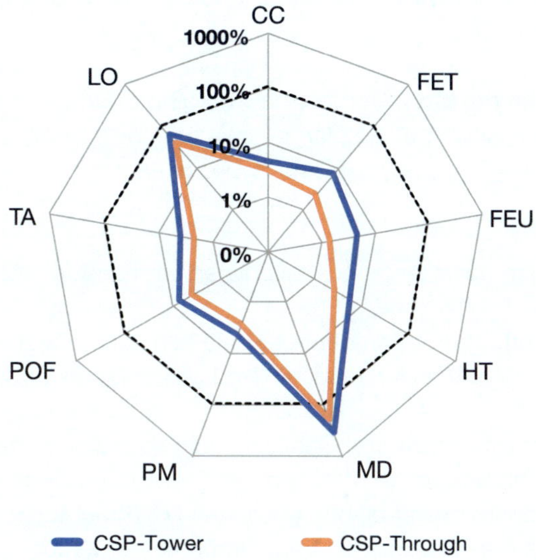

FIGURE 5

LCA results for Africa and Middle East CSP trough and tower systems by main components.
Abbreviations for the impact indicators are: CC-climate change; FET-freshwater ecotoxicity; FEU-freshwater eutrophication; HT-human toxicity; MD-metal depletion; PM-particulate matter formation; POF-photochemical oxidant formation; TA-terrestrial acidification; LO-land occupation.

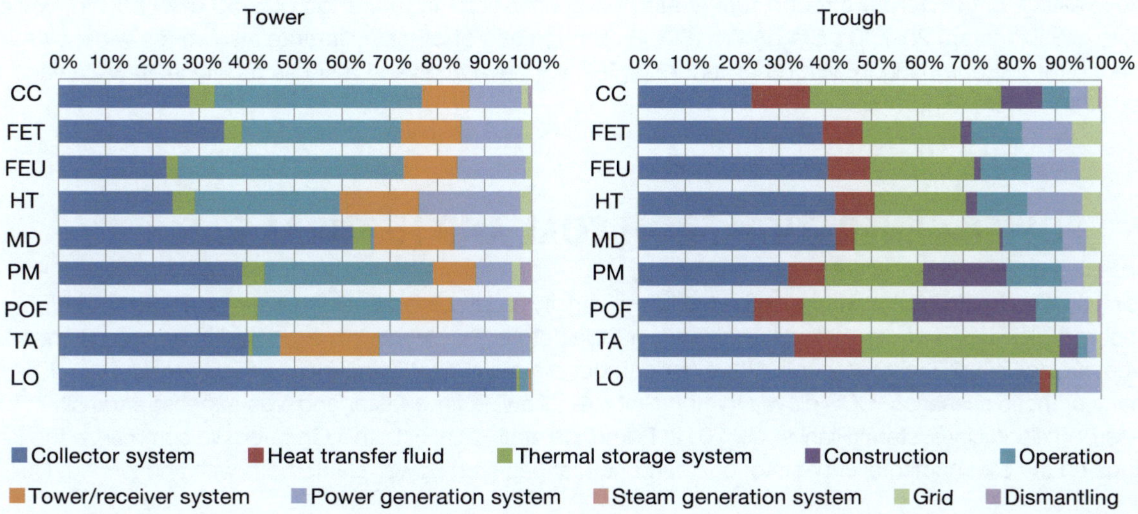

Land use: Our review shows that land use varies between 200-30,000 m^2/MW, or 0.04-6 m^2a/MWh. (Chapter 8.3.2.1.1)

Geological hazards: Geothermal energy production is associated with extensive extraction or circulation of geofluids and/or steam, large-scale and local manipulation of the shallow and deep ground. Landslides, subsistence, fractures, explosions and changes in natural seismicity have been connected to geothermal facilities.

Noise: High noise levels are associated with drilling and well testing.

Thermal effects: The amount of waste heat loss is around 4-10 times the amount of electricity generated, and is hence higher than for fossil fuel fired power plants of similar capacity.

Atmospheric emissions: Geofluids contain many contaminants. Pollutants such as H_2S, CO_2, and CH_4 are often discharged to the atmosphere. These non-condensable gases (NCG) are released from flash-steam and dry-steam power plants, because in contrast to steam, the gases do not condense at the turbine outlet. Emissions may also include trace amounts of mercury, ammonia, radium and boron.

Solid waste and water emissions: Liquid-dominated high temperature geothermal fields can result in significant waste of geothermal fluids. Critical contaminants of steam emissions, such as hydrogen sulfide (H_2S), boron (B), ammonia (NH_3), mercury (Hg) often occur in the fluids, as well as metals such as arsenic (As), lead (Pb), cadmium (Cd), iron (Fe), zinc (Zn), antimony (Sb), lithium (Li), barium (Ba) and aluminium (Al).

Water use: Water is used extensively in geothermal generation, especially for drilling, cooling, and to supplement steam production. The extent of cooling water use depends on the technology; air-cooled systems having a much lower water use, but also a lower efficiency and higher energy cost.

Life cycle results

LCAs available in the literature report fuel-related GHG emissions in the range of 6-50 gCO_2e/kWh, while fugitive emissions are 20-770 gCO_2e/kWh (8.4.4). The release of other pollutants also varies widely. As a reference for the comparison with other technologies, we have analysed a single facility in New Zealand and report results here (Figure 7).

3.6 POWER GENERATION FROM COAL AND NATURAL GAS

Fossil fuels are the dominant source of the world's electricity today. Given the long lifetime of mines, wells, transport facilities, and power stations and their versatile nature, fossil fuels are expected to remain an important source of electricity in the foreseeable future under most climate mitigation scenarios (Fischedick et al., 2011). In many of these scenarios, CCS plays an important role, allowing for a faster and less expensive transition to a low-carbon electricity system (Riahi et al., 2012; Edmonds et al., 2013). In this Chapter, we summarize the key findings on the environmental impacts of coal- and natural gas-fired power plants both with and without CCS.

State of coal and gas technologies

While coal and natural gas production and use are often viewed as mature technologies, there is in fact substantial scope for performance improvement. Until recently, the main driver for coal and gas technology development was securing energy supply and access; fuel costs and the decline of easily accessible sources were important variables.

FIGURE 6

Life cycle impacts for PV technologies in 2010 implemented in the OECD North America region, and normalized by the emissions associated with the present global power mix. Pollutant emissions are generally less than 10% of the global mix with the exception of freshwater ecotoxicity. Metal depletion is higher than the global mix. Thin-film cells perform better than silicon cells. Abbreviations for the impact indicators are: CC-climate change; FET-freshwater ecotoxicity; FEU-freshwater eutrophication; HT-human toxicity; MD-metal depletion; PM-particulate matter formation; POF-photochemical oxidant formation; TA-terrestrial acidification; LO-land occupation. Abbreviations of the technologies are: CdTe: Cadmium Telluride, CIGS: Copper Indium Gallium Selenide, Poly-Si: polycrystalline silicon. Ground refers to ground-mounted panel, roof to roof-mounted panels.

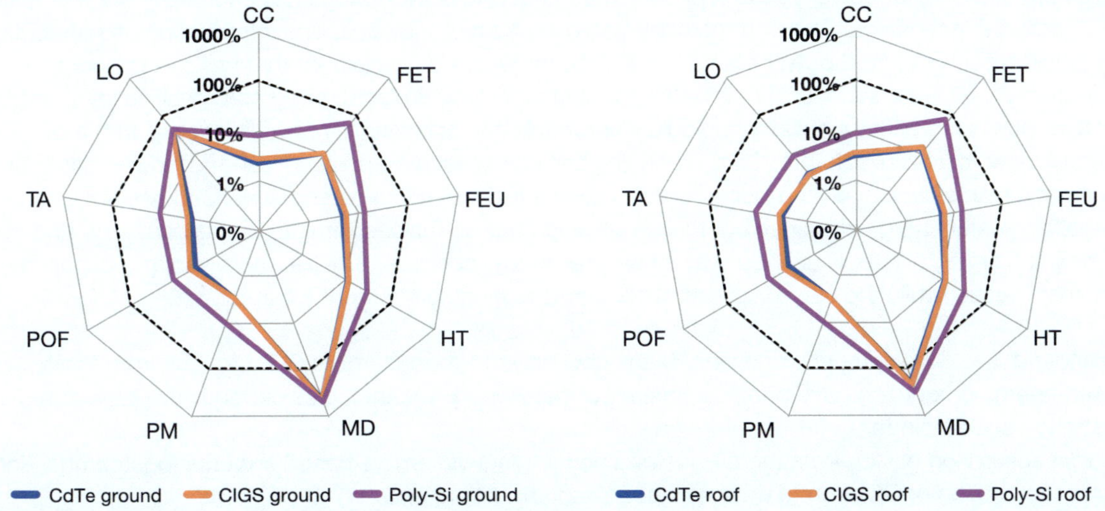

FIGURE 7

Environmental impacts for a 177 MW geothermal plant (Wairakei) relative to the OECD Pacific electricity mix of 2010. Abbreviations for the impact indicators are:
CC-climate change;
FET-freshwater ecotoxicity;
FEU-freshwater eutrophication;
HT-human toxicity;
MD-metal depletion;
PM-particulate matter formation;
POF-photochemical oxidant formation;
TA-terrestrial acidification;
LO-land occupation.

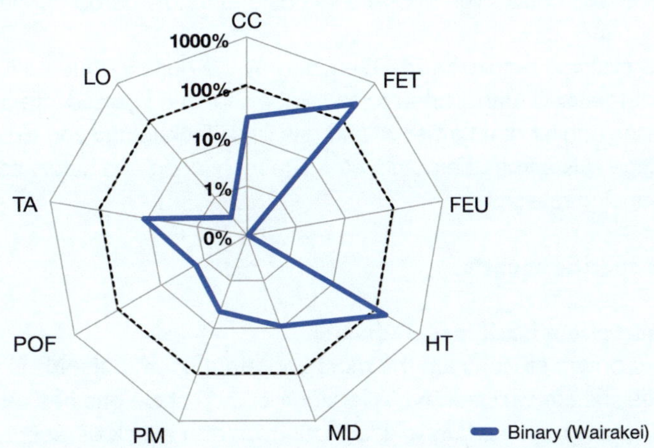

There have been significant advances in the production of fossil fuels with the hydraulic fracturing of shale for oil and gas production, horizontal drilling, deep-sea technology, coal seam methane extraction, and the mining of tar sands (Chapter 3.4.1), although many of these innovations are also associated with concerns about increased pollution. These new technologies have made accessible oil and gas resources that were previously considered uncommercial or technically impractical, raising the prospect of a continued increase in global CO_2 emissions (Rogner et al., 2012). Additional resources such as deep coal, Arctic gas or methane hydrates may also become accessible in future.

The maximum efficiency of power plants and combined heat and power plants has improved over the years, but not all efficient technologies are commercially viable in all circumstances. The introduction of supercritical and ultrasupercritical coal-fired power plants is a significant recent development that has raised energy efficiency from 35-37% for subcritical to 43-45% for ultrasupercritical plants. Integrated gasification combined cycle (IGCC) plants represent a new technological approach that achieves similar efficiencies, with the promise of further increases (Chapter 3.2). Technological advances in combined heat-and-power plants and polygeneration plants, which also produce cooling, include fuel-cell systems and advanced gas engines. Such systems offer greater energy efficiency but their application has so far been limited by the challenge and cost of matching the timing of supply and demand of several energy services. Smaller, more versatile units may yield fresh advances, while the use of fuel cells may increase electricity output (Chapter 3.2.4).

CCS is under active development worldwide. Many new ways to produce pure CO_2 streams from fossil fuel-based energy production have been discovered or developed. Technologies currently available at a demonstration scale include:

- chemical absorption of CO_2 from the off-gas of a power plant via amine-based solvents (postcombustion),
- physical adsorption of CO_2 from a synthesis gas (precombustion) and
- the combustion of fossil fuels with pure oxygen, producing CO_2 and water (oxyfuel).

These technologies are further explained in Chapter 3.3.1. Today, although CCS is currently used on a million-ton scale to clean natural gas and to produce commercial CO_2, it has only recently been implemented in a commercial power plant. The technology's greatest challenge lies in overcoming the combination of high investment costs, high operational costs, and low carbon prices.

CCS systems require that CO_2 emitted by the burning of fossil fuels is captured in as pure a form as possible, is compressed, transported to a storage site, and injected into a suitable deep geological formation, such as a saline aquifer or a former oil and gas field. Technology and experience with such injection and subsequent storage exists from the oil industry. The monitoring and safety assessment of large-scale CO_2 deposits is a focus of ongoing research.

Site-specific impacts

Impacts from fossil fuel production

Since power stations are the most important source of carbon emissions associated with fossil power, these emissions are well studied. A recent focus on shale gas has resulted in a wide range of estimates for fugitive methane emissions during gas production and a lack of recent, empirical data. Similarly, coal mine methane emissions vary widely across mines. Our life cycle calculations take into account recent estimates of fugitive emissions based on one source and are an increase on earlier estimates, especially for natural gas. Some fossil fuel production technologies, such as oil sands, also have high land and water requirements and impacts. (Chapter 3.4.1)

CO_2 transport and storage

Concerns about CO_2 transport and storage under high pressure include leakage, which undermines its mitigation effectiveness. Additional concerns associated with CO_2 leakage include the direct health hazard

FIGURE 8

LCA results for a 177 MW geothermal plant (Wairakei) indicate that toxic emissions, which are high relative to the reference mix, are from the operations of the power plant.

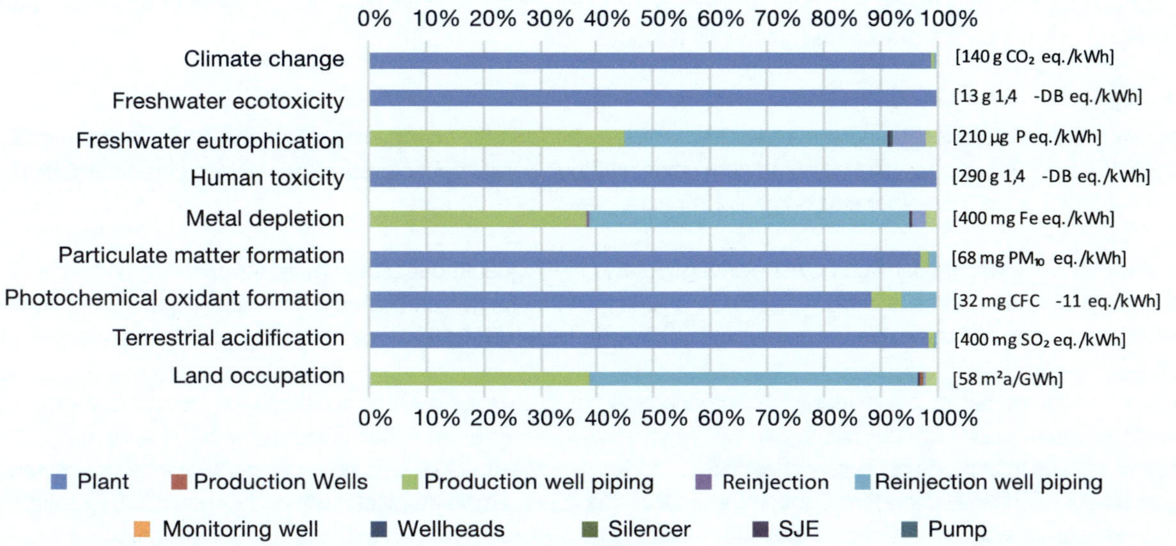

FIGURE 9

LCA results for fossil fuel fired systems modeled as if implemented in China and normalized to the existing global power mix. Abbreviations for the impact indicators are: CC-climate change; FET-freshwater ecotoxicity; FEU-freshwater eutrophication; HT-human toxicity; MD-metal depletion; PM-particulate matter formation; POF-photochemical oxidant formation; TA-terrestrial acidification; LO-land occupation. The technologies included are EXPC: existing pulverized coal; SCPC: supercritical pulverized coal; IGCC: integrated coal gasification combined cycle; NGCC: natural gas combined cycle; CCS: CO_2 capture and storage.

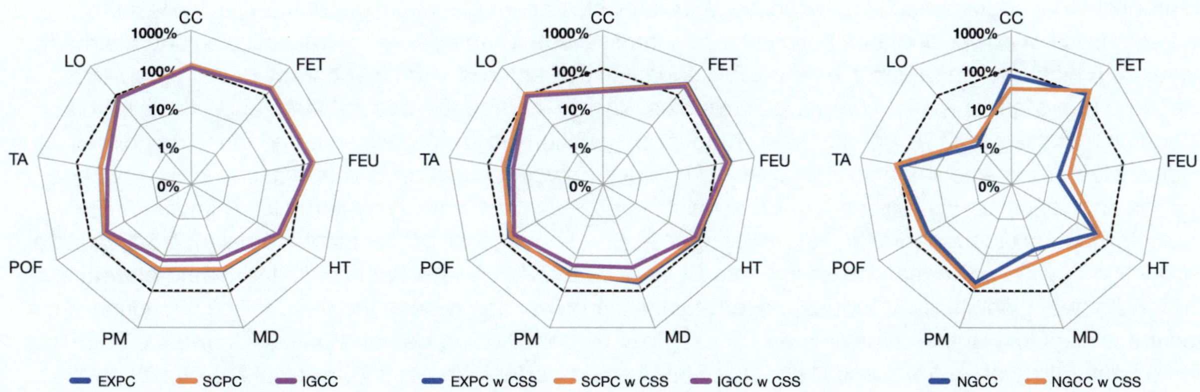

posed by locally high concentrations CO_2, and the potential mobilization of toxic heavy metals in the ground through the acidification of groundwater. CO_2 reacts at the storage site with geofluids and rocks, reducing the risk of leakage. The rate of these reactions depends on the geological conditions at the storage site. The highest risk of leakage is during CO_2 injection. Options have been proposed for monitoring, verifying and accounting for potential leaks. Similarly, proposals for technologies to seal leaks exist, some of which are based on existing solutions for leaky oil and gas wells (Chapter 3.5).

LCA results

A key finding from LCAs is that there is a clear trade-off between climate change mitigation and other environmental impacts (Chapter 3.8) of coal- and gas-fired generation. In other words technologies which reduce carbon emission from coal and gas increase other environmental impacts.

The study shows existing coal-fired power plants generally have higher impacts than more advanced supercritical and integrated gasification plants and much higher emissions than natural gas combined cycle plants (Figure 9). However the GHG emissions of modern power plants *with CSS* are between 22-26% those of existing coal fired power plants.

Furthermore, for the particulate matter and photochemical smog emissions which constitute the most important threats to human health, modern plants with CCS also show lower emissions than current coal fired plants, but higher emissions than modern plants without CCS. Modern plants with CCS also increase freshwater ecotoxicity and eutrophication compared to current plants without CCS. Comparing modern plants with and without CCS indicates that CCS increases environmental impacts (other than carbon) by 5-60% compared to the non-CCS alternatives.

NGCC plants have higher NO_x emissions than coal-fired plants, which poses a great risk of acidifying water bodies and soils. NO_x emissions also contribute to marine eutrophication, which is not shown here. As Figure 10 indicates, the most important contributors to environmental impacts are the operations of the power plant itself (for climate change, human toxicity, particulate matter formation and water use) and the extraction and refining of the fossil fuel (for land occupation, eutrophication, and freshwater ecotoxicity).

3.7 ELECTRICITY GRID AND ENERGY STORAGE

In this report, we have analyzed the environmental impacts of different electricity generation technologies on a kWh basis, according to the IEA scenarios. Energy resources, however, differ in their spatial and temporal distribution. The characteristics of resources and technologies for electricity generation, as well as the characteristics of power demand, have important implications for the design of transmission and distribution systems. A high fraction of variable renewable sources such as wind and solar energy poses an obvious challenge to system operation. Larger grid systems, energy storage, flexible demand, and/or the flexible operation of fossil-based power can help smooth out variations in supply. However, all these responses cause additional environmental impacts. The effect of different power sources on grid operations is very system specific and varies across regions and situations (Chapter 9.2 for electricity system characteristics). For example, various studies indicate that adjusting the operation of fossil power plants to balance the variable production of wind power can cause impacts as large as the life cycle impacts of installing and operating the wind power plant itself. At modest penetration, solar power can reduce the need for peak capacity as it generates electricity at the same time as, for example, air conditioning demand peaks (Chapter 9.3). Below, we give a brief outline of environmental impacts of the most important elements of a flexible electricity grid.

In the IEA scenarios, the investments in transmission are of a similar size to those in distribution. It is not clear that a mitigation scenario requires higher grid investments than a baseline scenario, as the mitigation scenario results in a lower total energy demand.

FIGURE 10

Contribution analysis for an integrated coal gasification combined cycle plant with pre-combustion CO_2 capture and a natural gas combined cycle plant with post-combustion CO_2 capture. Abbreviations for the impact indicators are: CC-climate change; FET-freshwater ecotoxicity; FEU-freshwater eutrophication; HT-human toxicity; MD-metal depletion; PM-particulate matter formation; POF-photochemical oxidant formation; TA-terrestrial acidification; LO-land occupation.

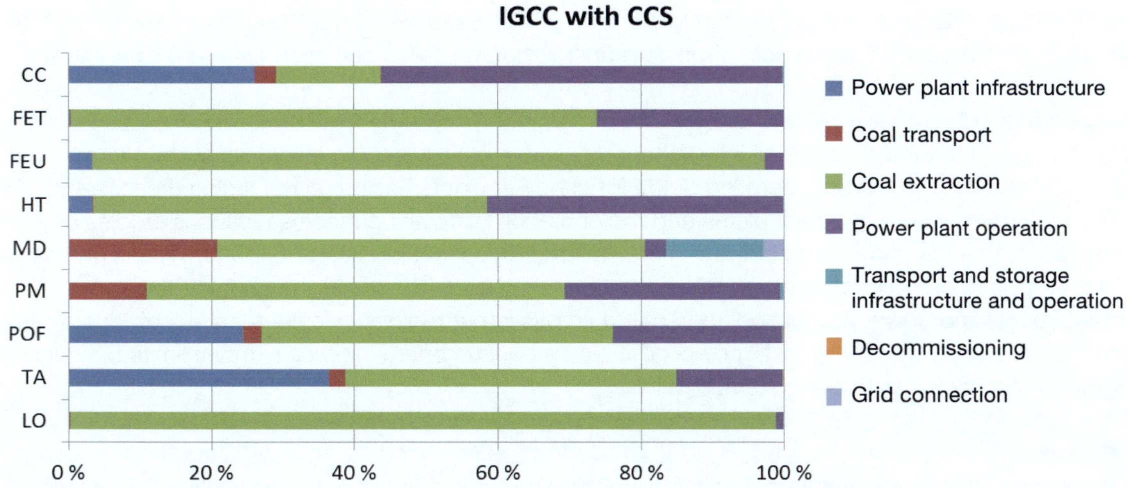

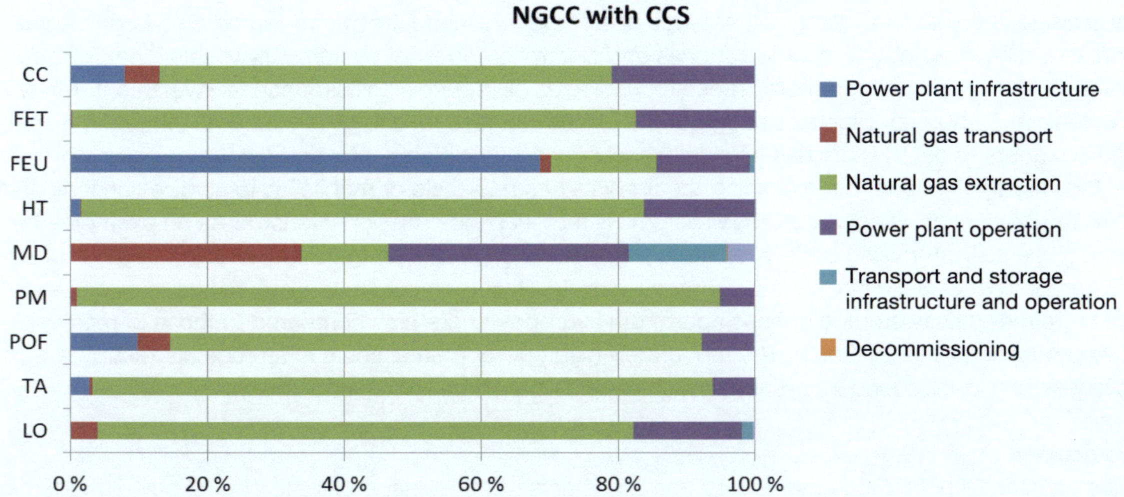

Electricity grid extension

Connecting larger areas of generation and demand can improve system operations and allow the integration of more renewables. High capacity, high voltage lines and cables can provide significant energy savings, allowing for more steady operation of power systems. All forms of electricity transmission incur losses, but these losses tend to be higher in systems with a weak transmission infrastructure. The construction of power lines, cables and transformer stations, however, causes a range of impacts both directly on habitat and wildlife and through the production of materials and equipment demanded.

Power lines take up land and are a cause of bird fatalities. A significant impact of electricity transmission is usually the power loss, which is often on the order of 1-3% for the high-voltage portion of the grid; losses in low-voltage distribution grid are commonly larger, 3% up to 40%. The electricity transmission infrastructure is also material intensive. In Norway, the construction of the transmission grid contributes approximately 1 gCO_2e per kWh of end-use electricity demand. A hypothetical grid for large-scale utilization of offshore wind power in the North Sea would add approximately 5 gCO_2e per kWh of power. Impacts from power transmission are generally low compared to impacts from power production, but they are not low enough to be ignored (Chapter 9.5.2.2). The impacts of power transmission on metal depletion are more significant.

Flexible operation of fossil power plants

The integration of substantial amounts of intermittent renewable energy into an electricity system dominated by fossil power requires the flexible operation of the fossil power plants, including managing the losses during the ramp-up and ramp-down of power plants and the operation of spinning reserves. Various studies of systems in North America and Europe indicate that this flexibility causes additional GHG emissions on the order of 15-70 gCO_2e per kWh of wind energy introduced into a grid. The larger the grid, the lower the costs, as the variability of wind power production aggregated across larger regions is lower than at individual sites. A fundamental challenge with using fossil power plants as a backup energy source is that it limits the share of very low-carbon technologies in the system.

Energy storage

Energy storage can deliver substantial benefits in stabilizing grid operations on all time scales, from seconds to months. Opportunities for effective electricity storage are limited, however. Pumped storage hydropower is the only technology widely used for large-scale energy storage today; it offers acceptable costs and efficient storage at suitable locations. Other types of storage foreseen for systems based on a large degree of intermittent renewable power include batteries and electrolysis/fuel cell systems, flywheels, compressed air storage, super-capacitors and more. These technologies all require significant capital investment. Many systems achieve 70-90% storage efficiency, but the losses increase as energy is stored on longer time scales, ranging from hours to days. There has been little analysis of the environmental and resource impacts of utility-scale energy storage options, but extending the analysis of small-scale or mobile systems gives an indication. Generally, the production of energy storage systems is material and emission intensive. As an example, the most environmentally promising battery technologies, lithium ion and sodium sulfide, emit in the order of 30-100 gCO_2e per kWh of electricity stored, over the life cycle. Based on our limited knowledge, environmental costs of current electricity storage options apart from hydropower are high compared to those of renewable electricity production (Chapter 9.5.3). The moderately high environmental costs of storage also limit the attractiveness of grid-independent systems and mini-grids based on PV or wind energy.

Flexible demand

There is a substantial potential to use energy demand that is not time-dependent to control power loads. For example, water heaters, district heating systems, refrigerator and freezers could use surplus electricity where it exists in a grid and so help to better match demand to variable supply. Other loads can be switched off at moderate costs. Smart grids and meters are one way to attain this goal. Some large industrial enterprises are already entering contracts that allow utilities to disconnect them in case of power shortages. Smart grids may make such options attractive to a much wider range of customers. Preliminary analysis indicates there are specific benefits from such strategies. However, the implementation of smart grids and meters is resource intensive, and little research exists to date on the environmental costs and benefits of flexible demand strategies.

4. COMPARATIVE RESULTS

4.1 CARBON MITIGATION EFFECTIVENESS

Five main low-carbon electricity technologies examined in this report can achieve life cycle carbon emissions of less than fifty grams of CO_2e per kilowatt hour (g/kWh): wind, PV, concentrated solar-thermal, hydro and geothermal power (Figure 1; Chapter 10.3.1).

This compares with 800-1000 g/kWh for standard coal-fired power generation and 600 g/kWh from natural gas combined cycle (NGCC) plants. With CCS, emissions from producing 1 kWh of coal and gas power drop to 200 grams.

The main sources of emissions in the life cycles of wind, PV and concentrated solar-thermal power are in the manufacturing and installation of the equipment. For onshore wind power, materials and manufacturing account for 80% of emissions, while for offshore wind, the joint contribution of installation, operations and decommissioning contribute is comparable with that of materials and manufacturing (Chapter 5.7.2). For concentrated solar thermal power, manufacture of the collector system accounts for 30-40% of the total greenhouse emissions (Chapter 6.5). For PV power, the manufacturing of modules contributes 45-85% of total GHG emissions, depending on the technology (Chapter 7.3.2).

FIGURE 11

Comparison of the life cycle carbon emissions of different electricity supply technologies, modelled for 1 kWh produced in Europe. For some technologies, substantial improvements are expected as a result of both technological improvement and the reduction of emissions in manufacturing due to cleaner energy. Abbreviations: CdTe – Cadmium telluride, CIGS – Copper indium gallium selenide, Poly-Si – Polycrystalline silicon, CCS – CO_2 capture and storage, IGCC – integrated gasification combined cycle, GB – gravity-based foundation.

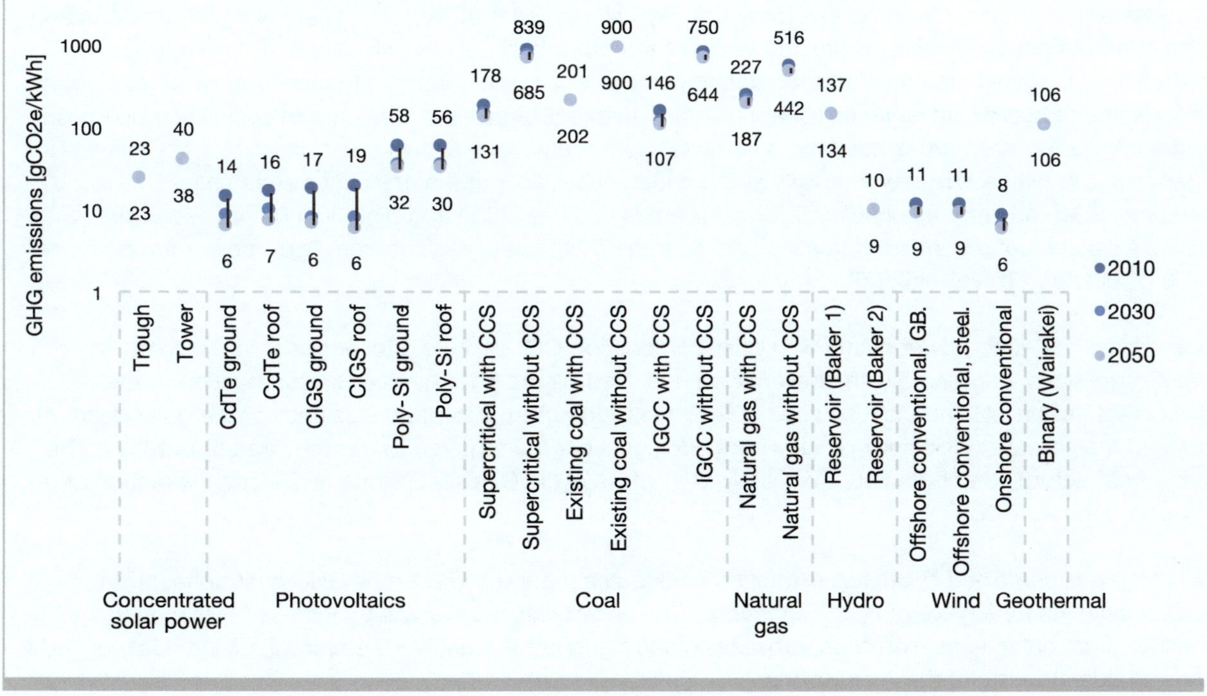

FIGURE 12

Human health impact of electricity production modelled for Europe in 2010. The measure is disability adjusted life years (DALY) per TWh of electricity generated following different damage pathways according to the ReCiPe (H) impact assessment method. Abbreviations: CdTe – Cadmium telluride, CIGS – Copper indium gallium selenide, Poly-Si – Polycrystalline silicon, CCS – CO_2 capture and storage, IGCC – integrated gasification combined cycle, GB – gravity-based foundation.

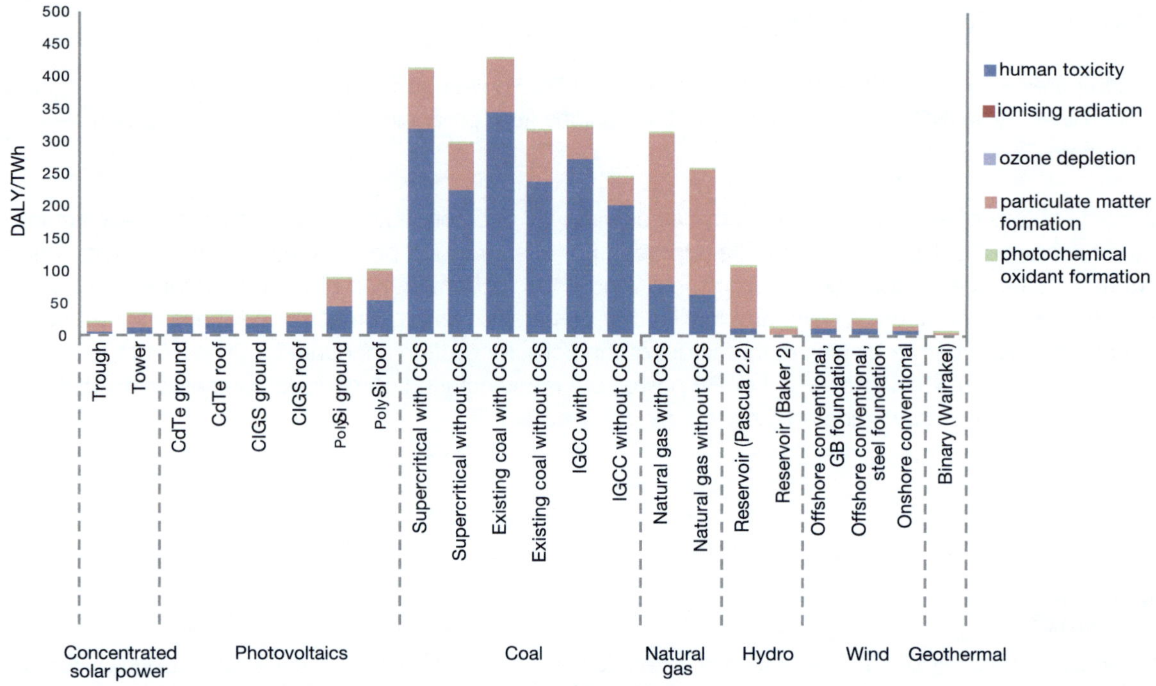

Hydropower plants generally have low fossil CO_2 emissions, most of which originate from the construction of the concrete dam and related infrastructure. On the other hand, there can be substantial ongoing generation of biogenic methane from the decomposition of organic matter in hydropower reservoirs, which offsets some of hydro's advantages (Chapter 4.4). The effect of dams on the carbon cycle is complex, so most analyses focus on gross methane emissions. Some reservoirs show no net emissions of GHGs, but large reservoirs with a low power density and a substantial inflow of biomass have substantial emissions, sometimes even as great as coal-fired power. There is wide variation and high uncertainty associated with these emissions, but global estimates indicate average methane emissions from hydropower are between 30-70 gCO_2e/kWh (Chapter 4.3.5).

Supercritical coal-fired power plants with post-combustion CO_2 capture offer scope for a 70% reduction in GHG emissions, compared with conventional coal fired plants (Chapter 3.4.3). Combustion at the power plant itself is still the main source of GHG emissions given a capture efficiency of the gas stream of 90% and the so-called efficiency penalty from additional energy required to run the capture process. The plant infrastructure contributes approximately 20% of life cycle GHG emissions, while coal extraction and transport contribute 15%.

For natural gas combined cycle, fuel extraction contributes around 65% of total GHG emissions, given the best estimate for fugitive methane emissions. The uncertainty and variability of these fugitive methane emissions from natural gas extraction, processing and transport is high. More empirical research is required to improve understanding of these emissions.

FIGURE 13

Ecosystem impacts of electricity production modelled for Europe in 2010. The impact is measured in species-year affected per 1000TWh of electricity following different damage pathways according to the ReCiPe (H) impact assessment method.

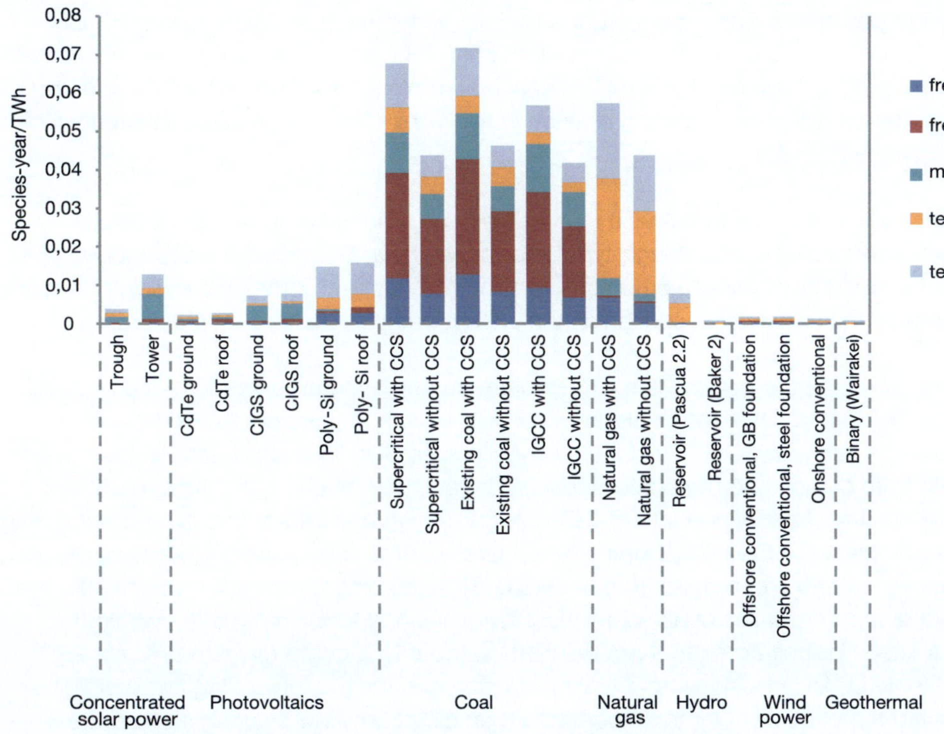

4.2 HUMAN HEALTH

The burning of fossil fuels and biofuels is the most important source of pollution-related human health issues. The World Health Organization's studies of the global burden of disease state that particulate matter from combustion is the most significant outdoor air pollution impact on human health, resulting in about 3.2 million premature deaths in 2010, while tropospheric ozone formed from air pollution was thought to cause 150,000 fatalities (Lim et al., 2012; Smith et al., 2012). A further 3.5 million deaths from indoor air pollution are due to respiratory infections and heart disease linked to particulate matter formed by products of incomplete combustion from biomass, coal and kerosene in primitive cooking and heating stoves in developing countries (Lim et al., 2012; Smith et al., 2012). Taken together, the annual death toll from air pollution is comparable with the annual death toll of World War II. Occupational health impacts, including accidents, also play a role in human health impacts from energy system, while the impacts from toxic pollution to water and soil are more uncertain.

Evaluated with ReCiPe midpoint and endpoint impact assessment (Goedkoop et al., 2008), low-carbon technologies perform as well as or better than modern fossil power plants with state-of-the-art pollution control (Figure 2). According to our assessment, health impacts from the toxic emissions of power generation are comparable to, and often larger than, impacts from particulate matter. This is true especially for fossil power plants, since metal leaching from mines continues for thousands of years and our assessment includes the toxic effects of this leaching. Such long-term releases are not yet considered by the WHO burden of disease studies, and there is substantial uncertainty about both low-dose chronic toxicity of metals and exposure avoidance by future generations. The other impact pathways on human health included in the study, namely photochemical oxidant (ozone) formation, ionizing radiation, and ozone depletion, have negligible impacts. The relatively low importance of photochemical oxidant formation is expected, as power plants usually contribute less to VOC and NO_x emissions than distributed sources such as transport.

The exposure of humans to particulate matter per unit of electricity generated from hydropower, PV, CSP, and wind power is an order of magnitude less than for modern coal- and gas power plants, with or without CCS, and two orders of magnitude less than for standard coal power plants. Particulate matter exposure is the impact pathway where we have the largest confidence in the results, and the results show renewable technologies perform substantially better.

Coal power is about four times more toxic to humans than gas power; among the renewable energy technologies, hydropower, onshore wind and trough-type solar power have the lowest toxicity scores. The high score for coal power is due to manganese emissions, which have not been examined in previous studies. These results should therefore be viewed with caution as the impact pathway deserves more scrutiny.

The amine-based solvents used in post-combustion CO_2 capture, degradation products from the capture process and compounds released during capture are all potentially toxic and therefore affect the overall toxicity rating of coal power plants with CCS. A major challenge in assessing the risks is that emissions and the composition of waste from carbon capture processes are not yet being made public. As a result, the understanding of the composition, toxicity and fate of CCS process emissions and products released during the waste treatment is incomplete (Da Silva and Booth, 2013). Under some circumstances, safety limits for toxic compounds in drinking water are exceeded (Karl et al., 2011). According to current assessments, the health risks posed by the reported releases of nitrosamines, nitramines and formaldehyde are within the range of health risks of toxic emissions from fossil power plants without CO_2 capture (Veltman et al., 2010; Da Silva and Booth, 2013). In LCAs, emissions from fuel production and the manufacturing and installation of the necessary equipment are of equal or larger importance than direct emissions during the capture process (Singh et al., 2011a). An increase of 40-80% in human toxicity impacts of fossil power plants with different CCS approaches has been reported, relative to their non-CCS counterparts. Particulate emissions from power plants with CCS are similar to those of similar plants without CCS, with differences ranging from a reduction of 10% to an increase of 20% (Singh et al., 2011a; Koornneef et al., 2012). The increase is due to the increased fuel requirements to run the CCS as well as equipment manufacturing and associated fuel chain emissions (Singh et al., 2011b; Koornneef et al., 2012). However, there is still a degree of technological uncertainty about the exact CCS solutions to be implemented and an insufficient understanding of emissions, reactions, and toxicity of the chemicals involved.

In this report, we do not include the potential human health impact from climate change in general. Other research suggests that the human health impact to be expected from climate change is more than the human health impact from particulate matter or toxic emissions (Singh et al., 2012). Since we aim to show the trade-off between climate mitigation and other environmental effects, we have elected not to combine the climate-related health effects with those arising from other mechanisms.

4.3 ECOSYSTEMS

The Millennium Ecosystem Assessment identifies habitat change, climate change, overharvesting (hunting and fishing), pollution (primarily nitrogen and phosphorus), and the introduction of alien species as the main threats to the Earth's biodiversity (Mooney et al., 2005). Ocean acidification is an important emerging concern, and pollution through acidifying and organic chemicals and heavy metals has caused regional or local impacts or impacts on particular species. Habitat change is the main driver of local and global species extinction today.

Climate change and ocean acidification – both linked to carbon emissions from burning of fossil fuels – are likely to substantially impact ecosystems by destroying the habitat of many species faster than these species can adapt or move (Emberson et al., 2012). Fossil fuel-based power systems also substantially increase atmospheric reactive nitrogen concentrations and mobilize phosphorus contained in coal, thereby contributing to the eutrophication of terrestrial, freshwater and marine ecosystems. Fossil fuel power plants further cause impacts through land use and toxic emissions such as mercury. Concerns about the biodiversity impacts of low-GHG energy systems mostly relate to habitat change caused by land use, water use, and the physical modification of the environment through structures such as dams, wind turbines, solar installations, and power lines. Pollution from mining, material processing and manufacture of the equipment contributes to eutrophication, acidification and toxic impacts. The quantitative assessment of such impacts, however, is complicated by the multitude of species and ecosystems to be protected and the difficulty of comparing the impacts on these ecosystems on a common scale. We were therefore unable to quantify all ecosystem impacts. Habitat change in particular is an issue that is very specific to site and project design parameters. We therefore discuss habitat impacts qualitatively and present land use requirements as an indicator for potential habitat change.

Fossil fuel-based power plants impact on the natural world through eutrophication (nitrogen and phosphorus pollution), acidification, toxic mine drainage, emissions of mercury and other toxic pollutants, climate change and ocean acidification (Chapter 3). Since similar pollution issues arise in the production of materials, an important consideration is the increased material requirements of renewable energy and carbon capture technologies. The comparison of life cycle impacts in Figure 13 indicates that only CCS leads to a modest increase in pollution-related ecosystem damage. Renewable energy technologies have significantly lower impacts than fossil power (Chapter 10). The emissions from material production for and the manufacture of renewable technologies are much lower than the combined emissions from mining, transport, and combustion of coal, as well as the waste treatment of ash.

Three of the CCS systems covered in this report use amine-based solvents that increase ammonia emissions and thus contribute to eutrophication and acidification. Increased fuel use causes further increases in eutrophication. Specific emissions from the carbon capture plant do not appear to be grounds for special concern, but it is important to pay attention to waste treatment. As with human health, the availability of emissions data related to capture plants is sparse.

For renewable electricity sources, the production of PV cells causes terrestrial and marine ecotoxicity. The pollution impacts from hydropower, wind power and CSP are small by comparison.

There are significant ecological concerns over habitat change resulting from the large-scale deployment of low-carbon technologies. The larger the land use, the greater the potential level of habitat change. The actual habitat change incurred, however, depends also on the specific project and site. For example, PV power may be produced in fertile valleys, pristine nature areas – or on rooftops and along highways. We apply land use rather than land use change as a generic indicator in this study, given that we do not investigate specific sites.

Figure 14 shows the intensive land use requirements for hydropower, coal power, CSP and PV. The lowest land use requirements are for power from natural gas combined cycle facilities, wind power and roof-mounted PV, where the roof area is not considered since the primary land use is attributed to the building itself. Bioenergy from dedicated plantations and forests, although not addressed in this assessment, entails even higher land use than hydropower. For direct land use associated with wind power, we consider only the area occupied by the windmill itself, its access roads and other installations, but not the land in between, because this land can be used as pasture, agricultural land or wilderness, with some restrictions. (If the entire wind park were used exclusively for power generation, land use would be on the order of 50-200 m^2a/MWh, which is substantial). For hydropower, the global average land use attributed to reservoirs is 100 m^2a/MWh (Barros et al., 2011) and is hence larger than the specific hydropower installations analyzed in this assessment.

FIGURE 14

Land occupation required for the production of electricity, Europe in 2010. For coal power, the dark green bar represents open pit mines (land use largely associated with the mine itself) and the total size of the bar reflects the land use associated with coal from underground mines. These underground mines use hard wood as structural material, which contributes most to land use.

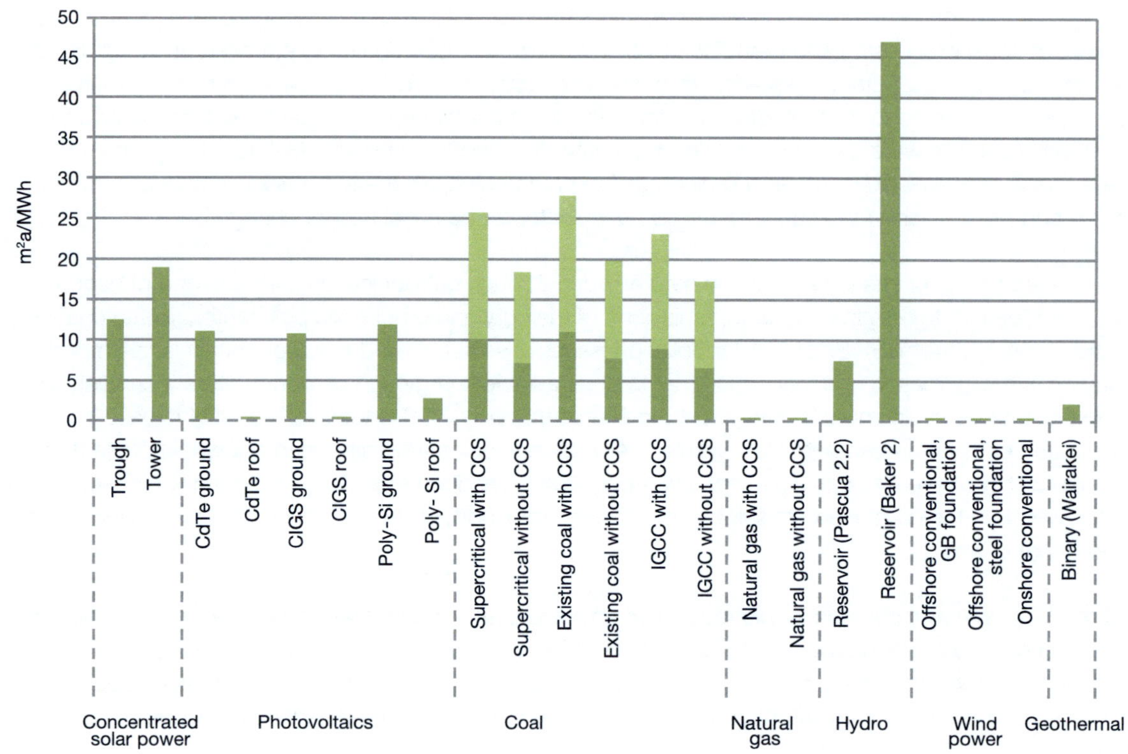

Not all land use is equivalent. The land use associated with open-pit coal mines or sealed surfaces of PV solar panels may have a greater ecological impact than the open water areas of hydroelectric reservoirs or the hardwood forest growing timber for underground coal mines. The ecological properties of the land during occupation vary significantly, as does the value of a site prior to its occupation.

Wind power and hydropower have specific ecological concerns. Wind power can cause injury and death to birds and bats through collisions (Chapter 5). Larger areas may thus become less suitable for particular species. Hydropower dams are migration barriers for fish and other aquatic species, and they change stream-flow and in-stream habitats (Chapter 4). River regulation also reduces flood plain habitat. Not all wind or hydro plants cause substantial impacts, however; some may in fact benefit local biodiversity. Offshore wind power creates new habitat for marine life, while creation of pondages and changes in flow regimes caused by some hydropower projects may benefit fish. Where sites and projects cause new pressures on particular species, it will be important to consider that these are additional to existing ecological pressures. There is, as yet, no clear understanding of the combined effects of these pressures, existing and new, on particular species. For example, the number of small bird deaths caused by collisions with wind power turbines is low compared to that caused by domestic cats, windows, or power lines, but large birds of prey are more frequently affected.

FIGURE 15

Bulk material and non-renewable energy requirements per unit power produced. Fossil technologies have high cumulative non-renewable energy demand (CED) and low bulk material requirements.

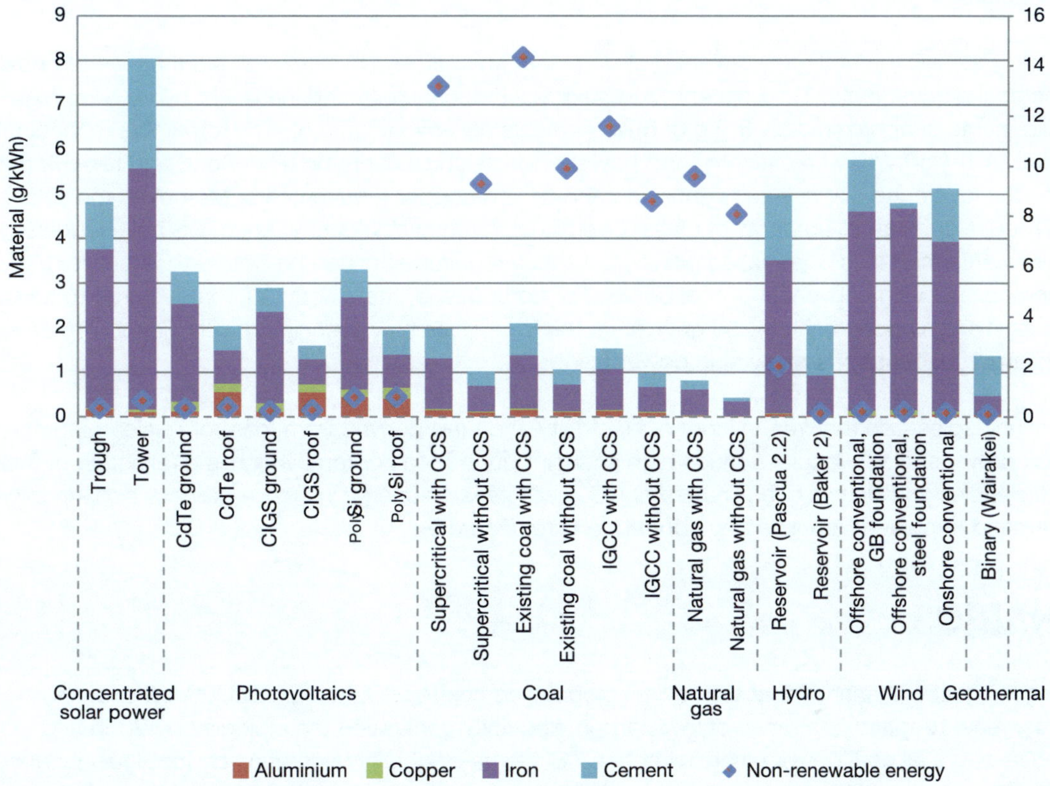

4.4 RESOURCES

The dependency of economies on finite fossil fuel reserves has been a concern since the Industrial Revolution (Jevons, 1866) and the current transition to less easily accessible resources is raising new and significant environmental issues. New technology is yielding access to plentiful fossil resources, albeit at higher prices and often, greater risk. While the peak in fossil fuel extraction is not considered imminent (Rogner et al., 2012), resources are finite and will eventually decline. Biomass was the first source of energy exploited by humans and, according to historians, the innovations of the industrial revolution came in response to a shortage of biomass for both energy and food. Land and water, which are the basis for growing biomass, are important resources for several of the clean technologies. Concerns have been raised recently about the availability of specialty minerals, such as rare-earth or transition metals, to provide permanent magnets for offshore wind power plants, or to manufacture concentrating PV solar cells (Andersson, 2000) and fuel cells (Kleijn and Van Der Voet, 2010).

Their increased use of land (see previous section), water and materials is often mentioned as a concern in the deployment of low-carbon technologies. The following resource indicators are considered in this report: the use of non-renewable energy; bulk materials such as cement, iron, aluminium, and copper; and an indicator for metal resource depletion. Often, there are technology-specific resources that may potentially be in low supply, such as rare earth metals for direct-drive wind turbines (Chapter 5.6), special metals for PV (Chapter 7.5.1), silver for CSP and PV (Chapter 7.4.1), and the availability of adequate storage space for CCS. Metals use in low-carbon technologies is addressed by the metal depletion indicator and discussed for specific technologies in the technology Chapters.

In terms of bulk materials, natural gas combined cycle plants and efficient hydropower plants generally have the lowest material requirements. The concentrated solar tower technology and inefficient hydro have high material requirements of approximately 8-9 g of bulk materials per kWh (Figure 5). The remaining technologies are in the range of 1-4 g/kWh, with offshore wind power and trough CSP on the higher end and roof PV and coal power on the lower end. PV has substantial aluminium and copper requirements. Moreover, the solar technologies require substantial amounts of glass. Overall, both renewable energy and CCS have higher material requirements than fossil fuel-based power, but these requirements can be limited to a four-fold increase in comparison with conventional coal power. For comparison, the amount of coal required to fuel a coal-fired power plant is approximately 250 g/kWh, so the total mass flow associated with a coal fired power plant is much larger than that of a renewable power plant.

Renewable technologies also require the input of electricity and fuels derived from fossil or nuclear sources. Naturally, fossil power stations require a much high energy input per unit output, which is a reflection of their conversion efficiencies and life cycle requirements. CO_2 capture is an energy intensive process and increases the energy demand of power production by about one third (Figure 5).

5. SCENARIOS

Like all climate mitigation scenarios that aim to limit global warming to 2°C, the IEA BLUE Map scenario foresees widespread adoption of a range of low-carbon electricity generation technologies by 2050 and the virtual phase-out of all coal power plants without CCS. Some electricity generation by fossil fuel power plants would be still exist in 2050, but most of it would be from power plants with CCS, and the remaining gas power plants without CCS would be used for fewer hours per year, serving mainly to balance variable renewable power production (Figure 16). As consequence of reduced coal use and CCS, greenhouse emissions associated with coal power would decline by 87% between 2010 and 2050.

The increasing market share of renewable electricity will also reduce the pollution impact per unit of electricity generated by a factor of two or more (Figure 16). In the face of continued growth in electricity supply, these improvements will enable us to stabilize particulate matter exposure and ecotoxic impacts on fresh water while at the same time reducing the emissions causing climate change and eutrophication. This downward trend contrast to the baseline scenario, where the increased use of coal and gas would lead to a proportionate increase in all its environmental impacts (Figure 16).

The mitigation scenario will lead to a reduction of the use of non-renewable energy resources and, surprisingly, also in land use. However the widespread deployment of low-carbon technologies also implies increased investment in infrastructure which leads to a greater demand for iron and steel, cement, and copper (Figure 16). Interestingly, the installation rate of new renewable power capacity in recent years is at a level that is, if sustained, consistent with the BLUE Map scenario (Figure 16). CSP and wind power plants will cause additional demand for cement and iron, while PV will lead to additional requirements for copper.

FIGURE 16

Impact indicators, resource demand and deployment characteristics of the investigated power generation technologies under the IEA BLUE Map scenario, consistent with the goal of limiting global warming to two degrees above pre-industrial level.

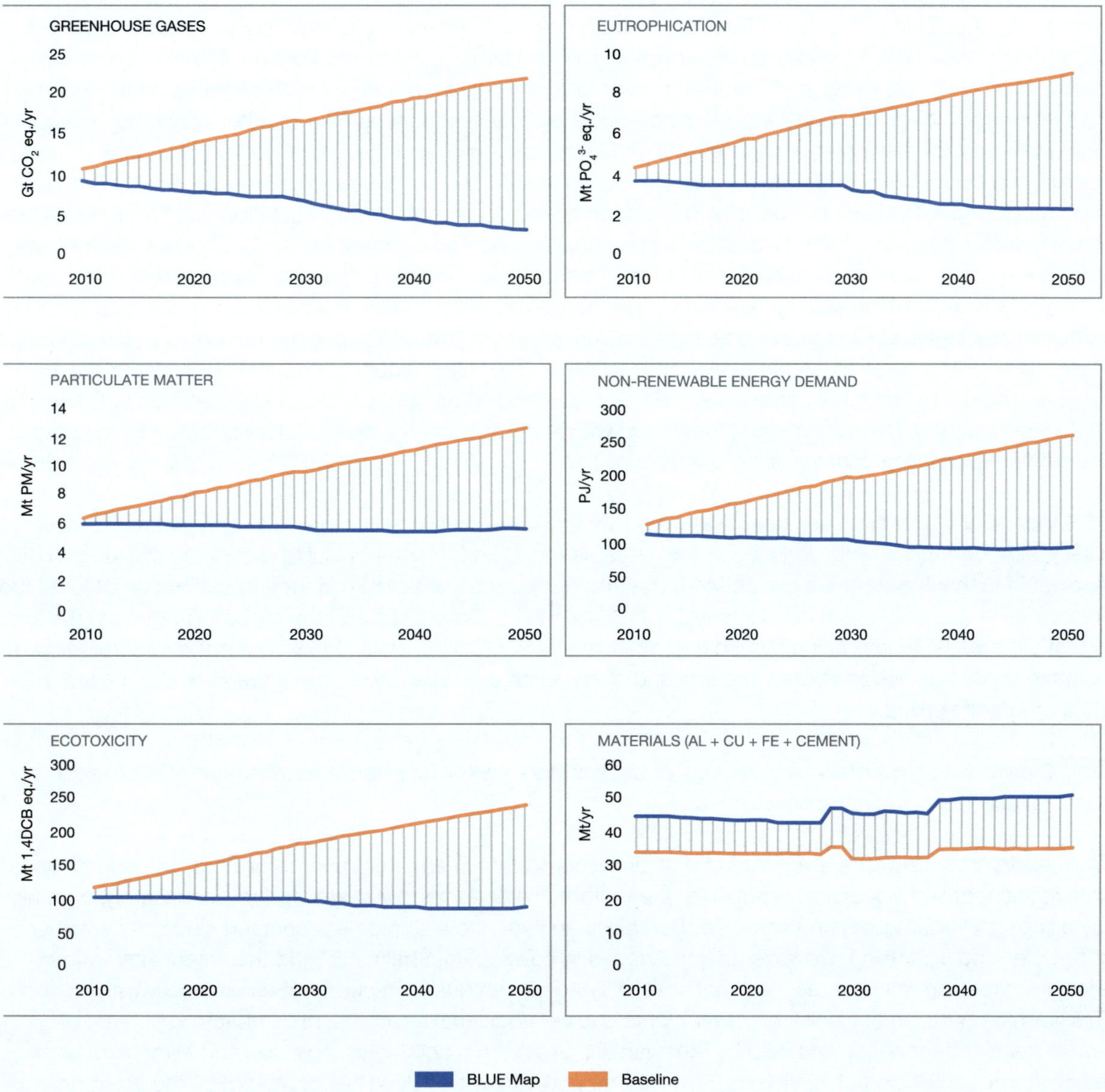

Source: Hertwich et al. under review

6. CONCLUSIONS

In replacing conventional fossil fuel-based power plants, renewable energy technologies offer substantial reductions in GHG and other pollutant emissions at the same time. The capture and storage of CO_2 from fossil fuel based power plants offers also substantial reduction in GHG emissions, but without the benefit of reducing other types of pollution. This report finds strong evidence that renewable electricity causes substantially less pollution than electricity from fossil fuels, including pollution causing eutrophication, acidification, particulate matter, photochemical smog, and various forms of toxicity (Table 2). Renewables also reduce dependence on finite reserves of fossil fuel. Renewable technologies, however, lead to a number of other concerns, principally their direct ecological impacts associated with land and water use, and their increased consumption of iron, cement and copper. Similar ecological impacts related to land and water use are also associated with fossil fuels: for example, land use by coal mining is similar in scale to that of wind and solar power. Fossil power plants use somewhat less water than do geothermal and concentrated solar power plants, but options such as air-cooling are now becoming available for all of these technologies. Proper project selection, design and operation will mitigate most adverse ecological impacts. The modest increase in iron and cement use associated with low-carbon technologies does not pose a serious problem given the availability of those resources and the relatively small share of total demand related to electricity systems. The use of copper and functionally important metals, however, may pose some concerns in the long term, depending on opportunities for substitution which are not yet fully understood. Overall, replacing fossil fuels with renewable energy offers a clear opportunity to reduce environmental pollution from electricity generation.

CCS technology also promises to substantially reduce GHG emissions compared to conventional power plants, although these reductions are not as large as those from most renewable technologies (Table 2). For renewable technologies, there is a concern that their use does not actually reduce the utilization of fossil fuels, but comes in addition. In power plants with CCS, the fuel is combusted and the carbon is stored in geological formation, thus no longer being available to other markets. CCS, however, leads to a moderate but uniform increase of most emission-related impacts and of resource use. In addition, the storage of CO_2 needs to be monitored and verified.

Table 2 provides an overview over findings of the literature review (qualitative assessment of ecological impacts) and the quantitative assessment.

While environmentally attractive, wind and solar resources are intermittent and do not provide a continuous or readily controlled electricity output. In some regions, peak demand is correlated to peak supply, e.g., air conditioning in hot regions and sunshine, but this is an exception. Customers demand electricity whether or not the wind is blowing. However the challenges in developing a balanced grid that integrates various energy sources are modest, as fossil-dominated systems can quickly respond to variable renewable supply. This flexible operation of fossil fuel power plants causes additional environmental impacts which are of similar magnitude to those imposed by renewables. Given the much higher pollution and climate impacts resulting from only using of fossil fuels, we find that grid integration does not compromise the environmental benefits of renewables in the medium term. However, integration challenges become more serious when variable renewables dominate the electricity mix. Building larger and stronger transmission grids, utilizing energy storage and flexible demand, and relying on a variety of uncorrelated sources of renewable energy are all promising response strategies. Indeed, grid integration challenges provide a persuasive rationale for use of concentrated solar power alongside thermal energy storage, wave power, tidal power, and offshore wind (which deliver energy at other points in time and with a higher capacity factor than onshore wind). Few assessments are currently available on the environmental impacts of power transmission and energy storage, but they indicate that strengthening and extending electricity grids has lower impacts than the forms of energy storage investigated. Further research and development will be needed to design integrated electricity systems with average emissions below 100 gCO_2e/kWh.

TABLE 2

Overview over the impacts of low-carbon technologies for electricity generation on climate, human health, ecosystems and resources, comparing state of the art power plants at well-suited locations. The reference is the current global mix.

	Climate	Human health	Ecosystem health	Resources
Wind	Low GHG (++)	Reduced particulate exposure (++) Potentially reduced human toxicity (--)	Bird and bat collisions (+=) Low ecotoxicity and eutrophication (=-)	High metal consumption (+=) Low water use and direct land use (==)
Photovoltaics	Low GHG (==)	Low PM (+=) Low HT (=-)	Low eutrophication and ecotoxicity (+-)	High metal use (+=) High direct land use for ground-based systems (++)
Concentrated Solar Power	Low GHG (==)	Low PM (=-) Low HT (=-)	Concern about heat transfer fluid (+=) Low eutrophication and ecotoxicity (+-)	High water use (++) High land use (++)
Hydropower	Low fossil GHG (++) High biogenic GHG from some dams (==)	Low air pollution impacts (=-)	Riparian habitat change (reservoir and downstream) (++)	Water use due to evaporation (+-) Land use for reservoirs (+=)
Geothermal power	Low fossil GHG (+-) Geogenic GHG for some types (=+)	Air and water pollution from geofluid flow in some sites (=-)	Aquatic habitat change/pollution (+=)	Cooling water use (+=)
Gas+CCS Coal+CCS	Low GHG (++) Substantial fugitive methane emissions (==) Concern about CO_2 leakage (-=)	Solvent-related emissions (==) High PM (==) High HT (++)	High eutrophication (++) High ecotoxicity (+=)	Increased fossil fuel consumption (++) Limited CO_2 storage volume (++)

Key to the assessment (##):

First symbol	(++) high agreement among studies	(==) moderate agreement	(--) low agreement
Second symbol	(+) robust evidence (many studies)	(=) medium evidence	(-) limited evidence

The key to future energy decisions lies in determining the right mix of technologies for the local or regional situation and policy objectives. This demands careful assessment of all the impact categories of the different energy alternatives, to avoid unintended negative consequences, and to achieve the most desirable mix of environmental, social and economic benefits.

The report shows that LCA is of central importance in determining the sustainability of different energy options. The Global Tracking Framework of the Sustainable Energy for All initiative[2] points out that sound criteria are needed to distinguish between different actions and technology choices in terms of their ultimate sustainability. These criteria will help ensure that overall sustainability goals are met and that actions are in line with global targets, such as the two degree warming target under the UN Framework Convention on Climate Change and the Aichi Biodiversity targets under the Convention on Biological Diversity.

This report lays the foundation for developing such sustainability criteria to support good decisions about the energy sources that will influence the whole human future.

7. LITERATURE

Andersson, B. A. 2000. Materials availability for large-scale thin-film photovoltaics. *Prog. Photovoltaics* 8(1): 61-76.

Arvesen, A. and E. G. Hertwich. 2011. Environmental implications of large-scale adoption of wind power: a scenario-based life cycle assessment. *Environmental Research Letters* 6(4): 045102.

Arvesen, A., C. Birkeland, and E. G. Hertwich. 2013. The importance of ships and spare parts in LCAs of offshore power. *Environmental Science & Technology* 47(6): 2948-2956.

Asmal, K., L. C. Jain, J. Henderson, G. Lindahl, T. Scudder, J. Cariño, D. Blackmore, M. Patkar, J. Goldemberg, D. Moore, J. Veltrop, and A. Steiner. 2000. *Dams and development - a new framework: The report of the World Commission on Dams*. London: Earthscan.

Barros, N., J. J. Cole, L. J. Tranvik, Y. T. Prairie, D. Bastviken, V. L. M. Huszar, P. Del Giorgio, and F. Roland. 2011. Carbon emission from hydroelectric reservoirs linked to reservoir age and latitude. *Nature Geoscience* 4(9): 593-596.

Bashmakov, I. A., T. Bruckner, Y. Mulugetta, H. Chum, A. D. L. V. Navarro, J. Edmonds, A. Faaij, B. Fungtammasan, A. Garg, E. Hertwich, D. Honnery, D. Infield, M. Kainuma, S. Khennas, S. Kim, H. B. Nimir, K. Riahi, N. Strachan, R. Wiser, and X. Zhang. 2014. Energy Systems. In *Climate Change 2014: Mitigation of Climate Change*, edited by O. Edenhofer, et al. Genevea: Intergovernmental Panel on Climate Change.

Bringezu, S., H. Schutz, M. O'Brien, L. Kauppi, R. W. Howarth, and J. McNeely. 2009. *Towards sustainable production and use of resources: Assessing Biofuels*. Paris: United Nations Environment Programme.

Burkhardt, J. J., G. A. Heath, and C. S. Turchi. 2011. Life Cycle Assessment of a Parabolic Trough Concentrating Solar Power Plant and the Impacts of Key Design Alternatives. *Environmental Science & Technology* 45(6): 2457-2464.

Da Silva, E. F. and A. M. Booth. 2013. Emissions from postcombustion CO_2 capture plants. *Environmental Science and Technology* 47(2): 659-660.

2 SE4All is an initiative launched by the UN Secretary General gathering support by a wide range of partners to reach three complementary objectives by 2030: universal access to modern energy services, doubling the share of renewables in the global energy mix, and doubling the global rate of improvement of energy efficiency. www.sustainableenergyforall.org

Demarty, M. and J. Bastien. 2011. GHG emissions from hydroelectric reservoirs in tropical and equatorial regions: Review of 20 years of CH 4 emission measurements. *Energy Policy* 39(7): 4197-4206.

EC. 2010. *Critical raw materials for the EU: Report of the Ad-hoc Working Group on defining critical raw materials*. Brussels: European Commission, DG Enterprise and Industry.

Edmonds, J., P. Luckow, K. Calvin, M. Wise, J. Dooley, P. Kyle, S. H. Kim, P. Patel, and L. Clarke. 2013. Can radiative forcing be limited to 2.6 Wm-2 without negative emissions from bioenergy AND CO_2 capture and storage? *Climatic Change* 118(1): 29-43.

Emberson, L., K. He, J. Rockström, M. Amann, J. Barron, R. Corell, S. Feresu, R. Haeuber, K. Hicks, F. X. Johnson, A. Karlqvist, Z. Klimont, I. Mylvakanam, W. W. Song, H. Vallack, and Z. Qiang. 2012. Chapter 3 - Energy and Environment. In *Global Energy Assessment - Toward a Sustainable Future*. Cambridge University Press, Cambridge, UK and New York, NY, USA and the International Institute for Applied Systems Analysis, Laxenburg, Austria.

EPIA. 2013. *Global Market Outlook for Photovoltaics 2013-2017*. European Photovoltaic Industry Association.

Fischedick, M., R. Schaeffer, A. Adedoyin, M. Akai, T. Bruckner, L. Clarke, V. Krey, I. Savolainen, S. Teske, D. Ürge-Vorsatz, and R. Wright. 2011. Mitigation Potential and Costs. In *IPCC Special Report on Renewable Energy Sources and Climate Change Mitigation*, edited by O. Edenhofer, et al. Cambridge (UK): Cambridge University Press.

Gibon, T., R. Wood, A. Arvesen, J. D. Bergesen, S. Suh, and E. G. Hertwich. 2015. A Methodology for Integrated, Multiregional Life Cycle Assessment Scenarios under Large-Scale Technological Change. *Environmental Science & Technology* 49(18): 11218-11226.

Goedkoop, M., R. Heijungs, M. Huijbregts, A. De Schryver, J. Struijs, and R. Van Zelm. 2008. *ReCiPe 2008: A life cycle impact assessment method which comprises harmonised category indicators at the midpoint and the endpoint level. First edition: Report I: Characterisation*. The Hague, NL: Dutch Ministry of the Environment.

Goodrich, A. W., M.; Noufi, R. 2011. CIGS Road Map. *NREL Technical Report (In preparation)*.

Hertwich, E. G., T. Gibon, E. A. Bouman, A. Arvesen, S. Suh, A. Ramírez, M. Vega Colorna, J. D. Bergesen, L. Shi, and G. A. Heath. under review. Resource requirements and environmental benefits of low-carbon electricity supply.

IEA. 2010. *Energy technology perspectives 2010: Scenarios and strategies to 2050*. Paris: OECD/IEA.

IEA. 2014. *World Energy Outlook 2014*. Paris: OECD/IEA.

Jevons, W. S. 1866. *The coal question: an inquiry concerning the progress of the nation, and the probable exhaustion of our coal mines*. London: Macmillan & Co.

Karl, M., R. F. Wright, T. F. Berglen, and B. Denby. 2011. Worst case scenario study to assess the environmental impact of amine emissions from a CO_2 capture plant. *International Journal of Greenhouse Gas Control* 5(3): 439-447.

Kleijn, R. and E. Van Der Voet. 2010. Resource constraints in a hydrogen economy based on renewable energy sources: An exploration. *Renewable and Sustainable Energy Reviews* 14(9): 2784-2795.

Koornneef, J., A. Ramírez, W. Turkenburg, and A. Faaij. 2012. The environmental impact and risk assessment of CO 2 capture, transport and storage - An evaluation of the knowledge base. *Progress in Energy and Combustion Science* 38(1): 62-86.

Kumar, A., T. Schei, A. Ahenkorah, R. C. Rodriguez, J.-M. Devernay, M. Freitas, D. Hall, Å. Killingtveit, and Z. Liu. 2011. Hydropower. In *IPCC Special Report on Renewable Energy Sources and Climate Change Mitigation*, edited by O. Edenhofer, et al. Cambridge, United Kingdom and New York, NY, USA: Cambridge University Press.

Lim, S. S.T. VosA. D. FlaxmanG. DanaeiK. ShibuyaH. Adair-RohaniM. AmannH. R. AndersonK. G. AndrewsM. AryeeC. AtkinsonL. J. BacchusA. N. BahalimK. BalakrishnanJ. BalmesS. Barker-ColloA. BaxterM. L. BellJ. D. BloreF. BlythC. BonnerG. BorgesR. BourneM. BoussinesqM. BrauerP. BrooksN. G. BruceB. BrunekreefC. Bryan-HancockC. BucelloR. BuchbinderF. BullR. T. BurnettT. E. ByersB. CalabriaJ. CarapetisE. CarnahanZ. ChafeF. CharlsonH. ChenJ. S. ChenA. T.-A. ChengJ. C. ChildA. CohenK. E. ColsonB. C. CowieS. DarbyS. DarlingA. DavisL. DegenhardtF. DentenerD. C. Des JarlaisK. DevriesM. DheraniE. L. DingE. R. DorseyT. DriscollK. EdmondS. E. AliR. E. EngellP. J. ErwinS. FahimiG. FalderF. FarzadfarA. FerrariM. M. FinucaneS. FlaxmanF. G. R. FowkesG. FreedmanM. K. FreemanE. GakidouS. GhoshE. GiovannucciG. GmelK. GrahamR. GraingerB. GrantD. GunnellH. R. GutierrezW. HallH. W. HoekA. HoganH. D. HosgoodD. HoyH. HuB. J. HubbellS. J. HutchingsS. E. IbeanusiG. L. JacklynR. JasrasariaJ. B. JonasH. KanJ. A. KanisN. KassebaumN. KawakamiY.-H. KhangS. KhatibzadehJ.-P. KhooC. KokF. LadenR. LallooQ. LanT. LathleanJ. L. LeasherJ. LeighY. LiJ. K. LinS. E. LipshultzS. LondonR. LozanoY. LuJ. MakR. MalekzadehL. MallingerW. MarcenesL. MarchR. MarksR. MartinP. McGaleJ. McGrathS. MehtaG. A. MensahT. R. MerrimanR. MichaC. MichaudV. MishraK. M. HanafiahA. A. MokdadL. MorawskaD. MozaffarianT. MurphyM. NaghaviB. NealP. K. NelsonJ. M. NollaR. NormanC. OlivesS. B. OmerJ. OrchardR. OsborneB. OstroA. PageK. D. PandeyC. D. H. ParryE. PassmoreJ. PatraN. PearceP. M. PelizzariM. PetzoldM. R. PhillipsD. PopeC. A. PopeJ. PowlesM. RaoH. RazaviE. A. RehfuessJ. T. RehmB. RitzF. P. RivaraT. RobertsC. RobinsonJ. A. Rodriguez-PortalesI. RomieuR. RoomL. C. RosenfeldA. RoyL. RushtonJ. A. SalomonU. SampsonL. Sanchez-RieraE. SanmanA. SapkotaS. SeedatP. ShiK. ShieldR. ShivakotiG. M. SinghD. A. SleetE. SmithK. R. SmithN. J. C. StapelbergK. SteenlandH. StöcklL. J. StovnerK. StraifL. StraneyG. D. ThurstonJ. H. TranR. Van DingenenA. van DonkelaarJ. L. VeermanL. VijayakumarR. WeintraubM. M. WeissmanR. A. WhiteH. WhitefordS. T. WiersmaJ. D. WilkinsonH. C. WilliamsW. WilliamsN. WilsonA. D. WoolfP. YipJ. M. ZielinskiA. D. LopezC. J. L. Murray and M. Ezzati. 2012. A comparative risk assessment of burden of disease and injury attributable to 67 risk factors and risk factor clusters in 21 regions, 1990-2010: a systematic analysis for the Global Burden of Disease Study 2010. *The Lancet* 380(9859): 2224-2260.

Mooney, H. A., A. Cropper, D. Capistrano, S. R. Carpenter, K. Chopra, P. Dasgupta, R. Leemans, R. M. May, P. Pingali, R. Hassan, C. Samper, R. Scholes, R. T. Watson, A. H. Zakri, and Z. Shidong, eds. 2005. *Ecosystems and human well-being: synthesis*, *Millennium Ecosystem Assessment*. Washington, DC: Island Press.

Riahi, K., F. Dentener, D. Gielen, A. Grubler, J. Jewell, Z. Klimont, V. Krey, D. McCollum, S. Pachauri, S. Rao, B. van Ruijven, D. P. van Vuuren, and C. Wilson. 2012. Chapter 17 - Energy Pathways for Sustainable Development. In *Global Energy Assessment - Toward a Sustainable Future*. Cambridge, UK and New York, NY, USA: Cambridge University Press.

Rogner, H.-H., R. F. Aguilera, R. Bertani, S. C. Bhattacharya, M. B. Dusseault, L. Gagnon, H. Haberl, M. Hoogwijk, A. Johnson, M. L. Rogner, H. Wagner, and V. Yakushev. 2012. Chapter 7 - Energy Resources and Potentials. In *Global Energy Assessment - Toward a Sustainable Future*. Cambridge University Press, Cambridge, UK and New York, NY, USA and the International Institute for Applied Systems Analysis, Laxenburg, Austria.

Singh, B., A. H. Strømman, and E. G. Hertwich. 2011a. Comparative life cycle environmental assessment of CCS technologies. *International Journal of Greenhouse Gas Control* 5(4): 911-921.

Singh, B., A. H. Strømman, and E. G. Hertwich. 2011b. Life cycle assessment of natural gas combined cycle power plant with post-combustion carbon capture, transport and storage. *International Journal of Greenhouse Gas Control* 5(3): 457-466.

Singh, B., A. H. Strømman, and E. G. Hertwich. 2012. Environmental Damage Assessment of Carbon Capture and Storage. *Journal of Industrial Ecology* 16(3): 407-419.

Smith, K. R., K. Balakrishnan, C. Butler, Z. Chafe, I. Fairlie, P. Kinney, T. Kjellstrom, D. L. Mauzerall, T. McKone, A. McMichael, and M. Schneider. 2012. Chapter 4 - Energy and Health. In *Global Energy Assessment - Toward a Sustainable Future*. Cambridge University Press, Cambridge, UK and New York, NY, USA and the International Institute for Applied Systems Analysis, Laxenburg, Austria.

Veltman, K., B. Singh, and E. G. Hertwich. 2010. Human and environmental impact assessment of postcombustion CO_2 capture focusing on emissions from amine-based scrubbing solvents to air. *Environmental Science and Technology* 44(4): 1496-1502.

Whitaker, M. B., G. A. Heath, J. J. Burkhardt, and C. S. Turchi. 2013. Life Cycle Assessment of a Power Tower Concentrating Solar Plant and the Impacts of Key Design Alternatives. *Environmental Science & Technology* 47(11): 5896-5903.

Woodhouse, M., A. Goodrich, R. Margolis, T. L. James, M. Lokanc, and R. Eggert. 2013. Supply-Chain Dynamics of Tellurium, Indium, and Gallium Within the Context of PV Module Manufacturing Costs. *IEEE Journal of Photovoltaics* PP(99): 1-5.

Woodhouse, M., A. Goodrich, R. Margolis, T. James, T. Barnes, T. Gessert, D. Albin, and R. Eggert. 2011. *Perspectives on the Pathways for Cadmium Telluride Module Manufacturers to Address Expected Increases in Tellurium Price (In Preparation)*. Golden, CO: National Renewable Energy Laboratory (NREL).

Chapter 1

Introduction

Lead authors: Edgar G. Hertwich, Thomas Gibon
Contributing author: Sangwon Suh

1.1 OBJECTIVE OF THIS ASSESSMENT

Anthropogenic greenhouse gas (GHG) emissions are the major cause of global warming (Stocker et al., 2013). The Intergovernmental Panel on Climate Change (IPCC) has identified electricity production as the single most important source of anthropogenic GHG emissions. Electricity production is responsible for 25 per cent of total emissions; this is approximately the same as the combination of deforestation, agriculture and other land use change, and is more than emissions attributed to industry (21 per cent) or transport (14 per cent) (IPCC, 2014). Climate change mitigation policy will likely increase the importance of electricity as an energy carrier, as electricity can be produced with lower emissions than most other energy carriers (Bruckner et al., 2014).

Substantial emissions mitigation policies have been proposed, and as a result, the IPCC Working Group III (WGIII) has investigated options for technological change and improvement of cost, feasibility and infrastructure requirements for existing technologies (IPCC, 2014). The International Resource Panel (IRP) study on priority products and materials identified fossil fuel extraction and use as a major source of not only GHG emissions, but also of most pollution-related environmental impacts, including eutrophication, acidification, particulate matter exposure, and toxicity (Hertwich et al., 2010). Technologies mitigating GHG emissions also require resources and cause various environmental impacts throughout their life cycle, regardless of whether they are cleaner energy technologies, such as renewable energy or nuclear power, or cleaning technologies, such as CO_2 capture and storage, or energy efficiency such as additional building insulation. Having higher investment costs, these cleaner energy technologies demand more materials, manufacturing and construction activities than conventional fossil fuel-fired power plants. Some scientific studies suggest substantial impacts from the construction of renewable power plants. In addition, there has recently been a focus on the availability of the minerals used in energy technologies. Moreover, the public is clearly concerned about the impacts on wildlife and visual effects of some technologies. Media reports suggest substantial impacts from land use and question the ability of novel technologies to provide a substantial fraction of electricity demand. There is hence a need to better understand the environmental and resource impacts of different energy technologies, both at a per-delivered-kWh basis in a direct comparison with each other, and their role in a large-scale implementation scenario.

While fossil fuel power plants have large combustion-related emissions, the environmental impacts of renewable power plants are mostly connected to the manufacturing and installation of the power conversion devices. The assessment should hence consider all relevant impacts from the extraction of the materials used to construct or operate the power plants to the plant's final decommissioning.

Substantial investments in low carbon electricity generation are required in order to meet the climate stabilization target of limiting global average temperature rise to 2°C, set forth by the international community in the Cancún agreement, or the more ambitious Paris agreement. The purpose of this report is to inform decision makers about technology selection and the design of energy projects. Energy scenarios analysed by the IPCC clearly indicate that the investment in several low carbon energy supply technologies is required to meet climate targets. No single technology will meet all needs, but at the same time, not all available technologies are required. Decision makers hence face a technology choice. In making this choice, decision makers should take into account more than technical feasibility and economic costs. They should also consider implications for human health, ecosystems, and resources. Often, decisions are not principal decisions about whether or not to employ a specific technology, but what energy project to realize. Project selection and design choices can substantially influence the extent of environmental impact. The purpose of this report is hence both to provide an understanding of the general environmental characteristics of different power production technologies and to identify issues that vary on a project-by-project basis so that decision makers can institute routines which ensure that the selection and design of energy projects limits environmental impacts.

With this study, we aim to provide a consistent, comparative assessment of the environmental consequences and resource requirements of electricity generation technologies with low GHG emissions, known as low carbon technologies. The study adopts a life cycle perspective that accounts for the environmental impacts and resource requirements associated with the extraction and transport of the fuels, as well as the construction, operation, maintenance and dismantling of the power plants (Verbruggen et al., 2011). Life cycle assessment is the method of choice for assessing and comparing the environmental impacts of products. It is further described in Chapter 2. This study is part of an ongoing effort to investigate resource requirements and environmental impacts of climate mitigation technologies addressing both the supply of and the demand for energy. The IRP has already commenced a follow-up report on demand-side mitigation measures. The current assessment consists of a systematic review of the literature, an extended assessment of consistent life cycle inventory (LCI) data collected by teams of experts, and a modelling of the global environmental pressure resulting from introducing technologies following widely used scenarios. The review encompasses literature addressing site-specific, ecological impacts. The assessment extends scenario-based life cycle assessment (LCA) (Arvesen and Hertwich, 2011; Singh, 2011; Viebahn et al., 2011) to include, for the first time, changes in the upstream energy mix used for power plant construction as a result of the implementation of the investigated technology. The assessment is global in scope and has a very modest level of geographical detail, representing the world in nine regions following the International Energy Agency's (IEA) energy scenario model. As a result, we do not assess local, site-specific ecological impacts in a comprehensive manner. Rather, we use a simple literature review as the basis to identify issues to be considered when planning new installations.

The IPPC WGIII has reviewed the life cycle GHG emissions from and the expected future costs of different mitigation technologies. It has also provided information on some selected air pollutants, without assessing the environmental or health impacts of these species (Bruckner et al., 2014; Sathaye et al., 2011). Currently available GHG emissions from different energy technologies are from many discrete studies that have specific assumptions, and hence inherently lack comparability due to incompatible assessment principles, system boundaries and scenario assumptions. The present study goes beyond a review of the LCA literature to produce consistent life cycle inventories assessed within a single life cycle assessment model.

Substantial deployment of low carbon energy technologies will change the economy-wide energy mix and lower its carbon intensity. Most published life cycle assessments do not address the potential impact of such changes. The present study investigates the widespread deployment of clean technologies following a mitigation scenario and thus accounts for the effect of changes in the electricity mix on the environmental impact of manufacturing new energy conversion devices.

1.2 SCOPE

1.2.1 TEMPORAL AND GEOGRAPHICAL SCOPE

The assessment addresses the wide scale adaptation of low carbon technology for electricity generation, investigating one particular scenario with low GHG emissions, the IEA BLUE Map scenario, and compares this scenario with the Baseline scenario. The scenarios, taken from the IEA Energy Technology Perspectives (IEA, 2010), are global in scope, represent the world in nine regions, and provide information on the adaptation and characteristics of the technologies in 2030 and 2050. The BLUE Map scenario is a climate change mitigation scenario moving towards the 2°C target and requires stringent climate policies. The Baseline scenario does not assume any additional policy adoptions and sets the world on a pathway towards a global temperature increase 5-6°C.

1.2.2 TECHNOLOGY SCOPE

The assessment is based on a comparison of clean technologies that are relevant for climate mitigation with conventional fossil fuel power plants. The technology choice was strongly influenced by the IEA's Energy Technology Perspectives (IEA, 2010), and aimed to cover relevant power sources (Figure 1.1).

The comparative analyses includes coal- and gas-fired power both with and without CO_2 capture and storage (CCS), hydropower, wind power, photovoltaic power, concentrating solar power (CSP), and geothermal power. These particular technologies were selected because they all play an important role in future energy scenarios. We did not address nuclear power due to unexplained divergences by one order of magnitude in the results of different assessments (Lenzen, 2008; Warner and Heath, 2012). We also omit bioenergy because we did not have access to a sufficiently detailed land use model nor to scenarios concerning nutrition and urbanization; these are key aspects of bioenergy and would substantially influence the conclusions made.

FIGURE 1.1

Average annual electricity capacity additions to 2050 needed to achieve the BLUE Map scenario
Dark sections indicate historical production capacity, and blue sections indicate additional capacity. Note that in recent years, the rate of installation of new PV capacity has increased rapidly and has now reached the level indicated as necessary.

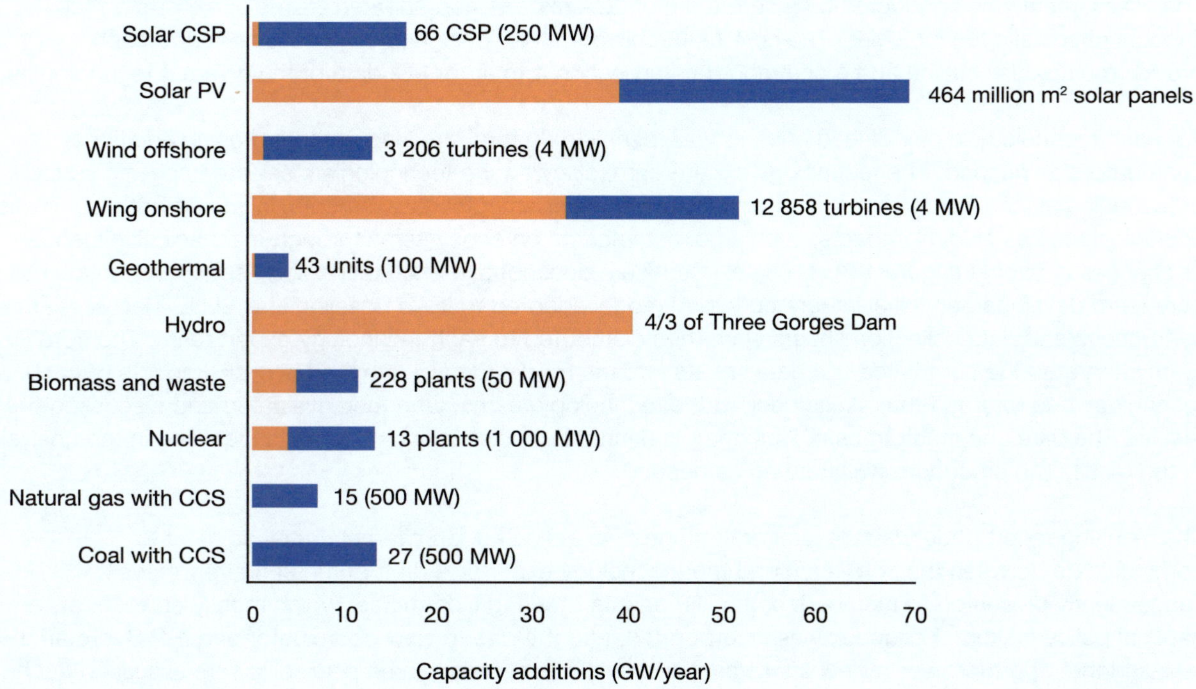

Source: REN21, 2014; IEA, 2014

1.2.3 SCOPE OF THE INVENTORY

The report addresses life cycle steps from the extraction of resources to the dismantling and removal of the power plant, but not recycling or waste treatment processes. The exclusion of this part of the end-of-life is justified by the long lifetime of the power plants and the poor availability of technology descriptions, as well the uncertainty regarding what waste treatment would be available so far in the future. The life cycle assessments presented in the report combine information on the inputs of material and energy from process-based life cycle databases with inputs of services from input-output tables, following the methodology of hybrid LCA (Lenzen, 2002; Suh et al., 2004). In this manner, the inventories are more complete than most LCAs.

1.2.4 IMPACT ASSESSMENT

In this report, we seek to address environmental impacts as identified in the scientific literature. The quantitative analysis focuses on those issues easily quantified based on current methods of life cycle impact assessment. The Chapters also identify site-specific issues for which no life cycle impact assessment methods are available, such as bird collisions with wind power plants or groundwater issues potentially arising from the geological storage of CO_2. The life cycle impact assessment is based on the ReCiPe 1.3 method, addressing both midpoints reflecting specific environmental mechanisms, such as the contribution to freshwater ecotoxicity and terrestrial acidification, and endpoints describing the damages to human health and ecosystems (Goedkoop et al., 2008).

1.3 ASSESSMENT PROCEDURE

The assessment was conducted by a core team of LCA experts and affiliated teams of experts on the impacts of specific technologies. The core team collaboratively designed an assessment approach, providing a chapter outline and a common data collection format for LCI data of the different technologies.

For each technology, a pair of lead authors was identified. In turn, the lead authors drew in additional contributors as needed. The technology expert team reviewed the technologies, selected specific subsets of technologies and issues to address, and reviewed environmental assessments, including both LCAs and ecological studies of local impacts, such as direct impacts on ecosystems through mechanical influence, or changes in habitat (Figure 1.2). These elements are described in each technology-specific chapter. The core team developed an initial integrated hybrid model adapted to IEA's Baseline and BLUE Map scenarios. Instructions and data collection sheets were then distributed to each technology expert team. The core team integrated the completed LCI data sheets into the model to calculate the life cycle impacts of each technology and returned the results back to the technology teams, who then evaluated and discussed the results. The data and result transfer was complemented by a discussion of aspects identified in the analysis of the result. The procedure was iterated as needed.

The resulting report underwent several draft stages. As indicated, an internal draft was used to communicate between the core team and the technology experts. A draft of the entire report was subsequently presented to members of the IRP and its Steering Committee, which mostly consists of national policy makers. Feedback was incorporated, and the peer review coordinator and IRP Secretariat released the report for peer review after Panel approval. On average, three peer reviewers assessed each chapter. In addition, three reviewers received the entire report, with the request to review, in particular, the logic of the report and the Introduction and results chapters. Sections of the report were also published as scientific journal papers that were submitted during the course of this work (Arvesen and Hertwich, 2012; Bayer et al., 2013; Bergesen et al., 2014; Hertwich, 2013; Hertwich et al., 2015; Corsten et al., 2013).

FIGURE 1.2

Flow chart of the assessment procedure for this report

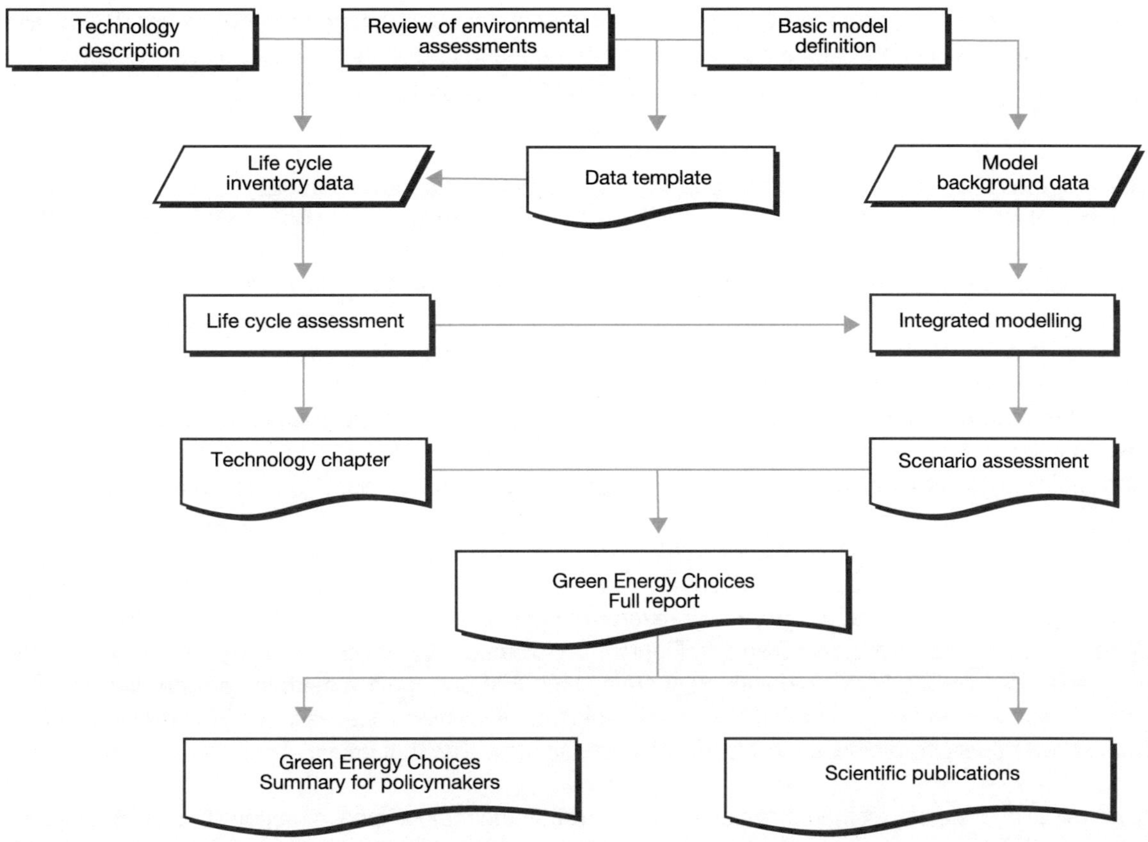

1.4 PREVIOUS WORK

Studies have long addressed life cycle energy use (Boustead and Hancock, 1979; Herendeen et al., 1979), environmental impacts and resource requirements of specific energy technologies from a life cycle perspective. Specific assessment challenges have been noted, particularly system boundary issues, allocation, and the "apples and oranges" comparison of different emissions species resulting from their varying environmental impacts (Holdren, 1982; Holdren et al., 1980). To compare different environmental pressures, impact assessment methods have been developed (Udo de Haes et al., 2002). To address system boundary issues and questions of scope (Majeau-Bettez et al., 2011), hybrid LCI methods were developed. The most systematic and influential effort of data collection and assessment of energy technologies were produced by the Swiss Inventory of Energy Technologies, which developed into the widely adopted ecoinvent database (Frischknecht and Jungbluth, 2007). Similar efforts were made in other countries, particularly in Japan and the United States. Only recently have efforts been made to systematically compare technologies using LCA results (Jacobson, 2009; Lenzen, 2010; Pehnt, 2006). Such efforts may be problematic as they usually begin as disparate studies conducted using inconsistent system definitions and arbitrary assumptions or descriptions of local circumstances, such that the results obtained from a particular study are not directly comparable to that of other studies (Farrell et al., 2006). A comprehensive collection and review of existing LCA results for the IPCC Special Report on Renewable Energy (SRREN) has been performed (Sathaye et al., 2011). SRREN also includes a discussion of water and land use, air pollution and a range of ecological aspects of renewable energy sources, and as such is a very current and constructive document. The analysis of life cycle results focused on GHG emissions for which the result values varied widely (Figure 1.3). The large range reflects not only uncertainty, but also variability and differing system boundaries (Figure 1.3).

As a result of these problems, the U.S. National Renewable Energy Laboratory (NREL) initiated a comparison and harmonization project, conducting a systematic meta-analysis of existing assessments of life cycle GHG emissions, following a prominent example for bioethanol (Farrell et al., 2006). This work led to the publication of a special issue of the *Journal of Industrial Ecology* in mid-2012 (Heath and Mann, 2012; Brandão et al., 2012). This harmonization focused on adjusting assumptions, such as solar radiation intensity in photovoltaic technologies, in different studies for the same technology and ensured that the assessments are consistent and reflect similar assumptions. The NREL harmonization studies consider only GHG emissions. In its fifth assessment report (AR5), the IPCC presented an overview of the results of the harmonization studies, the raw literature results included in the SRREN, and a contribution analysis to indicate the relative importance of emissions associated with the power plant, the fuel supply and the power plant and fuel chain infrastructure (Figure 1.3).

Some of the findings of the present study were already included in the work for the AR5. In particular, the numbers for gas power plants and hydropower reservoirs were adjusted for new findings on methane emissions in the fuel chain (Chapters 3 and 4). Figure 1.3 indicates that for coal- and gas-fired power plants, the direct emissions of CO_2 from the power plant are so high that only fugitive methane emissions in the fuel supply are of some importance, while producing the required installations causes low GHG emissions. For renewable power apart from bioenergy, it is the installations, however, that cause all the GHG emissions.

Two reviews of life cycle assessments of electricity generation technologies have recently been published in the peer-reviewed scientific literature (Masanet et al., 2013; Turconi et al., 2013). These reviews focus on a similar range of technologies and include power from lignite, bioenergy, and nuclear power. Masanet et al. (2013) also address ocean energy. Both reviews struggle with the fact that the LCA literature usually does not contain the full life cycle inventories and case studies use different impact assessment methods that are not comparable. Both reviews solve this problem by focusing on a few environmental indicators. Turconi et al. (2013) list air pollutants, while Masanet et al. (2013) also include water use, primary energy consumption, solid waste generation and land use. The different indicators, however, are not always taken from the same LCA studies. Additional reviews focus on single indicators such as land use (Fthenakis and Kim, 2009) and water use (Fthenakis and Kim, 2010). The reviews indicate the need for publishing complete life cycle inventories that can be utilized in reviews and meta-analyses, not only summary results as is common practice today.

The IPCC SRREN contains a discussion of site-specific ecological impacts of power generation (Sathaye et al., 2011). Like the present work, this is a qualitative discussion based on a literature review.

FIGURE 1.3

Estimates of life cycle GHG emissions (g CO_2eq/kWh) for broad categories of electricity generation technologies

Published by the IPCC in AR5 (Bruckner et al., 2014). The figure shows both the assessments assembled for the IPCC SRREN (labelled in grey) and assessments assembled for the AR5 (labelled in black). The SRREN values reflect literature numbers, while the AR5 ranges harmonized assumptions (narrow bars). The contribution analyses (wide bars) are from individual studies and shows the importance of direct emissions from the power plant, methane emissions from the supply chain (gas and coal) and the reservoir (hydropower), and other life-cycle emissions (infrastructure and supplies). For biomass-based electricity, SRREN indicates the possibility of negative GHG emissions, derived from assessments that credit combined heat and power plants with avoiding emissions from fossil-fuel based heat production.

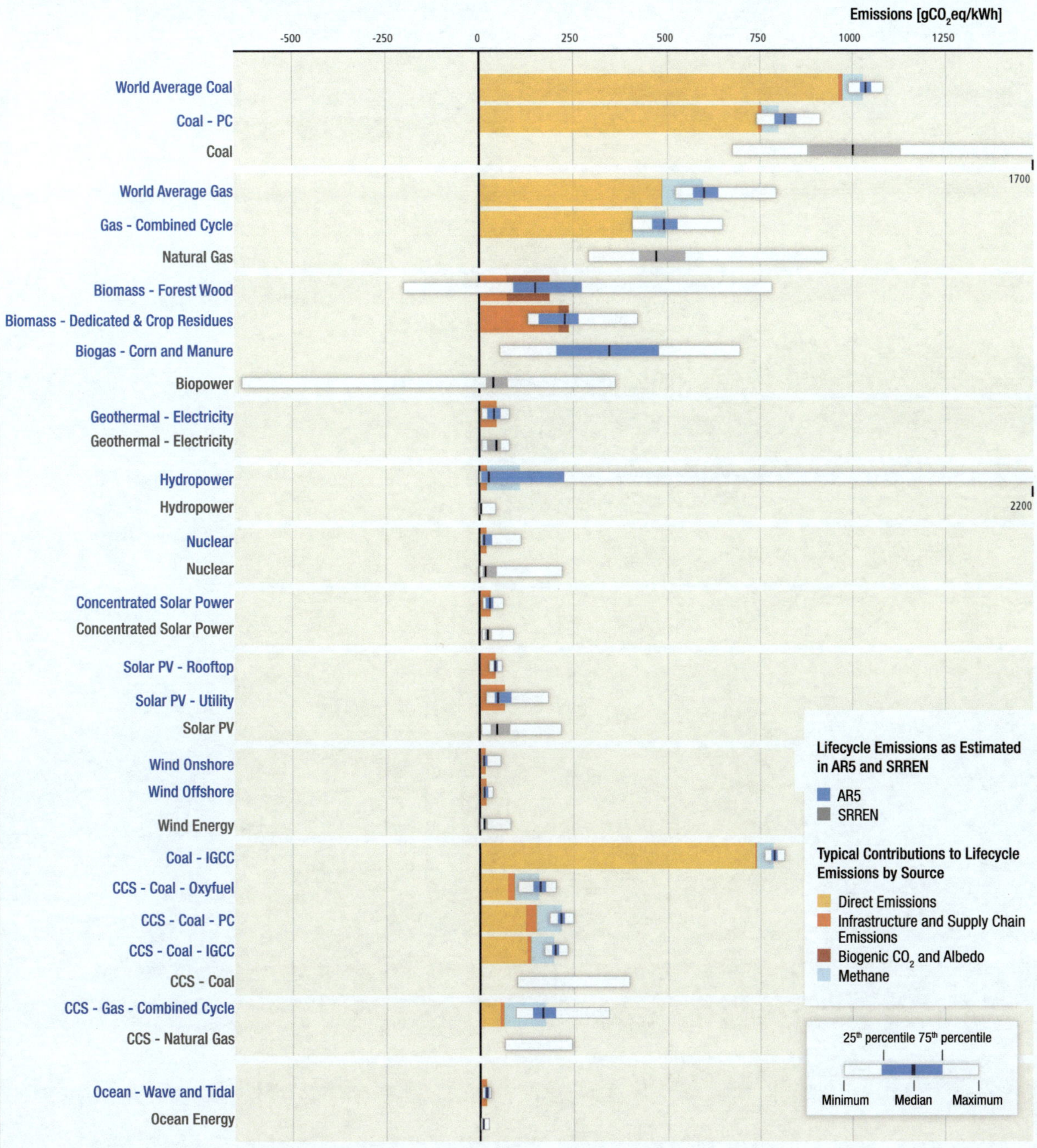

1.5 REFERENCES

Arvesen, A. and E. G. Hertwich. 2011. Environmental implications of large-scale adoption of wind power: A scenario-based life cycle assessment. *Environmental Research Letters* 6(4): 045102.

Arvesen, A. and E. G. Hertwich. 2012. Assessing the life cycle environmental impacts of wind power: A review of present knowledge and research needs. *Renewable and Sustainable Energy Reviews* 16(8): 5994-6006.

Bayer, P., L. Rybach, P. Blum, and R. Brauchler. 2013. Review on life cycle environmental effects of geothermal power generation. *Renewable and Sustainable Energy Reviews* 26: 446-463.

Bergesen, J. D., G. A. Heath, T. Gibon, and S. Suh. 2014. Thin-Film Photovoltaic Power Generation Offers Decreasing Greenhouse Gas Emissions and Increasing Environmental Co-benefits in the Long Term. *Environmental Science and Technology* 48(16): 9834-9843.

Boustead, I. and G. F. Hancock. 1979. *Handbook of industrial energy analysis*. Chichester, UK: Ellis Horwood.

Brandão, M., G. Heath, and J. Cooper. 2012. What Can Meta-Analyses Tell Us About the Reliability of Life Cycle Assessment for Decision Support? *Journal of Industrial Ecology*.

Bruckner, T., I. A. Bashmakov, Y. Mulugetta, H. Chum, A. D. L. V. Navarro, J. Edmonds, A. Faaij, B. Fungtammasan, A. Garg, E. Hertwich, D. Honnery, D. Infield, M. Kainuma, S. Khennas, S. Kim, H. B. Nimir, K. Riahi, N. Strachan, R. Wiser, and X. Zhang. 2014. Energy Systems. In *Climate Change 2014: Mitigation of Climate Change. Contribution of Working Group III to the Fifth Assessment Report of the Intergovernmental Panel on Climate Change*, edited by O. Edenhofer, et al. Cambridge, United Kigdom and New York, NY, USA: Cambridge University Press.

Corsten, M., A. Ramírez, L. Shen, J. Koornneef, and A. Faaij. 2013. Environmental impact assessment of CCS chains – Lessons learned and limitations from LCA literature. *International Journal of Greenhouse Gas Control* 13: 59-71.

Farrell, A. E., R. J. Plevin, B. T. Turner, A. D. Jones, M. O'Hare, and D. M. Kammen. 2006. Ethanol Can Contribute to Energy and Environmental Goals. *Science* 311(5760): 506-508.

Frischknecht, R. and N. Jungbluth. 2007. *ecoinvent 2*. Dübendorf: Swiss Centre for Life Cycle Inventories.

Fthenakis, V. and H. C. Kim. 2009. Land use and electricity generation: A life-cycle analysis. *Renewable and Sustainable Energy Reviews* 13(6-7): 1465-1474.

Fthenakis, V. and H. C. Kim. 2010. Life-cycle uses of water in U.S. electricity generation. *Renewable and Sustainable Energy Reviews* 14(7): 2039-2048.

Goedkoop, M., R. Heijungs, M. Huijbregts, A. De Schryver, J. Struijs, and R. Van Zelm. 2008. *ReCiPe 2008: A life cycle impact assessment method which comprises harmonised category indicators at the midpoint and the endpoint level. First edition: Report I: Characterisation*. The Hague, NL: Dutch Ministry of the Environment.

Heath, G. A. and M. K. Mann. 2012. Background and Reflections on the Life Cycle Assessment Harmonization Project. *Journal of Industrial Ecology*.

Herendeen, R. A., T. Kary, and J. Rebitzer. 1979. Energy Analysis Of The Solar Power Satellite. *Science* 205(4405): 451-454.

Hertwich, E. G. 2013. Addressing Biogenic Greenhouse Gas Emissions from Hydropower in LCA. *Environmental Science & Technology* 47(17): 9604-9611.

Hertwich, E. G., E. van der Voet, M. Huijbregts, S. Suh, A. Tukker, P. Kazmierczyk, M. Lenzen, J. McNeely, and Y. Moriguchi. 2010. *Environmental impacts of consumption and production: priority products and materials, International Resource Panel Report*. Paris: UNEP.

Hertwich, E. G., T. Gibon, E. A. Bouman, A. Arvesen, S. Suh, G. A. Heath, J. D. Bergesen, A. Ramirez, M. I. Vega, and L. Shi. 2015. Integrated life-cycle assessment of electricity-supply scenarios confirms global environmental benefit of low-carbon technologies. *Proceedings of the National Academy of Sciences of the United States of America* 112(20): 6277-6282.

Holdren, J. P. 1982. Energy hazards: What to measure, what to compare. *Technology Review* 85(3): 33-38, 74-75.

Holdren, J. P., G. Morris, and I. Mintzer. 1980. Environmental Aspects of Renewable Energy Sources. *Annual Review of Energy* 5: 241-291.

IEA. 2010. *Energy Technology Perspectives 2010: Scenarios and Strategies to 2050*. Paris, France: OECD/IEA.

IEA. 2014. *Energy Technology Perspectives 2014: Harnessing Electricity's Potential*. Paris, France:

IPCC. 2014. *Climate Change 2014: Mitigation of Climate Change. Contribution of Working Group III to the Fifth Assessment Report of the Intergovernmental Panel on Climate Change*. Edited by O. Edenhofer, et al. Cambridge, United Kingdom and New York, NY, USA: Cambridge University Press.

Jacobson, M. Z. 2009. Review of solutions to global warming, air pollution, and energy security. *Energy and Environmental Science* 2: 148 - 173.

Lenzen, M. 2002. A guide for compiling inventories in hybrid life-cycle assessments: Some Australian results. *Journal of Cleaner Production* 10(6): 545-572.

Lenzen, M. 2008. Life cycle energy and greenhouse gas emissions of nuclear energy: A review. *Energy Conversion and Management* 49(8): 2178-2199.

Lenzen, M. 2010. Current state of development of electricity-generating technologies: A literature review. *Energies* 3(3): 462-591.

Majeau-Bettez, G., A. H. Strømman, and E. G. Hertwich. 2011. Evaluation of Process- and Input–Output-based Life Cycle Inventory Data with Regard to Truncation and Aggregation Issues. *Environmental Science & Technology* 45(23): 10170-10177.

Masanet, E., Y. Chang, A. R. Gopal, P. Larsen, W. R. Morrow, R. Sathre, A. Shehabi, and P. Zhai. 2013. Life-Cycle Assessment of Electric Power Systems. *Annual Review of Environment and Resources* 38(1): 107-136.

Pehnt, M. 2006. Dynamic life cycle assessment (LCA) of renewable energy technologies. *Renewable energy* 31(1): 55-71.

REN21. 2014. *Renewables 2014 Global Status Report*. Paris, France: REN21 Secretariat.

Sathaye, J., O. Lucon, A. Rahman, J. Christensen, F. Denton, J. Fujino, G. Heath, S. Kadner, M. Mirza, H. Rudnick, A. Schlaepfer, and A. Shmakin. 2011. Renewable Energy in the Context of Sustainable Energy.

In *IPCC Special Report on Renewable Energy Sources and Climate Change Mitigation*, edited by O. Edenhofer, et al. Cambridge, United Kingdom and New York, NY, USA: Cambridge University Press.

Singh, B. 2011. Environmental evaluation of carbon capture and storage technologies and large scale deployment scenarios. PhD thesis, Department of Energy and Process Engineering, Norwegian University of Science and Technology, Trondheim.

Stocker, T. F., D. Qin, G.-K. Plattner, M. Tignor, S. K. Allen, J. Boschung, A. Nauels, Y. Xia, V. Bex, and P. M. Midgley, eds. 2013. *Climate Change 2013: The Physical Science Basis. Contribution of Working Group I to the Fifth Assessment Report of the Intergovernmental Panel on Climate Change*. Edited by IPCC. Cambridge, United Kingdom and New York, NY, USA: Cambridge University Press.

Suh, S., M. Lenzen, G. J. Treloar, H. Hondo, A. Horvath, G. Huppes, O. Jolliet, U. Klann, W. Krewitt, Y. Moriguchi, J. Munksgaard, and G. Norris. 2004. System Boundary Selection in Life-Cycle Inventories Using Hybrid Approaches. *Environ. Sci. Technol.* 38(3): 657-664.

Turconi, R., A. Boldrin, and T. Astrup. 2013. Life cycle assessment (LCA) of electricity generation technologies: Overview, comparability and limitations. *Renewable and Sustainable Energy Reviews* 28: 555-565.

Udo de Haes, H. A., O. Jolliet, G. Finnveden, M. Goedkoop, M. Hauschild, E. G. Hertwich, P. Hofstetter, W. Klöpffer, W. Krewitt, E. W. Lindeijer, R. Mueller-Wenk, S. I. Olson, D. W. Pennington, J. Potting, and B. Steen. 2002. *Life Cycle Impact Assessment: Striving towards Best Practice*. Pensacola: Society of Environmental Toxicology and Chemistry.

Verbruggen, A., W. Moomaw, and J. Nyboer. 2011. Glossary. In *IPCC Special Report on Renewable Energy Sources and Climate Change Mitigation*, edited by O. Edenhofer, et al. Cambridge, United Kingdom and New York, NY, USA: Cambridge University Press.

Viebahn, P., Y. Lechon, and F. Trieb. 2011. The potential role of concentrated solar power (CSP) in Africa and Europe-A dynamic assessment of technology development, cost development and life cycle inventories until 2050. *Energy Policy* 39(8): 4420-4430.

Warner, E. S. and G. A. Heath. 2012. Life Cycle Greenhouse Gas Emissions of Nuclear Electricity Generation: Systematic Review and Harmonization. *Journal of Industrial Ecology* 16(SUPPL.1): S73-S92.

Chapter 2

Method description

Lead authors: Thomas Gibon, Edgar G. Hertwich, Joseph Bergesen, Sangwon Suh

2.1 INTRODUCTION

This report presents an assessment of the impact of key power plant technologies on human health, ecosystem health, and resources, using a life cycle approach. Life cycle assessments were conducted using an integrated model capable of modelling impacts on a regionally disaggregated level, thereby reflecting region-specific technologies in the nine regions of the International Energy Agency (IEA) Energy Technology Perspectives (ETP) model (IEA, 2010). The total emissions resulting from the widespread implementation of the assessed low carbon technologies were assessed by combining the life cycle inventories (LCI) produced in this assessment with the deployment foreseen in the IEA ETP BLUE Map scenario (IEA 2010), which is consistent with the goal of limiting global warming to 2°C. We compared the resulting global emissions rates and resource use with a deployment of energy technologies foreseen in the IEA ETP Baseline scenario. This chapter presents a description of the general procedure and the methods employed in this study. A more detailed description is available in the literature (Gibon et al., 2015).

2.2 METHOD DEVELOPMENT

2.2.1 LIFE CYCLE ASSESSMENT

Life cycle assessment (LCA) is a method to assess the environmental impacts and resource requirements associated with products. LCA characteristically accounts for several life cycle stages and a wider set of environmental pressures. The exact scope varies. A product life cycle is commonly defined as including the life cycle stages of resource extraction, the materials refining, product manufacturing, distribution and transport, operation and maintenance, and disposal (end-of-life). The system providing not only the product itself but also operational inputs, maintenance and disposal is called the product system. There is an international standard for LCA (ISO 14040) and a related standard exists for carbon footprints (ISO 14064) (Finkbeiner et al., 2006).

The basic principle of LCA is to connect the environmental impact resulting from the production, distribution and disposal processes of the products in question. A common challenge occurs in LCA when a specific process produces different fractions that may be used as unique products. In most LCA studies, the input requirements and environmental flows of a specific production process are allocated in equal portion to the product output of that process. This work also adopts this strategy for multi-output processes. LCA is hence a procedure that assigns the responsibility for environmental impacts to the individual products, i.e., the user of the product.

For the purpose of LCA, it is common to define a functional unit, which describes the unit of analysis, such as the quantity and duration of the service delivered. For power production, a common functional unit is a kWh or a million kWh of electricity produced. The term does not imply that all kWh of electricity are equal, this is rather used as a basis of comparison. An alternative would be to define a specific temporal profile of electricity demand that different combinations of technology can achieve. Such analyses are yet uncommon.

An LCA consists of a goal and scope definition, an inventory of emissions and natural resource use, also called environmental interventions, an assessment of the potential impacts of the environmental interventions, and results interpretation. The goal and scope definition defines the study. The life cycle inventory (LCI) is built upon a model of the interconnection of the different life cycle stages, with each production process specified in terms of intermediate products, inputs and outputs as well as resource inputs and waste/pollution outputs. Life cycle inventories can be more or less comprehensive, depending on how many of the many input requirements and economic processes as well as environmental interventions are considered. Standard databases provide descriptions of common processes, representing the work of many person-years of analysis. They are indispensable in conducting LCAs.

Since process-based LCI databases do not encompass most industrial, agricultural, commercial or residential activities of an economy (Majeau-Bettez et al., 2011), LCA practitioners sometimes source input for such activities from input-output tables. An input-output table provides a complete description of the economy at a rougher level of detail. Studies combining inputs in physical and monetary terms are called hybrid life cycle assessment (HLCA) (Suh et al., 2004; Strømman et al., 2006). In HLCA, process LCA complements input-output with its accuracy and detail, which provide more comprehensive modelling and analysis. It addresses LCA's weaknesses related to system boundaries issues (Lenzen, 2001).

Our economy uses many different resources and produces thousands of pollutants. Typical life cycle inventories list hundreds of these environmental interventions. Impact assessment methods have been developed to assess the importance of the environmental interventions associated with any specific product system. Impact assessment commonly proceeds by multiplying the inventoried amount of emissions or resource extraction with a characterization factor quantifying the magnitude to which a unit of emissions or resource extraction contributes to an identified environmental issue. Each of these characterization factors is derived from the sometimes complex modelling of the environmental mechanisms by which the flow contributes to an environmental problem. In this study we use ReCiPe, an impact assessment methodology commonly used in Europe. ReCiPe addresses the contribution of pollutants to identified environmental problems, such as global warming or human toxicity, called midpoint level impacts, and further the human health damage caused by both climate change and human toxicity effects.

In LCA, one often distinguishes between a foreground system, which describes the production processes under study and contains the product-specific data collected for the life cycle of the product, and a background system, which describes generic inputs required by many foreground systems. For the latter, several LCI databases with generic data exist. As new energy technologies become widely adopted, they become part of the background and their level of implementation eventually will impact the life cycle inventories of any product.

For this report, we integrate scenario modelling and HLCA to comprehensively assess the environmental impacts of new energy technologies. Scenario models project the composition of future energy mixes by estimating the extent to which the different new energy technologies will be adopted and the resulting technology mix for each energy carrier.

The HLCA set-up is similar to earlier scenario work for CO_2 capture and storage (Singh, 2011) and wind power (Arvesen and Hertwich, 2011); a commonly used process-level LCI database (ecoinvent 2.2) is combined with an input-output model (EXIOBASE) (Frischknecht and Jungbluth, 2007; Wood et al., 2014). Hence, inputs to the foreground system can be either physical inputs from the process LCI database or economic inputs from the input-output database. In this work, we go a step further by feeding results from the different foreground systems of the respective new energy technologies into the generic process-level LCI database.

In order to model a prospective LCI for low carbon technologies, it is necessary to predict at least four types of changes. First, the progressing energy and material efficiency of key processes in the inventory database and key sectors of the economy should be incorporated into the model. For that purpose, one might rely on scenario analyses provided by authorities such as the IEA or prospective literature. Second, advancements and direct changes of the product or process being modelled may be anticipated by the extrapolation of historical data, scenario analyses or expert opinion. In this study, the authors chose to rely on data provided by experts who in turn rely either on literature or their own judgment. A disaggregated energy sector ensures that these changes can be modelled at a detailed level. Third, an important structural change in the background data is the share of renewable energy in the global electricity mix. Finally, considering general changes in energy and resource efficiency of the global economy is a necessary but delicate step. Since the inputs to any economic sector vary over time, there is a need to model changes in energy and material inputs to all economic sectors. The details of these modifications are described in Chapter 2.2.2. We describe how exogenous energy scenarios, the different energy technology foreground systems, the background LCI database, and the multiregional input-output table are combined to facilitate scenario modelling.

2.2.2 DATA SOURCES FOR THE INVENTORY MODELLING

Figure 2.1 shows how different data sources were utilized to construct representative life cycle inventories for the various technologies for each of nine world regions and the years 2010, 2030, and 2050 as represented in the IEA energy scenarios. A description of the methods and the underlying inventory data for technologies has been published (Hertwich et al., 2014).

Energy scenarios

While the method developed for this study is generic, its implementation uses two scenarios taken from the IEA ETP 2010: the Baseline and BLUE Map scenarios. The BLUE Map scenario defines renewable energy penetration levels in each region required to achieve a 50% reduction of energy-related CO_2 emissions over the period 2005-2050. In conjunction with the LCA literature, it is possible to use these scenarios to estimate the greenhouse gas (GHG) emissions and other environmental impacts of the entire electricity supply for the years leading to 2050. This estimate is then refined using an iterative approach by using the scenario-based model results described in this report to determine more accurate life cycle impacts for the different energy technologies. Additionally, the IEA ETP (IEA, 2010) estimates energy and resource efficiency pathways for important and impactful sectors of the economy. These data can be adapted to modify the emissions and resource inputs of these sectors in both process LCA and input-output databases.

Background life cycle inventory

Reliance on physical process data is the traditional way of evaluating the environmental profile of an activity, when a list of physical inputs and outputs of materials and energy is available. The impact of technology development described in terms of improved physical efficiency, direct emission reduction, variations in resource use, loss reduction or enhanced use of recycled material and recycling rates can be modelled by modifying these parameters in a process LCI. For example, the stressor list available in the process-based ecoinvent 2.2 database with 580 air pollution species is more comprehensive than that in EXIOPOL, which contains only 20 air emissions (Frischknecht and Jungbluth, 2007). ecoinvent 2.2, updated in 2010, is used as a background for the present situation, with modifications made to the electricity mixes for consistency with the underlying assumptions of the input-output table energy mixes.

Modifications brought to the ecoinvent 2.2 database concerns energy mixes and key processes. For the first, electricity mixes representing the nine regions were added. For the second, "key processes" were modified because these processes have been identified as dominant impact contributors in the life cycle assessment of energy technologies. Modifications were based on the realistic-optimistic scenario for LCIs of the New Energy Externalities Development for Sustainability (NEEDS) project (ESU and IFEU, 2008). The realistic-optimistic scenario achieves 440 ppm atmospheric CO_2 and has been considered as the closest match to BLUE Map, since it is based upon actual data on best available techniques and reasonable efficiency trends.

Two specific modifications were made to individual database processes. The first modification was a mere adaptation of the key processes to future scenarios. The second adaptation was carried out as far as background energy mixes are concerned, for specific processes. This adaptation is applied only when necessary, i.e., when foreground systems require processes that occur in a specific energy context. As ecoinvent offers a majority of processes specific to Switzerland or Europe, electricity inputs from these processes were substituted by another specific regional energy mix whenever needed.

As far as energy mixes are concerned, IEA power generation projections for 2030 and 2050 have been considered, utilizing, where available, the LCIs from this study to describe the inventories of technologies in the mix. Wave and tidal energy were not represented.

Modification of fugitive methane emissions from fossil fuel production

Fugitive methane emissions from natural gas and coal mining have been revised. It has been found that ecoinvent 2.2 underestimates methane emissions from natural gas mining, and overestimates them for the coal mining processes, according to Burnham et al. (2011). The following table presents how the new emission values have been derived and how the adaptation has been made on ecoinvent 2.2. Regional variation is kept, although adjusted so the new mean value matches the mean value reported in the literature.

Background multiregional input-output table (hybrid life cycle assessment)

HLCA can take advantage of the latest advances in input-output: among others, multiregional input-output models (Tukker and Dietzenbacher, 2013). By definition a multiregional input-output model is able to cover both the *global* and the *regional resolution* aspects of such a model.

We utilize the EXIOBASE database from the EXIOPOL project (Tukker et al., 2013). EXIOPOL consists of a 44-region world input-output table with 129 economy sectors for the year 2000. The first step is transforming the 44-region table to a 9-region table. No direct hierarchy exists between the two regional classifications, i.e., each region in the 44-region set is not contained in exactly one of the other set's regions; some regions might therefore be double counted. The rest-of-the-world region in EXIOBASE, representing approximately 150 countries, was broken down into different subregions that will fall under the nine-region world classification, in a procedure similar to (Stadler et al., 2014). As a first approximation, we use the relative output shares of the countries belonging to the rest-of-the-world group from the GTAP database to estimate the fraction of total economic flows attributable to each of them. Further, the electricity supply was modified to reflect generation mixes specified by IEA.

Emissions matrix

Emissions per sector are also likely to change, due to improved efficiency and external policy pressure on pollutant emissions. The small range of emissions considered in EXIOBASE embodies mainly GHGs, heavy metals and particulate matter. These substances are controlled, reported and regulated. To estimate the future evolution of national emissions, we have assumed continuity with the historical evolution of most of these pollutants in Europe. The model thus relies on the assumption that future emissions per euro will decrease as pollution control technologies improve and regulations become stricter worldwide, at the same pace as it has been in Europe for a couple of decades. To project these potential changes in the model, historic emission trends in the EU27 from 1990 to 2009 are extrapolated (Gibon et al., 2015). The pollutants in question are cadmium, carbon monoxide, dioxins, HCB, HCH, mercury, ammonia, non-methane volatile organic carbons, NO_x, lead, PCB, PM10, PM2.5, SOx, and total PAH. Other parts of the world are assumed to undergo a similar reduction in emission factors as Europe did as a result of the Convention on Long-Range Transboundary Air Pollution (CLRTAP) (European Environment Agency, 2013). With the notable exception of copper emissions and arsenic emissions, these pollutants cover a majority with the most important environmental stressors used in EXIOPOL that contribute to this project's stated impact categories. The best possible technique to adapt this data to our model is the following: pollutant emissions were normalized by the total GDP of the EU27 countries during 1990-2009 in order to adjust for

TABLE 2.1

New emission factors per unit of natural gas extracted, in g CH_4/m^3 gas, for the different regions addressed in ecoinvent 2.2.

	Average	Ratio (ecoinvent = 1)	Germany	Algeria (at production onshore)	the Netherlands	Russia	North America	Great Britain	the Netherlands	Norway	Algeria (at liquefaction plant)	Europe	North America
in g per	m³	-	m³	m³	m³	m³	m³	m³	m³	m³	m³	m³	m³
ecoinvent	1.4	1.0	0.26	1.2	0.19	3.1	1.4	0.00015	0.19	0.066	0.33	0.33	8.8
Burnham	**13**	**9.0**	**5.4**	**25**	**3.9**	**63**	**3.6**	**0.0031**	**3.9**	**1.3**	**6.7**	**6.8**	**22**

In bold, the values that have been chosen. Reprinted with permission from Burnham et al., 2011.

Source: Burnham et al., 2011.

TABLE 2.2

New emission factors per unit of coal extracted, in g CH_4/m^3 gas, for the different regions addressed in ecoinvent 2.2.

	Average, underground	Average, open pit	Ratio, underground (ecoinvent = 1)	Ratio, open pit (ecoinvent = 1)	Australia	China	Eastern Asia	Eastern Europe	Latin America	North America	Russia	Western Europe	South Africa
in g CH_4 per	kg	kg	-	-	kg	kg	kg	kg	kg	kg	kg	kg	kg
ecoinvent	-	-	1.0	1.0	2.7	17	3.0	8.2	0.16	3.0	9.2	14	3.5
Burnham	**7.7**	**1.2**	**1.2**	**0.18**	**0.7**	**9.8**	**1.7**	**4.7**	**0.014**	**0.88**	**3.8**	**7.9**	**1.2**
Underground					32%	100%	97%	100%	0%	42%	67%	100%	50%
Open pit					68%	0%	3%	0%	100%	58%	33%	0%	50%

In bold, the values that have been chosen.

Source: Burnham et al., 2011. Reprinted with permission from Burnham, et al., 2011 changes in economic output that could increase or decrease overall emissions. Compound decrease was then assumed for the emissions in each sector. For every substance, emission levels were modelled for the 1990-2009 time period and, on this basis, extrapolated to 2050. Finally, improvement factors were derived from this extrapolation. This method is a first approximation of what can be achieved under "business-as-usual" efforts in pollutant control regulations.

Time series

In developing a time series model for scenarios in LCA, it is first necessary to model endogenous change dictated by the scenario that is being followed. The example in this report follows the IEA BLUE Map scenario, which dictates the amount of energy generation by source in a given year. Thus, the life cycle impacts of a specific energy technology in the baseline year will change the overall impact of energy consumption in the economy in the following year.

TABLE 2.3

GDP variation figures used to extrapolate demand vectors and flow matrix (IEA, 2010)

GDP variation	2000-2007	2007-2030	2030-2050
China	191%	375%	210%
India	162%	406%	192%
OECD Europe	115%	142%	115%
OECD North America	117%	162%	132%
OECD Pacific	115%	135%	141%
Other Developing Asia	132%	215%	168%
Economies in Transition	160%	211%	201%
Africa and Middle East	135%	259%	174%
Latin America	122%	185%	164%

2.2.3 LIFE CYCLE IMPACT ASSESSMENT

The results of the LCA are calculated using the ReCiPe impact assessment method (Goedkoop et al., 2008). ReCiPe offers a comprehensive set of characterization factors for 18 "midpoint indicators" and 3 "endpoint indicators". Midpoint indicators describe the common environmental mechanism caused by the presence of a certain group of compounds emitted to a variety of receiving compartments such as freshwater, ocean, low-density populated area, and soil. Examples of midpoint indicators include climate change, toxicity and eutrophication. Endpoint indicators compile and weight a selection of midpoint indicators that collectively describe the damage caused to humans, ecosystems and resources. ReCiPe proposes three perspectives, analogous to scenarios: egalitarian (precautionary), hierarchical (following most common policy principles) and individualist (accounting for short-term interest and technological optimism). Since no specific assumptions were taken on the time horizon and uncertainty of certain characterization factors, the hierarchical perspective was therefore used in this report. This approach relies on widely used metrics, such as the GWP_{100} indicator, to facilitate comparison with literature.

FIGURE 2.1

Flowchart of the different flows of information and data in the model.

Green arrows represent base data, purple arrows represent external information that modify these base data. Figures are solely shown for illustration purposes.

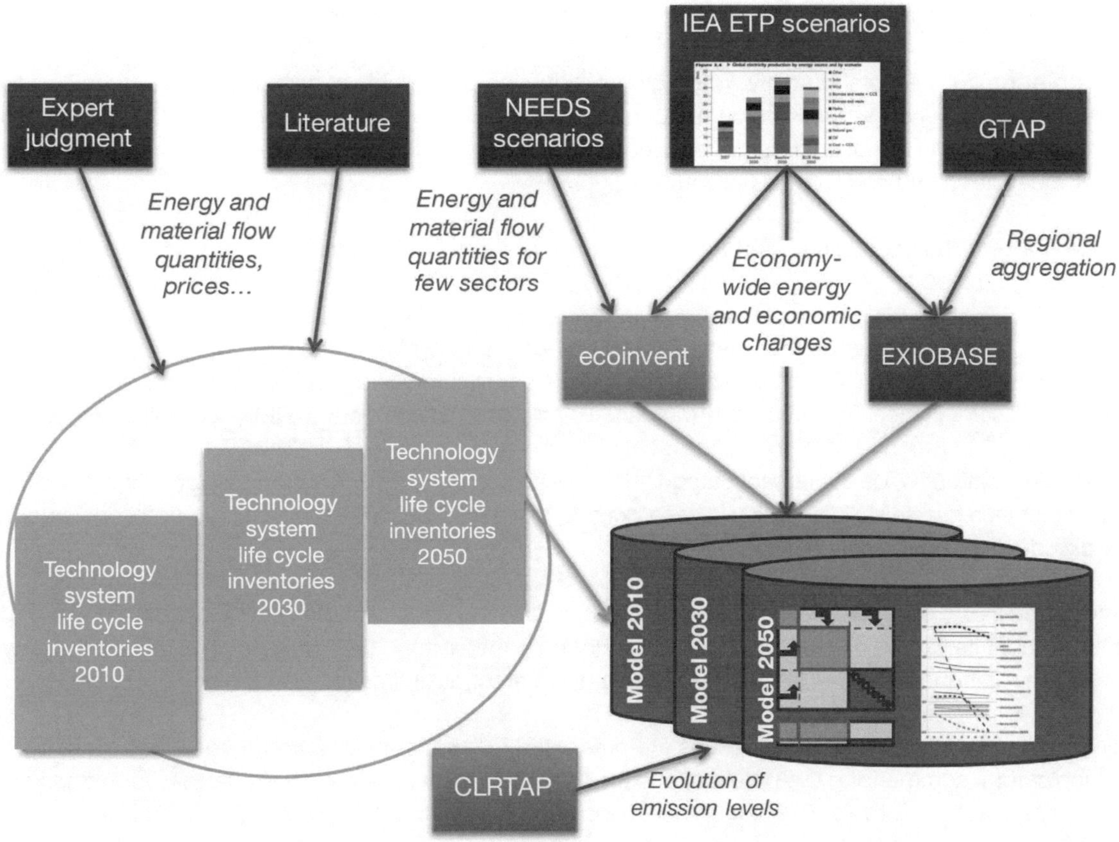

Technology comparisons are based on several impact categories. Of the 18 midpoint indicators in ReCiPe (Goedkoop et al., 2008), 9 are presented in the analysis. The following paragraphs describe these indicators.

Climate change (CC) is the impact category quantifying the global warming potential due to GHG emissions. These emissions increase radiative forcing in the atmosphere over various time horizons. The reference unit for this indicator is kg CO_2 equivalents. Typical GHG include carbon dioxide, methane and dinitrogen oxide.

Freshwater eutrophication (FEU) is the impact category quantifying the response of freshwater environments to the addition of nutrients such as phosphates and nitrates resulting from human activities. A common negative effect of eutrophication is the decrease of available oxygen (hypoxia) in aquatic environments and eventual loss of animal life. This indicator is measured in kg PO_4^{3-} (phosphate ion) equivalents. Typical eutrophying substances are fertilizers based on nitrogen, phosphorous or potassium.

Human toxicity (HT) quantifies the toxic potential of compounds in the human body. The characterization factors for this impact category are a combination of fate and exposure factors that represent the behaviour of each compound in the given set of compartments. The reference unit for this indicator is kg 1,4-dichlorobenzene (1,4-DB) equivalents emitted to urban air.

Freshwater ecotoxicity (FET) is measured roughly in the same manner as HT, and quantifies the toxicity to living organisms other than humans, in 1,4-DB equivalents emitted to freshwater.

Mineral depletion (MD) aims at quantifying the global reduction of available mineral resources, based on the United States Geological Survey reports on the current available reserves of most ore types. Characterization is derived from the modelling of cost damage, specifically the marginal cost increase per kg extracted. The indicator is measured in kg iron (Fe) equivalents.

Particulate matter (PM) accounts for all particulate matter emissions. The reference unit is kg PM_{10} (up to 10 μm diameter) emitted to air.

Photochemical oxidant formation (POF) is an impact category that evaluates the contribution of individual substances to ozone formation. Ozone is in turn a health hazard to humans causing inflamed airways and lung damage. The indicator is measured in kg non-methane volatile organic compound (NMVOC) emitted to air.

Terrestrial acidification (TA) is the impact category that measures the atmospheric deposition of inorganic substances that increase soil acidity, which is hazardous to plant species. The reference unit for the indicator is kg sulfur dioxide (SO_2) equivalents emitted in soil.

Land occupation (LO) is the sum of all agricultural and urban land directly and indirectly occupied by a system throughout its life cycle. It is measured in m^2a (square meter-annum), a quantity that represents how much of an area (in square meters) is occupied over a given amount of time (in years).

In addition to these midpoint indicators, results are presented in "human health" and "ecosystem diversity" endpoint indicators, summarizing the potential effect of a system from a damage perspective, for human and non-human species, respectively. The human health indicator is measured in "disability-adjusted life years" (DALY), which is a measure of overall disease burden expressed as the number of years lost due to ill-health, disability or early death[1]. The unit for the ecosystem diversity indicator is the "loss of species during a year", or species.year, which measures species extinction rate.

2.3 MODEL SETUP

FIGURE 2.2

Schematic describing the setup of the hybrid inventory model (see page 65 below)

The foreground contains the flows among processes for which the IRP analysts collected data (green area). Inputs, e.g. of materials, from the LCI background databased are traced in the light orange area of the foreground column. Inputs from the wider economy are traced in the purple area of the foreground column. The background columns have input of electricity from the foreground system, displayed in the top row of the background columns.

Source: Strømman et al., 2006; Suh and Huppes, 2005

1 http://www.who.int/healthinfo/global_burden_disease/metrics_daly/en/

A
Foreground
LCI background
LCI background
IO background
EU
US
EU
US
EU
US
PV
proc 1
proc n
PV production mix
proc ...
sect 1
sect n
Functional unit ->
CdTe system
Poly-Si system
αCdTe,EU kWh/FU × βCdTe,EU kWh/kWh
αPoly-Si,EU kWh/FU × βPoly-Si,EU kWh/kWh
αCdTe,US kWh/FU × βCdTe,US kWh/kWh
αPoly-Si,US kWh/FU × βPoly-Si,US kWh/kWh
αCdTe,US kWh/FU × βCdTe,US kWh/kWh × γCdTe,US kWh/MEUR
αPoly-Si,EU kWh/FU × βPoly-Si,EU kWh/kWh × γPoly-Si,EU kWh/MEUR
αCdTe,US kWh/kWh * x kWh/MEUR
βPoly-Si,US kWh/kWh * x kWh/MEUR
0
V
Profits etc.
Gross capital formation
Jobs
F
Global
Ecoinvent classification of stressors
EXIOPOL classification of stressors

2.4 REFERENCES

Arvesen, A. and E. G. Hertwich. 2011. Environmental implications of large-scale adoption of wind power: A scenario-based life cycle assessment. *Environmental Research Letters* 6(4): 045102.

Burnham, A., J. Han, C. E. Clark, M. Wang, J. B. Dunn, and I. Palou-Rivera. 2011. Life-Cycle Greenhouse Gas Emissions of Shale Gas, Natural Gas, Coal, and Petroleum. *Environmental Science & Technology* 46(2): 619-627.

ESU and IFEU. 2008. *New Energy Externalities Developments for Sustainability (NEEDS) - LCA of background processes*. Ulster, CH: ESU Services.

European Environment Agency. 2013. *European Union emission inventory report 1990-2011 under the UNECE Convention on Long-range Transboundary Air Pollution (LRTAP)*. Luxembourg: European Commission.

Finkbeiner, M., A. Inaba, R. Tan, K. Christiansen, and H.-J. Klüppel. 2006. The New International Standards for Life Cycle Assessment: ISO 14040 and ISO 14044. *The International Journal of Life Cycle Assessment* 11(2): 80-85.

Frischknecht, R. and N. Jungbluth. 2007. *ecoinvent 2*. Dübendorf: Swiss Centre for Life Cycle Inventories.

Gibon, T., R. Wood, A. Arvesen, J. D. Bergesen, S. Suh, and E. G. Hertwich. 2015. A Methodology for Integrated, Multiregional Life Cycle Assessment Scenarios under Large-Scale Technological Change. *Environmental Science & Technology* 49(18): 11218-11226.

Goedkoop, M., R. Heijungs, M. Huijbregts, A. De Schryver, J. Struijs, and R. Van Zelm. 2008. *ReCiPe 2008: A life cycle impact assessment method which comprises harmonised category indicators at the midpoint and the endpoint level. First edition: Report I: Characterisation*. The Hague, NL: Dutch Ministry of the Environment.

Hertwich, E. G., T. Gibon, E. Bouman, A. Arvesen, S. Suh, G. Heath, J. D. Bergerson, A. Ramirez, M. V. Coloma, and S. Lei. 2014. Integrated life cycle assessment of electricity supply scenarios confirms global environmental benefit of low-carbon technologies. *Proceedings of the National Academy of Sciences of the United States of America*: accepted for publication.

IEA. 2010. *Energy technology perspectives 2010: Scenarios and strategies to 2050*. Paris: OECD/IEA.

Lenzen, M. 2001. Errors in conventional and input-output-based life-cycle inventories. *Journal of Industrial Ecology* 4(4): 127-148.

Majeau-Bettez, G., A. H. Strømman, and E. G. Hertwich. 2011. Evaluation of Process- and Input–Output-based Life Cycle Inventory Data with Regard to Truncation and Aggregation Issues. *Environmental Science & Technology* 45(23): 10170-10177.

Singh, B. 2011. Environmental evaluation of carbon capture and storage technologies and large scale deployment scenarios. PhD thesis thesis, Department of Energy and Process Engineering, Norwegian University of Science and Technology, Trondheim.

Stadler, K., K. Steen-Olsen, and R. Wood. 2014. The 'rest of the world' - estimating the economic structure of missing regions in global multi-regional input-output tables. *Economic Systems Research* 26(3): 303-326.

Strømman, A. H., C. Solli, and E. G. Hertwich. 2006. Hybrid life-cycle assessment of natural gas based fuel chains for transportation. *Environmental Science & Technology* 40(8): 2797-2804.

Suh, S. and G. Huppes. 2005. Methods for life cycle inventory of a product. *Journal of Cleaner Production* 13(7): 687-697.

Suh, S., M. Lenzen, G. J. Treloar, H. Hondo, A. Horvath, G. Huppes, O. Jolliet, U. Klann, W. Krewitt, Y. Moriguchi, J. Munksgaard, and G. Norris. 2004. System Boundary Selection in Life-Cycle Inventories Using Hybrid Approaches. *Environmental Science & Technology* 38(3): 657-664.

Tukker, A. and E. Dietzenbacher. 2013. Global multiregional input–output frameworks: an introduction and outlook. *Economic Systems Research* 25(1): 1-19.

Tukker, A., A. de Koning, R. Wood, T. Hawkins, S. Lutter, J. Acosta, J. M. Rueda Cantuche, M. Bouwmeester, J. Oosterhaven, T. Drosdowski, and J. Kuenen. 2013. EXIOPOL - development and illustrative analyses of detailed global multiregional, environmentally extended supply and use tables and symmetric input-output tables. *Economic Systems Research* 25(1): 50-70.

Wood, R., K. Stadler, T. Bulavskaya, S. Lutter, S. Giljum, A. de Koning, J. Kuenen, H. Schütz, J. Acosta-Fernández, A. Usubiaga, M. Simas, O. Ivanova, J. Weinzettel, J. Schmidt, S. Merciai, and A. Tukker. 2014. Global Sustainability Accounting—Developing EXIOBASE for Multi-Regional Footprint Analysis. *Sustainability* 7(1): 138-163.

©Alexander Khodarev/

Chapter 3

Fossil fuels and carbon dioxide capture and storage

Lead author: Andrea Ramírez
Contributing authors: Peta Ashworth, Bhavik Bakshi, Joule Bergerson, Anne Maree Dowd, Liang Shih Fan, W. Michael Griffin, Takeshi Kuramochi, Heather McLean, Sanjay M. Mahajani, J. Marcelo Ketzer, Yu Qian, Evert Bouman, Thomas Gibon, Edgar G. Hertwich

3.1 INTRODUCTION

The aim of this chapter is threefold. First, it aims to provide a systematic overview of the fossil fuels spectrum and the technologies used for fossil fuel-based power production, including their current status and key constraints. Secondly, this chapter aims to provide a review of the potential environmental impacts and trade-offs reported in literature. Energy is central to addressing the major challenges of the twenty first century: climate change, poverty, economic and social development. Historically, most of the world's energy requirements have been supplied by fossil fuels (about 81 per cent of the world's primary fuel mix in 2010) and it is expected that they will continue to play a major role in the coming decades. For instance, in its 2013 *World Energy Outlook*, the International Energy Agency (IEA) indicates that fossil fuels (oil, coal and natural gas) will remain the dominant sources of energy until 2035 with shares of about 80 per cent in the *Current Policies Scenario* and 64 per cent in the 450 ppm scenario[1] (IEA, 2013).

However, the combustion of fossil fuels results in emissions of CO_2, nitrogen oxides (NO_x), sulfur oxides (SO_x), particulates, volatile organic compounds and heavy metals such as mercury. The IEA (2013) estimates that energy-related CO_2 emissions reached a record 31.2 billion tons in 2011, representing around 60 per cent of global greenhouse gas (GHG) emissions (measured on a CO_2-equivalent basis). The Global Carbon Project reported an estimate of 34 billion tons in 2013 (GCP, 2014). Besides decreasing CO_2 emissions, the energy sector also faces significant challenges in controlling and decreasing non-CO_2 emissions. Contributions of fossil fuel combustion to environmental issues such as acidification, eutrophication and health impacts have been extensively analysed in literature (GEA, 2012). Nowadays, there is increasing agreement that there is no single option that can achieve this level and therefore an extensive portfolio of energy initiatives must be developed and deployed together (GEA, 2012; IEA, 2012a).

The focus of this chapter is on the potential impact of carbon dioxide capture and storage (CCS) on the environmental performance of fossil fuel power production. The impacts are examined both at the facility level and from a lifecycle perspective. Finally, the chapter presents the inventory used for the integrated assessment of technologies which is at the core of the present report.

1 The *current policy scenario* government policies that had been enacted or adopted by mid-2012 continue unchanged. In the *450 ppm scenario*, policies are adopted that have a 50 per cent chance of limiting the global increase in average temperature to 2°C in the long term, compared with pre-industrial levels (IEA, 2013).

3.2 TECHNOLOGY DESCRIPTION: FOSSIL FUEL-BASED POWER PLANTS

3.2.1 PULVERIZED COAL FIRED POWER PLANT

In a power plant, coal can be burned in a variety of combustor or boiler types such as pulverized coal combustors, stoker combustors, or fluidized bed combustors (Stultz and Kitto, 1992). The heat energy released by the combustion is used to generate high-pressure, high-temperature steam that drives a steam turbine system to generate electricity. Currently, pulverized coal-fired (PC) power plants account for more than 90 per cent of the electricity generated from coal (Miller, 2004). Figure 3.1 illustrates a schematic flow diagram of a typical PC power plant.

Here, coal is first transported from the storage facility to the pulverizer that grinds the coal into a fine powder. The powdered coal is then conveyed to the boiler, which is a large enclosed combustion chamber. Preheated air and coal powder are mixed and then introduced into the burner nozzle to enhance mixing. The air-fuel mixture ignites and combusts once introduced to the boiler, resulting in temperatures exceeding 1,500°C (2,800°F). The heat carried by high-temperature effluent gases from the boiler then is extracted to produce superheated steam to drive high-pressure steam turbines. The steam discharged from these high-pressure steam turbines is then directed to the intermediate pressure and low pressure steam turbines for the subsequent generation of additional electricity. The low grade steam discharged from the low pressure steam turbines is condensed by cooling water prior to entering the next steam generation cycle. Cooling towers release the heat removed in the condensation step to the environment.

Although the underlying concept seems simple, modern PC power plants present the following challenges to be addressed: enhancement of energy conversion efficiency, effective control of hazardous pollutants emission, and CCS. When CO_2 emission control is required, the components, configuration and operating conditions of the coal combustion plant directly affect the design and performance of the CO_2 emission control systems, which must be integrated to the plant.

3.2.2 SUPERCRITICAL PULVERIZED COAL-FIRED POWER PLANT

An increase in combustion process efficiency reduces coal consumption, pollutant emissions, and potentially the cost of electricity generation. The first generation of coal-fired power plants constructed in the early 1900s converted only eight per cent of the energy contained in coal to electricity (Yeh and Rubin, 2007). Although significant improvements in plant efficiency have been made since then by operating plants at higher temperatures and pressures, the efficiency is constrained by the corrosion resistance of materials. Most current operating PC power plants have energy conversion efficiencies ranging from 33-37 per cent (Ansolobehere et al., 2007). More advanced plants operating at even higher temperature and pressure, called supercritical pulverized coal-fired power plants, reach efficiencies of 37-40 per cent (Ansolobehere et al., 2007). Advances in materials such as super alloys, increased environmental concerns and the rising cost of coal during the last two decades have stimulated the revival of the advanced supercritical technology, particularly in Europe and Japan. This technology now dominates the construction of new coal-fired power plants.

3.2.3 ULTRA SUPERCRITICAL PULVERIZED COAL-FIRED POWER PLANT

Recent advances in coal combustion technologies are highlighted by the generation of ultra supercritical (USCPC) steam conditions that can achieve even higher process efficiencies. The ultra supercritical condition refers to operating steam-cycle conditions above 565°C (1,050°F). The higher pressure and temperature of the steam generated from existing ultra supercritical power plants results in an energy conversion efficiency of more than 43 per cent (Bugge et al., 2006). The global ongoing research and development activities on advanced USCPC boilers are expected to result in plant efficiencies reaching 50 per cent and higher (IEA, 2012c). In 2012 the first ultra supercritical power plant in the United States (the John W. Turk Jr. Coal Plant) came online. This plant is operated by American Electric Power (AEP) and it has specially designed chrome- and nickel-based super alloys. The adoption of these advanced metals increases the plant's construction cost by five per cent in comparison to a similarly-sized supercritical

FIGURE 3.1

Simplified schematic diagram of a pulverized coal combustion process for power generation

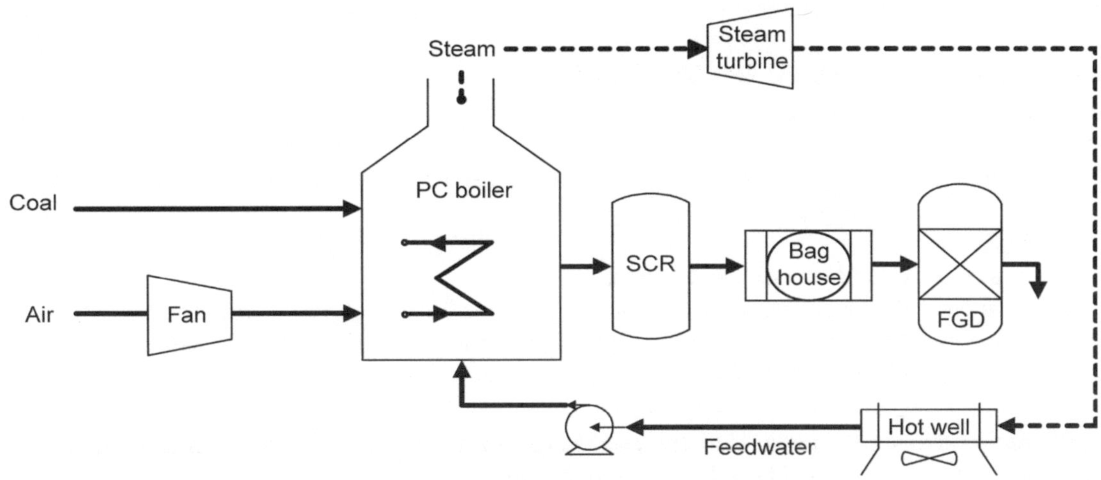

PC: pulverized coal; SCR: selective catalytic reduction; FGD: flue gas desulfurization; bag house: filter used to control particulate matter; - - - steam

FIGURE 3.2

Flow diagram of an IGCC power plant

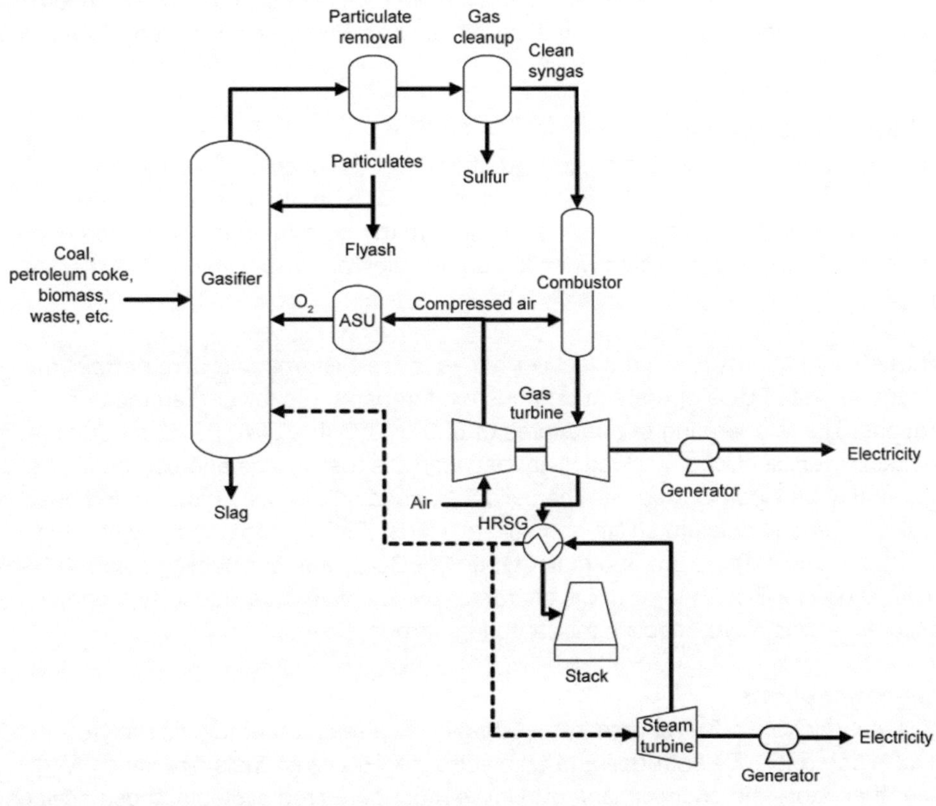

ASU: air separation unit; HRGS: heat recovery steam generator

power plant. However, since a USCPC plant can be up to 11 per cent more efficient than a supercritical plant, the new plant will reduce the coal consumption per unit electricity generated and thus also produce electricity with reduced emissions intensity.

3.2.4 ADVANCED COAL-FIRED CONFIGURATIONS

While typical coal-fired configurations today are implemented throughout the world, extensive research into advanced coal-fired systems is underway to improve the efficiency of fuel conversion and to enhance the environmental controls in a new generation of coal-fired configurations. These advanced technologies are alternatives for the today's coal-fired configurations.

3.2.4.1 Integrated gasification combined cycle

Integrated gasification combined cycle (IGCC) technology combines coal gasification processes with a combined cycle employing a gas turbine and steam turbine (Figueroa et al., 2008). This advanced approach first uses oxygen and steam to convert the solid fuel to a gaseous fuel known as synthetic gas (syngas) that consists of carbon monoxide, hydrogen, carbon dioxide, and water. The syngas is then processed to remove particulate and sulfur contaminants. The purified gaseous fuel is then burned in a gas turbine to produce electricity. The residual heat from the gas turbine exhaust is used to generate more steam, which combined with the steam from the gasifier drives steam turbines to generate more electricity. A schematic of the IGCC process is shown in .

The IGCC system has several advantages over conventional coal combustion: it requires smaller volumes of syngas, can use any type of coal, produces less waste, and consumes fewer resources such as water. IGCC units also have higher efficiencies available at the gas turbine. In addition, the IGCC system can capture the CO_2 component in the syngas prior to combustion. The syngas can also be used to produce petrochemicals and refining products. However, IGCC systems are complex and require care in design and construction. IGCC demonstration systems are operating worldwide. In June 2013, Duke Energy started a commercial unit in Edwardsport, Indiana.

3.2.4.2 Fluidized bed combustion systems

Fluidized bed combustion (FBC) systems are coal-fired configurations that are characterized by the suspension of solid fuel during combustion. The FBC system comprises of a bed of solid fuel materials such as coal and limestone and the introduction of combustion air from the bottom. The solid materials are entrained and become "fluidized," behaving as a fluid when the combustion air passes through the bed at a sufficiently high flow rate.

Unlike pulverized coal-fired configurations, FBCs are fuel flexible as these systems can burn petroleum coke, biomass, and any type of coal. FBCs operate at combustion temperatures lower than those of pulverized coal configurations. These operating temperatures of 800-900°C (US EPA, 2010) produce fewer NO_x emissions while maintaining efficient chemical reactivity between the fuel source and combustion gas. Sorbents such as limestone may be incorporated into the solid fuel; such substances capture and absorb sulfur dioxide (SO_2) emissions, forming calcium sulfite and sulfate solids. FBC systems thereby eliminate the need for auxiliary pollutant control. There are two major types of FBC systems: bubbling fluidized beds (BFB) and circulating fluidized beds (CFB). The former is characterized by lower gas velocities, coarser particles and higher pressures, and is more effective for capturing carbon dioxide.

3.2.4.3 Advanced coal power plants

Fuel cells convert chemical energy from a fuel source into electricity via a reaction at the electrolyte. Similar to batteries, fuel cells comprise of electrodes and an electrolyte. However, fuels directly convert the chemical energy rather than store the chemical energy. Unlike most coal-fired systems, these devices form usable energy such as electricity and heat without combustion. This technology typically combines hydrogen, the fuel source, and oxygen from air in a low-emission fuel conversion configuration. Recently, fuels cells have been integrated with typical fuel power systems such as coal-fired gasification technologies to improve the thermal efficiencies.

Solid oxide fuel cells (SOFC) are fuel cells where the electrolyte is a solid oxide material or ceramic to facilitate oxygen transfer from cathode to anode. SOFC can be combined with a gas turbine (GT) to produce electricity in a hybrid power generation plant. Solid oxide fuel cells-gas turbine (SOFC-GT) combined cycles operate the SOFC at higher pressures to improve the efficiency of typical SOFCs. In these combined systems, coal or other hydrocarbon fuel is gasified to fuel the SOFC. The GT converts the rejected thermal energy from the SOFC to additional electricity. The SOFC produces around 80 per cent of electrical power as compared to the turbine system which produces around 20 per cent of electricity (US DOE, 2004a). SOFC-GT configurations have the potential for higher thermal efficiencies. SOFC-GT can achieve 75-80 per cent of higher heating value (HHV) for fuel-to-electricity efficiency (Brouwer, 2006). This hybrid system development is limited by the durability and cost-effectiveness of the SOFC material, and is the subject of much on-going research.

3.2.5 CONVENTIONAL NATURAL GAS-FIRED CONFIGURATIONS

3.2.5.1 Natural gas combined cycle

NGCC is an advanced power generation technology that improves the fuel efficiency of natural gas. Most new gas power plants in North America and Europe are of this type. Although coal is the cheapest fossil fuel available, it also is the most polluting. Natural gas is a comparatively cleaner fuel, with the primary products of combustion containing very little sulfur and nitrogen oxide impurities (NPCC, 2002). The design of NGCC facilities is similar to that of a coal-fired IGCC plant, with the exception of lowered conditioning requirements for natural gas. Natural gas primarily consists of methane (CH_4) and is combusted with excess air in a pressurized combustion chamber. The combustion gases pass through a gas turbine to generate electricity. The resulting gases are sent to a heat recovery steam generator (HRSG) that produces steam that passes through a steam turbine to produce electricity. The combined cycle thus ensures that both heat energy and the pressure of the gas stream is used for electricity production. .

NGCC systems eliminate the problem of ash-handling and air pollution control units for other polluting compounds. These systems are complex and require detailed and robust design considerations. The more expensive infrastructure required for gas transport is also a significant factor in the overall economic feasibility of the process. NGCC plants are expected to play a major role in meeting the world energy demand in the near future.

3.2.5.2 Advanced combined cycle

An SOFC stack can be coupled with a cascaded humidified advanced turbine (CHAT) for high efficiency power generation from natural gas. Figure 3.3 compares various electricity generation technologies in terms of process efficiency and generation capacity. As indicated in the figure, proton exchange membrane or polymer electrolyte membrane (PEM) fuel cells and solid oxide fuel cells (SOFC) are more efficient for small capacity power generation in comparison to conventional internal combustion engine (ICE). For large scale or centralized power plants, the advanced combined cycle system (SOFC+GT, ST or CHAT/SOFC) is a promising option. The improved power generation efficiency obtained by the CHAT/SOFC process mainly results from the system integration between and SOFC and combined cycle systems. In such system, the exhaust gases from SOFCs, containing considerable amount of leftover fuel, are utilized. This renders the full fuel conversion in the SOFC stack less important, and thus decreases the size and the cost of the SOFC stack. CHAT/SOFC process has been demonstrated with 50 per cent electrical efficiency with natural gas, and 85 per cent thermal efficiency for cogeneration (Brouwer, 2006).

3.2.5.3 Decentralized power and heat generation from conventional fossil fuels

Distributed generation of electricity (DG) is expected to become increasingly important in the future for energy supply infrastructure, particularly in future electric utilities in economies with deregulation (Ogden, 2002). DG stations are generally smaller than 100-150 MW_e (Ackermann et al., 2001) and combined heat and power generation (CHP) is one of the major applications of DG due to its high overall energy efficiency. A study conducted by the IEA suggests that in G8+5 countries, which account for more than two-thirds of global primary energy consumption, the share of CHP in electricity generation may increase from 11 per cent in 2005 to 24 per cent in 2030 in a scenario with a pro-CHP policy regime (IEA, 2008).

CHP can be defined as "the sequential of simultaneous generation of multiple forms of useful energy (usually mechanical and thermal) in a single, integrated system" (US EPA, 2014). In a centralized power plant, a significant amount of low/medium grade heat is contained in the stack gas. This heat, however, is often not utilized and is vented to the atmosphere; centralized power plants are often located far from heat consumers and the transport and distribution of heat over a long distance results in significant energy losses. These losses often make the use of cogenerated heat economically infeasible. CHP plants make the use of cogenerated heat by being installed near the heat consumers.

3.2.5.3.1 Applications

There are three categories of CHP applications: industrial, commercial/institutional, and district heating and cooling (DHC) (IEA, 2008). An overview is presented in Table 3.1.

FIGURE 3.3
Advantage of fuel cell over other technologies

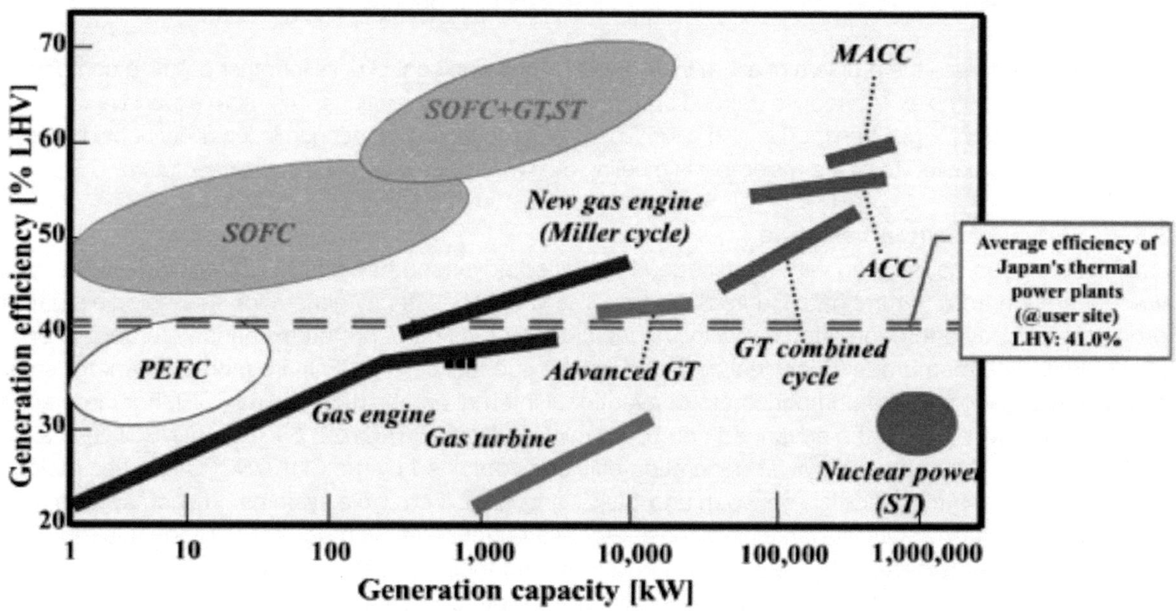

GT: gas turbine;
SOFC: solid oxide fuel cell;
PEFC: polymer electrolyte fuel cell;
ACC: Advanced Combined Cycle;
MACC: More Advanced Combined Cycle.

Source: Murazeki, 2009

TABLE 3.1

Overview of CHP applications

Feature	CHP - Industrial	CHP – commercial/ institutional	District heating and cooling
Typical customers	Chemical, pulp and paper, metallurgy, heavy processing (food, textile, timber, minerals), brewing, coke ovens, glass furnaces, oil refining	Light manufacturing, hotels, hospitals, large urban office buildings, agricultural operations	All buildings within reach of heat network, including office buildings, individual houses, campuses, airports, industry
Ease of integration with renewable and waste energy	Moderate to high (particularly industrial energy waste streams)	Low to moderate	High
Temperature level	High	Low to medium	Low to medium
Typical system size	1 – 500 MWe	1 kWe – 10 MWe	Any
Typical prime mover	Steam turbine, gas turbine, reciprocating engine (compression ignition), combined cycle (larger systems)	Reciprocating engine (spark ignition), Stirling engines, fuel cells, micro-turbines	Steam turbine, gas turbine, waste incineration, CCGT
Energy/fuel source	Any fuel, including industrial process gases	Liquid or gaseous fuels	Any fuel
Main players	Industry (power utilities)	End users and utilities	Include local community ESCOs, local and national utilities and industry
Ownership	Joint ventures/third party	Joint ventures/third party	From full private to full public, including utilities, industry and municipalities
Heat/electricity load patterns	User and process specific	User specific	Daily and seasonal fluctuations mitigated by load management and heat storage

Industrial heat supply

Industry has been a major user of CHP for decades. Energy-intensive sectors such as food processing, pulp and paper, chemicals, metal- and oil refineries represent more than 80 per cent of the total global electric CHP capacities (IEA, 2007). CHP is attractive to industry actors for two main reasons. Firstly, heat demand of these industrial sectors is high and is not subject to daily and seasonal fluctuations. Secondly, industrial plants have operations and maintenance personnel competent to manage CHP systems (IEA, 2007). While industrial systems over 1 MW_e account for the vast majority of global CHP capacity, many smaller scale industrial sites also use CHP systems that are similar to those used in commercial and institutional buildings (IEA, 2007).

Commercial and institutional heating

The use of CHP in commercial and institutional buildings has increased steadily in recent years, largely due to technical improvements and cost reductions in small-scale pre-packaged systems. Moreover, these buildings often have significant energy costs in addition to balanced and constant electricity and heating and cooling loads, making CHP a cost-effective option to reduce their carbon footprint (IEA, 2008). Residential micro-CHP technologies are also being developed and sold to individual households. Micro-CHP may become a mass-market CHP product if fully competitive and reliable products can be released on the market (IEA, 2008).

Combined heat and power district heating and cooling (CHPDHC)

Space and water heating require low and medium temperature heat. For this purpose, low grade heat from CHP plants, industrial processes and waste incineration is often used. District cooling also uses low grade heat to drive absorption chillers (IEA, 2008). District cooling is becoming an increasingly popular alternative to conventional electricity- or gas-driven air conditioning systems (IEA, 2008). District cooling systems reach efficiencies that are five to ten times higher than those of typical electricity-driven air conditioning systems because they make use of resources that would otherwise be wasted or difficult to use (Euroheat & Power, 2011; IEA, 2008). Because of large energy losses encountered during CHPDHC distribution and the high installation costs of heating and cooling distribution networks, population density is an important factor for the cost effectiveness of CHPDHC. CHPDHC is also considered to be a viable way to introduce renewable energy resources into heat and electricity sectors (IEA, 2008) because it does not require high quality fuel such as natural gas to achieve high overall system energy efficiency.

3.2.5.4 Combined heat and power technologies

Table 3.2 presents an overview of CHP technologies. Gas turbine power generators, particularly NGCC, and coal-fired steam turbine generators are also widely used for centralized power generation, whereas gas and oil engine generators and fuel cells are mainly used for distributed generation.

TABLE 3.2

Overview of combined heat and power (CHP) technologies

CHP system	Typical capacity [MW_e]	Power efficiency[1)]	Overall efficiency[1]	Typical heat-to-power ratio	Fuel types	Uses for thermal output
Gas turbine	0.5-40 (NGCC: ~250)	24-40%	77-83%	0.5-2	natural gas, biogas, propane, oil	Heat, hot water, low and high pressure steam
Reciprocating engine	0.01-5	24-44%	77-88%	1-2	natural gas, biogas, propane, landfill gas	Hot water, low pressure steam
Steam turbine	0.5-250	16-42%	84-88%	3-10	All types	Low and high pressure steam
Fuel cells	0.005-2	33-66%	72-88%	0.5-1	H_2, NG, propane, methanol	Hot water, low and high pressure steam

1. Efficiency values reported in higher heating value (HHV) are converted to lower heating value (LHV) terms using a multiplication factor of 1.05 for coal and 1.1 for other fuels.

Source: Kuramochi et al., 2011, based on various sources (IEA, 2008; IEA GHG, 2007; IPCC, 2005; US EPA, 2014). Energy efficiency values are presented in LHV terms.

The following descriptions on CHP generator technologies are based on the characterization of technologies provided in the United States Environmental Protection Agency's (US EPA) CHP Catalogue (US EPA, 2014), unless otherwise stated.

Boiler with steam turbines

The high-temperature, high-pressure steam produced in the boiler is expanded in a turbine to generate electricity. Some of the steam discharged from the turbine can be used in turn to supply useful heat to consumers. Because of the simplicity of this system, a wide variety of fuels can be used. The costs of a complete boiler/steam turbine CHP system is relatively high on a per kW basis in comparison to other generator technologies because of their high heat-to-power ratios (HPR), the size of the equipment, the complexity of the fuel and steam handling systems and the custom nature of most installations. Steam turbine CHP systems are therefore typically used in medium to large scale industrial or institutional facilities with high thermal loads and where solid or waste fuels are readily available for boiler use.

3.2.5.4.1 Gas turbines with heat recovery

Gas turbine generators are available in a wide range of sizes, from 500 kW up to 300 MW. Although gas turbines can operate on a variety of fuels, most generally operate on gaseous fuel and use liquid fuels as a backup. Gas turbines can be used in two general CHP configurations: (1) a single gas turbine and a heat recovery steam generator (HRSG) (simple GT-CHP) and (2) combined cycle CHP, in which high pressure steam is generated in HRSG and used partially for additional power generation using a steam turbine and partially for useful heat supply. Gas turbines are well suited for CHP because their high temperature exhaust can generate high quality process steam at high pressure and temperature conditions reaching up to 83 bar and 480°C. Simple cycle CHP applications are common in smaller installations, typically less than 40 MW, while combined cycle CHP installations can be 250 MW or larger. Microturbines, which are available in sizes from 30 kW to 250 kW, can burn a wide variety of gaseous and liquid fuels, even those with high sulfur content. Microturbine CHP systems operate similarly to larger gas turbine CHP systems, but usually generate only hot water as the useful heat product. Microturbine generation is a rather new technology; it entered field-testing in 1997 and the first commercial units began service in 2000.

3.2.5.4.2 Reciprocating engines

There are two common types of reciprocating engines used in CHP applications: spark ignition (SI) and compression ignition (CI)[2]. SI engines use spark plugs to ignite a compressed fuel-air mixture within the cylinder and they are available in sizes up to 5 MW. CI engines, also called diesel engines, operate on diesel fuel or heavy oil and are among the most efficient simple-cycle power generation options in the market. The main strengths of reciprocating engines for CHP applications are that they start quickly, follow load well, have good partial load efficiencies, and generally have high reliabilities. Reciprocating engines are well suited to applications with hot water or low-pressure steam demands.

3.2.5.4.3 Advanced combined heat and power technologies: fuel cells

Fuel cells use an electrochemical process to convert the chemical energy of a fuel, often hydrogen, into water and electricity. In CHP applications, heat is generally recovered in the form of hot water or low pressure steam, depending on the type of fuel cell and its operating temperature. There are currently five types of fuel cells under development: (1) phosphoric acid (PAFC), (2) proton exchange membrane (PEMFC), (3) molten carbonate (MCFC), (4) solid oxide (SOFC), and (5) alkaline (AFC). PAFC systems are commercially available in two sizes, 200 kW and 400 kW, and two MCFC systems are commercially available, 300 kW and 1,200 kW. MCFC and SOFC, which are still in the pilot phase, can be scaled up to multi-MW_e size. The installed costs of fuel cell systems are still high today and thus, the most cost-effective applications of fuel cell systems are for CHP applications.

2 SI engines use petrol or gasoline as fuel. They use an Otto cycle where fuel combustion occurs at a constant volume. In this engine a spark is used to initiate the burning process as petrol has a high self-ignition temperature. CI engines use diesel as fuel. They use a diesel cycle in which the combustion occurs at a constant pressure contrary to petrol, diesel has low self-ignition temperature and therefore no spark is needed as the ignition of fuel occurs due to the compression of the air-fuel mixture

3.2.6 UNCONVENTIONAL FOSSIL FUELS

3.2.6.1 Oil sands

The first commercial oil sands project began production in 1967, producing 12,000 barrels per day (bpd) in the western Canadian province of Alberta (CRS, 2008). The industry stagnated for several decades until high oil prices and technological advances prompted significant expansion. Oil sands production from the area was approximately 1.5 million bpd in 2010, and it is expected that production will increase by 150,000 bpd in 2012 (CGES, 2011).

Canadian oil sands reserves are estimated at 175 billion barrels, third after only Saudi Arabia and Venezuela in terms of global oil reserves (US EIA, 2011b). There are oil sands reserves outside of Canada, however, with the exception of Venezuela, they are less extensive, and none of these reserves have been nor are expected to be developed within the next decade (CRS, 2008). This section focuses on developments and issues related to the Canadian oil sands due to their relative maturity.

The oil sands are a mixture of sand, water, and bitumen; bitumen is a heavy oil which does not flow under ambient conditions due to a high viscosity. It is hence more difficult to recover and process than conventional crude oil. For deposits up to 75m in depth, surface mining is the preferred method of recovery. In this process, oil sands are excavated and transported in trucks to separation facilities, where the bitumen is separated from the sand using hot water. Currently, all surface-mined bitumen is sent to an upgrader for further processing, while the residual water and other waste materials, consisting mostly of sand and unrecovered bitumen, are sent to tailings ponds. In situ recovery methods are more suitable for deeper deposits where surface mining is impractical. For deposits with lower bitumen viscosities, the oil sands mixture is pumped out of vertical wells, a method known as cold heavy oil production with sand (CHOPS); otherwise, thermal techniques are required (CRS, 2008).

The two most common thermal in situ methods are cyclic steam stimulation (CSS), and steam assisted gravity drainage (SAGD), both of which involve pumping steam into the ground to reduce the viscosity of the bitumen and separate it from the sand so that it can be recovered by pumping. The SAGD process involves the drilling of two horizontal wells. Steam is continuously injected into the upper well, mobilizing the surrounding bitumen and allowing it to flow into the lower well. It allows for a higher bitumen recovery rate than CSS: up to 70 per cent versus 25-30 per cent, respectively (CRS, 2008). A key indicator of the efficiency of a thermal in situ project is the steam-to-oil ratio (SOR), a measure of the amount of water in the form of steam required to produce one barrel of oil. Although many analyses of SAGD operations assume an SOR of 2.5 for a well-performing SAGD operation (Lacombe and Parsons, 2007; Toman et al., 2008), some projects operate at slightly lower SORs, while others operate at much higher SORs. The government of Canada's well-to-wheel model for transportation fuels, GHGenius, is frequently used to examine the energy use and GHG emissions associated with various fuel pathways. GHGenius assumes default SOR values of 3.2 and 3.4 for SAGD and CSS, respectively (Brandt, 2012). However, limited explanation is provided to justify the choice of these parameters (Charpentier et al., 2009). Once recovered using in situ techniques, bitumen is either upgraded to synthetic crude oil (SCO) or is diluted with a lighter fuel. This dilution allows pipeline transport of the bitumen to a refinery for further processing into petroleum products, such as gasoline and diesel.

While surface mining accounts for approximately 55 per cent of current oil sands production, about 80 per cent of the Canadian oil sands reserves will require in situ methods (ERCB, 2008). Although a considerable number of existing in situ projects use CSS, SAGD is the preferred technology for new in situ projects (NEB, 2006). The oil sands industry is a key driver of economic growth in Canada, and more focus is being put towards developments to overcome their environmental challenges. With oil sands production expected to reach 4.21 million bpd by 2020, their significance is increasing in the global energy context (CAPP, 2011).

3.2.6.1.1 Emerging technologies in the oil sands

The recent rapid expansion of the oil sands industry has induced significant investment in the development of new technologies targeted towards overcoming some of the technical, economic, and environmental challenges that face the industry. Current investments in emerging technologies generally focus on new in situ methods that produce bitumen at lower cost by reducing or eliminating the natural gas needed for steam production. These initiatives are either modifications to the existing SAGD process to reduce the SOR, or new methods that remove the requirement for steam altogether. By decreasing the need for natural gas, which is one of the main contributors to GHG emissions associated with the production of oil sands products; these technologies have the potential to offer substantial reductions in the industry's GHG emissions. Many emerging technologies are also suitable for use in reservoirs where the physical characteristics of the reservoir make existing technologies unfeasible. Technologies currently in development that are considered promising include: solvent processes, in situ combustion, and electric heating.

Improvements to surface mining operations generally focus on incremental changes to site design and operation. The primary aim of these improvements is to improve the efficiency of the operation to reduce energy use, costs and GHG emissions. Heat integration, where the waste heat from one process is used in another, has also been identified as a means of reducing demand for external energy sources. Heat integration is also applicable to in situ operations, where large volumes of both steam and hot water are required for bitumen production. Steam processes reduce costs by lowering energy requirements. These processes involve modifications to the existing SAGD method to improve recovery efficiencies by changing well configuration and placement. The performance of these emerging steam processes at the full industrial scale remains uncertain (Bergerson and Keith, 2010).

Solvent processes use similar well configurations to SAGD; however, some or all of the steam typically required for SAGD is replaced with a solvent that reduces the viscosity of bitumen in the reservoir. When the bitumen is produced, a portion of the injected solvent is recovered and can be re-injected. In addition to reductions in energy and natural gas demand for bitumen production, solvent processes may also reduce or eliminate water use. A hybrid steam-solvent approach can be taken, where both steam and solvents are injected into the reservoir to increase bitumen production. It is still uncertain whether solvent recovery and recycling rates that would make this technology viable at an industrial level can be achieved. Large quantities of solvents lost in the reservoir may also negatively affect local ecosystems.

In situ combustion is an alternative in situ technology that has been under development for many years and is now moving toward commercial scale. With in situ combustion, air is injected into the reservoir to prompt combustion or gasification of the heavy portion of the petroleum in the reservoir. One advantage of this process is the potential for bitumen to be partially upgraded by the combustion process within the reservoir, reducing the need for upgrading and further processing once the bitumen is recovered. Although in situ combustion may result in higher bitumen recovery rates than current in situ methods, GHG emissions may be higher as natural gas is replaced with a heavier fuel.

Electro-thermal processes are also being considered as an alternative to steam injection. One such technology, the electro-thermal dynamic stripping process (ET-DSP), places a grid of electrodes in the ground surrounding a central extraction well (McGee and McDonald, 2009). An electrical current is passed through these electrodes. The current flows through the water in the formation and heats the surrounding bitumen. Pilot applications of this technology show promise. Depending on the GHG-intensity of the energy source used to provide electricity, ET-DSP may result in considerable emissions reductions.

Many technologies related to bitumen production are still in development, though several of these technologies have been found to perform favourably at the pilot scale. Although the main incentive for the development of these technologies has been to reduce capital and operating costs, these new technologies also offer the potential to provide access to reservoirs that would otherwise be inaccessible or uneconomical using existing technology. The full-scale performance of these technologies remains uncertain, but they may potentially also provide significant environmental benefits, including reduced GHG emissions.

3.2.6.2 Unconventional natural gas

Unconventional gas includes shale gas, tight gas and coal bed methane. These sources of natural gas represent the largest portion of American gas reserves and potentially represent a major source of long-term natural gas supplies worldwide. Unconventional is simply defined as production that uses extraordinary means to extract a fossil energy resource. For natural gas the "extraordinary means" usually means horizontal drilling and hydraulic fracturing.

Conventional oil and gas deposits consist of porous reservoirs (sandstone or carbonate) "capped" by an impervious layer (structural or stratigraphic trap) that allows oil, natural gas, and water to accumulate. These deposits of oil and/or gas are generally small in area. When multiple fluids (gas, oil and water) are present they separate based on density and fluid mobility within the reservoir. Most past gas and oil production and the vast majority of current production is from conventional reservoirs. The hydrocarbons recovered from these reservoirs are termed conventional gas or oil.

An unconventional deposit is usually defined by contrasting it to the conventional reservoir definition. An unconventional deposit has oil and/or gas distributed throughout the reservoir rock and limited migration. The formation can spread over vast areas, and because of low permeability, unconventional methods to extract the hydrocarbon are required. Thus, unconventional natural gas is simply gas produced from an unconventional reservoir.

Below the three major sources of unconventional gas sources are discussed. They include coal bed methane (CBM), tight gas, and shale gas.

3.2.6.2.1 Coal-bed methane

CBM is a gas consisting predominantly of methane but can contain small amounts of heavier hydrocarbons and varying amounts of non-hydrocarbon gases. Coal-bed methane is associated with coal seams throughout the world, although development of this resource occurs predominantly in the United States. There is increasing activity related to CBM in Canada, Australia, India and China. Coal seams are both the source rock and reservoir for this resource. The permeability of coal is low and gas production generally requires dewatering and/or fracturing of the coal to mobilize the gas (Rogner, 1997). Methane is produced during the coal maturation process.

Coal can store large quantities of methane, estimated at 6 or 7 times as much gas as an equal volume of rock in a conventional reservoir (USGS, 2000). The CBM is found in a free gas phase, dissolved in water, and adsorbed to the surface of the coal. Water is produced along with methane during the maturation process and permeates throughout the coal seam. Thus, considerable volumes of water varying in quality are often co-produced with CBM. This water must be removed to reduce hydrostatic pressure and thereby release gas from the coal surface, so during the early phases of production, levels of water production are high (USGS, 2000). The water is commonly saline but in some areas it can be potable. Produced water disposal can be a concern, and the disposal must be done in an environmentally responsible manner.

3.2.6.2.2 Tight gas

Tight gas is produced from sandstone and carbonate reservoirs (Ambrose et al., 2008) with very low permeability corresponding to less than 0.1 millidarcys; this categorises tight gas as unconventional (Aguilera and Harding, 2008). There is some debate as to the characterization defining tight gas formations. Schmoker (2005) considers tight gas as continuous gas accumulation that has a large areal extent with indistinctly defined boundaries (Schmoker, 2005). The formations are associated with conventional reservoir rocks (Schenk and Pollastro, 2002). On the other hand, Shanley et al. (2004) believe that the resource is simply poor quality reservoir rock in conventional traps (Shanley et al., 2004). Regardless of the exact physical definition of this resource, many nations consider it an extension of conventional gas and lump tight gas statistics with the conventional resources, making it difficult to accurately assess reserves and resources (IEA, 2012b).

3.2.6.2.3 Shale gas

Shale gas is mainly a dry gas extracted from shale formations. The gas has methane content exceeding 90 per cent,, but in some cases, may contain heavier hydrocarbons such as natural gas liquids (US DOE, 2009). Shale is a sedimentary rock that has low permeability that the industry previously considered a barrier to oil and gas migration (Boyer et al.,, 2006), ranging from 0.01 to 0.00001 millidarcy (US DOE, 2009). This low permeability requires the use of horizontal drilling and hydraulic fracturing to induce economic flows of gas. The development of these techniques led to the first large-scale shale gas production exploiting the Barnett Shale in North-Central Texas in the 1980s and 1990s (US EIA, 2011a). A number of other shale formations are being rapidly developed in the United States. The current "hot topic" play is the Marcellus shale formation in northeastern United States. All of this development activity has led the U.S. Energy Information Agency (EIA) to declare that, "The development of shale gas plays has become a 'game changer' for the American natural gas market." In the United States, shale gas production accounts for 35 per cent of the total national gross production of 29.5 trillion cubic feet in 2012 (US EIA, 2013b).

Shale gas has been developed mainly in the United States. There has, however, been some production in Canada, amounting to about 4 billion cubic feet in 2008. This is expected to reach 169 billion cubic feet by 2012 (NEB, 2010). In Europe, there are some early exploration efforts in the UK and Poland. A report from the EIA (2013a) indicates that the technically recoverable shale gas resources is in the range of 7299 to 7795 trillion cubic feet with only 10-15 per cent of this reserves being located within the United States (US EIA, 2013a). The potential for shale gas production is large and as the EIA suggests, it could be a game changer, but the use of unconventional production processes increases per well costs. The EIA (2011c) projects that wellhead prices will remain relatively constant at approximately 4 USUS$/million cubic feet through 2015 and thereafter rise, exceeding the 6 US$ mark towards 2035 (US EIA, 2011c). This puts pressure on shale gas production since the break-even price for gas production in most shale formations is between 4 to 6 US$/million cubic feet (Medlock III et al., 2011). The presumably low margins make drilling efficiency and the presence of oil and natural gas liquids essential to profitability.

Drilling activity in the Marcellus play demonstrates the impact of the presence of natural gas liquids on the economics of shale gas. Wet gas present in the western portion of Marcellus can be captured above a realized value, which takes into account of all components in the gas stream, of 7 US$/million cubic feet of gas produced (Gue, 2010). There is sustained activity in southwest Pennsylvania and increasing activity in the Unita shale in Eastern Ohio, where wet gas is found. This comes at the expense of drilling in the drier portions of the Marcellus.

3.2.6.3 Production technology

3.2.6.3.1 Drilling

As suggested in the definition, unconventional gas requires unconventional extraction processes. In most cases, this includes the use of horizontal drilling and hydraulic fracturing. Some formations have been exploited using conventional vertical wells, but require hydraulic fracturing in order to produce economical flows of gas. Historically, most oil and gas production wells were drilled vertically and contacted the reservoir directly below the drill site. Due to the limited thickness of unconventional gas formations, a vertical well has only minimal contact with the gas-containing layer. With the advent of horizontal, or directional, drilling, wells deviating from the vertical could be drilled. This increases the exposure length of the wellbore to the reservoir and enables the drilling of multiple wells from a single location.

Although drilling is typically accomplished with a conventional rotary drilling rig, some innovative designs available in the market aim to reduce costs. These innovations include walking rigs that can move about the well pad to facilitate the drilling of multiple wells and rigs that can be taken apart in sections (Kulkarni, 2010).

A well consists of a number of steel casings (defined as any pipe that is cemented into place) that protects the well from collapse and prevents migration of fluids into or out of the well. The casing diameter gets progressively smaller as well depth increases. Cement and casing integrity is important to the overall safety of drilling, completion and production from a well.

The final step of the drilling process is completion. If the lateral portion is cased, explosive charges are used to penetrate the casing. These perforations permit the transfer of fluids from the reservoir to the wellbore. In a horizontal well, there are multiple perforations along the horizontal production interval. Alternatively, a pre-slotted non-cemented liner can be installed (US EIA, 1993).

3.2.6.3.2 Hydraulic fracturing

Hydraulic fracturing uses a fluid, typically water that is pumped into the reservoir at a rate that exceeds the ability of the formation to accept the flow in a radial flow pattern (US DOE, 2004a). As resistance to flow increases, the pressure in the wellbore increases until it induces fractures in the low permeability reservoir. As the fracture propagates, the fluid carries proppant of fine-grained sand or ceramic material into the fractures to hold the fractures open and facilitate gas migration. The ideal fracturing fluid must be compatible with the formation rock and fluids, generate enough pressure to create wide fractures, transport the proppant into and down the full length of the fracture, decompose to a low viscosity fluid for clean-up, and be cost effective (US DOE, 2004b).

The pressures achieved during fracturing can be as high as 8000 psi and can induce fractures as far as 3000 feet from the well bore (Kerr, 2010). In a horizontal well, up to 3-5 million gallons of water can be required. The fracturing job consists of 4 stages: an acid stage, a pad stage, a prop sequence stage, and a flushing stage. The acid stage acts as a preparation phase as it clears cement debris, removes carbonate minerals from the reservoir face, and opens small fractions near the wellbore. The pad stage injects a plug of slickwater without proppant to fill the wellbore and facilitate flow into the formation. The slickwater can contain a number of chemicals depending on the characterization of the well, including biocides, scale inhibitors, chelating agents, corrosion inhibitor, oxygen scavengers, friction reducing agents (hence the term slickwater) and viscosity modifiers (URS Corporation, 2011).

The prop sequence consists of injecting slickwater with the proppant. The status and the success of hydraulic fracturing are monitored through use of microseismic fracture mapping, simulation modelling, and tiltmeter analysis (US DOE, 2009). Finally there is a flushing phase, which uses freshwater to flush excess proppant from the wellbore.

3.2.6.4 Water-soluble gas

Water soluble gas (WSG), is a type of the unconventional natural gas that is dissolved in water under high pressure and mainly consists of methane and some non-hydrocarbon gases (Wang et al., 2008). WSG reservoirs have been found in many countries such as America, Italy, Hungary, the Philippines, Iran and Japan, among others (Battino, 1984). The total global reserves are estimated to be about 34,000 trillion m^3 (Zhang, 1995), which is more than ten or even hundred times larger than that of conventional natural gas and is only exceeded in size by the reserves of natural gas hydrates. Many oil-gas basins are rich in WSG, for instance, some American basins along the Gulf of Mexico, basins in north and south Caspian, basin of West Siberian, Volga-Ural Basin and Azov-Cuban Basin (Zhou et al., 2011). Other countries, such as China and Japan, also have WSG reserves. Table 3.3 shows the geographic distribution and size of major water-soluble gas reservoirs.

WSG mainly consists of hydrocarbons, which are typically methane, nitrogen and sour gas, among other components (Chen et al., 2006). Generally, it can be divided into different categories if the volume percentage of certain component exceeding 50 per cent, and then further divided into different types according to whether the value exceeding 25 per cent. Unlike conventional natural gas, WSG is characterized by low reserve abundance and production, and sensitivity to the surrounding environment, which makes exploitation and utilization difficult.

TABLE 3.3

Major distribution and resource of water soluble gas in the world

Area	Resource (in trillion m³)
Gulf of Mexico Basin	2,699
South Caspian Basin	2,590
West Siberian Basin	1,000
North Caspian Basin	980
Volga-Ural Basin	140
Azov-Cuban Basin	180
China	19
Japan	0.739-0.887

3.2.6.5 Natural gas hydrates

Natural gas hydrates are crystalline solids composed of water and gas. The gas molecules (guests) are trapped in water cavities (host) formed by hydrogen-bonded water molecules (Sloan and Koh, 2008). They form under moderately high pressure and at temperatures approaching the freezing point of water.

The vast amounts of gas clathrate hydrates occurring in nature present a new potential energy resource, and offer one possible solution to the world's energy concerns. According to a recent US DOE report, the potential hydrate reserves have been suggested to approach around 400 million trillion cubic feet (TCF), which is orders of magnitude higher than the currently proven global gas reserves of 5500 TCF (Rath and Marder, 2007). Therefore, global energy assessments of gas from hydrated deposits vary widely, from exceeding all conventional gas resources, to even surpassing the amount of all hydrocarbon energy including coal, oil, and natural gas combined (Grace et al., 2008). Over 220 gas hydrates deposits (GHD) have been found worldwide in permafrost regions onshore and in ocean-bottom sediments at water depths exceeding 450 meters.

The interest in hydrates as an energy resource grows worldwide from North America (United States and Canada), Asia (India, Japan, Korea, Taiwan, and China), and Australasia (New Zealand). The countries with limited domestic resources such as India and Japan are investing significant financial and technical resources in exploration programs along their coastlines that are far exceeding the efforts in North America (Koh et al., 2009).

Production methods for energy recovery from arctic hydrated deposits include depressurization, thermal and geothermal stimulations, inhibitor injection, and carbon dioxide–methane exchange, where carbon dioxide is sequestered within the hydrate framework during methane production.

3.2.6.6 Underground coal gasification

Underground coal gasification (UCG) utilizes coal that otherwise could not be exploited; there is no other known technique that can extract such reserves. Early studies suggest that the use of UCG could potentially increase global extractable reserves by as much as 600 billion tons. The commercial development of UCG began in the 1930s in the Soviet Union; there are currently some commercial units that have been in operation for approximately 50 years, such as that in Angren, Uzbekistan. Feasibility studies and demonstrations are being conducted worldwide in the United States, United Kingdom, Russia, China, South Africa, New Zealand, Canada and India, among others. The efforts made in various countries at both research and pilot levels are well documented in literature (Khadse et al., 2007; Shafirovich and Varma, 2009).

UCG is the process of in situ conversion of coal into combustible synthesis gas (syngas), which can be used either as a fuel or as a chemical feedstock. UCG offers a number of environmental and other benefits over conventional mining, and is therefore proposed to be the coal utilization technique of the future. It eliminates the need for mining and also eliminates the need for specialized coal processing equipment and gasification reactors, thereby reducing the required capital investment significantly. The cost of syngas produced by UCG can be as low as 50-66 per cent of that from a surface gasifer (Friedmann et al., 2009). Other benefits of UCG include increased worker safety, avoidance of surface disposal of ash and coal tailings, low dust and noise pollution, low water consumption, larger coal resource exploration and low methane emissions to the atmosphere.

3.2.6.6.1 Process

A schematic of UCG process is shown in Figure 3.4. UCG in its most general form consists of a pair of process wells, one injector and one producer, drilled from the surface into the coal seam at a specified distance apart from each other. After making the wells, a highly permeable channel is created in order to establish the link between the two wells within the coal seam. Various configurations of well connections and drilling techniques are possible and are reviewed elsewhere (Khadse et al., 2007). Once a permeable link of desired size is developed, air or a mixture of steam and oxygen is injected at high rate and high pressure into one of the wells. Gasification occurs when a mixture of air or oxygen with steam, forced into the coal seam through the injection well, reacts chemically with coal. As the reaction proceeds, a cavity consisting of coal, char, ash, rubble, and void space, is created underground. This consumes a bulk of the coal producing a combustible gas mixture containing CO, CO_2, CH_4, H_2, H_2S and other non-gaseous substances such as H_2O, char, tars, etc. The pressure maintained in the cavity is normally less than the hydrostatic head such that water flows into the cavity instead of allowing dangerous gases to escape to the surface. The product gas is then cleaned, treated and used for power generation or as a chemical feedstock. The successful application of such a process would provide a low to medium heating gas (88.23-264.70 kJ/mol), depending on whether air or a mixture of oxygen and steam is used. As in surface gasification, the removal of H_2S and NH_3 compounds in the syngas is relatively less expensive compared to the removal of SO_x and NO_x produced during combustion. The UCG product gas is expected to contain relatively more hydrogen and carbon dioxide. The ENN project in Wulanchabu is aimed at methanol production whereas Linc Energy's project at Chinchilla will make Fischer-Tropsch liquids (Friedmann et al., 2009).

3.2.6.6.2 Process engineering aspects

UCG is an inherently transient process and maintaining a uniform product gas composition is the key requirement from the viewpoint of its downstream applications. Hence, a reliable modelling tool that predicts the product gas composition and the state of cavity will be integral to the success of a UCG installation. The sensitivity of the performance to the controllable parameters such feed flow rate, feed composition and operating pressure must be thoroughly studied. The simultaneous occurrence of a number of exothermic and endothermic reactions, heat and mass transfer effects, non-ideal flow patterns in the cavity and thermo-mechanical failure of the coal leading to crack development and spalling makes the modelling and simulation of a UCG installation a challenging task. Though notable efforts have been made to this effect, there is still a tremendous scope for researchers to develop empirically validated models that will represent different events in the UCG process (Aghalayam, 2010).

3.2.6.6.3 Hurdles

The main environmental issues associated with UCG are groundwater contamination and surface subsidence. Careful site selection, such as ensuring the coal seam is at sufficient depth with no aquifers in the nearby area will possibly avoid these problems. Another major obstacle is potentially adverse public perceptions and reactions, which could either stop or delay proposed installations. A case study on public perception in UK (Shackley et al., 2006) recognized the potential of UCG as a secure future energy source. They have discussed potential benefits to the local community, potential risks, the role of CCSCCS, and links to the hydrogen economy. It is also recommended that an open, transparent and counselling process of decision-making is necessary and that UCG should be developed at a remote site, preferably on land, before initiating UCG projects in coal seams close to populated areas. Public perception is, however, not an issue inherent to UCG alone and it is discussed in more detail later in the chapter.

FIGURE 3.4

Schematic of underground coal hydrogasification (UCHG)

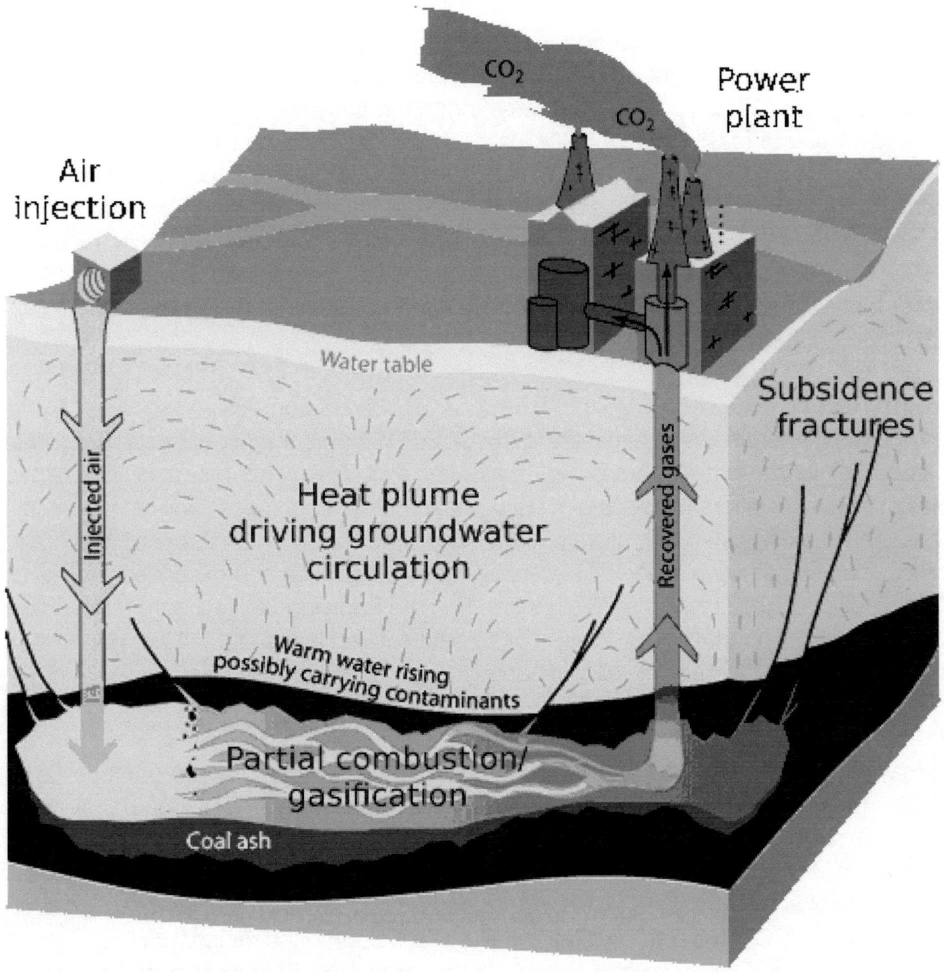

Produced by the Free Range Energy Beyond Oil Project version 1, October 2011.
web: http://fraw.org.uk/pubs/e11.html

3.3 TECHNOLOGY DESCRIPTION: CARBON DIOXIDE CAPTURE, TRANSPORT AND STORAGE

Carbon dioxide capture, transport and storage (CCS) technology entails the capture of CO_2 from large anthropogenic sources, transport of the CO_2 to an underground storage reservoir and long-term isolation from the atmosphere (IPCC, 2005). CCS is an interesting technology for climate change mitigation because it allows the continued use of fossil fuels while reducing their CO_2 emissions intensity. Since fossil fuels are expected to dominate the world's primary energy supply for the decades to come, CCS can play a key role on decarbonising the energy and industrial sectors. The IEA estimates, for instance, that in a BLUE Map scenario of 450 ppm CO_2-eq by volume, CCS contributes about 21 per cent of emissions reductions by 2050 (IEA, 2012b). Similar findings have been reported by studies examining the role of CCS at the regional (Odenberger and Johnsson, 2010; Strachan et al., 2011) and national level (Remme et al., 2011; van den Broek et al., 2010).

The CCS value chain can be coarsely divided into four parts: capture, compression, transport and storage. In the last decade, the most attention has been paid to the CCSCCS components. Pilot capture plants are under development, and CO_2 injection projects are being undertaken and closely monitored. In 2011, 74 large-scale integrated projects[3] existed in different stages of development. Fourteen of these projects are either in operation or construction and have a combined CO_2 storage capacity of over 33 million tons a year. The other projects are in earlier implementation stages such as identification, evaluation, definition and execution (Global CCS Institute, 2011). The adoption of CCS needs to increase significantly in the following years. Figure 3.5 indicates the number of projects and the amounts of CO_2 that must be sequestered in the 2015-2050 period in order to reach the 21 per cent reduction on CO_2 emissions described in the IEA scenario.

3.3.1 CAPTURE TECHNOLOGIES

There are three main routes considered for CO_2 capture: post-combustion capture, pre-combustion capture and oxyfuel combustion. These routes can use one or more separation technologies (see Figure 3.6). In chemical or physical absorption, solvents are used to capture the CO_2. The CO_2 is then released through changes of temperature or pressure. Almost all near- and mid-term post-combustion capture processes under development are absorption-based (Global CCS Institute, 2011). Adsorption processes use materials with high surface areas such as zeolites to separate CO_2 from gas mixtures by taking up CO_2 onto the material surface. The CO_2 is released by changes in temperature or pressure. In membrane processes, CO_2 is separated by using membranes which allow passing the CO_2, or other components, through a membrane wall. In order for the CO_2 to pass through the membrane wall, the partial pressure of the CO_2 must be higher on one side of the membrane than the other side. This is obtained by means of pressurizing the flue gas, applying a vacuum or a combination of both. In a cryogenic process, CO_2 is separated through condensation at extremely low temperatures.

3.3.1.1 Post-combustion capture

As its name indicates, in this system, CO_2 is captured from the flue gases produced after fossil fuels or biomass are burned (Figure 3.7). As the CO_2 is captured after combustion, the technique could be used for retrofitting power plants. A detailed description of coal and natural gas power plants are provided in Chapter 3.2. The flue gas is close to atmospheric pressure and CO_2 is relatively dilute with concentrations of 4-8 per cent by volume in natural gas-fired and 12-15 per cent by volume in coal-fired power plants. Given these conditions, chemical absorption is considered likely the first generation technology that will be adopted in power plants. The technology consists of three steps. First, the flue gas is cleaned of contaminants such as NO_x, ash and SO_2 via selective catalytic reduction, electrostatic precipitator and flue gas desulfurization unit, respectively, in order to minimize solvent degradation and cost. In the second step, the cleaned flue gas is sent to an absorber where

3 Large scale integrated projects were selected based on one of these criteria: a) have not less than 80 per cent of 1 million tons per annum of CO_2 captured and stored annually for coal-fired power generation; and b) not less than 80 per cent of 0.5 million tons of CO_2 captured and stored annually for other emissions-intensive industrial facilities (including natural gas-fired power generation) (Global CCS Institute, 2011).

FIGURE 3.5

Number of CCS projects required to achieve a 19 per cent reduction of CO_2 gases by 2050 (IEA, 2009b)

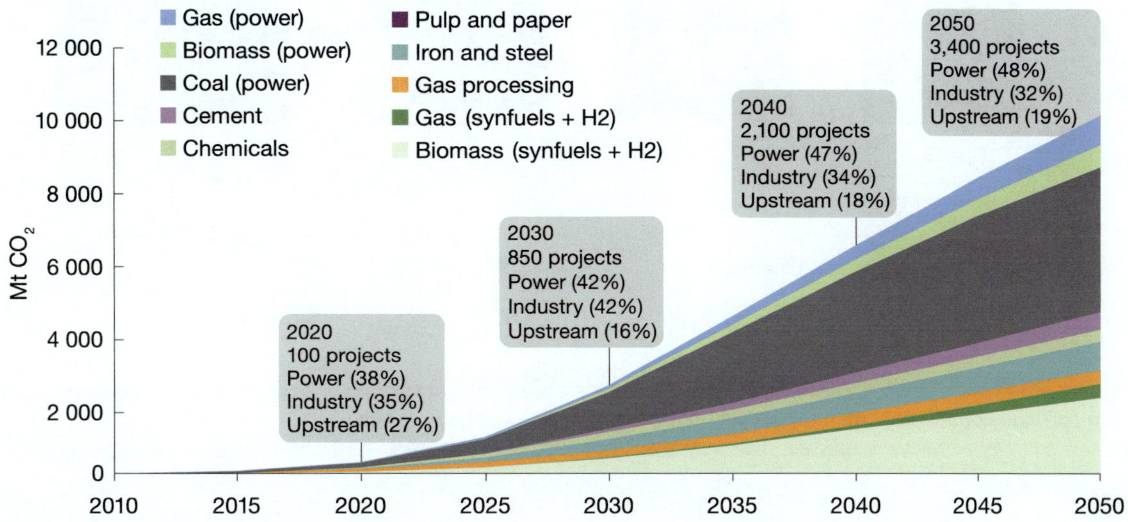

FIGURE 3.6

Options for CO_2 capture. Reprinted with permission from (Rao and Rubin, 2002).

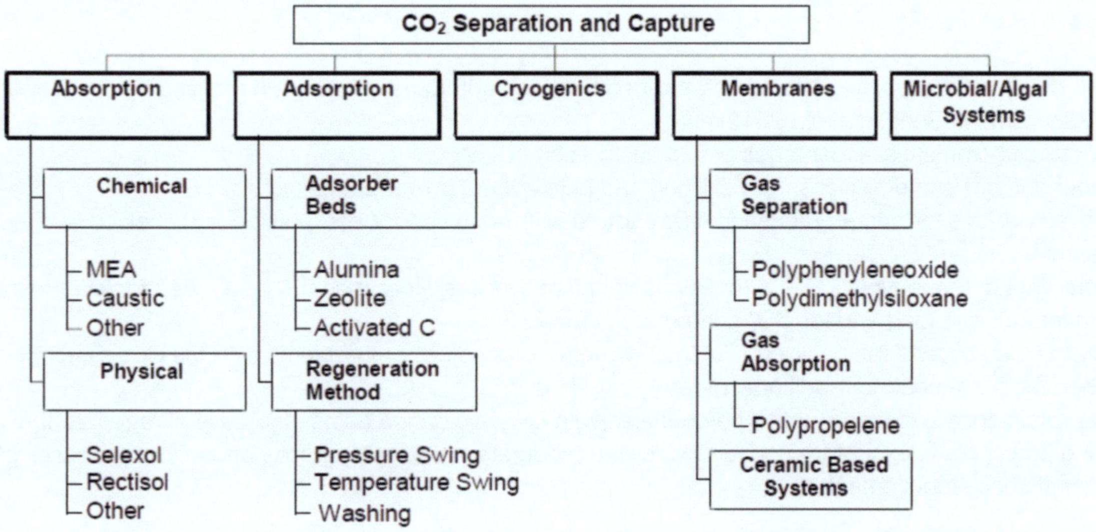

FIGURE 3.7

Process flow diagram of a post-combustion process adapted from Rubin (2008) and NETL (2010b)

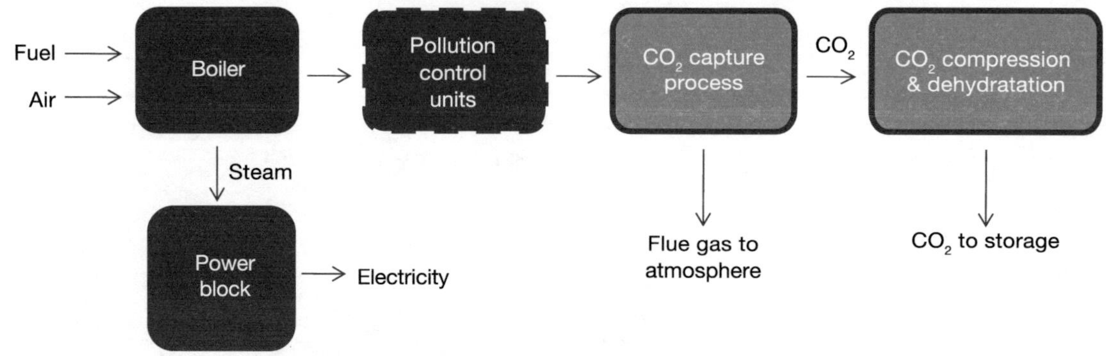

Note that pollution control units are required only in the case of coal-fired power plants.

a chemical solvent absorbs 90 per cent or more of the CO_2. Finally, the CO_2-loaded solvent is pumped into a regenerator unit called a stripper where heat is applied to release the CO_2 from the solvent. The steam supplying the required heat originates from the steam turbine. The solvent is returned to the absorber, while the CO_2 is dehydrated, compressed and piped to an underground storage location.

The chemical solvents available for industrial CO_2 capture are aqueous solutions of alkanolamines (monoethanolamine (MEA), dimethylethanolamine (DMEA)), sodium hydroxide (NaOH) and ammonia (NH_3). Amine-based chemical solvents such as aqueous MEA have been used to remove acid gases such as CO_2 and H_2S from natural gas streams and to produce food-grade CO_2 for use in beverages and other products (NETL, 2010b) for more than 60 years. As a result of these solvents' mature role in CO_2 removal processes, they are the most studied solvents considered for CO_2 capture applications.

The main drivers of the growing interest in post-combustion capture have been summarized by the Global Energy Assessment (Benson et al., 2012) as:

- Slow rate of commercial acceptance of fuel gasification (IGCC);
- Number and scale of emissions from existing and planned PC power plants;
- Improved designs for post-combustion CO_2 capture with more vendor competition and choices of chemical solvents;
- Minimal impact to the typical NGCC or PC power plant process other than the large need for low pressure steam for CO_2 stripping and for CO_2 compressor power;
- Ability to easily bypass the back-end flue gas scrubber process when problems with the CO_2 system occur or when there is a need for additional peaking power; and,
- Lower total capital expenses (not to be confused with CO_2 avoidance costs) and ease of retrofit to the existing power plant, except for accounting for the moderately high net capacity and efficiency losses plus additional space requirement.

3.3.1.2 Pre-combustion capture

In this concept, CO_2 is removed prior to combustion (Figure 3.8). It is applied to integrated gasification combined cycles (IGCC), which have been described in Chapter 3.2.4.1. Pre-combustion capture is, in fact, adapted IGCC technology with the addition of shift reactors to convert CO to CO_2. As with post-combustion capture, removal of sulfur is required before the capture process. The gas entering the CO_2 capture unit is at relatively higher

FIGURE 3.8

Schematic process flow diagram of a pre-combustion capture process adapted from Rubin (2008) and (NETL 2010b)

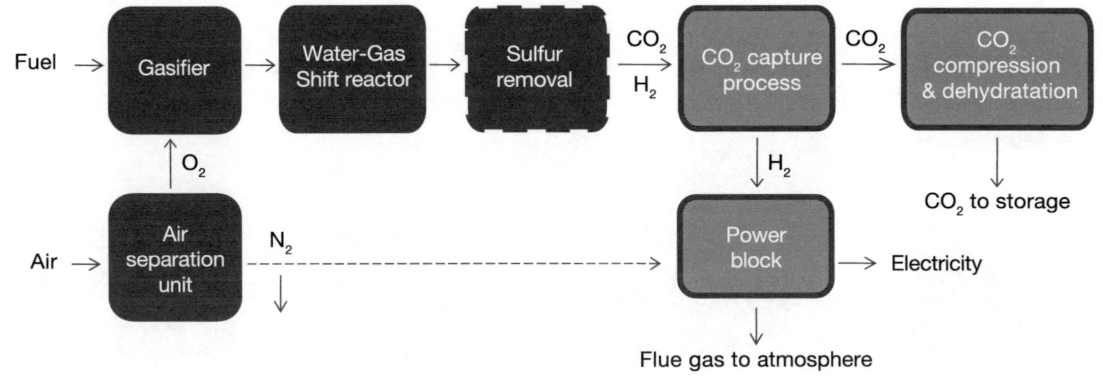

Note that sulfur removal units will be needed in the case of coal-fired power plants. If the ASU unit is located in situ, N_2 can be used for turbine NO_x control

partial pressure and concentration than in post-combustion processes (over 30 bar, approximately 40 per cent CO_2 concentration), making the separation of CO_2 from the hydrogen-rich flue gas easier.

In pre-combustion capture, the sulfur-free flue gas passes through an absorber where the CO_2 is captured via physical absorption. Selexol and Rectisol are the most commonly used physical solvents. Rectisol is used most frequently in processes synthesizing chemicals, because these processes produce a cleaner syngas, including the removal of heavy metals (Falcke et al., 2011). Selexol appears quite frequently in the literature considering CO_2 capture (Falcke et al., 2011; IPCC, 2005). This sorbent absorbs acid gases, in our case, CO_2, at high pressure. The CO_2 is released from the sorbent in a stripper at lower pressure and higher temperature. Less steam is required for sorbent regeneration than in the case of the chemical absorption process used in post-combustion capture. The stripped CO_2 is dried and compressed for transport and storage, while the hydrogen-rich gas is sent to the power block as a fuel to produce electricity.

3.3.1.3 Oxyfuel combustion

Oxyfuel capture involves the combustion of a fuel in oxygen rather than air, thereby producing a smaller volume of flue gas containing a much higher concentration of CO_2 (Liu and Shao, 2010). This implies that large amounts of oxygen must be produced, as about 2.5 times more pure oxygen is required in comparison to pre-combustion capture. One of the principal drivers for oxyfuel technology is the capability to reach near-zero emissions primarily in terms of CO_2, but also of other pollutants such as NO_x, SO_x, and particulates (Scheffknecht et al., 2011). A schematic process diagram is shown in Figure 3.9. There are three main basic components in oxyfuel combustion units: the air separation unit, the boiler and air quality control, and the CO_2 purification unit. The flue gas consists mainly of water vapour, high concentrations of CO_2, excess O_2, which is needed to ensure complete combustion of the fuel and small traces of pollutants such as NO_x. Note that part of the flue gas is recycled in order to control the boiler temperature. The CO_2 is separated from the water by cooling and condensing the flue gas. The CO_2 is then compressed and transported for underground storage.

Oxyfuel combustion technologies (without capturing CO_2) are used, for instance, in metallurgical and glass industries. However, there are no full-scale power plants in operation. Currently, there are a number of pilot-scale facilities around the world, typically ranging in size between 0.3–3.0 MW_t (Scheffknecht et al., 2011). A number of demonstration projects ranging in size from 30 MW_t to 300 MW_e have been proposed.

FIGURE 3.9

Schematic oxyfuel combustion process adapted from Rubin (2008) and NETL (2010b)

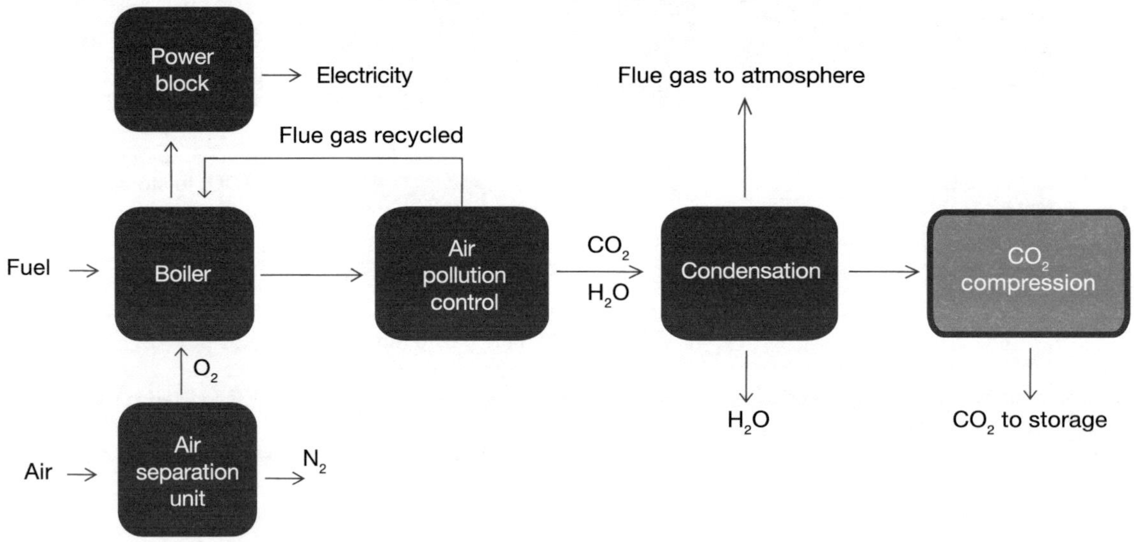

FIGURE 3.10

Comparison of efficiency penalties induced by CO_2 capture from power generation after harmonization of findings provided by different studies in the literature (Finkenrath, 2011)

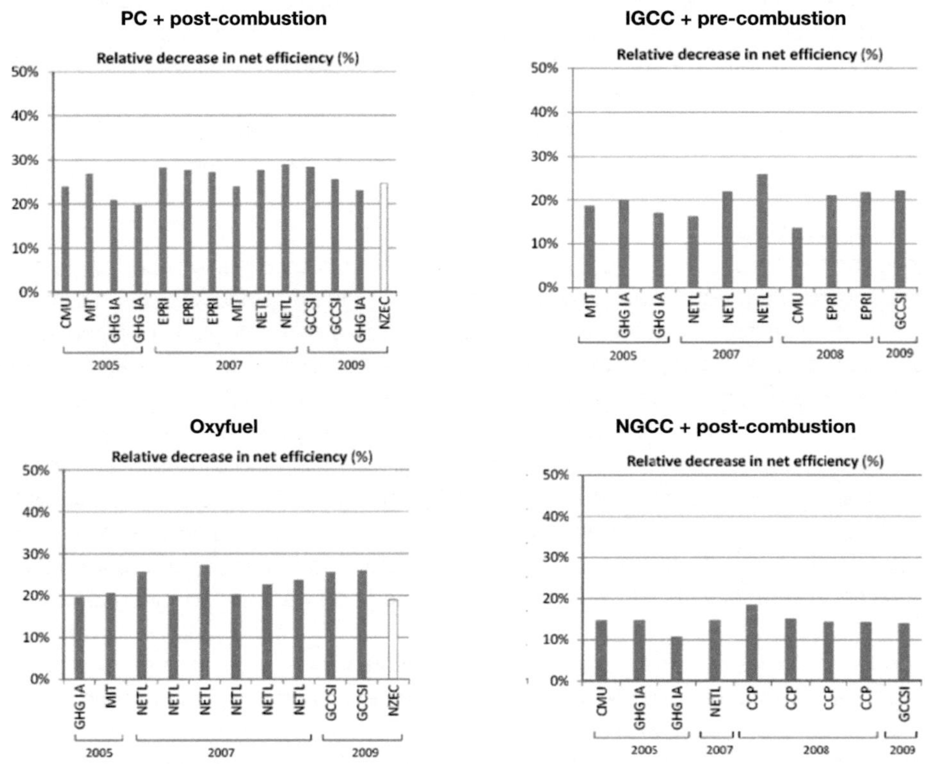

FIGURE 3.11

Differences between CO_2 capture and CO_2 avoidance

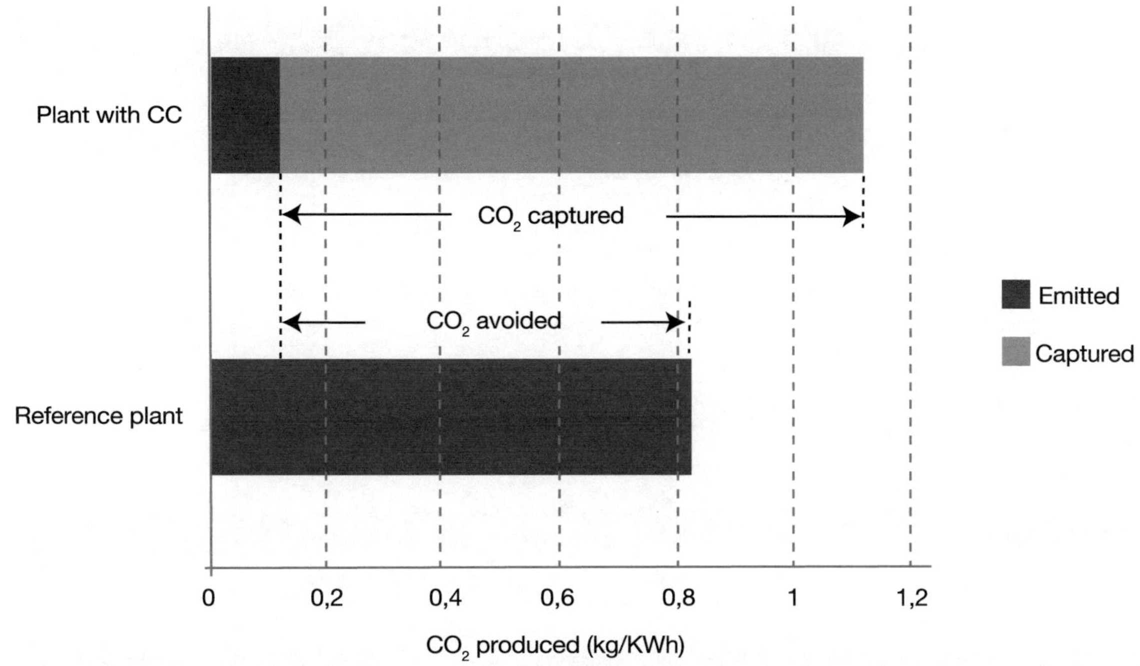

Reprinted with permission from Rao and Rubin, 2002. Copyright 2002, American Chemical Society

3.3.1.4 Energy penalty

Capturing and compressing the CO_2 at power plants affects their power efficiency regardless of the technology used. Lower plant efficiency means that more fuel is needed to generate the same amount of electricity output and requirements for materials and water along the chain increase (see e.g., Chapter 3.6.2). A study published by the IEA (Finkenrath, 2011) compares the effects of CO_2 capture on the net efficiency of new-build commercial power plants by harmonizing results of studies previously published in the literature (See Figure 3.10). The study reports an average net efficiency decrease of 25 per cent, ranging between 24-29 per cent for PC with MEA-based post-combustion capture; NGCC with post-combustion results in a 15 per cent decrease in net efficiency, ranging between 11-19 per cent; for IGCC with CO_2 capture, the average value reported is 20 per cent with a range of 14-26 per cent and for oxyfuel plants with CCS, 23 per cent with a range between 19-27 per cent.

The decrease in net efficiency is attributable to several factors. In post-combustion capture, over half of the efficiency reduction is due to the steam used for solvent regeneration, about 40 per cent is due to electricity needed to operate fans, pumps, and CO_2 compressors, while the rest is caused by the power loss due to steam extraction. The reported efficiency losses due to CO_2 capture in IGCC are lower than those in PC due to the higher CO_2 partial pressure in IGCCs, which requires a less energy-intensive physical solvent scrubbing. The lowest efficiency losses are reported for NGCC with post-combustion capture due to lower solvent-regeneration heat requirements as less CO_2 has to be captured from the flue gas. In the case of oxyfuel, the reduction in efficiency is mainly caused by the energy consumed by the oxygen production unit itself, which is responsible for about 60 per cent of efficiency reduction. Power for compression causes about 30 per cent of the efficiency reduction while CO_2 compressors account for the final 10 per cent.

TABLE 3.4

Levelised cost of electricity (LCOE) and the CO_2 avoidance costs for hard coal-fired power plants by state of commercial deployment

		Levelised electricity costs (LCOE) €/MWh	CO_2 avoidance cost €/TCO$_2$
Low fuel cost 2.0 €/GJ			
Reference case - no capture	State of the art	44.4-44.6	-
Hard coal PF post-combustion capture	FOAK early commercial	65.9-68.5	32.1-36.0
	NOAK early commercial	62.9-65.9	27.5-32.1
Hard coal IGCC with pre-combustion capture	FOAK early commercial	70.2-75.3	38.6-46.7
	NOAK early commercial	66.3-70.2	32.5-38.6
Hard coal PF oxycombustion	FOAK early commercial	71.3-81.9	40.5-56.6
	(Reference plant) NOAK early commercial	(39.1) 58.5-64.3	29.1-38.2
Moderate fuel cost 2.4 €/GJ			
Reference case - no capture	State of the art	48.1-48.3	-
Hard coal PF post- combustion capture	FOAK early commercial	70.3-72.9	33.3-37.2
	NOAK early commercial	67.2-70.3	28.5-33.3
Hard coal IGCC with pre- combustion capture	FOAK early commercial	74.7-80.0	39.8-48.3
	NOAK early commercial	70.5-74.7	33.3-39.8
Hard coal PF oxycombustion	FOAK early commercial	76.0-86.7	42.1-58.2
	(Reference plant) NOAK early commercial	(42.8) 63.0-69.1	30.5-39.9
High fuel cost 2.9 €/GJ			
Reference case - no capture	State of the art	52.7-52.8	-
Hard coal PF post-combustion capture	FOAK early commercial	75.9-78.5	34.7-38.8
	NOAK early commercial	72.6-75.9	29.7-34.7
Hard coal IGCC with pre-combustion capture	FOAK early commercial	80.2-85.9	41.2-50.3
	NOAK early commercial	75.8-80.2	34.4-41.2
Hard coal PF oxycombustion	FOAK early commercial	82.0-92.6	44.2-60.2
	(Reference plant) NOAK early commercial	(47.4) 68.7-75.1	32.2-42.0

FOAK: first of a kind; NOAK: nth of a kind

Source: ZEP, 2011

TABLE 3.5

Levelised cost of electricity (LCOE) and the CO_2 avoidance costs for natural gas fired power plants by state of commercial deployment

		Levelized electricity costs (LCOE) €/MWh	CO_2 avoidance cost €/TCO_2
Low fuel cost 4.5 €/GJ			
Base Reference case - no capture	State of the art	47.2	
Natural gas CCGT post- combustion capture	FOAK early commercial	73.7	91.8
Opti Reference case - no capture	State of the art	45.5	
Natural Gas CCGT Post Combustion Capture	NOAK early commercial	64.0	65.9
Moderate fuel cost 8.0 €/GJ			
Base Reference case - no capture	State of the art	71.9	
Natural Gas CCGT Post Combustion Capture	FOAK early commercial	103.5	109.7
Opti Reference case - no capture	State of the art	69.3	
Natural Gas CCGT Post Combustion Capture	NOAK early commercial	91.5	79.0
High fuel cost 11.0 €/GJ			
Base Reference case - no capture	State of the art	93.0	
Natural Gas CCGT Post Combustion Capture	FOAK early commercial	129.0	125.0
Opti Reference case - no capture	State of the art	89.7	
Natural Gas CCGT Post Combustion Capture	NOAK early commercial	115.1	90.2

FOAK: first of a kind; NOAK: nth of a kind. Base reference refers to a conservative case while Opti is a case which includes technology improvements, refined solutions and improved integration.
Source: ZEP, 2011

3.3.1.5 Costs of carbon dioxide capture

A broad range of costs for CO_2 capture technologies is reported in the literature. Assumptions about the type of fuel, design, operation, and financing of the power plant, which capture technology is applied as well as assumptions on the performance of the CO_2 capture technologies, and the level of technological development (demonstration, first of a kind, N^{th} of a kind) vary among the studies resulting in different cost values being reported.

When assessing cost figures reported in the literature, a distinction must be made between the cost per unit of CO_2 captured and per unit of CO_2 avoided. Figure 3.11 presents a graphical representation of the difference between capture and avoidance. The additional use of heat and electricity induced by CO_2 capture processes results in a reduction of the efficiency of the power plant, which translates into increased coal consumption per kWh and therefore, additional CO_2 production per kWh. CO_2 avoided is the difference between the emissions produced by a reference plant without capture technology and the plant with capture. The amount of emissions avoided is smaller than the amount of CO_2 captured and as a result, the cost per ton avoided is greater than the cost per ton CO_2 captured.

Several studies have attempted to harmonize costs for the different technologies. A report published by the IEA (Finkenrath, 2011) for newly built early commercial plants reports expected capture costs of 58 US$ per ton CO_2 avoided, with a range of 40 to 74 US$/ton for PC with MEA-based post-combustion capture, 80 US$ per ton CO_2 avoided with a range of 60 to 128 US$/ton for NGCC with post-combustion, 43 US$ per ton CO_2 with a range of 26 to 62 US$/ton for IGCC with CO_2 capture and 52 US$ per ton CO_2 with a range of 35 US$/t to 72 USUS$/t for plants with oxyfuel combustion. A study published by the European Technology Platform for Zero Emission Fossil Fuel Power Plants (ZEP, 2011) estimated the costs of CO_2 capture technologies for different type of power plants at three levels of fuel costs. Results for coal fired power plants and natural gas fired power plant are shown in Table 3.4 and Table 3.5, respectively.

3.3.2 CARBON DIOXIDE TRANSPORT

3.3.2.1 Pipelines

Oil and natural gas are commonly transported by pipeline. There are already about 6,000 km of installed pipeline transporting CO_2 from natural and some anthropogenic sources to be used in enhanced oil recovery (EOR) operations, mainly in the USA. CO_2 can be transported in a pipeline as a gas, a liquid, a supercritical fluid, or in a two-phase flow. Given the low density of CO_2 in the gaseous phase and relatively high pressure drop experienced during pipeline transportation, this phase is considered economically infeasible for long-distance pipeline transport. For small-scale transport, short-distance transport to a trunk line, the gaseous phase could be a suitable option (Serpa et al., 2011; Skovholt, 1993). The supercritical phase (>7.38 MPa, 31°C) allows for the most efficient transportation of CO_2, because an increased mass per unit volume can be transported since supercritical CO_2 has the density of a liquid but the compressibility and viscosity of a gas.When considering transport conditions, the avoidance of two-phase flow is recommended for a number of reasons:

- It may cause cavitation in the pipeline, which will decrease the strength of the pipeline.
- When liquid-phase CO_2 turns into a gas, it causes turbulence, which can agitate the liquid-phase CO_2 and thus damage the pipeline
- The density of gaseous CO_2 is multiple times lower than that of liquid CO_2, which significantly reduces the transport capacity.
- Two-phase flow requires special equipment, particularly compressors and pumps.

A key factor for CO_2 transport is to guarantee that the fluid is completely dry and free of water. In the presence of water, CO_2 forms a weak acid known as carbonic acid (H_2CO_3). Carbonic acid can lead to corrosion rates exceeding 10 mm/y depending on the CO_2 partial pressure, temperature and the presence of impurities. Other contaminants such as H_2S, NO_x or SO_2 will also form corrosive acids in combination with free water (DNV, 2008). Another consequence of free water presence in the CO_2 flow is the potential formation of hydrates. Hydrates can cause localized damage in the pipeline, thereby reducing transport capacity, as well as pipe blockages and mechanical damage to plant components. The maximum water concentration for CO_2 transport reported in the literature varies between 50 and 600 ppm for the FEED studies, the CCS Longannet project (UK) and the ROAD project (NL) specify a water content of less than 50 ppm by volume basis. The FEED of the Kingsnorth CCS Project (UK) reports a water content of 24 ppm by volume for normal operation.

In terms of pipeline material, based on economic considerations and experience in the field, it is recommended to use carbon steel pipelines for long distance transport (DNV, 2008). Nevertheless, for short distances, higher-grade pipelines can be considered for 'wet' transport (Seiersten and Kongshaug, 2005). Table 3.6 shows an overview of main materials that can be used for CO_2 transport and their requirements.

TABLE 3.6

Suitability of different pipeline materials for CO_2 transport

	Carbon steel	13% Cr steel	Polymer-coated	Duplex/high alloy steels
Source	(DNV 2010; Seiersten and Kongshaug 2005)	(Choi et al. 2010; DNV 2010; Seiersten and Kongshaug 2005)	(Tabe et al. 2000)	(Seiersten and Kongshaug 2005); (DNV 2010)
Dry conditions	Good, with limited impurities	Good, depending on impurities	Not investigated	Good
Wet conditions	High corrosion rate	Depending on impurities	Not investigated, inhibits hydrate formation	Good, depending on impurities
Relative material cost[1]	1	2	?	>4

[1] As compared to carbon steel

Although transport by pipeline is a relatively mature technology, compared to some of the CCSCCS technologies, there are still major challenges that need to be faced:

- The amount of CO_2 to be transported will require a vast new international and domestic pipeline network that must be constructed in a relatively short period of time;
- CO_2 is a fluid with unusual properties compared to other fluids transported by pipeline. The critical point of pure CO_2 is close to the pressures/temperatures during pipeline operation requiring more stringent design and operation if two-phase flow is to be avoided;
- Impurities in the CO_2 streams can affect the thermodynamic behaviour of CO_2 and the transport capacity of the pipeline; impurities can also influence ductile fracture propagation and induce corrosion;
- Pipelines transporting CO_2 will require careful control during decompression in order to avoid a rapid cooling of the flow, which would result in formation of solid CO_2.

3.3.2.2 CO_2 pipeline costs

CO_2 transport costs have been estimated to contribute relatively little to the total costs in the CCS value chain; these have been assessed in two reports to be between 8-15 per cent (McKinsey & Company, 2008) or 1-3 per cent (WorleyParsons and Schlumberger, 2011) of total costs. Despite this relatively low share, transport costs can play a key role in the development of optimal CCS chains. The capital investments are large, especially if economies of scale are to be exploited. Key cost determiners of pipeline construction costs are diameter, operating pressures, distance and terrain. Other factors, including pipe material, climate, labour costs, competition among contracting companies, safety regulations, population density and rights of way, may cause construction costs to vary significantly from one region to another (Knoope et al., 2013). There are several models available in literature to estimate the capital costs of CO_2 pipelines. An overview of the cost estimated using some of the models is shown in Figure 3.12. Note that for distances longer than 150-200 km, booster stations will be required to make up for pressure drop and keep the flow in dense phase. These costs are not included in Figure 3.12. It is also important to highlight that since data on CO_2 pipelines is not easily available in the available literature, most cost models use information from natural gas pipelines, particularly historical cost figures. Since CO_2 will be transported at higher pressures, the costs of CO_2 pipelines are likely to be higher than those estimated by the models.

FIGURE 3.12

Capital costs predicted by various diameter-based models in million €2010/km for a pipeline 25 km in length on flat agricultural terrain (plotted as lines) and capital cost estimations in literature, for actual planned or realized projects (plotted as markers)

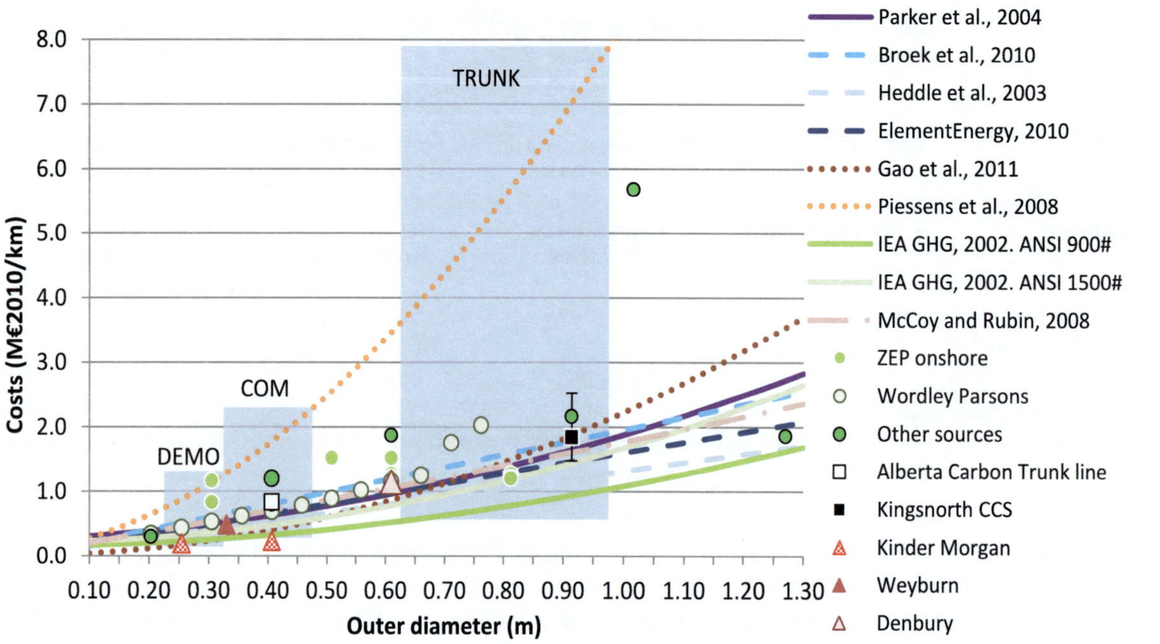

The shaded areas represent the ranges for three different base cases (demo: mass flow of 50 kg/s; COM: 150 kg/s; TRUNK: 750 kg/s).

Source: Knoope et al., 2013

3.3.3 CARBON DIOXIDE UNDERGROUND STORAGE

Subsurface storage of CO_2 involves both technical and non-technical issues. The latter are mainly related to legal and regulatory aspects, and public acceptance, especially regarding local communities in the vicinity of a storage complex. Technical aspects include storage capacity, long-term integrity of reservoir, caprock, and wellbore materials, among others. Some of the technical aspects related to CO_2 injection in geological reservoirs will be reviewed below.

In order to store CO_2 in a geological reservoir, CO_2 must be first pumped through an injection well, which is in many aspects similar to oil and gas production wells. However, the wellbore materials used during well completion, such as casing and cement, should be chosen while taking into consideration the chemical reactions that may occur due to the high concentration of CO_2 encountered within the well. CO_2 can be injected either in a gas or supercritical state, which is a state where CO_2 has gas-like viscosity and therefore has high mobility, and has a high density, typical of liquids, which is advantageous for maximizing storage capacity. This state is reached when the temperature is above 31.1°C and the pressure is higher than 73.9 bar. Since pressure and temperature both increase with depth in the Earth's crust, with geothermal and hydrostatic gradients, averaging an increase of approximately 30°C and 100 bar per kilometre, respectively, at approximately 800 m depth, CO_2 is likely to be found in a supercritical state. CO_2 storage is, therefore, likely to occur in reservoirs at depths greater than 800 m in the lithosphere.

FIGURE 3.13

Carbon dioxide (CO_2) storage options in petroleum fields, saline aquifers, and coal deposits

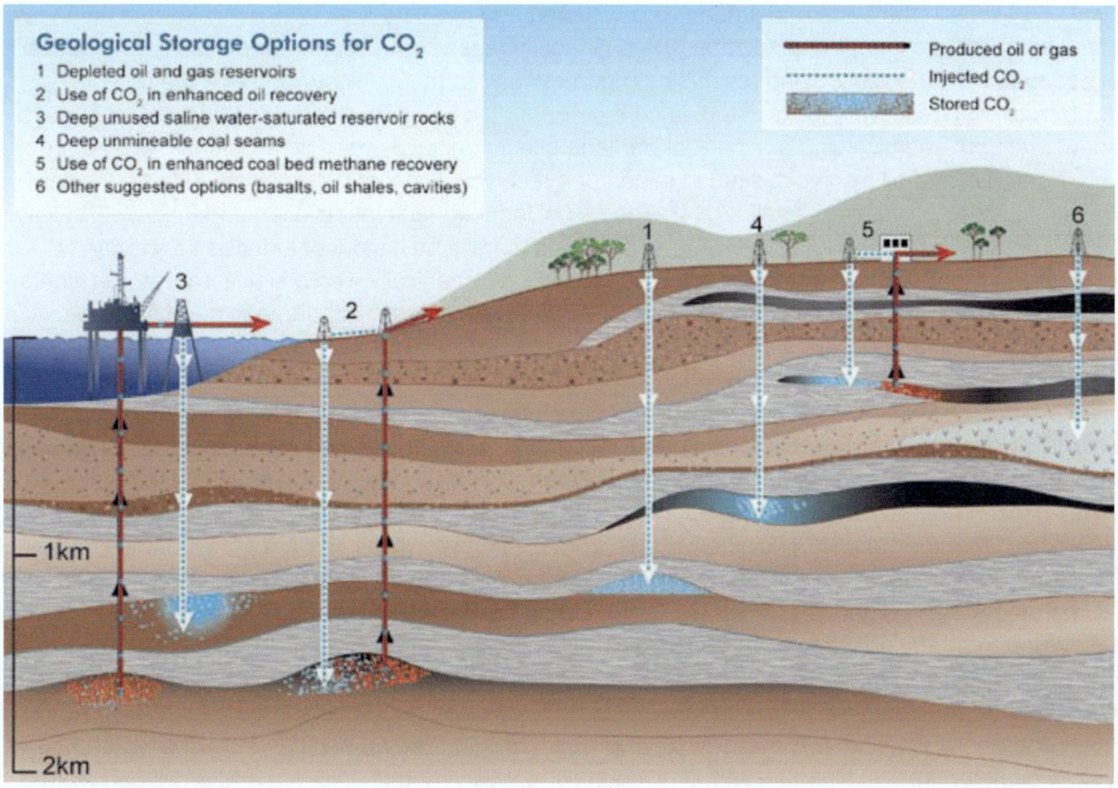

Source: IPCC, 2005

Once injected in the reservoir, the storage will be similar to a fossil fuel system. A storage system is composed of a reservoir and a caprock, and may or may not contain a trap, that is, a volume of reservoir encased by the caprock. A geological reservoir is a porous and permeable rock, often a sedimentary rock that contains fluids such as water, oil and gas, CO_2 or H_2S, and can safely store CO_2 for geological periods of time, or longer than 1,000 years, at a minimum. Porosity is the space within rock matrix, such as that between grains of sand and/or rock fractures that contain fluids, which will be partly occupied by CO_2 during storage. A reservoir is preferentially a sedimentary rock such as sandstone or limestone, or less commonly, a fractured metamorphic or igneous rock such as basalts (Figure 3.13).

3.3.3.1 Types of reservoirs

There are three main groups of reservoirs that are likely to store CO_2: petroleum fields, saline aquifers and coal deposits. The first two are already in their commercial phase and the latter, although proven in pilot scale, still needs to be demonstrated in larger scale.

3.3.3.1.1 Petroleum fields

Reservoirs of oil and gas fields are appealing targets for CO_2 storage as the trapping efficiency is proven, as these reservoirs held hydrocarbons for millions of years. Also, these units are often well studied by oil companies, with plenty of data available. Moreover, the injection of CO_2 has been already carried out in many oil fields for many years , particularly in the United States, in order to improve the oil/gas recovery rates in a method known as enhanced oil recovery (EOR). Storing CO_2 in oil and gas fields may, however, imply geochemical alterations in the system that need to be studied and predicted, as they may alter the reservoir, caprock, fault and wellbore integrity. In addition, the changes in reservoir pressure, temperature, and stress state, that accompany CO_2 injection and reservoir filling, may tend to mechanically damage the reservoir-caprock system or even reactivate faults. Studies conducted on the integrity of reservoir, caprock and faults, though, show that both chemical and mechanical damage effects generally pose little leak risk, at least in the case of former gas and oil reservoirs sited in sandstone reservoir systems in tectonically stable regions (e.g. Hangx et al., 2013; Pluymakers et al., 2014). The risk of induced fault reactivation and (micro)seismicity is also low and not increased by injection of CO_2, especially if there is no history of induced seismicity during hydrocarbon production (Samuelson and Spiers, 2012). Mechanically weak, highly porous carbonate reservoirs such as chalks are more susceptible to coupled chemical-mechanical damage and should be evaluated individually (Liteanu et al., 2013).

Estimation of storage capacity in oil or gas fields is the most straightforward of all potential reservoir types, as the petroleum industry gathers many data from these reservoirs before and during exploration, to establish the precise amount of petroleum reserves and resources. Furthermore, oil and gas fields have discrete volumes with fairly well-defined boundaries, which allows for a more precise assessment. In a simplistic approach, it is possible to assume that the volume of fluids produced from a reservoir will be similar to the volumes that can be used for storage and that injection of CO_2 in a petroleum field will be carried out until the original reservoir pressure is restored. On the other hand, petroleum production history and methods in a reservoir will influence directly the available storage space for CO_2 (Ketzer et al., 2012).

Storage capacities of petroleum reservoirs can be calculated based on geometrical parameters, namely reservoir area and depth, physical and hydrodynamic properties such as CO_2 density and water saturation, and other data gathered from exploration activities, such as recovery factors, injected and produced water volumes (Bachu et al., 2007). The global storage capacity estimate for petroleum fields is 920 billion tons CO_2 (IEA, 2012b), and is based on the volumes of known petroleum reserves in the world.

3.3.3.1.2 Saline aquifers

Saline aquifers are porous and permeable rock formations containing highly saline water, characterized as brines with salinity similar or higher than seawater, e.g., ca. 35 g/L. Permeability of saline aquifers reservoir should be sufficiently high so as to allow constant injection of millions of tons of CO_2 per year for the time duration of the project, generally a few decades. Since the injected CO_2 will displace the original fluid, low permeability may cause clogging and excess reservoir pressure that may result in hydraulic fracturing of the reservoir. Similar to a petroleum reservoir, a saline aquifer must have a continuous, overlying caprock with low permeability, with a minimum presence of faults and fractures over the range of the estimated storage area. Moreover, this formation must resist the extra (above natural) hydraulic pressure experienced during the injection phase. This makes saline aquifers more susceptible to induced fault motion and (micro) seismicity during the injection phase than exhausted hydrocarbon reservoirs, especially if in a tectonically stressed geological setting where faults make already be close to activating (Zoback & Gorelick, 2012). Nonetheless, management of injection pressure can be used to avoid such effects.

The great advantage of saline aquifers over other storage reservoirs is their enormous theoretical capacity and widespread distribution. On the other hand, there is much less information and data available for saline aquifers, since the economic incentive for their study is much lower, resulting in less accurate capacity estimates (Bachu et al., 2007). Therefore, global CO_2 storage capacity estimations of deep saline formations have only been roughly estimated thus far; these estimates fall between 1 and 10 trillion tons CO_2 (IPCC, 2005).

Coal Fields

Coal traps CO_2 mainly by adsorption; the gas molecule is bound to the solid coal surface by Van der Waals forces. As in petroleum fields, CO_2 storage in coalbeds may be economically viable because coalbed methane can be produced as free gas as a by-product of the storage process. In this case, injected CO_2 will be adsorbed preferentially, displacing the naturally occurring methane from the coal matrix, which can be produced through wells, a technique known as enhanced coalbed methane recovery (ECBM). CO_2 storage in coal beds is still in the early stages of development, compared to petroleum fields and saline aquifers; at present, some ECBM demonstration projects have been deployed.

Storage capacity in coalbeds can be estimated by analogy with coal bed methane (CBM) reserve estimation, which implies the determination of initial gas in place (IGIP) and producible gas in place (PGIP) for a given coal seam (Bachu et al., 2007). The global storage capacity estimate for coal fields is 200 billion tons CO_2 (IPCC, 2005).

3.4 CARBON DIOXIDE STORAGE AND SAFETY RISK

Safety of CO_2 storage is a key element for CCS. CO_2 leakage can result in potential environmental impacts at both the global and local level by releasing CO_2 back in the atmosphere - making CCS a less effective mitigation measure - and impacting ecosystems, animal and human health. Figure 3.14 depicts potential leakage pathways. Based on analogous experiences from the natural gas storage industry, the three largest sources of risk for CO_2 storage are considered to be (Benson et al., 2012):

- Inadequate site selection
- Leakage from wells
- Leakage through undetected faults or fractures in the storage reservoir seal

Leakage to adjacent geological formations may cause geochemical reactions and mobilization of potentially polluting elements such as heavy metals. A study of the IEA GHG (2011) identifies three potential mechanisms of CO_2 leakage that could result on negative impacts on groundwater resources, namely: (i) leakage of buoyant CO_2 from storage site into potable aquifers: potential impacts include acidification and mobilization of other substances, such as heavy metals; (ii) displacement of brine from deeper storage formations into potable aquifers and (iii) disruptions of aquifer flow systems and groundwater discharge patters. Furthermore, environmental impacts are also expected if (elevated concentrations) of CO_2 contacts soil and/or the sea floor. In the first case, CO_2 could affect the soil pH and impact the chemistry of nutrients, trace metals and plant growth (Bachu, 2008; Saripalli et al., 2003). Leakage of CO_2 into the depth of the sea floor, could affect local pH and marine ecosystems (Bachu, 2008). The risk of leakage changes over time and is expected to be the greatest during injection. Figure 3.15 depicts a schematic risk profile over time. The injection of CO_2 can increase reservoir pressure, which depending on the injection pressure, could lead to migration of CO_2 either through existing pathways or by inducing fracturing or fault reactivation. Because pressure will decrease after the active injection phase and CO_2-trapping mechanisms will increase over time[4] (e.g., solubility and mineral trapping), the risks posed by CO_2 storage are expected to decrease in time (Bachu, 2008). Operation and the period just after injection stops are therefore considered the most critical periods (though risk of leakage after these periods is still present).

Appropriate monitoring, verification and accounting programs (MVA) are therefore a key requirement of CCS projects as they play a key role to help meet the goals for safe, secure and verifiable permanent

4 These are mechanisms that keep the supercritical CO_2 securely stored, for instance by structural trapping (the CO_2 cannot go beyond the impermeable layer of caprock), residual trapping (the CO_2 is trapped in the porous space in the formation) and solubility trapping (CO_2 dissolves in the salt water already present in the porous rock).

FIGURE 3.14

Potential escape pathways for stored CO_2 in a geological formation

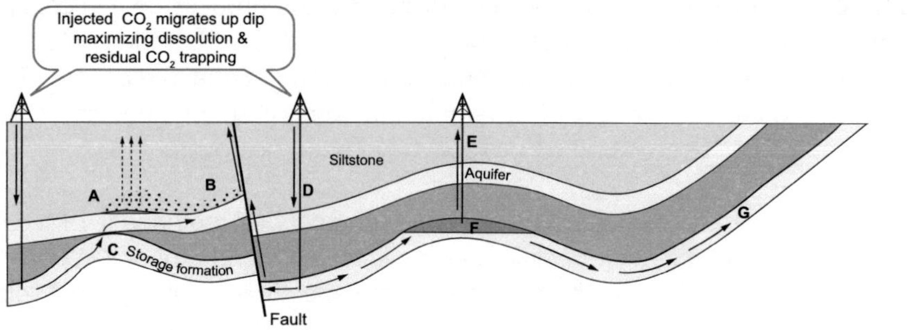

Potential Escape Mechanisms

A. CO_2 gas pressure exceeds capillary pressure & passes through siltstone	B. Free CO_2 leaks from A into upper aquifer up fault	C. CO_2 escapes through 'gap' in cap rock into higher aquifer	D. Injected CO_2 migrates up dip, increases reservoir pressure & permeability of fault	E. CO_2 escapes via poorly plugged old abandoned well	F. Natural flow dissolves CO_2 at CO_2 / water interface & transports it out of closure	G. Dissolved CO_2 escapes to atmosphere or ocean

Remedial Measures

A. Extract & purify ground-water	*B. Extract & purify ground-water*	*C. Remove CO_2 & reinject elsewhere*	*D. Lower injection rates or pressures*	*E. Re-plug well with cement*	*F. Intercept & reinject CO_2*	*G. Intercept & reinject CO_2*

Source: IPCC, 2005

FIGURE 3.15

Risk evolution for an underground CO_2 storage

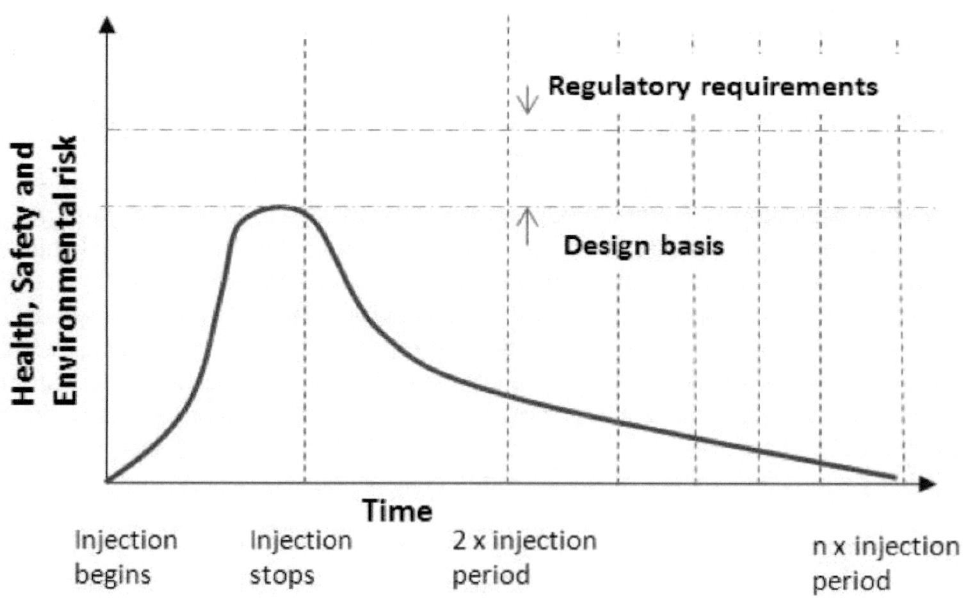

Adapted from: Benson et al., 2012

CO_2 storage (NETL, 2013 (see Figure 3.16). They are required to commence pre-injection to establish baseline levels and continue through operational and post-closure phases. MVA is especially important to avoid slow gradual leakage to go undetected for long periods of time as this has the greatest potential to cause broad scale environmental damage (Bachu, 2008; Damen et al., 2006). Monitoring techniques for CO_2 storage projects are similar to the technology in use in oil and gas exploration, natural gas underground storage and radioactive and industrial waste disposal (Heemann et al., 2011). The choice for appropriate monitoring methods and instruments will depend on the storage complex characteristics and regulatory requirements (Ketzer et al., 2012). Monitoring techniques can be targeted to the subsurface (reservoir, overburden, storage complex, aquifers) or surface (soil, water column of lakes and sea, air). Some techniques use changes in physical properties (e.g., acoustic, thermal, electrical) or chemical composition (e.g., tracers, stable and radioactive isotopes, etc.) to detect and measure CO2 in the reservoir, and within or out the storage complex. The IEA-GHG provides an online monitoring selection tool which contains up-to date overview of 40 monitoring techniques and case studies to support the design of monitoring protocols (for more information see: http://www.ieaghg.org/index.php?/Monitoring-Selection-Tool.html)

Besides monitoring programs, there is also need for correction actions to stop leakage in case it occurs. Correction actions will vary depending on the leakage pathway. The level of experience is quite different between repairing leakages through wells or through caprock formations, faults or fracture zones (Réveillère and Rohmer, 2011). In the first case, standards techniques from the oil and gas industry can be applied, defective well elements can be replaced or the well can be closed and abandoned. In the second case, there is lack of experience on repairing CO_2 leaks. Current research focuses on the creation of chemically or microbiologically induced barriers that change the hydraulic properties within or above the pathway and the creation of hydraulic barriers that counter the hydraulic gradient that drivers the flow up in the leak (Esposito and Benson, 2011; Réveillère and Rohmer, 2011). Research is also being conducted on how to treat contamination caused by leakage (e.g., in situ treatment, pump and treat technologies) (IEA GHG, 2011).

FIGURE 3.16

Role of monitoring, verification and accounting (MVA) on reducing uncertainties and risk of CO_2 storage.

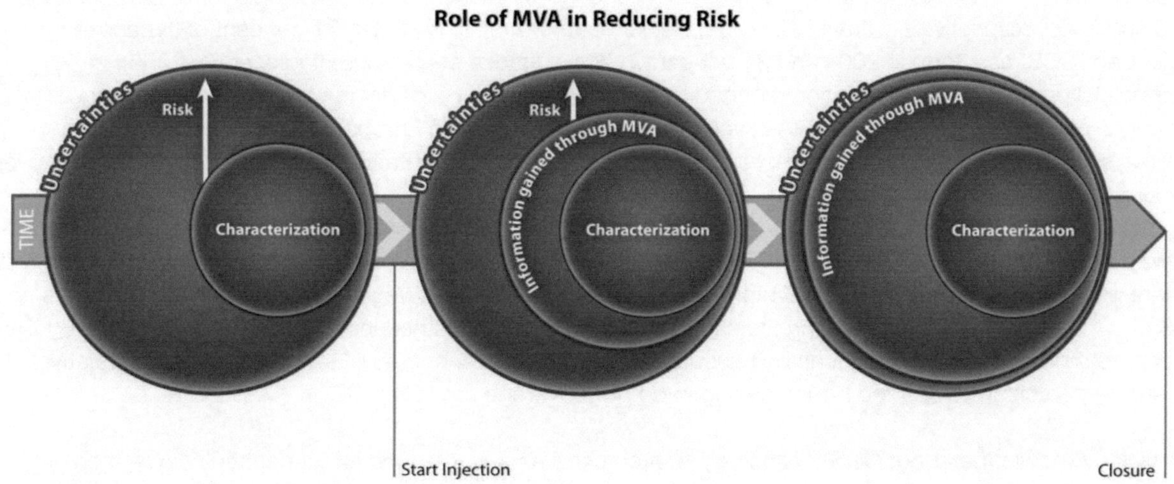

Source: NETL, 2013

3.5 SOCIAL ACCEPTANCE

Societal acceptance of CCS technology has been identified as an obstacle for its development and deployment at the commercial scale (Ashworth and Cormick, 2011). At the same time, if we are to successfully mitigate the release of large volumes of carbon dioxide from the atmosphere, the adoption of CCS becomes critical. One might assume then, that if we convince the public of the necessity of mitigating carbon dioxide, CCS will be more widely accepted. However, this does not appear to be the case in reality. Early attempts to deploy CCS technology have been delayed and even brought to a halt as a result of public opposition (Bradbury and Wade, 2010; Desbarats et al., 2010; Feenstra et al., 2010; Voosen, 2010).

Despite the promise of new technologies solving complex global challenges, they often have high uncertainty associated with them (Scheufele and Lewenstein, 2005). Such uncertainty raises fears regarding potential health and environmental problems and other negative social, moral and ethical consequences (PCAST, 2005). For example, concerns have been raised about the potential health risks of nanotechnologies, genetically modified foods and nuclear power, which have ultimately affected how these technologies are accepted in society (Ashworth and Cormick, 2011; Bergstein, 2008; Chang, 2008); this is no less true for CCS technology.

3.5.1 PERCEPTIONS OF CCS TECHNOLOGY

International research results have shown that CCS is often perceived negatively by the general public. Reasons most frequently cited for opposition to a project include concerns regarding the fear of leakage and the resulting impacts on human health and local ecosystems, the effects on local groundwater located near the storage sites, the likely impact on local housing prices, and a fundamental resistance to the continued exploitation of fossil fuel-based industries, which have a legacy of negative environmental impacts (Ashworth et al., 2009a; de Best-Waldhober et al., 2011; Oltra et al., 2010). In the developing world, there is an additional concern that CCS does not improve energy security but rather reduces it because CCS is still a technology in development. These perceived risks have, in some instances, led to the rejection of CCS as a mitigation strategy (Oltra et al., 2010).

3.5.2 AWARENESS AND KNOWLEDGE OF CCS

While media coverage about CCS has generally increased and become more positive (Hansson and Bryngelsson, 2009), studies from around the world indicate that although public awareness is improving, overall the general public has low levels of knowledge about both CCS as a technology and the environmental concerns that it addresses (Ashworth et al., 2009b; Sharp et al., 2009).

Results from national surveys show that 4, 5 and 17 per cent of Americans recognized the term 'CCS' in 2003, 2006 and 2009, respectively (Curry et al., 2005; Reiner et al., 2006). In contrast, 31 per cent of Japanese recognized 'CCS' as a term in 2004 and 34 per cent in 2010; (Itaoka et al., 2009; Itaoka et al., 2005). In Australia, 18 per cent of the population reported no knowledge of CCS, 22 per cent moderate knowledge and only 2 per cent a high knowledge (Ashworth et al., 2010a). Finally, in Europe, as part of a Fossil Energy Coalition's (FENCO-ERA NET) project, researchers conducted six representative national surveys, surveying over a thousand individuals in the Netherlands, UK, Germany, Norway, Greece and Romania (Reiner, 2010).

Norway and the Netherlands showed the least number of individuals who had never heard of CCS. This result is more likely given the exposure of CCS in these two countries; the Norwegian government fully supports CCS and there are Norwegian CCS projects underway, while in the Netherlands, controversy surrounding the Barendrecht project may have potentially raised awareness of the technology. However, in general, relatively few survey participants reported having seeing in de pth coverage regarding CCS.

This lack of knowledge about CCS technology is often cited as a key reason for its negative perception among the general public. Some researchers (see e.g. de Best-Waldhober et al., 2009; Duan, 2010; Itaoka et al., 2005) suggest that increased support may be gathered for the technology if the public is more fully informed of its potential to mitigate GHG emissions. Itaoka et al.'s (2009) research supports this claim.

They found that individuals who claim to have more knowledge about CCS perceive the technology more favourably. However, research has also shown that when these people are presented with neutral information concerning CCS, favourability declines, further reinforcing the limited knowledge members of the general public have about CCS and the potential problems associated with a low knowledge base.

3.5.3 ROLE OF INFORMATION DISSEMINATION ON ATTITUDES TOWARDS CCS

Researchers have also explored the role that a trusted and credible information source plays in influencing public attitudes towards CCS. Research has found that information from sources that do not have a vested interest in the technology is more likely to be trusted and have a positive influence on attitudes towards CCS (Ter Mors et al., 2010; Terwel et al., 2009; Tokushige et al., 2007). The credibility of the information source was also supported in the findings of earlier work in Australia where they compiled expert opinions from authorities in several differing fields concerning a range of energy technologies (Ashworth et al., 2011). When the information was presented as a consensus reached by experts from diverse backgrounds and interests such as environmental non-government organizations, coal industry representatives and researchers, the conclusion was considered objective and more trustworthy (Ashworth et al., 2010a).

Researchers in Australia have also experimented with various approaches to engage the general public on the topic of climate and energy to understand how these influence individual attitudes. They have used citizens' panels, small group workshops, participatory action research and large group processes. Their results have consistently shown that when the public is engaged on the topic of climate and energy technologies, with CCS being presented as an element in a portfolio of options for low carbon energy, support increases (Ashworth et al., 2010a). Recently, the large group process was replicated in Calgary, Canada with a group of 74 and a similar outcome was achieved. When asked how strongly they agreed (7) or disagreed (1) with CCS as a mitigation option, the mean support for CCS increased from 4.53 to 5.41 (Einsiedel et al., 2011).

However, the process was also repeated in Amsterdam, the Netherlands, this time with a group of 111, and in this case, support declined over the course of the workshop from a mean of 4.2 to 3.7 (Brunsting et al., 2011). However, it is difficult to draw conclusions as to why the result was so different in the Dutch study. The obvious conclusion may be the influence of the political landscape around Barendrecht and the withdrawal of support by government for any onshore storage projects.

Examples of participants' concerns about CCS from the Dutch workshop were related to uncertainties about the technology. For example (Brunsting et al., 2011) quoted "If something explodes on the North Sea, then..." and, "If you put CO_2 underground, there is a chance that they start drilling in the future at that same place, which would release the CO_2 in the atmosphere. That would be a big problem".

These findings also support the findings from the international research of Bradbury and colleagues in the United States, who found that context and legacy issues were important factors for the acceptance of projects (Bradbury et al., 2009).

3.5.4 LEARNING FROM PROJECTS

In their comparative study across several of the early Department of Energy regional partnerships in the United States, Bradbury et al. (2009) report that in many cases, social factors including socio-economic status, desire for compensation, benefits to the community and past experience with industry and government in the local area were of greater factors in CCS acceptance than the perceived risks of the technology itself. In fact, participant concerns regarding fairness and trust were key determinants of perceptions about CCS technology in the communities and regions studied. Such procedural justice issues were also found to be critical for successful deployment of a range of energy technologies in a study by Desbarats et al. (2010). Priorities included information transparency and active stakeholder input to and influence on the process. It was also considered important to have a project representative available who could be contacted at any time if concerns were raised. The successful engagement of Total SA with their Lacq project in France reinforces the importance of the public consultation process and treating community members with respect. In their example, an open and transparent engagement of several organization representatives led to a successful outcome for the project (de Marliave, 2009). The Lacq project injected in total 51,000 metric tonnes of CO_2 between 2010 and 2013, when the pilot finished. A more recent comparative study of CCS projects has identified a number of key criteria that should be considered in order to achieve successful adoption. These criteria are described below:

- The extent to which key government and development team members are aligned in terms of support for, and coordination of, the project; without alignment, projects are unlikely to move forward.
- The deployment of communication experts as part of the project team from the outset; communication and engagement are considered equally important to the technological and geological components.
- The consideration of social context relative to site selection and project design and implementation. Taking this step ensures an understanding of the local history of projects in the area and acknowledges that current happenings in the local community will influence the outcome of the current CCS project.
- Finally, the degree of flexibility in framing the project and adjusting the project implementation strategy. Flexibility has been identified as critical to allow time for affected stakeholders to provide input into the process (Ashworth et al., 2010b).

Related to the last criterion that addresses flexibility is the timing of community engagement activities for CCS projects. The research found that timing can have a decisive influence on the acceptance of a project. Early engagement with the community such as that undertaken in the CO_2CRC Otway case, has emerged as the best approach to facilitate meaningful participation and to instil a sense of empowerment within the community. Evidence has shown that announcing project plans prior to public engagement has contributed to significant conflict between stakeholders in a number of cases, including Barendrecht. Meaningful dialogue amongst project developers, national and local authorities and the public is essential well before the finalisation of project plans (Ashworth et al., 2010b).

3.5.5 CONCLUSION

The evidence clearly suggests that there is a compelling reason for CCS project proponents to willingly engage and partner with communities early in a project's life. However, experience also demonstrates that many companies do not consider it an imperative to undertake public engagement to gain a social license to operate. Rather, some companies appear to regard an operating license as a right. Regulators must therefore address the challenge of balancing and encouraging the necessary base of community engagement without restricting and framing conversations to the point where they are meaningless for both the communities and project proponents.

A critical consideration will be to carefully evaluate the regulatory processes that have been used for early CCS projects, to identify what has been successful, highlight and analyse areas of improvement. In this instance, there is an opportunity for government, industry and community to cooperate in order to better understand what will comprise a rigorous regulatory regime for successful public engagement and consultation for CCS projects. Such reflective and participatory processes early in the life of CCS deployment should help to set the stage for best practice to become the norm in this area. Inclusivity, project flexibility and transparency of process will be critical if CCS is to be successfully deployed at a scale that is sufficient to achieve significant reductions in GHG emissions.

3.6 EMISSIONS AND ENVIRONMENTAL IMPACTS

This section provides an overview of the potential environmental impacts of fossil fuel combustion both with and without CCS throughout their life cycle. GHG emissions resulting from natural gas production and processing have been an active topic of investigation for a number of years. In some cities efforts have already been done to substitute natural gas for coal and using natural gas in niche applications where methane's clean burning attributes for criteria pollutants could provide human health benefits, such as in intercity buses and vehicle use in confined spaces.

3.6.1 FOSSIL FUEL EXPLOITATION

3.6.1.1 Conventional fossil fuels

There are significant environmental issues associated with the exploration, production, refining and distribution of fossil fuels. These issues are related to the leakage or spillage of gas, oil, coal dust and acid mine drainage, to impacts of infrastructure establishment and operation, as well as to land and water use. The fossil fuel production process requires energy and materials. No thorough analysis of either the environmental impacts or GHG emissions of the fossil fuel industry at the global scale could be found. The last comprehensive review of these issues for life cycle assessment was, to our knowledge, conducted for ecoinvent version 1.1 in 2004, and a partial update is available in ecoinvent version 2.0 (Dones et al., 2007b; Dones et al., 2007a). These data are being used in our current modelling work. In general, air emissions from fuel combustion processes dominate those of upstream processes, but the production of fuels can dominate other environmental pressures such as land use and toxic emissions. A discussion and review of these issues is beyond the scope of this report. Recent research indicates that the magnitude and variability of fugitive methane emissions from coal mines and oil and gas fields have been underestimated in the past. The discussion below will focus on this issue. For information on environmental impacts from coal mining and oil exploration we refer to (US EPA, 2008; Greb et al., 2006; UNEP, 2007; Weng et al., 2012).

GHG emissions data published by the IEA indicate that in 2008, "other energy industries" (comprising all energy industries except for power generation) were responsible for direct emissions of 1.5 billion tons CO_2 (IEA, 2012b); the consumption of electricity by these industries causes 0.5 billion tons CO_2 (IEA, 2012b). In comparison, total global emissions from fuel combustion were 30 billion tons CO_2. Coal and peat were responsible for 0.28 billion tons, and oil and gas for 1.2 billion tons (IEA, 2012b). Reported global methane emissions from oil and gas extraction and delivery lie in the order of 1.5 billion tons CO_2-eq (Alsalam and Ragnauth, 2011; IEA, 2010), those from coal mines to the order of 0.6 billion tons (Alsalam and Ragnauth, 2011), compared to a global total methane emissions of 11 billion tons CO_2-eq. These emissions estimates are based on an emission factor approach following the 2006 IPCC guidelines (IPCC, 2006).

3.6.1.1.1 Coal bed methane

Coal contains a gas consisting partly of methane that leaks during mining. Coal seam gas has typically posed a safety hazard to mining operations. Coal bed methane is formed through two mechanisms: biogenic formation from the bacterial degradation of coal and residual biomass and thermogenic formation from the volatilization of shorter-chained components (Moore, 2012; Song et al., 2012). Biogenic generation is dominant in lower rank coal from peat and lignite through sub-bituminous to highly-volatile bituminous coal, while the thermal gas formation processes dominate in higher-rank coal from medium-volatile bituminous coal to anthracite. The concentration of coal seam gas is typically around 2-6 m^3 per ton raw coal but can reach up to 20 m^3/ton (Moore, 2012). It depends not only on the rank of the coal but also on potential leakage prior to extraction, such as losses through tectonic movements (Hou et al., 2012).

Depending on the specific geological formation at the site, some coal bed methane may be extracted for energy production before the mining (Karacan et al., 2011), offering a significant safety benefit (Packham et al., 2011). In the absence of methane extraction, seam gas is vented to protect mine workers. Various technologies exist to utilize or oxidize low-concentration methane from mine ventilation (IEA, 2009a; UNECE, 2010).

Estimates of methane emissions from coal mining operations have traditionally relied on generic emission factors developed by the IPCC for national emissions inventories. Measurements for underground mines are based on volume and methane concentration measurements in ventilation air (Su et al., 2011). Open pit mines have presented a special challenge for emissions estimation. Emission factors were based on selected concentration measurements and inverse air quality modelling. New measurement techniques have been developed based on the gas content of the coal and rock and models for its volatilization (Saghafi, 2012). Saghafi (2012) reports emission factor estimates for Australian coal vary between 0.1 and 3.3 kg CH_4 per of ton of coal produced. Su et al. (2011) report on methane recovery and release from five mining areas in

China, where operators have begun capturing coal bed methane. The drainage gas rates range from 4 to 38 kg CH_4 per ton of coal produced, with capture to the order of 30-40 per cent foreseen in the future. China's average national emissions correspond to 5 kg CH_4/ton coal. Similar ranges have been found in the United States (Venkatesh et al., 2012). With modern coal-fired power plants, fugitive emissions of coal seam gas can hence vary from 1-300 g CO_2-eq/kWh, with likely average values around 30 g CO_2-eq/kWh. It should be emphasized, however, that the high variation indicates a significant opportunity for avoiding coal purchases from highly emitting mines and for general mitigation of coal bed methane.

3.6.1.1.2 Emissions from oil and gas extraction and distribution

Reported global emissions of methane from oil and gas extraction and distribution lie in the order of 1.5 billion tons CO_2-eq (Alsalam and Ragnauth, 2011; IEA, 2010), although this estimate is likely highly uncertain. Sources indicate that there is a high level of variability. Lower emission factors are reported for one-third of global production controlled by multinational companies (OGP, 2010). These numbers, however, hide large uncertainty and variability. This uncertainty has only recently been called to our attention in context of conflicts about shale gas and unconventional natural gas (Cathles III et al., 2012; Howarth et al., 2011). Reviewing others' work, Weber and Clavin (2012) report ranges of 10-20 g CO_2-eq/MJ gas for both shale gas and conventional gas in the United States, with slightly higher values for conventional gas, while earlier sources suggested slightly higher emissions for shale gas (Jiang et al., 2011; Stephenson et al., 2011). These emissions depend deeply on what practices are implemented in the field (Weng et al., 2012).

A recent analysis of hydrocarbon concentrations in the Colorado Front Range (Pétron et al., 2012) presents questions about the reliability of the emissions factor approach. The emissions levels that would explain observed elevated concentrations are about twice as high as those derived from an emissions-factor approach, with little overlap of the uncertainty ranges. Emissions correspond to about 4 per cent of the methane produced, but these estimates are also disputed (Cathles III, 2012).

As the large and easily exploitable oil and gas fields enter later stages of production, more energy is required to produce additional oil and gas (Gagnon et al., 2009; Hertwich et al., 2008; Murphy and Hall, 2010). Increasing oil prices prompt a shift to unconventional sources such as heavy oils, oil sands, shale gas and oil, or synfuel production from coal, all of which have higher input requirements and higher emissions. A lack of analysis and empirical work focusing on environmental aspects of fossil fuel production hampers the assessment of the current situation and likely future development.

3.6.1.2 Unconventional fossil fuels: case of unconventional gas

The environmental impacts of unconventional fuel extraction are poorly understood. Emissions associated with synthetic crude production from oil sands are higher than those from most conventional oil resources (Charpentier et al., 2009), and these emissions are related to extra energy requirements, fugitive emissions from venting and flaring (Johnson and Coderre, 2011), and land use change (Rooney et al., 2012).

Impacts are evident when a mishap occurs, water is polluted, or property destroyed. Michaels et al. (2010) describe a variety of publicized incidents (Michaels et al., 2010). However, single events cannot be easily generalized. Additionally, some serious environmental impacts can be subtle and require scale and time to emerge. Table 3.7 shows a list of the potential environmental issues related to unconventional gas development and production. Many are covered in detail below. Although the list is extensive, it is by no means exhaustive. Also, many may dispute the inclusion of one or more items. The message is simply that there is potential for adverse effects from unconventional gas development.

TABLE 3.7

Potential environmental impacts of unconventional natural gas

Resource degradation:
• Extensive water use • Water quality deterioration • Wastewater handling • Release of naturally occurring radioactive material (NORM) and technically enhanced NORM (TNORM) • Increased GHG emissions and air quality deterioration
Ecological impacts:
• Habitat destruction and fragmentation • Increased erosion and sedimentation of surface waters • Reduction of ecosystem sustaining water flows and downstream effects
Human health and safety impacts:
• Induced seismic activity • Increased truck traffic and infrastructure deterioration • Fire and other accidents • Handling and disposal of hazardous and radioactive materials • Methane migration and accumulation in buildings

3.6.1.2.1 Land use and habitat impacts

Unconventional gas formations cover vast areas. Even with the use of horizontal drilling, of which a single well can replace multiple vertical wells, thousands of well pads must be drilled to fully exploit a reservoir. As an example, the Marcellus shale gas play covers between 140,000 and 250,000 km^2 (Kargbo et al., 2010) and the economically viable region lies under southern New York, much of Pennsylvania, northern West Virginia, and western Ohio in the United States. In Pennsylvania alone, an estimated 42,000[5]-60,000[6] wells may be drilled (Johnson et al., 2010). Simply assuming 6 to 10 wells per well pad (NYSDEC, 2011) results in potentially 7,000-10,000 well pads distributed throughout the state. Each well pad causes direct land disturbance, land used for pad development and on-going operations. There is additional land required for ancillary services such as access roads, gathering lines, gas treatment facilities, and new transmission lines.

Jordaan et al. (2009) and Johnson (2010) investigated the land use of conventional and unconventional gas, respectively. Their results are summarized in Table 3.8. Jordaan et al. (2009) studied the overall land use associated with Canadian oil sands production and conventional natural gas that is used to extract oil sand. They not only measured the direct land disturbance but also included the important ecological impact of "edge effects". It is a well-known ecological concept that adjacent lands, especially in forested areas, can be impacted from disturbance. The disturbance creates new edges within "interior ecosystems," and this can impact habitat of sensitive flora and fauna. The accumulative impacts of multiple disturbances can also result in habitat fragmentation.

Jordaan et al. (2009) estimated that a conventional well requires 3 ha per well, with 1.2 ha being used for well pad development and 1.8 ha required for other ancillary development. Johnson (2010) explored the land use footprint of drilling and production in the Pennsylvanian portion of the Marcellus shale and also included the impacts of edge effects. They estimated the impacted area was about 12 ha per well pad. Assuming 6 to 10 wells per pad, this would equate to 1.2 to 2 ha per well, a reduction from the single vertical wells explored

5 Calculated based on USGS (2011) estimate of 84 trillion cubic feet of gas in the Marcellus and the average well ultimate recover for unconventional gas of 0.5 to 3.5 (2.0 most likely) (Weber and Clavin, 2012). Calculated based on USGS (2011) estimate of 84 trillion cubic feet of gas in the Marcellus and the average well ultimate recover for unconventional gas of 0.5 to 3.5 (2.0 most likely) (Weber and Clavin, 2012).

6 Based on the study author's assessment of investment and academic data about Marcellus shale development potential.

by Jordaan et al. (2009). Viewed in this way, the use of horizontal drilling reduces the industry footprint. A single well or even single well pad's impact is not particularly the issue but rather the accumulative land use of the developing industry and that associated with mature production. A distinguishing characteristic of unconventional gas plays is the large continuous areal extent of the resource. Johnson (2009) estimates that between 7,000-16,000 Marcellus well pads could be developed by 2030. At this level of development, land use likely will not scale linearly since ancillary services will be optimized by the sharing of these services amongst multiple well pads. Linear scaling, however, provides a general impression of magnitude. Using the data in Table 3.8, 84,000 to 192,000 ha could be used directly and indirectly over time. Johnson (2010) projects that two-thirds of these well pads will be developed in forested land, corresponding to approximately 14,000-128,000 ha, raising the concern for habitat destruction and fragmentation.

TABLE 3.8

Land use for production of natural gas

	Land use		
	Unconventional gas[1]		Conventional gas[2]
	(ha/well pad)	(ha/well[3])	(ha/well)
Area cleared for well pad	1.2	0.1 to 0.2	1.2
Ancillary infrastructure	2.4	0.2 to 0.4	1.8
Edge effects	8.3	0.8 to 1.4	*
Total[4]	12	1.2 to 2	3

*not available
[1] Source: (Johnson et al., 2010); [2] Source: (Jordaan et al., 2009) Jordaan et al. used the same 100m factor to value edge effects as Johnson et al., but did not break these out and are aggregate in the values presented. [3] Assuming 6 to 10 wells per pad; [4] Totals may not sum up due to rounding

3.6.1.2.2 Erosion from land use

Surface disturbance due to the construction of well pads, roads, pipelines, and ancillary facilities can cause increased runoff or sediment transport and lead to stream sediment deposition, increased nutrient transport impacting eutrophication, and detrimental impacts on aquatic life (Williams et al., 2008). Site erosion management techniques are available to reduce impacts and are routinely practiced by the industry (Veil, 2010). However, only limited data is available to benchmark these impacts or to evaluate the efficacy of mitigation efforts. Williams et al. (2008) studied three well sites near Denton, Texas and compared runoff results to two undeveloped reference sites. The well sites were described as having inner gravel pads and an outer disturbed area resulting from construction. After construction the outer areas were graded and left to re-vegetate naturally, which is common practice for the industry in that area. The entire developed area was slightly sloped to facilitate drainage. The well sites all had mulch berms. The authors found that sediment transport could be 49 times higher in the developed sites compared to undisturbed sites but over time, sediment transport at the well sites decreased. The authors suggested this reflected a "site stabilization effect" where the more easily mobilized material was transported and removed from the site soon after development. The outer disturbed area was determined to be the source for most of this sediment. The practice of leaving areas to re-vegetate naturally is not universal, however. In Pennsylvania, for example, all such portions of a well site must be planted with grasses in order to stabilize and prevent erosion (PADEP, 2008). No studies were found that evaluated the efficacy of this approach but logic would dictate that this should result in reduced erosion. In another study, Entrekin et al. (2011) monitored seven streams in the

Fayetteville shale area that had various levels of development based on well density, measured as wells/unit area. They monitored stream turbidity and found a correlation between increased development activities and increased turbidity. The study described the results as only being preliminary evidence and there was no description of specific well site activities (Entrekin et al., 2011).

These two studies suggest that both site development-related erosion and the resulting runoff impact surface water quality. It is critical to consider the impact from other potential development activities that might occur in the region over time, and if mitigation techniques can reduce these impacts to a satisfactory extent. The key question is if the impact from that of other potential development activities that might occur in the region over time and whether mitigation techniques can satisfactorily reduce these impacts.

3.6.1.2.3 Water use and water use impacts

Water is necessary for all drilling, regardless of whether the well is vertical or horizontal or the production is conventional or unconventional. The large quantities of water required to exploit unconventional gas, especially for hydraulic fracturing, has led to widespread concern. In this section, we discuss the consumptive water use in drilling in the form of drilling mud and in well completion, or hydraulic fracturing.

Drilling mud is required for pressure control, lubrication, maintaining the wellbore integrity, and removal of drill cuttings from the borehole. The cuttings are removed, the drilling mud re-circulated and chemistry adjusted as required during drilling (CRS, 2009). In drilling a horizontal well, the mud also provides the power and cooling required for the down-hole motor and bit and is the transmission medium for down-hole sensor readings. Drilling muds are generally classified as water-based, oil-based or synthetic oil-based (NYSDEC, 2011), and their composition can vary. A short description of the composition of these drilling fluids may be found in the appendix of (Acharaya et al., 2011). Drilling horizontal wells can require between 400–4,000 m^3 of water per well (Gregory et al., 2011). Veil (2010) estimates a slightly lower range of 230–4,000 m^3, depending on the drilling fluid composition and the depth and length of the horizontal sections. The author suggests an average figure of 300 m^3. Used drilling muds are "reconditioned" for re-use, thereby reducing water needs.

Hydraulic fracturing requires an estimated 7,600 to 15,000 m^3 of water per well (US DOE, 2009). The New York State Department of Environmental Conservation (NYSDEC, 2011) has estimated a larger range of water use of 9,000 to 30,000 m^3 of water per well. Carter et al. (2011) reviewed publicly available data provided by the Pennsylvania Department of Conservation and Natural Resources for horizontal Marcellus wells completed between 2005 and 2010 and found that on average, 11,000 m^3 of water, the midpoint of the DOE range stated above, was used for fracturing (Carter et al., 2011). They found a range of 95 to 60,000 m^3. None of these studies provides a reason for the large range in extremes.

Water requirements for hydraulic fracturing of coal bed methane are estimated to be much lower than those for shale gas production. These have been estimated at between 95 to 1,300 m^3 per well (Holditch, 1993; Palmer et al., 1993). The water issue associated with coal-bed methane is the large amount of produced water that must be removed from the seam before the methane can desorb from the coal. Disposal of this water is a concern. In any event, this water is usually taken from surface water sources or municipal treatment works (Gregory et al., 2011).

Water use impacts are regional and are variable. After an initial period of development where the industry uses only fresh water, re-circulation of the produced water returns to the surface during flowback (sometimes called flowback water) has become the norm in some shale plays such as Marcellus. This re-use of water has reduced water consumption and disposal quantities. Not all of the flowback may be recycled, however. For instance, Acharaya et al. (2011) reported that re-use might be restricted to water recovered early in the flowback period. This water has a chemical composition similar to the original hydraulic fracturing fluid and is more suitable for re-use.

Also, not all injected water is recovered; an estimated 20-90 per cent remains in the shale formation. This range represents the extremes of the ranges presented by (Gregory et al., 2011) and (CRS, 2009). Thus, make-up water must meet the demands of additional hydraulic fracturing. This makes extrapolation of values to estimate total water use for the development of a shale reservoir over time difficult. For instance, for a single pad using 11,300 m^3 per well for hydraulic fracturing, scaling linearly the values for single well statistics would results on a requirement of 68,000 m^3 of freshwater and disposal of 34,000 m^3 flowback water. However, if early flowback from the first five days of recovery is re-used, the requirements would decrease to 54,000 m^3 freshwater, of which 20,000 m^3 would require disposal.

After all necessary completions have been performed on a well pad, some operators transfer the water for use at other drilling sites, particularly if transfer costs are less than overall disposal costs. Efforts are being made to develop methods to increase re-use and to find alternative sources of water rather than relying on and potentially straining local freshwater source.

Water use impacts are region specific. Water consumption can affect surface water or ground water if it exceeds the capacity of the system. Since water for hydraulic fracturing can be large as previously described, management of consumption is important. Reducing flow can impact downstream uses, aquatic biota via exposure, temperature effects, concentrate existing pollutants, and exacerbate drought impacts (NYSDEC, 2011). Groundwater withdrawal rates that are higher than the system's recharge rate will result in lowering the aquifer level. Although the immediate impact might be area wells running dry, groundwater ultimately re-charges rivers, streams, and lakes. The decreased aquifer level can reduce water levels in the rivers, lakes and streams, resulting in the same issues as direct surface water withdraws. The importance of water use may not be apparent at the level of a single well pad, but it can become significant over time and across entire river basins.

Water, the formulated drilling muds and fracturing fluids (see next section), and flowback water must be stored in tanks or impoundments at the well pad during operations. Also, water production from the well continues throughout the lifetime of the well. The water is disposed of differently on a regional basis. In dry areas, the water can be evaporated and the remaining solids transported to a landfill. In areas where evaporation is not practical, the water is injected into underground disposal wells (Clark and Veil, 2011). Some development is going into water treatment technologies to reduce the need for injection such as the use of reverse osmosis, thermal distillation and crystallization, etc. At this point, all of these water treatment technologies have limitations related to handle high total dissolved solids or economics (Gregory et al., 2011). It should be pointed out that after the produced water from flowback is recovered, most shale continue to produce small amounts of water (2-8 m^3/day) for the life of the well (Gregory et al., 2011). This water is separated from the gas stream and is stored in tanks that are periodically emptied and transported by truck for disposal.

3.6.1.2.4 Chemicals in hydraulic fracturing fluid

The chemical constituents of hydraulic fracturing fluid have been widely cited (US DOE, 2009). The water-based fluid usually contains quartz sand or ceramic material as proppants, gels to increase viscosity and loss, acids to clean the wellbore, biocides to control microbial growth, scale inhibitors, corrosion inhibitors, surfactants to enhance flow characteristics and a variety of other chemicals based on company preferences and water- and site-specific characteristics (Kargbo et al., 2010; US DOE, 2009). The exact fluid composition is considered proprietary but some companies have provided lists of components. As quoted from a United States Congressional investigation, “Between 2005 and 2009, the fourteen oil and gas service companies used more than 2,500 hydraulic fracturing products containing 750 chemicals and other components” (Waxman et al., 2011). From that same report, two of the most used chemicals included methanol, a hazardous air pollutant, and 2-butoxyethanol (2-BE), a chemical that can cause hemolysis and spleen, liver and bone marrow damage. It should be noted that EPA recently found 2-BE in drinking water wells in Pavillion, Wyoming (US EPA, 2010). The report further describes chemicals that are known or suspected human carcinogens, regulated under the Safe Drinking Act, and hazardous air pollutants.

Colborn et al. (2011) investigated chemicals used in general natural gas operations. Although the authors did not clearly define which chemicals were derived from which process due to study design constraints and the large number of chemicals involved, it was apparent that many products and specific chemicals were poorly characterized. In some cases, only the trade names of the product could be found and composition was unavailable. In other cases, products reported only 1 per cent of their components, and many reported chemicals had no Chemical Abstracts Number available to aid in characterization (Colborn et al., 2011).

The array of chemicals found in fracturing fluids should be no surprise. There are different levels of purity requirements for chemicals used in food or cosmetics versus industrial formulations. Thus, contaminants left over from the production process many times accompany the active ingredient in an industrial formulation.

Concern about unintended release of these fluids and the subsequent contamination of surface waters and aquifers is widespread. This concern is usually focused on hydraulic fracturing fluid and the water returned to the surface during the flowback period. Fracturing fluids change in composition over the course of the fracturing process and during flowback as the water entrains and solubilizes components of leftover drilling muds and minerals from the reservoir (Acharaya et al., 2011). Alley et al. (2011) analysed the composition of "produced waters[7]" from 377 coal bed methane samples, 137 tight gas sands samples and 541 shale gas samples (Alley et al., 2011). The report summarizes results from these analyses according to common water quality parameters. The authors identified constituents of concern based on whether a specific constituent concentration exceeded the criteria from the guidelines of United States Environment Protection Agency water quality criteria for surface discharge.[8] The criteria were selected to protect the aquatic biota from toxic concentrations of constituents in the receiving system. Shale gas-produced water exceeded the guidelines in 10 of 12 criteria where data were available for comparison,[9] tight gas exceeded 10 of 13 criteria, and coal bed methane exceeded 14 of 18. It is clear from this analysis that the chemical compounds present in drilling muds and fracturing fluids should not be released to surface waters or allowed contact with groundwater. The potential for such release as the industry expands and matures is unknown but has occurred.

3.6.1.2.5 Storage and disposal

Increased seismicity

Operations related to unconventional gas production may potentially cause "induced seismicity". For a recent review see (NRC, 2012). The most obvious of these operations is hydraulic fracturing, but less obvious is deep well injection, which is the major disposal method for shale gas wastewater consisting of flowback and produced water. Recently, the Oklahoma Geological Survey investigated and documented fifty earthquakes between 1.0 and 2.8 magnitude associated with a particular hydraulic fracturing event. Although the event met many of the requirements proposed by (Davis and Frohlich, 1993) to show causality, the author remained sceptical of the relationship (Holland, 2011).

There is concern that the injection of large amounts of fluid into the subsurface can stimulate earthquakes (Nicholson and Wesson, 1990). In the 1990 USGS report, the authors reported documented earthquakes due to deep well injection in five American states and possible events in three others (Nicholson and Wesson, 1990). Most of the documented cases of fluid injection-related earthquakes were associated with secondary oil recovery, or water flooding. The postulated link was attributed to injecting fluids such as water at high pressure into reservoirs of low permeability.

7 These waters contained flowback samples as well but were not distinguished in the analysis.

8 The authors addressed FAO guideline for agriculture uses and peer reviewed toxicity for specific species. The EPA guidelines are highlighted here due to their particular relevance to unintended releases.

9 The dataset was not specifically designed to match all of EPA criteria. The criteria analysed for shale gas included pH, alkalinity, nitrate, phosphate, Al, B, Ba, Cl, Cu, Fe, Mn and Zn. For tight gas the criteria assessed are pH, alkalinity, oil and grease, ammonia – N, As, Cd, Cl, Cr, Cu, Fe, Mn, Ni, and Zn. For coal-bed methane: pH, alkalinity, nitrate, phosphate, ammonia – N, Al, As, B, Ba, Cd, Cl, Cr, Cu, Fe, Hg, Mn, Ni, and Zn are assessed.

The USGS study (Nicholson and Wesson, 1990) also found two documented instances where deep well injection for waste disposal caused earthquakes[10]. These events occurred in Ashtabula, Ohio and Denver, Colorado. In the Ashtabula case, a series of small shallow earthquakes at 1.8 km depth where triggered, with the largest being of 3.6 magnitude. The Denver incident occurred at the Rocky Mountain Arsenal where fluid was injected into an impermeable formation and caused the largest known injection-induced earthquake at that time of the report; the earthquake was of 5.5 magnitude and caused an estimated half a million dollars of damage.

More work is needed in the area to come to a scientific consensus regarding the relationship between subsurface fluid injection and earthquakes, in general. If unconventional operations that rely on hydraulic fracturing and deep well injection for wastewater disposal are indeed responsible for the reported increases in seismic activity, these risks need to be assessed in order to ensure public safety.

Gas migration

Gas in drinking water wells has been reported. Methane is not considered a human health hazard, but in confined spaces, it can present an explosion risk. In 1983, Harrison reported drinking water wells in northwestern Pennsylvania that had been impacted by natural gas drilling in tight gas sands (Harrison, 1983). The wells showed turbidity from entrained gas bubbles soon after drilling and completion; a home owner demonstrated the presence of gas in their well water by successfully lighting a flame at the end of a garden hose, and a pump house door was blown off from the accumulation of gas. These incidents are so prevalent that the Pennsylvania Department of Environmental Protection issues a fact sheet concerning mitigation of methane accumulation in water wells (PADEP, 2004).

Osborn et al. (2011) look at the relationship between shale gas production and methane presence in residents' drinking water wells. They found a statistical correlation between the concentration of gas and the distance from a producing gas well. The authors provide chemical isotope analyses that supported contention that the gas was thermogenic (Osborn et al., 2011). (Molofsky et al., 2011) use a larger database than Osborn et al. (2011) and find that methane concentrations in water wells were directly related to being located in lowland areas. They were unable to consistently correlate high methane concentrations in groundwater and proximity to producing natural gas wells. The isotopic data demonstrated that the thermogenic methane originating from Upper and Middle Devonian deposits overlay the Marcellus shale. In their discussion, they showed that the Osborn et al. (2011) isotopic data closely matched the methane from the Upper and Middle Devonian layers. With these conflicting results, it is difficult to draw definitive conclusions regarding the source of the gas.

3.6.1.2.6 Greenhouse gas emissions

In general, the GHG emissions associated with natural gas production, transmission, distribution and use come from fugitive methane emissions and fuel combustion (Jaramillo et al., 2007). A number of GHG life cycle analyses calculating these emissions throughout the natural gas life cycle have been published (Ally and Pryor, 2007; Arteconi et al., 2010; Kim and Dale, 2005; Odeh and Cockerill, 2008; Okamura et al., 2007; Venkatesh et al., 2011). However, unconventional gas production requires unconventional methods. The concern for increased emissions from horizontal drilling and hydraulic fracturing has stimulated a recent spate of GHG life cycle analyses (Burnham et al., 2011; Howarth et al., 2011; Hultman et al., 2011; Jiang et al., 2011; NETL, 2011a; Stephenson et al., 2011). The studies examined generic shale gas plays (Burnham et al., 2011; Howarth et al., 2011; Hultman et al., 2011; Stephenson et al., 2011), or specific plays such as Marcellus or Barnett. Also, these studies looked at a variety of end uses for the natural gas, such as electricity generation, and transportation. The reports all make different modelling assumptions that ultimately lead to some variability in results. Figure 3.17 shows the well to plant gate (through the transmission system) from each of the studies. The study results of Howarth et al. (2011) were converted to 100 year global warming potential (GWP) values for comparison purposes.[11]

10 Holland (2011) covers these examples and a number of confirmed and possible cases. At this point it represents the best review of this area.

11 Howarth et al. (2011) argues that using 20 GWP values from Shindell et al. (2009) is more appropriate as it provides a better picture of the short term impacts of methane emissions which have short atmospheric residence time compared to CO_2 but greater radiative forcing.

FIGURE 3.17

GHG emissions from unconventional and conventional gas production

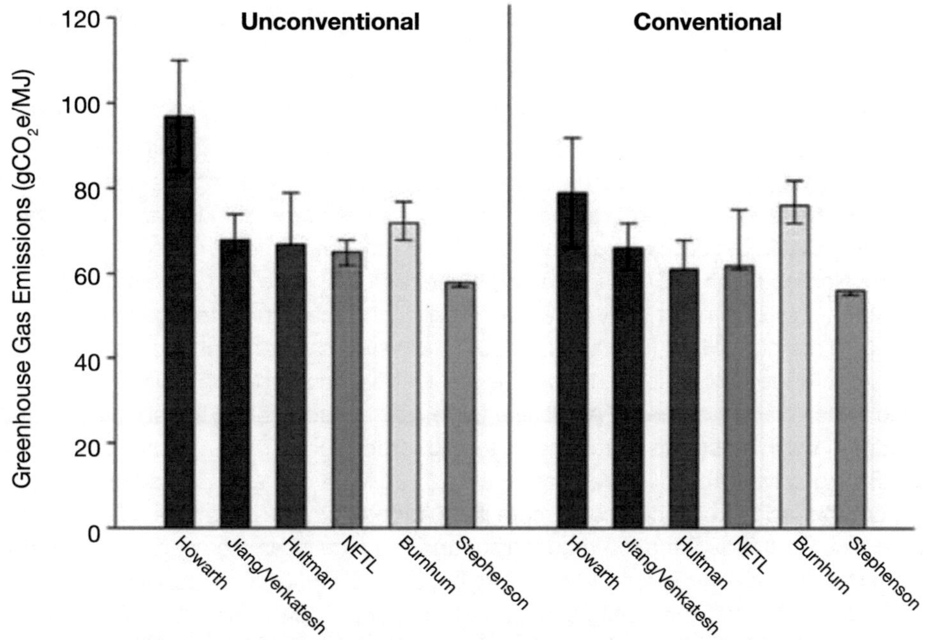

Reprinted from Weber and Clavin, 2012. Copyright 2012, American Chemical Society

Weber and Clavin (2012) review the studies shown in Figure 3.17. They reconciled differences in upstream data and assumptions and conducted a Monte Carlo uncertainty analysis of the carbon footprint of both shale and conventional natural gas production. They found the "likely" upstream "carbon footprint" natural gas production, regardless of whether it was conventional or unconventional, to be largely similar, with overlapping 95 per cent confidence intervals of 11.0–21.0 g CO_2CO_2-eq/MJ for shale gas and 12.4–19.5 g CO_2-eq/MJ for conventional gas. The upstream emissions represent less than 25 per cent of the total life cycle emissions from providing heat, electricity, transportation services, or other functions from natural gas.

3.6.2 POWER PLANT OPERATION

3.6.2.1 Emissions to air

The combustion of fossil fuels results in emissions of CO_2, NO_x, SOx, particulates (PM), volatile organic compounds (VOC) and heavy metals such as mercury. The amounts of CO_2 discharged to the atmosphere are mainly dependent on the carbon content of the fossil fuel, and, to a lesser extent, on the efficiency of the thermodynamic cycle. Emissions of non-CO_2 substances depend not only on fuel characteristics but also on specific conditions such as type of technology, combustion, operation and maintenance conditions, size and age of the facility, and emission control policy. An overview of emission factors for power plants with and without CCS reported in literature is shown in Table 3.9. Gas-fired power plants produce fewer CO_2 emissions and, generally, negligible levels of SOx and particulates while the levels of NO_x are 30-60 per cent of coal-fired power plants.

As Table 3.9 indicates, the deployment of CO_2 capture technologies can change the emission profiles of power plants at the plant level:

- In post-combustion concepts with amine solvents, for instance, SO_2 and PM emissions are expected to be very low at the plant level since the concentration of SO_2 needs to be limited in order to avoid solvent degradation and low levels of PM are necessary in order to assure a stable capture process. The emission of particulate matter from natural gas fired cycles can be considered negligible. With regard to NO_x, NO_2 needs to be removed since it can react with amine-based solvents, causing degradation. However, NO_2 accounts for only approximately 5-10 per cent of the total NO_x emissions, making the net impact of its removal fairly minor. The effect of CO_2 capture on the emission levels of mercury is under discussion. Some studies, such as (Nie, 2009) and (Korre et al., 2010) use mercury removal rates in the order of 76-80 per cent in the post-combustion (MEA) capture unit. Other studies, such as (Cui et al., 2010), however, indicate such removal rates would most likely be attained only if the mercury entering the CO_2 capture unit was present as oxidized mercury (Hg^{2+}) (Lee et al., 2009)[12]; however, the concentration of this oxidized form of mercury is expected to be limited since the flue gas first passes through FGD units[13] prior to CO_2 capture. Finally, due to its volatility, some MEA will also be lost by evaporation in the absorber. These losses are minimized by adding washing stages at the top of the absorber column. MEA losses are reported in the range of 1.6 to 3.1 kg solvent/ton CO_2 (Veltman et al., 2010). MEA can also degrade into products such as ammonia, organic acids, oxidants, carbamate salts, nitrosamines, nitramines, etc. The degradation products formed depend on the degradation mechanism and the type of amine used[14]. MEA degradation rates reported in the literature are in the range of 0.29 kg/t CO_2 to 0.73 kg/t CO_2, for flue gas containing approximately 3 per cent CO_2 and 5 per cent O_2 (Goff and Rochelle, 2004). Currently, R&D efforts are being conducted by different research groups around the world in order to better understand the potential impact and level of MEA degradation products.

- Studies in literature for pre-combustion concepts report both lower and higher NO_x values, compared to similar plants without CO_2 capture (Davison, 2007; IEA GHG, 2006; NETL, 2007). The levels of NO_x reported are dependent on the assumed gas turbine performance and the efficiency penalties. Pre-combustion is expected to further lower PM emissions in IGCC plants, which have low PM emissions even without CO_2 capture units, due to high removal efficiencies in the gas cleaning section (NETL, 2007). Mercury removal is considered easier and more cost-effective in IGCCs than in PCs because mercury can be removed from the syngas at elevated pressure prior to combustion. Furthermore, syngas volumes are much smaller than flue gas volumes in comparable PC cases. High levels of removal efficiencies of 90-95 per cent are attainable with pre-sulfide carbon beds in the syngas stream (NETL, 2009). At the moment, little information is available in literature on the interaction of impurities, e.g., mercury with solvents such as selexol, nor is it clear from the literature whether the CO_2 capture unit would further lower mercury emissions.

- In the case of coal oxyfuel combustion with CO_2 capture, NO_x emissions are expected to be low due to three mechanisms: reduction of fuel NO_x and inhibition of thermal NO_x since the nitrogen supplied by the air is removed, decomposition of NO_x in the recycled flue gas, and reaction of recycled NO_x and char (Croiset et al., 2000; Hu et al., 2001; Okazaki and Ando, 1997). SO_2 emission levels are reported lower than in PCs due to the high concentration of SO_2 inside the furnace since there is no dilution in

12 High temperatures in coal combustion in a PC vaporize the mercury in the coal to form gaseous elemental mercury (Hg^0). Subsequent cooling of the combustion gases and interaction of the gaseous Hg^0 with other combustion products result in a portion of the Hg being converted to gaseous oxidized forms of mercury (Hg^{2+}) and particle-bound mercury. The Clean Air Mercury Rule (CAMR) in the U.S. indicates environmental targets for mercury emissions levels in the range of 20 x 10-6 lb/MWh (PC/bituminous coal and IGCC) to 175 x 10-6 lb/MWh (PC/lignite coal) (NETL, 2009f).

13 Wu et al. (2010) compared the speciation of mercury in PC power plants with and without a selective catalytic reduction unit (SCR). Their results show that after the FGD unit, the share of Hg^0 dominates the total concentration of Hg (about 70-80 per cent).

14 MEA can degrade due to contact with oxygen or metal ions in the flue gas, high temperatures in the reboiler and stripper and reactions with oxidized nitrogen in the atmosphere in processes called oxidative, thermal and atmospheric degradation, respectively.

nitrogen, which enhances sulfur retention in fly ash, called ash sulfation (Hu et al., 2001; Scheffknecht et al., 2011). However, a study made by Fleig et al., (2009) based on model simulations, indicates that SO_3 concentrations are significantly higher than the concentrations seen in conventional PCs15. Finally, PM emissions are estimated in literature to be lower per kWh, compared to conventional pulverized coal-fired power plants. No information has been found in publicly available literature about the fate of trace heavy metals in oxyfuel combustion.

- With regard to pre-combustion capture in IGCCs unit, low SO_x emissions are expected since high level of recovery of sulfur compounds is expected from the Claus unit (see Table 3.9).

TABLE 3.9

Average, minimum and maximum emission factors for energy conversion concepts with and without CO_2 capture as reported in the literature

Capture technology	Conversion Technology	CO_2 g/kWh	NO_x mg/kWh	SO_2 mg/kWh	NH_3 mg/kWh	VOC mg/kWh	PM mg/kWh
No capture	IGCC	766 (694-833)	229 (90-580)	64 (40-141)	..	..	**28 (27-29)**
	NGCC	370 (344-379)	168 (90-262)	..	..	..	..
	PC	826 (706-1004)	374 (159-620)	414 (100-1280)	7 (3-10)	10 (9-11)	**39 (7-51)**
Oxyfuel combustion	GC	10 (0-60)	..	..	..	..	..
	NGCC	8 (0-12)	..	..	..	..	..
	PC	47 (0-147)	172 (0-390)	25 (0-98)	..	..	**3 (0-10)**
Post-combustion	NGCC	55 (40-66)	188 (110-275)	..	6 (2-19)	..	..
	PC	143 (59-369)	537 (205-770)	9 (1-13)	209 (187-230)	..	**52 (9-74)**
Pre-combustion	GC	21 (0-42)	..	..	..	..	..
	IGCC	**97 (71-152)**	**209 (100-550)**	**28 (10-51)**	..	..	**34 (34-35)**

Source: Koornneef et al., 2010. The ranges report the minimum and maximum values found in 171 studies.
The emissions factors are based on various fuels and power plant configurations and performance.
Post-combustion capture includes capture with amine-based solvents and chilled ammonia.

15 The increase in SO_3 is reportedly caused by three properties of the oxy-fuel process (Fleig et al., 2009). a) The oxidizer in oxyfuel combustion contains SO_2, which increases the amount of sulfur present during combustion. b) The oxidizer in oxyfuel combustion has a higher concentration of O_2, which decreases the volume flow through the furnace and, thus, increases the concentration of SO_3 and c) the change from N_2 to CO_2 increases the SO_3/SO_2 ratio.

3.6.2.2 Impacts to water

Thermoelectric power generation is an important user of fresh water resources. To understand the relationship between water and energy from fossil fuels, two concepts are of importance: water withdrawal, where water is taken from a source and returned to the same source, and water consumption, which is the use of water that is not returned to the source. The rates of water withdrawal and consumption depend on the type of cooling strategy used. For instance, in an open loop process, often called once-through, cooling water is taken from a water body such as a river or the sea and is passed through a condenser and discharged back to the same water body. In this case, there is a high rate of water withdrawal and relatively low water consumption. If a closed loop is used instead, such as cooling towers, cooling ponds, there is low water withdrawal and relatively higher water consumption.

Vassolo and Döll (2005) made a first estimate of the global annual water withdrawal and consumption attributable to thermoelectric power[16]. Although they use 1995 as reference year due to data availability, their results provide an indication of the global water use as a consequence of fossil fuel combustion: 401 billion m^3 of water was withdrawn in 1995 by fossil fuel power generation while 11 billion m^{3}were consumed (Vassolo and Döll, 2005). More recent estimates are provided at the country level. For instance, the United States Geological Survey (USGS) estimated that in 2005, US thermoelectric power plants withdrew approximately 41 per cent of freshwater sources, corresponding to about 541 million m^3 per day, primarily for cooling needs[17], while water consumption was estimated at 23.5 million m^3/day (Kenny et al., 2009). Water withdrawals by Chinese thermoelectric plants are estimated between 49 (PGA 2010) and 67 billion m^3 (Yu et al., 2011), corresponding to 37-50 per cent of the total industrial water withdrawal, respectively. Chinese water consumption induced by thermoelectric power is estimated at about 7 billion m^3 (PGA, 2010). Shares of water withdrawals of up to 70 per cent are reported for several European countries: Germany (73 per cent), Belgium (72 per cent), Netherlands (59 per cent), Poland (72 per cent), and Spain (23 per cent) (Eurostat, 2011)

Water is used at power plants to generate steam that drives steam turbines, for cooling the exhaust steam and for other operations including ash disposal, emissions control and potable use. In a cooling tower system, raw water use is dominated by the makeup requirements for the cooling tower, ranging between 80-95 per cent of total water use, depending on the technology. For once-through systems, over 90 per cent of the raw water is used by cooling of the steam turbine. Water losses, reported as water consumption, can be classified as process losses, flue gas losses and cooling water losses[18]. For systems with cooling towers, cooling water losses have the largest share at 75 per cent. These losses are mainly attributed to water evaporation. Process losses are higher in the IGCC plants due to additional water requirements induced by the gasification process and the water-gas shift reactor. In the case of flue gases, the FGD units for PC power plants result in significant water losses.

Cooling water requirements are affected by plant size, the energy source used, cooling technology, plant efficiency, FGD type, ambient temperature, and whether carbon capture technologies are deployed. Table 3.10 provides an overview of water withdrawal and consumption factors reported in the literature for different types of power plants. The type of cooling also affects the efficiency of the power plant. This is mainly due to differences in the cooling temperatures that can be achieved, which in turn affects the pressure of the steam turbine condenser and the steam turbine efficiency. Differing geographical areas adopt different cooling technologies. In the U.S., for instance, 48 per cent of coal power plants use wet cooling towers, 39 per cent use once-through systems, 12.7 per cent use cooling ponds and only 0.2 per cent use dry systems (NETL, 2010b) while in China,

16 The authors assessed 63,590 power plants using an average water consumption factor of 4 m^3/MWh (withdrawal) and 1.33 m^3/MWh for cooling towers and 180 m^3/MWh (withdrawal) and 0.65 m^3/MWh for once-through flow cooling systems.

17 In addition to the fresh water withdrawal, the survey also reports additional 58.1 billion gallons of saline water withdrawals per day, which corresponds to 95 per cent of total saline withdrawal.

18 Note that water losses in once-through cooling plants are generally not accounted for since these types of technologies return the water used for cooling to its source, such as a local river or lake. Some studies have, however, indicated that temperature rises of 10-15°C might be expected in the receiving water body, which could cause additional evaporation in the water body (EPRI, 2002; EPRI, 2000). It has been argued that since this evaporation is caused by the use of the resource by the plant, these evaporative water losses should be allocated to the power plant.

FIGURE 3.18

Water use in three different types of power plants with and without CCS

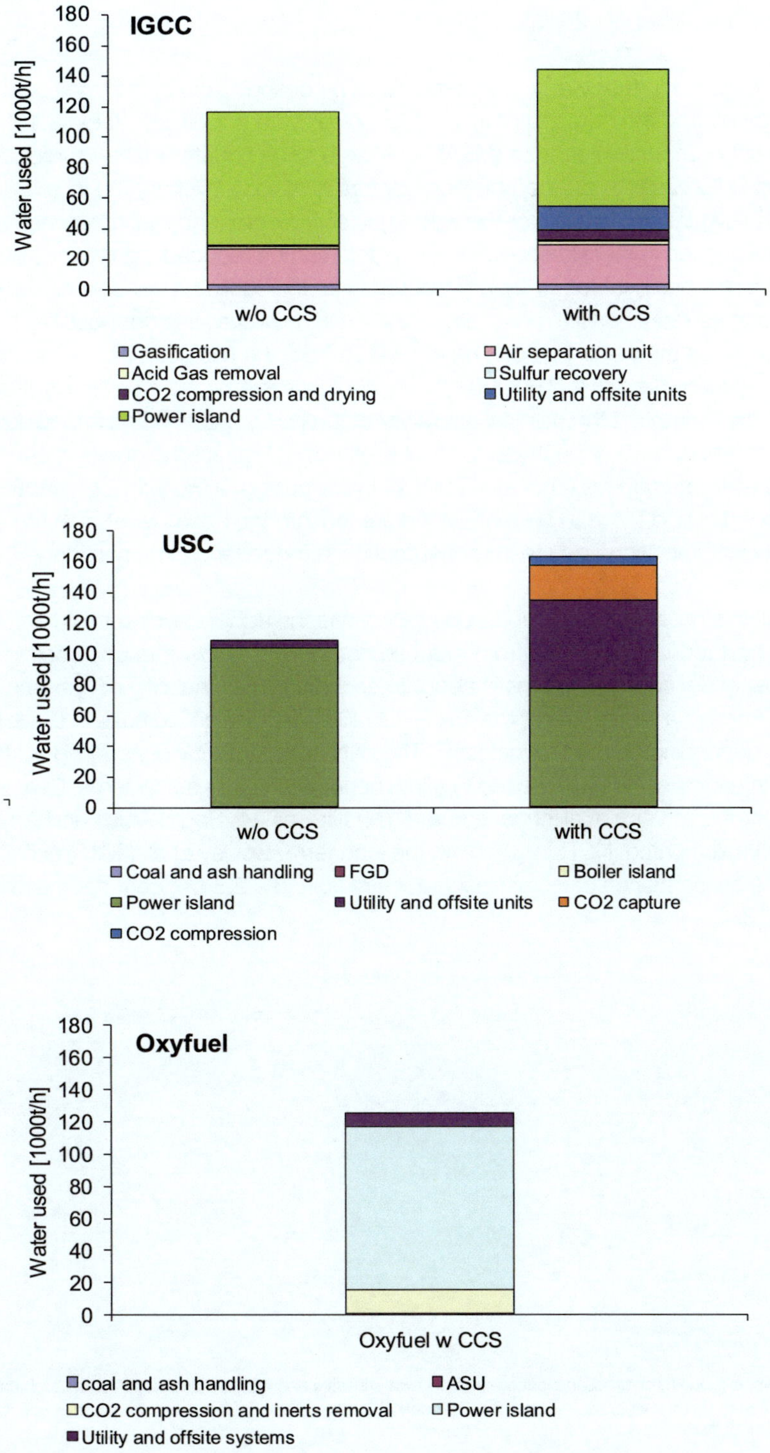

USC: Ultra supercritical power plant; IGCC: Integrated Gasification Combined Cycle.
Source: IEA GHG, 2011

currently 90 per cent of the coal power plants use once-through systems[19]. Davies et al. (2013) indicate that since most old power plants, which mainly use once-through systems, will be replaced in the coming years, a shift towards wet cooling towers can be expected, resulting in a significant decrease in water withdrawal and an increase in water consumption (Davies et al., 2013).

CO_2 capture technologies will increase further water requirements as a consequence of the additional fuel used to compensate for the energy penalty and the demand of the CO_2 capture system itself (see Table 3.10). For instance, coal-fired power plants with post-combustion capture (MEA) have large cooling water make-up requirements, while increased water demand in IGCCs with pre-combustion capture is mainly driven by the increased cooling load required to further cool the syngas and steam for the water gas shift reactor and the increased auxiliary load (NETL, 2009). Water consumption in thermoelectric power plants can be reduced significantly by using dry cooling systems that use convective heat rather than evaporation as the cooling mechanism. For PCs and NGCCs, water consumption can be reduced to almost zero. For IGCCS, however, it is not possible to eliminate the net raw water consumption since the water that can be recovered from the flue gas does not suffice to satisfy the entire water demand of the power plant (IEA GHG, 2011). Dry cooling systems will result in significantly lower water requirements; however, they will affect the thermal efficiency of the power plant[20]. The impact is even larger when CO_2 capture technologies are applied since these technologies not only affect the power island but also compression intercoolers and ASU intercoolers (Zhai et al., 2011). In the case of a PC with CO_2 capture (MEA), the use of air to cool down the sour gas and the lean solvent, which are fed into the absorber at high temperatures, will also lead to an increase of solvent circulation and steam consumption in the regeneration section.

Dry cooling systems also result in a net increase in capital costs because the additional cost related to the air cooled condenser is greater than the cost reduction realized from reduced cooling water flow rate and cooling tower duty (NETL, 2010b). The capital cost of dry cooling systems is about three to five times that of wet systems (Yang and Lant, 2011; Zhai et al., 2011). For a coal power plant, the IEA GHG (2011) reports an increase in the total investment cost of between 4-8 per cent, depending on the technology[21]. The penalty induced by dry cooling systems in conjunction with the increase in capital costs has resulted in a low penetration of these systems. Only when cooling water supplies are severely limited, such as in California, some regions in China[22], South Africa and Australia, dry cooling systems are being implemented (NETL, 2011b). In the long term, Dooley et al. (2013) expect a larger penetration of dry systems as a consequence of increasing water stress, and a 2-5 per cent decrease in the cost of dry systems (Dooley et al., 2013).

19 Note that consumptive water use will vary due to changes in location- and season-sensitive operating parameters such as ambient temperature; (Zhai et al., 2011) Zhai et al. report that makeup water requirements for wet cooling tower systems increase by more than 10 per cent when the ambient temperature increases from 15-25°C.

20 The performance of a dry cooling system is limited by the ambient dry-bulb air temperature. Since dry-bulb temperatures are higher than corresponding wet-bulb temperatures, the performance of dry cooling systems will be less than once-through or recirculating wet systems.

21 The range applies to both, power plants without CCS (dry versus wet cooling) and power plants with CCS (dry versus wet cooling)

22 In China, dry cooling has been adopted for many new plants. It has been reported that dry cooling had been installed on more than 35000 MW of new plant as of 2008 (NETL, 2011b). Indicative penetration of dry cooling in other countries is: 12445 MW (USA), 15456 MW (Europe), 12250 MW (Africa and Middle East); 4195 MW (Asia), 705 MW (Australia) (SPX, 2009).

TABLE 3.10

Water withdrawal (l/MWh) and water consumption (l/MWh) in power plants with and without CCS

Fuel	Cooling system	Boiler type	Water use				Power plant efficiency		Source
			Without CCS		With CCS		Without CCS [%]	With CCS [%]	
			Water withdrawal	Water Consumption	Water withdrawal	Water consumption			
Coal	Once-through	USC	139,927	104	240495	410	44	34.8	(IEA GHG, 2011)
	Wet-cooling tower	SC	2,400	1,685	4379	3,085	38.3	26.4	(Zhai et al. 2011)
				1,900		2,700	42	32	(Smart and Aspinall, 2009)
			2,168	1,721	4,091	3,152	39.3	28.4	(NETL ,2010b)
	Air cooled	SC	0	0	0	0	42.1	32.6	(IEA, GHG 2011)
				100		800	40	29	(Smart and Aspinall, 2009)
			313		2,660				(NETL, 2010b)
	Hybrid system	SC	480	235.9	2,714.98	1,820.15	NS	NS	(Zhai et al., 2011)
NGCC	Wet-cooling tower	NA		850		1,000	52	39	(Smart and Aspinall, 2009)
		HRSG	961	738	1,878	1408	50.2	42.8	(NETL, 2010b)
	Air cooled	CCGT		100		250	50	37	(Smart and Aspinall, 2009)
		HRSG	22		1,366		42.8		(NETL, 2010b)
IGCC	Once-through	NA	146,932	126	185,187	411	38	31.5	(IEA GHG, 2011)
	Wet-cooling tower			1,785		2194	38.2	32.92	(Ikeda et al., 2007)
		NS	1,229	1,658	2,567	2,012	39.7	31	(NETL, 2010b)
	Air cooled		0	0	0	39	35.7	28.9	(IEA GHG, 2011)
	Air cooled	CoP	537		1,252				(NETL, 2010b)
	Once-through				226,120	63	35.4	32.7	(IEA GHG, 2011)
Coal-oxyfuel	Wet-cooling tower					3,668		31.94	(Ikeda et al., 2007)
	Air cooled					945		29.89	(Ikeda et al., 2007)

UC: ultra supercritical; SC: supercritical; NGCC: natural gas combined cycle; IGCC: integrated gasification combined cycle; HRSG: heat recovery steam generator; CCGT: combined cycle gas turbine

3.6.2.3 Waste and by-products generated by fossil fuel power plants

The primary solid wastes of fossil fuels power plants without CCS are bottom ash and fly ash both of which are produced during the combustion process. Bottom ash is composed of agglomerated particulates that are too large to be carried in the flue gas. It is collected by impinging on the furnace walls or falling through open grates to an ash hopper at the bottom of the furnace (US EPA, 2010). Fly ash, which is a finer ash material, is removed from the plant exhaust gases by electrostatic precipitators and wet scrubber systems. Boiler slag is the molten bottom ash. The ashes differ in characteristics depending on the content of the fuel burned, but in general, over 90 per cent of both ashes are composed of silicon, aluminium, iron and calcium in both elemental and oxide forms. In addition to ashes, waste is also produced by the FGD units; this consists predominantly Ca-SOx compounds. A wet FGD produces waste in the form of an ash, unreacted lime, calcium sulfate and calcium sulfite slurry. A dry FGD generates a mix of unreacted solvent, sulfur salts and fly ash. Table 3.11 shows an overview of the type of waste produced each type of power plant according to different literature sources.

Coal power plants are considered one of the largest sources of industrial wastes worldwide. The American Coal Ash Association (ACAA), for instance, reported that in the United States, 52 million tons of fly ash, 14 million tons of bottom ash, 1.7 million tons of boiler slag and 32 million tons of FGD waste were generated from coal power plants in 2012 (ACAA, 2013). The total amount of waste produced in EU15 in 2008 has been estimated at 56 million tons and over 100 million tons for EU 25 (Ecoba-Eurelectric, 2011), while annual coal waste generation in China and India is reported in the order of 200 and 100 million tons, respectively (ADB, 2009; Senapati, 2011). Besides ashes, other gypsum and sulfur are produced as well. Gypsum, sulfur and ash wastes, however, can, be used as feedstock in other industrial processes. On average, it is estimated that about 47 per cent of the wastes are used in other processes in the United States (ACAA, 2013), while the percentage in Europe is larger, at 56 per cent (Ecoba, 2006). Hazardous waste is generated in coal-based power plants mainly from the use of activated carbon and other enhanced sorbents for reducing emissions of mercury to air.

The deployment of CO_2 capture technologies increases the amounts of waste produced. Koornneef et al. (2012) report a relative annual 18 to 100 per cent increase in waste, depending on the technology (Koornneef et al., 2012). Although the impact differs by source and technology type, a general trend is observed with post-combustion capture consuming substantially more solvent than IGCC-CCS power plants. In the case of post-combustion capture with MEA, waste is also produced in the amine reclaimer. This waste contains MEA, heat stable salts, solid precipitates, and small amounts of absorption solvents, heavy metal corrosion inhibitors and about 30 per cent by weight of water. Reclaimer waste is considered hazardous and its treatment includes metal removal and incineration. The waste could also be disposed of in a cement kiln where the waste metals become agglomerated in the clinker.

Note that the studies reported in Table 3.11 assume low slip rates. However, Thitakamol et al. (2007) indicate that the quantity of reclaimer waste will vary with the slip stream. In a case study, they report 14.9 kg waste per ton of CO_2 captured in a facility with 2 per cent slip compared to 3.7 kg per ton of CO_2 in a plant with 0.5 per cent slip (Thitakamol et al., 2007). The amounts of reclaimer waste will change depending on the solvent type. (NETL, 2010b) indicates that the use of advanced solvents instead of MEA significantly reduces the amount of solvent lost and waste production. Davidson (2007) found similar results; he reports 2.63 g/kWh of solvent waste for a MEA-based process and 0.26 g/kWh for a process using a novel solvent (MHI KS- 1).

It is important to note that the data provided in Table 3.11 deals with the most typical concepts in CCS, namely post-combustion MEA, pre-combustion and oxyfuel, information on the implications of novel concepts and solvents such as chemical looping, potassium carbonate, chilled ammonia and amino acids is scarce in the literature. Smith et al. (2009), for instance, indicate that using potassium carbonate as a solvent could result in the formation of nitrates, nitrates, sulfates and sulfites that have potential commercial value. Yeh and Bai (1999) indicate that a by-product of the chilled ammonia concept is ammonium sulfate, which could be recovered and be used as a fertilizer. None of these studies, however, provide a quantitative estimate of the amount of by-product generated.

TABLE 3.11

Example of values published in literature regarding waste streams and by-products originating from fossil fuel power plants with and without CO_2 capture

Power plant	Bottom ash, fly ash & slag	Gypsum	Sulfur	Hazardous waste	Reference
PC	58.8				(Rubin et al., 2007)
PC		9.08			(Koornneef et al., 2008)
PC MEA		11.9		2.1	(Koornneef et al., 2008)
PC MEA				4.4	(Korre et al., 2010)
USP MEA	48.9	19.1		2.63	(Davison, 2007)
USP MHI	48.3	18.8		0.26	(Davison, 2007)
Oxyfuel	48				(Davison, 2007)
IGCC without CC	55.8		3.48	0.02	(Davison, 2007)
IGCC with CC	32.4		7	0.005	(Rubin et al., 2007)
NGCC MEA				0.94	(Rubin et al., 2007)
NGCC MEA				1.19	(Davison, 2007)
NGCC MHI				0.20	(Davison, 2007)

Values expressed in g/kWh.

3.6.3 FINDINGS FROM LIFE CYCLE ASSESSMENTS

During the last decades, several environmental studies have assessed the potential impacts of CCS. Early studies (Waku et al., 1995) (Summerfield et al., 1995), based their assessments on mass and energy balances. Later studies have used LCA methodologies to assess the impacts throughout the whole chain. An early review of LCA studies for CCS was first presented in (Hertwich et al., 2008) and subsequently updated in (Singh et al., 2011a), (Zapp et al., 2012), and (Corsten et al., 2013). The reviews highlight differences in terms of the technologies assessed, detail in processes modelled, completeness of the life cycle inventory and emissions included in the assessments. This section presents an overview of the results shown by the LCAs in the literature as well as insights gained in previews overviews. The potential environmental impacts of less mature capture technologies such as membranes, solid sorbents, and chemical looping have been evaluated in much less detail and will not be examined in detail in this chapter. The focus will be placed on post-combustion capture with MEA, pre-combustion capture and oxyfuel combustion. The following environmental impact categories will be assessed: global warming, eutrophication, acidification, toxicity and photochemical oxidation.

Focus of the assessment, functional unit and reference system

The main focus of the studies found in the literature is on power plants using relatively mature CO_2 capture concepts such as post-combustion capture with chemical absorption (MEA), pre-combustion capture and oxyfuel combustion. The functional unit is fundamental for the understanding of LCA results and provides a common basis for comparison of results from different systems or studies. The functional unit in this report is 1 kWh of electricity delivered to the grid. Note that the studies used different technological parameters such as states of commercialization, efficiencies and efficiency penalties to define the technologies examined. In this chapter, both absolute and relative changes in the difference between a power plant with CCS and a power plant without CCS as defined in each study are reported. Note that since the studies have not been fully harmonized due to differences in system boundaries and lack of background data, the relative values provide a more accurate picture of the changes induced by CCS while the absolute values are to be used as an indication of the emissions levels.

TABLE 3.12

Overview of studies examined in this chapter

Author	Fuel	Type of CO_2 capture technology
(Akai et al., 1997)	Coal; natural gas	IGCC-Selexol; LNG-MEA; MFCC-MEA
(Bauer et al., 2008)	Coal; natural gas	PC-MEA; oxyfuel; NGCC-MEA
(Carpentieri et al., 2005)	Biomass	Biomass-MEA
(Doctor et al., 2001)	Coal	IGCC-Selexol
(Sundkvist et al., 2004)	Natural gas	NGCC-MEA; AZEP membranes
(Khoo and Tan, 2006b)	Coal	PC-MEA; PC-cryogenics; PC membranes; PC-PSA
(Khoo and Tan, 2006a)	Coal	PC-Mineralization
(Koornneef et al., 2008)	Coal	PC-MEA
(Korre et al., 2010)	Coal	PC-MEA; PC-other solvent
(Lombardi, 2003)	Coal; natural gas	IGCC-Amines; NGCC-MEA
(Markewitz et al., 2009)	Coal	PC-MEA
(Modahl et al., 2012)	Natural gas	NGCC-MEA
(Odeh and Cockerill, 2008)	Coal; natural gas	PC-MEA; IGCC-Selexol; NGCC-MEA
(Pehnt and Henkel, 2009)	Coal	PC-MEA; IGCC-Selexol; oxyfuel
(Rao and Rubin, 2002)	Coal	PC-MEA
(RECCS, 2008)	Coal; natural gas	PC-MEA; IGCC-Rectisol; oxyfuel; NGCC-MEA
(Schreiber et al., 2009)	Coal	PC-MEA
(Spath and Mann, 2004)	Coal; biomass; natural gas	PC-MEA; NGCC-MEA; Biomass-MEA
(Svanes, 2008)	Coal	PC-MEA; PC-other solvent
(Thitakamol et al., 2007)	Coal; natural gas	PC-MEA; PC-other solvent; NGCC-MEA
(Tzimas et al., 2007)	Coal; natural gas	PC-MEA; IGCC-Selexol; NGCC-MEA
(Viebahn et al., 2007)	Coal	PC-MEA; IGCC-Rectisol; oxyfuel
(Waku et al., 1995)B	Coal	IGCC-Selexol; LNG-MEA
(Weisser, 2007)	Coal	PC-MEA; IGCC-Selexol
(IEA GHG, 2006)	Coal; natural gas	PC-MEA; IGCC-Amines; NGCC-MEA
(Nie, 2009)	Coal	PC-MEA; oxyfuel
(NEEDS, 2009)	Coal; natural gas	PC-MEA; oxyfuel; NGCC-MEA
(Singh et al., 2011a)	Natural gas	NGCC-MEA
(Singh et al., 2011b)	Coal; natural gas	PC-MEA;NGCC-MEA; partial oxidation; oxyfuel

IGCC: integrated gasification combined cycle; NGCC: natural gas combined cycle; PC: pulverized coal; MFCC: molecular fractionation with conjugated caps; MEA: monoethanolamine; AZEP: advanced zero emission plant; PSA: pressure swing adsorption; LNG: liquefied natural gas.

3.6.3.1 Impacts in selected categories

3.6.3.1.1 Global warming potential

The main motive for deploying CCS technologies is the reduction of CO_2 emissions. The impact of CCS on GWP is therefore an intuitive impact category analysed by nearly all studies. Figure 3.19 shows a comparison of the GWP values reported for fossil fuel-fired power plants both with and without CCS. CCS significantly reduces the GWP of fossil fuel-fired power plants. The literature indicates a decrease in GWP over the life cycle of a PC with CCS in the order of 65-84 per cent relative to similar plants without CCS. In absolute terms, GWPs of PC without CCS are reported in the range of 690 to 1,100 g CO_2-eq/kWh. For PC power plants with post-combustion capture using MEA, the range is 79-275 g CO_2-eqeq/kWh. The studies also indicate that in PC plants without CCS, direct emissions from the power plant account for about 80-95 per cent of the total while in PC plants with CCS, this share is much lower at 43-60 per cent.

FIGURE 3.19

Global warming potential of power plants with and without CCS

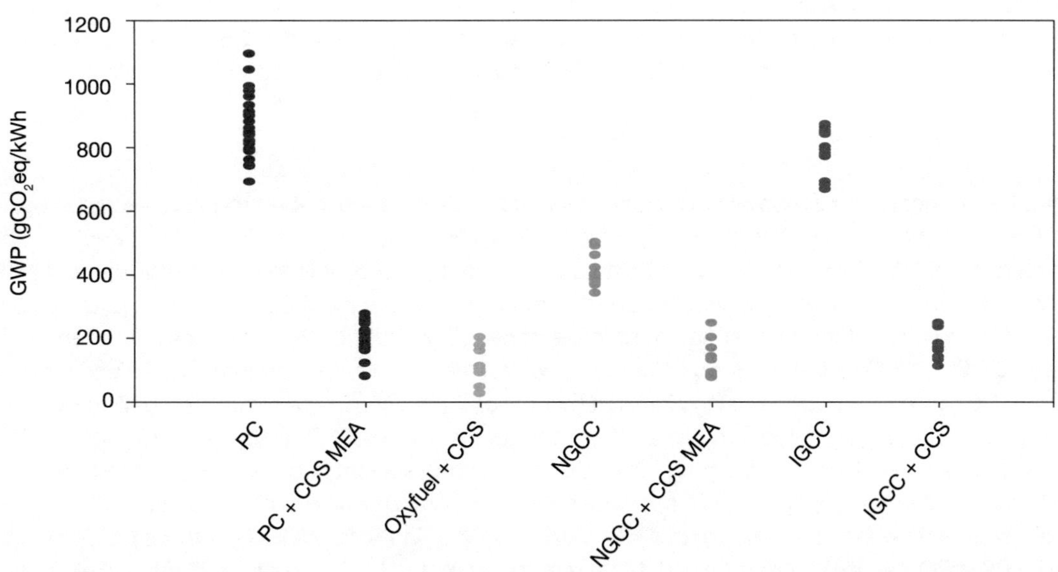

As reported by life cycle assessments in literature

Deployment of CCS results in a decrease in GWP in IGCCs with CCS in the order of 68-87 per cent. The absolute GWP values of IGCCs without CCS range from 666 to 870 g CO_2-eq/kWh, and with CCS from 110 to 245 g CO_2-eq/kWh. The GWP of IGCCs using an amine solvent to capture CO_2 instead of Rectisol or Selexol is reported by only one study (IEA GHG, 2006). The calculated value of 235 g CO_2-eqeq/kWh falls within the range of the GWP reported for PCs using MEA. For oxyfuel power plants with CCS, GWPs are reported in the range of 25-176 g CO_2-eqeq/kWh. Since there are no oxyfuel plants considered without CCS, most studies use a PC plant as a reference. In such a case, implementation of oxyfuel with CCS results in a 78-97 per cent relative decrease in GWP.

Finally, a relative decrease in the GWP of NGCCs with post-combustion capture using MEA is reported in the range of 51-80 per cent. In absolute terms, this corresponds to 340-499 g CO_2-eqeq/kWh for NGCCs without CCS and 75-245 g CO_2-eqeq/kWh for NGCCs with CCS. Remarkably, there is a relatively lower decrease reported for NGCC plants than for PC power plants, which is due to assumptions made on methane leakage from upstream transport (Corsten et al., 2013). The management of upstream methane emissions therefore plays a key role in the performance of NGCC chains.

3.6.3.1.2 Eutrophication potential

Eutrophication is generally associated with the environmental impacts of excessively high nutrient levels that lead to shifts in species composition and increased biological productivity. Eutrophication potential (EP) values reported in the literature are depicted in Figure 3.20. The range for PC power plants without CCS is 0.04-0.29 g PO_4^{3-}-eq/kWh, while the EP of PC power plants with post-combustion capture using MEA varies from 0.06 to 0.30 g PO_4^{3-}-eq/kWh. The results indicate that post-combustion capture in coal power plants leads to an increase in their EP. The level of the change differs however by literature source, with some sources reporting an increase of about 19 per cent while others report significantly larger values, with the largest of these values corresponding to a relative increase of 170 per cent compared to PCs without CCS. In terms of direct vs. indirect contributions, the largest share, ranging from 55-92 per cent, in PC plants *without* CCS is allocated to the operation of the power plant itself; most of the total EP stems from direct impacts. Although only a handful of studies provide values for the individual chain steps for PC plants with CCS, these studies indicate the increase in EP to be mainly caused by emissions from MEA production and degradation. Higher NO_x emissions from coal production and ship transport as a result of the energy penalty induced by CO_2 capture also plays a role, albeit a minor one, on the increased EP.

The reported EPs of coal-based *oxyfuel power plants with CO_2 capture* are in the range 0.01 to 0.094 g PO_4^{3-}-eq/kWh. If compared with PC without CCS, the EP is significantly lower, at 43 to 78 per cent. This may be due to the reduced NO_x formation in oxyfuel plants in comparison to coal-fired power plants; most NO_x originates from the oxidation of atmospheric nitrogen, a reaction that is minimized in oxyfuel plants since a pure oxygen stream is used for combustion. In the case of *IGCCs without CCS,* EP values in the range of 0.025 to 0.21 g PO_4^{3-}-eq/kWh are reported in literature, which are in the same range as the values reported for PCs. For *IGCCs with CO_2 capture using Selexol or Rectisol,* EP values in the range of 0.035 to 0.18 g PO_4^{3-}-eq/kWh are reported, corresponding to a 30-40 per cent increase of EP relative to similar power plants without CCS. CO_2 can also be captured in an IGCC using an amine solvent such as in PCs. As shown in Figure 3.20 , the EP is in the upper range of the values reported for PCs using amines to capture CO_2. The 40 per cent increase reported is relative to an IGCC without CCS. Finally, for *NGCCs without CCS,* EP values of 0.01 and 0.09 g PO_4^{3-}-eq/kWh are reported. Note that the lowest value in the range is most likely due to the exclusion of CO_2 transport and storage from the system boundaries chosen by the study (IEA GHG, 2006). For *NGCCs with post-combustion capture using MEA,* the two reported values are 0.02 and 0.11 g PO_4^{3-}-eq /kWh, corresponding to a relative increase in EP of 21 per cent and 35 per cent,, respectively. The studies do not discuss the cause of the increase, but as in the case of PCs with MEA, it is most likely due to the NH_3 emissions from MEA production and degradation.

3.6.3.1.3 Acidification potential

Acidification is caused by the emission of acid-forming substances, which change pH conditions in ecosystems and contribute to, for example, fish mortality and damage to forests and buildings. Figure 3.21 shows a compilation of the acidification potential (AP) reported in the literature for power plants with and without CCS. The AP range reported for *PC plants without CCS* varies from 0.39 to 2.76 g SO_2-eqeq/kWh. For *PC plants with post-combustion capture using MEA,* the AP is in the range of 0.34-2.10 g SO_2-eq/kWh. The relative change in AP between plants with and without CCS ranges from -23 per cent to +91 per cent. As explained in Chapter 3.6.2.1, PCs with CO_2 capture require a high removal of SO_2 and NO_2 to avoid degradation of the solvent. However, due to the energy penalty, more coal is required to generate the same amount of electricity, resulting in increasing SO_2 and NO_x emissions during the production and transport of coal. Although NO_2 is removed prior to the capture

FIGURE 3.20

Eutrophication potential of power plants with and without CCS

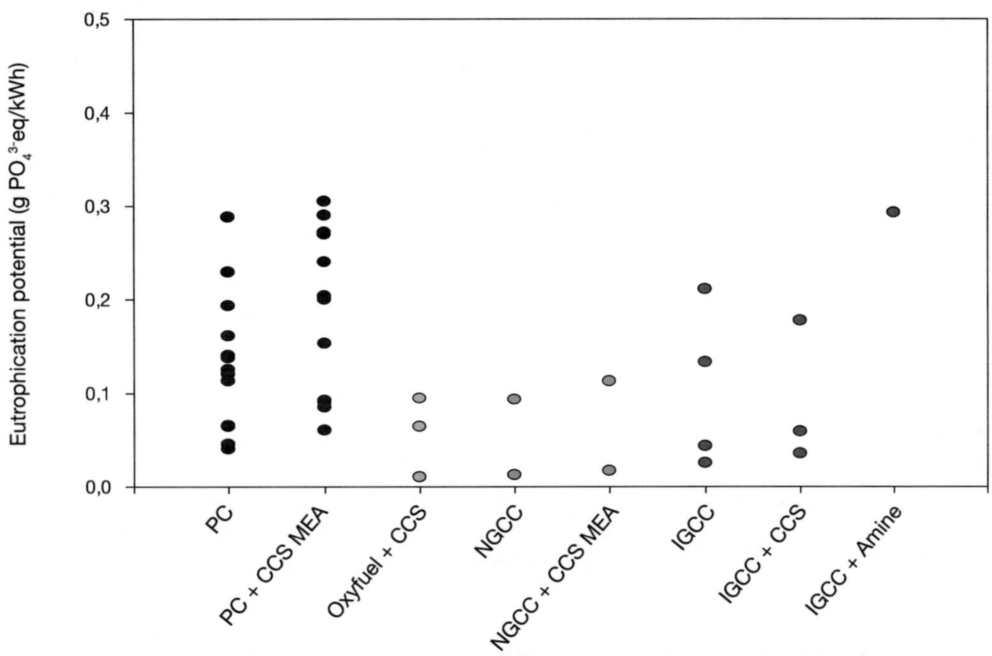

FIGURE 3.21

Acidification potential of power plants with and without CCS

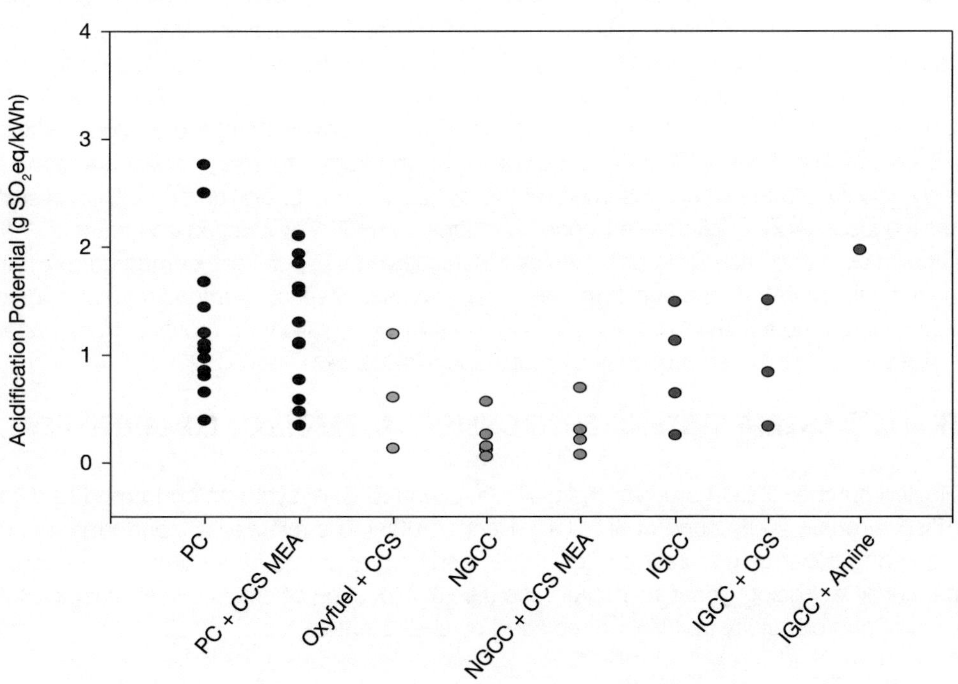

process, its removal has a limited effect since NO_2 only accounts for about 5-10 per cent of the total NO_x formed. An additional contributor to the AP of PCs with CCS using MEA is the emission of NH_3 due to MEA production and degradation. The share contribution of these emissions to the AP of power plants with CCS is reported in the order of 30-40 per cent. Some PC studies however, do not include these emissions in their system boundaries, resulting in lower AP values and in some cases, relative decreases in AP compared to power plants without CCS.

AP values of oxyfuel plants with CO_2 capture are reported in the range 0.13-1.19 g SO_2-eqeq/kWh. In oxyfuel power plants, SO_2 emissions per kWh are expected to decrease as the reduced flue gas volume leads to higher SO_2 concentrations, which are likely to increase the removal efficiency of SO_2 in FGDs. Also, the NO_x formation is expected to be lower as NO_x formation is suppressed when combustion occurs in an atmosphere with reduced nitrogen quantities. The energy penalty, however, will result in increased emissions of SO_2 and NO_x from coal production and transport. Of the three studies that are examined this technology, two report a relative decrease of 38 per cent and 80 per cent and one an increase of 40 per cent in AP values.

For IGCCs without CO_2 capture, AP values are reported in the range 0.25-1.5 g SO_2-eqeq/kWh and for IGCCs with pre-combustion CO_2 capture using Selexol or Rectisol, in the range 0.33-1.5 g SO_2-eqeq/kWh. For IGCC with MEA, reported emission are 2 g SO_2-eq/kWh. In all cases, this corresponds to a 32 per cent relative increase in AP. Finally, compared to PCs, the range of *NGCCs without CCS* is lower, at 0.06-0.56 g SO_2-eqeq/kWh. The sulfur content of natural gas is very low and thus SO_2 emissions are lower for natural gas-fired power plants without CCS. The implementation of post-combustion CO_2 capture using MEA results in a relative increase of 23-26 per cent.

3.6.3.1.4 Toxicity

In LCA, four different kind of toxicity categories are reported: human toxicity potential (HTP), which refers to the impact of toxic substances on human health in the air, water and soil; freshwater aquatic ecotoxicity potential (FAETP) refers to the impact of toxic substances on aquatic ecosystems; terrestrial ecotoxicity potential (TETP), which is the impact of toxic substances on terrestrial ecosystems and marine aquatic ecotoxicity potential (MAETP), which refers to the impact of toxic substances on marine ecosystems.

Figure 3.22 to Figure 3.25 show the values reported in the literature. Note that for most of the technologies, there are too few studies examining toxicity to allow the drawing of robust conclusions. We thus limit ourselves in this report to discussing PCs with and without CCS. In the case of HTP, the values reported in the literature go from a 30 per cent decrease to 260 per cent increase compared to similar PC plants without CCS. MEA production is reported as the main contributor to human toxicity, primarily because of the ethylene oxide emissions from MEA production. For FAETP, relative increases ranging from 8 to 256 per cent are reported. In this case, the energy penalty and the steel consumed for the production and operation of the CCS system are indicated as the main causes. Additionally, the ethylene emissions from MEA production contribute to the FAETP. The TETP values shown both decreases (-34 per cent) and increases (42-57 per cent) compared to plants without CCS. As in the case of FAETP, the energy penalty and increased steel consumption are considered the main drivers. The study reporting a relative decrease in FAETP assumes significant removal of trace metals in the CO_2 capture unit. The validity of this assumption is currently under discussion (see 3.6.2.1).

3.6.4 DECENTRALIZED ENERGY SYSTEMS: ENVIRONMENTAL EFFECTS OF COMBINED HEAT AND POWER

There is a rather limited number of LCA studies on fossil fuel-powered CHP systems performed to date (Bauer and Heck, 2009; Fischer et al., 2008; Lund et al., 2010; Pehnt, 2008). The relative CO_2 performance of CHP technologies compared to separate generation of electricity and heat depends on a number of factors such as choice of prime mover technology, heat-to-power ratio (HPR), allocation of emissions to electricity and heat, and the source of grid electricity that will be replaced by CHP plants.

CHP saves direct fuel consumption compared to separate production of heat and power. However, the degree to which GHG emissions are reduced is largely case-specific. A "cradle-to-grave" LCA analysis on

FIGURE 3.22

Human toxicity potential of fossil fuel power plants with and without CCS

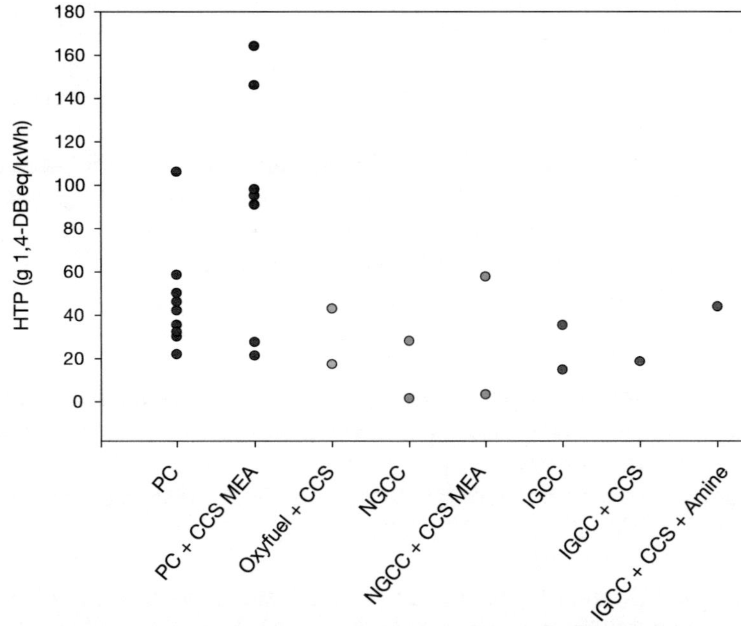

FIGURE 3.23

Freshwater aquatic ecotoxicity potential of fossil fuel power plants with and without CCS

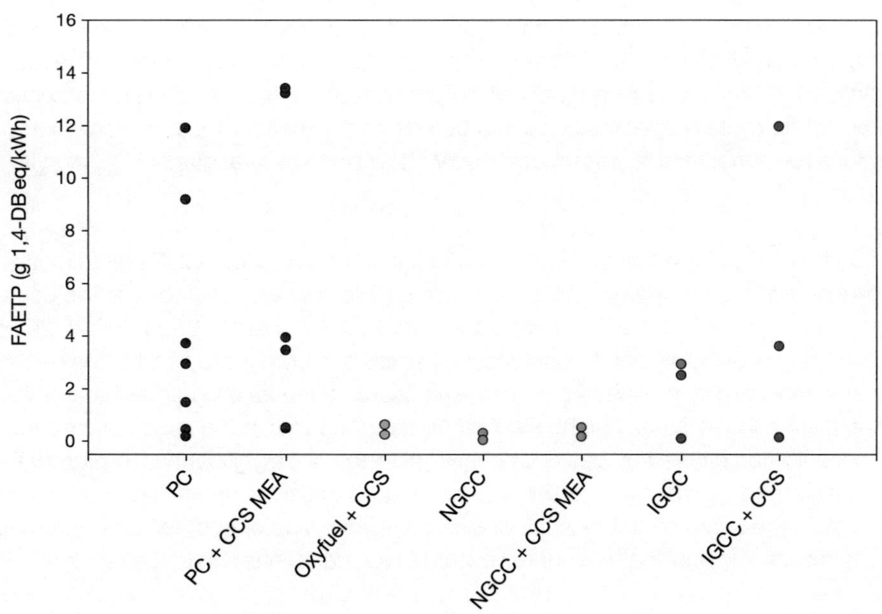

various residential and district heating CHP technologies under German conditions has been performed by Pehnt (2008). Results of this study indicate that both micro-CHP and district heating CHP are superior not only to a reference case using a German average electricity CO_2 emission factor, but also to a reference case with electricity from a state-of-the-art NGCC[23] facility. In another study, Fischer et al. (2008) investigate the performance of fuel cell CHPs. Their results indicate that fuel cell CHPs are not advantageous regarding GHG emissions compared to conventional prime movers. This is partly due to the assumed high heat-to-power ratio (HPR) of 1.7:1, which does not maximize the high electrical conversion potential of fuel cells. Strachan and Farrell (2006) stated in their study of an American case that at high HPR of about 2:1, the combustion-based CHP technologies have an emissions profile comparable to that of fuel cells because low electricity demand relative to the heat demand reduces the importance of the inherent efficiency of the prime mover (Strachan and Farrell 2006). Note, however, that in real-life applications, the GWP reduction potential of CHP may be smaller than the values published in the literature because an economically optimized configuration does not necessarily maximize its technical, and consequently environmental, performance.

3.6.4.1 Air pollution

It has often been argued that CHP facilities contribute to environmental relief due to their decentralized nature and high overall efficiency (Pehnt, 2008). However, the introduction of CHP technologies does not necessarily lead to improved air quality. The emission performance of CHP plants on air pollutant emissions compared to separate generation of electricity and heat depends, in addition to the factors already discussed, on the emission control performance and the location of CHP plants.

Regarding acidifying emissions such as NO_x, SO_2 and NH_3, the performance of natural gas-based CHP plants depends on the prime mover technology used and the application of emission control measures. Figure 3.26 shows a comparison of these emissions between large CHP and conventional electricity production plants without CHP in a German case study. The figure shows that fuel cells and Stirling engines with the innovative burner reduce acidification impact, while gas engines may increase acidifying emissions due to less efficient emission control systems compared to that for large-scale centralized power plants. In the same paper, Pehnt (2008) also shows that the use of oil-fired reciprocating engine CHP plants lead to a seven-fold increase in acidifying emissions compared to gas-fired reciprocating CHP plants.

The fact that CHP plants can emit more acidifying pollutants than the separate generation of electricity from NGCC without CHP and heat from a gas-fired condensing boiler is also suggested by (Allison and Lents, 2002). Furthermore, the authors report a breakdown of acidification potential by contribution substance. The results show that NO_x is the largest contributor, accounting for about 60 per cent of the total acidification potential, followed by SO_2. The majority of life cycle NO_x emissions is attributable to the fuel combustion from CHP plant operation, while SO_2 emissions are almost entirely from fuel supply chain since natural gas combustion itself emits little SO_2.

Some authors have indicated that the prevalence of district, residential and commercial CHP plants may aggravate the urban air pollution and health problem because these CHP plants are located near the consumers whereas large scale power plants are often further from populated areas (Canova et al., 2008; Pehnt, 2008). The literature seems to agree that gas engine CHP plants increase NO_x emissions both locally and globally, but the local environmental impacts depend largely on specific local characteristics such as orography, meteorological conditions and design of the plant such as stack height. Pehnt (2008) reports that in the case of Germany, the increase in local NO_x concentration caused by gas engine CHP plants is not significant according to the local legislation. The author performed a dispersion calculation for a hypothetical residential area with relatively critical weather conditions (e.g., large share of stable weather situations, low wind speed) flat topography and urban housing structure. The results showed that the annual average NO_2 concentration in the residential area

23 These results are based on a functional unit of 1 kWh electricity and on the avoided burden approach, in which the cogenerated heat is credited with an alternative generation route. This approach allocates all the benefits of cogeneration to electricity generation

FIGURE 3.24

Terrestrial aquatic ecotoxicity potential of fossil fuel power plants with and without CCS

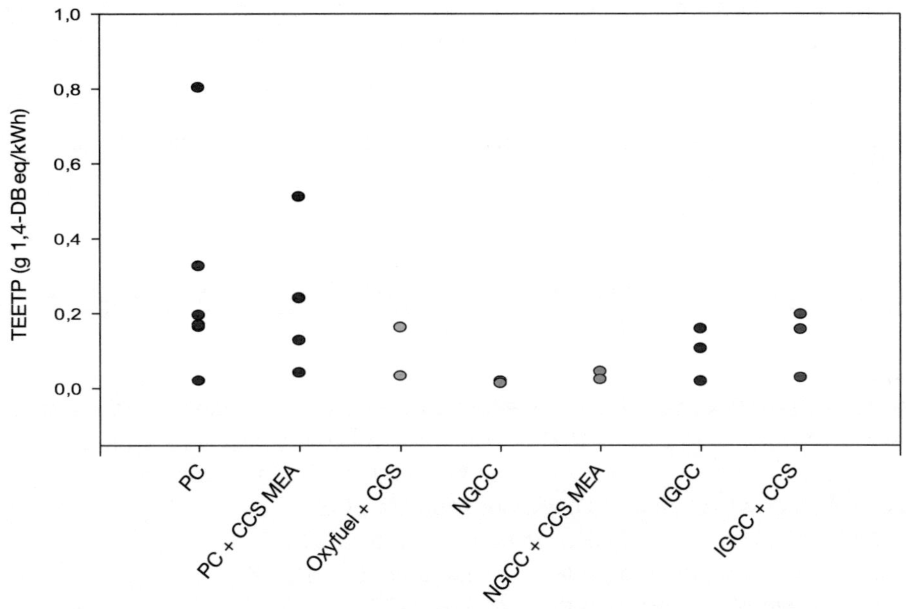

FIGURE 3.25

Marine ecotoxicity potential of fossil fuel power plants with and without CCS

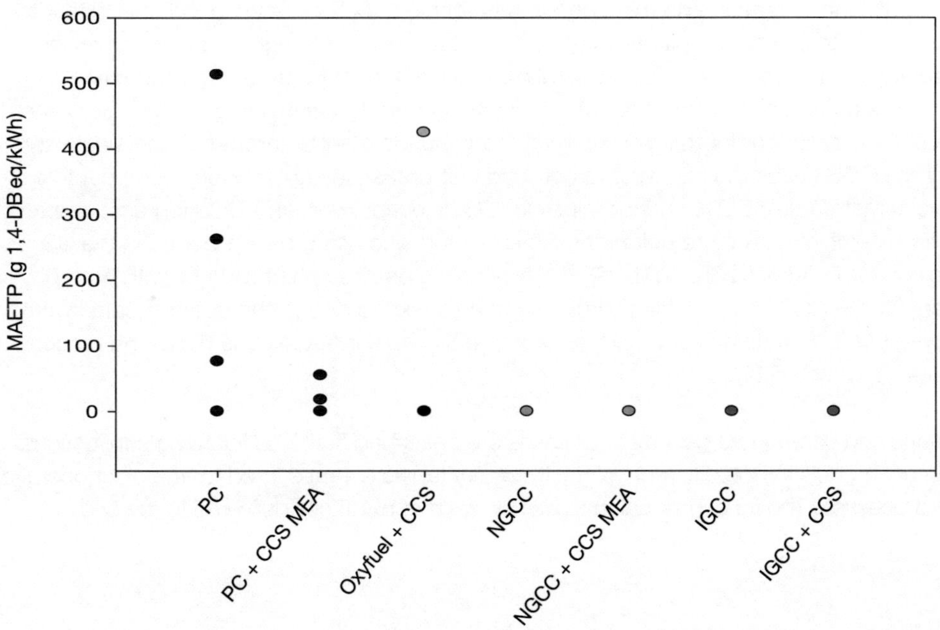

increases by 0.6 µg/m³ while the national limit is 40 µg/m³. The results also show that the maximum short-term NO_2 concentration (by Pehnt defined as concentration that should not be exceeded in more than 18 hr) is found to be generally below 7 µg/m³, where the national short-term limit is 200 µg/m³. The author concludes that such an increase in NO_x concentration does not create serious additional environmental impacts. A study by Canova et al. (2008) draws similar conclusions. The authors investigated the changes in both global and local emissions of NO_x for microturbine and gas engine CHP plants in the Italian context. The results (Figure 3.27) show that there is a significant difference between the changes in global and local emissions (scenario 1 versus scenario 2, and scenario 3 versus scenario 4), but the overall conclusion remains the same: microturbine CHP plants reduce emissions and gas engine CHP plants increase emissions.

3.7 LIFE CYCLE INVENTORIES USED IN THE INTEGRATED ASSESSMENT

In this report, hybrid life cycle assessment is employed to evaluate the environmental impacts of low carbon electricity producing technologies (see Chapter 3.6). This section describes the life cycle inventories for four different types of fossil fuel plants which are used for the integrated modelling. These power plants include post-combustion for coal and natural gas fired power plants and coal pre-combustion capture:

- Sub-critical pulverized coal fired power plant (EXPC) without and with CCS
- Supercritical pulverized coal fired power plant (SCPC) without and with CCS
- Integrated gasification combined cycle power plant (IGCC) without and with CCS
- Natural gas combined cycle power plant (NGCC) without and with CCS

The inventories have been set up according to a general structure, and adopting a cradle-to-gate perspective. The term cradle-to-gate in this chapter implies that potential environmental impacts are assessed only for the life cycle up and until the delivery of the electricity to the grid (which is the main product of the power plant). Electricity transport and distribution to the end-users as well as the necessary infrastructure needed for distribution is outside the system boundaries of the analysis (see Figure 3.28). Note however that the unit processes considered in the foreground and background describe the full life cycle of the unit processes, e.g., coal mine construction, coal mine operation and mine decommissioning are all included in the coal extraction unit process. For the cases describing power plants without CCS, the following foreground unit processes are included: fossil fuel extraction, transport, power plant operation and plant infrastructure. Fuel extraction unit processes include both infrastructure and operation, but the plant infrastructure (including construction) and plant operation are intentionally split in order to investigate the contribution of infrastructure to the electricity production process. For the CCS cases the following foreground unit processes are included: CO_2 capture and compression on-site infrastructure, CO_2 transport pipeline, CO_2 injection well, and CCS (on-site) operation. Unless otherwise specified, plant infrastructure unit processes are modelled using data reported in the US National Energy Technology Laboratory (NETL, 2010b), which contains detailed plant designs with both physical and cost data, necessary for the HLCA input. This source has been selected due to the detailed data inventory presented; transparency in the assumptions used and the fact that the data input used is within the ranges found in the literature (see Chapter 3.6.3).

In this section, assumptions and parameters general to all cases are described first. The following subsections briefly describe the inventory models for the different plants. Inventory tables are presented for the plant operation and CCS operation unit processes. The functional unit used is one kWh of electricity delivered to the grid.

FIGURE 3.26

Life cycle assessment of acidifying emissions (NO_x, SO_2, NH_3) of micro cogeneration technologies

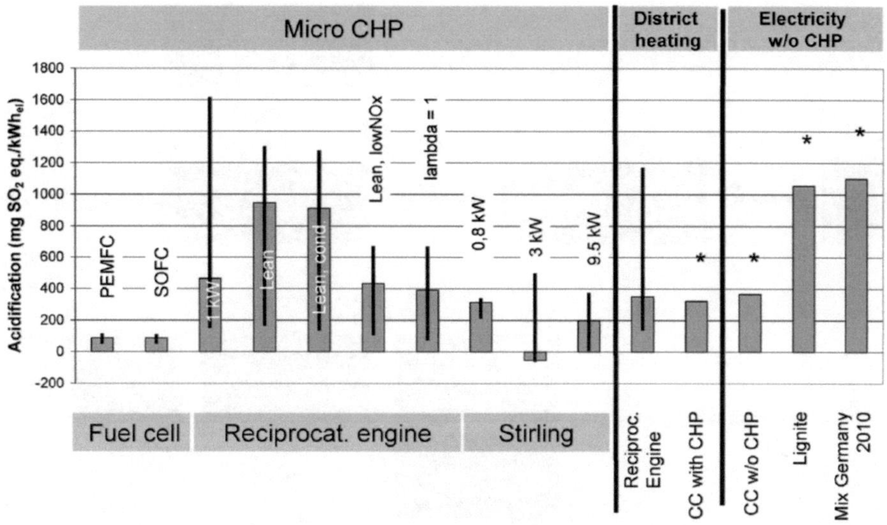

Source: Pehnt, 2008. LCA as compared to compared to large CHP and conventional electricity production without CHP in a German context. *No data for bandwidth available. Functional unit 1 kWh electricity at low voltage level, co-produced heat is credited ("avoided burden").

FIGURE 3.27

Case study scenarios: NO_x emission balances per kWh electricity produced.

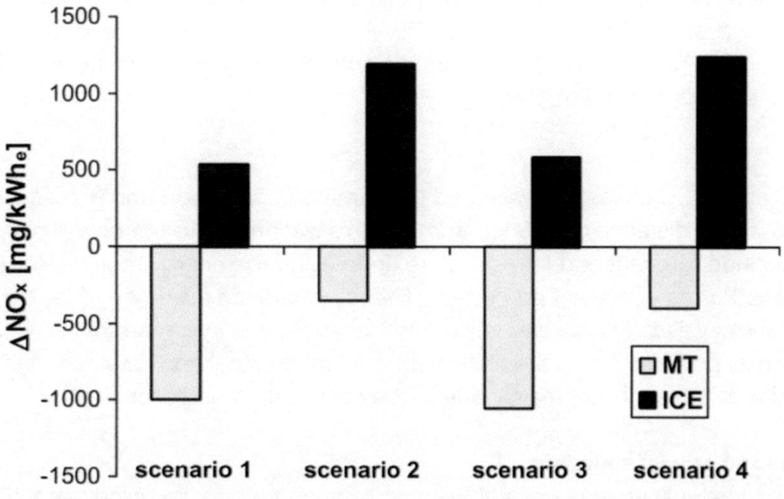

MT: microturbine; ICE: internal combustion engine; Scenario 1: global emission balance, full-load emission factors; Scenario 2: local emission balance, full-load emission factors; Scenario 3: global emission balance, emission factors at partial load; Scenario 4: local emission balance, emission factors at partial load.

Source: Canova et al., 2008

FIGURE 3.28

System boundaries with the foreground LCA unit processes for a power plant with and without CCS

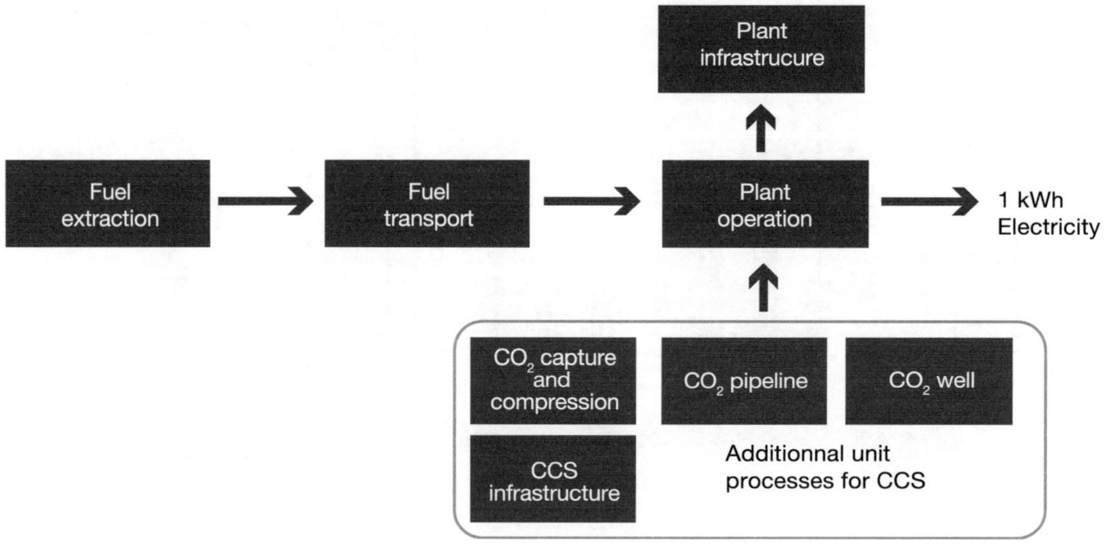

3.7.1 GENERAL APPROACH

Plant capacity, lifetime and capacity factor

Table 3.13 shows the main characteristics of the power plants. A lifetime of 30 years was assumed for all cases. The power plant location is North America and North American hard coal (Illinois n6) is used for operation. Please note that the energy density and carbon content of fossil fuels can vary regionally. We have chosen to model the North American supply chain in the foreground in order to increase comparability between interregional results. The influence of varying energy and carbon content associated with the fossil fuel supply chain is discussed by Bouman et al. (2015). All values, including efficiencies reported in this section, are based on HHV. Plant infrastructure, land use and financing costs are included in the inventory and grouped over different sectors in the EXIOBASE.

Coal transport

For the EXPC, SCPC and IGCC cases it is assumed the same coal transport unit process. Coal is assumed to be transported by rail over a distance of 330 km from the excavation site to the power plant (NETL, 2010a). Coal transport data include the transport infrastructure (trains) and the energy required for transport (NETL, 2010a) and are modelled using ecoinvent processes. The rail tracks are assumed to be constructed and are not included in the inventory. During coal extraction and transport, it is assumed that no coal is lost. The main components of the diesel powered trains are aluminium, chromium steel and steel, and the environmental emissions associated with transport are mainly due to the combustion of diesel.

Allocation of water use and process emissions

Total water demand can be split up in raw water withdrawal and internally recycled water. For simplicity, it is assumed here that all consumed water is eventually evaporated for cooling duties. In this report, total water withdrawal is modelled distinguishing between process water discharge and water consumption. For simplicity, it is assumed that water is obtained from an unspecified natural origin, even though NETL specifies the water sources for the power plant.

TABLE 3.13

General power plant characteristics

	EXPC	SCPC	IGCC	NGCC
Net power output without CCS (MW)	550	550	629	555
Net power output with CCS (MW)	550 [a]	550 [a]	497	474
Capacity factor	0.85	0.85	0.80	0.85
Net plant efficiency (with CCS)	36.8% (26.2%)	39.3% (28.4%)	42.1 % (31.2%)	50.2% (42.8%)
CO_2 capture efficiency	90%	90%	90%	90%
Flue gas desulfurization (FGD) efficiency	98%	98%[b]	Sulfur captured in Selexol process	Low sulfur fuel
Selective catalytic reduction (SCR) efficiency	86%	86%	N/A	90%
Particulate matter (PM) removal efficiency	99.8%	99.8%	Particulate removal by cyclone and barrier filter	N/A
Hg reduction efficiency	90%	90%	95%	N/A

Source: NETL, 2010b
a: the nominal net output for the EXPC and SCPC cases was maintained at 550 MW for the cases with CCS. This is done by increasing the boiler and turbine/generator sizes to account for a larger auxiliary load due to the carbon dioxide capture process. For the IGCC and NGCC cases, the plant size was kept constant, leading to a lower net power output. b: the efficiency of the FGD is the same in both cases, however in the CCS case the flow from the FGD unit passes through an extra unit in order to reduce degradation of the solvent in the capture unit.

As noted previously in this chapter, the addition of CCS increases the water and fuel use of the power plant significantly, and subsequently the related emissions. However, water use and emissions data is only available for total operation, i.e., for the plant operation and CO_2 operation unit processes (shown in Figure 3.28 together). The breakdown between plant operation and CO_2 operation was not reported because additional use or emissions due to CCS do not necessarily occur in the CO_2 capture section but also in the power island. In order to allocate water use and emissions between plant operation and CO_2 capture the following approach was carried out: as data is available for both the power plants without and with CCS system, the ratio between the plant efficiency of the power plant without and with CCS is used to calculate the plant emissions of the plant operation unit process for the plant with CCS, based on the emissions of the plant without CCS. The difference in the totals will be the emissions associated with the CO_2 operation process. Water use allocation is done in a similar way. This disaggregation could lead to slight misrepresentation of separate contributions of the foreground processes plant operation and CCS operation, but does not affect total emissions.

TABLE 3.14

Inventory for the subcritical pulverized coal (EXPC) plant operation unit process

EXPC Plant operation		without CCS	with CCS	Unit	Reference
INPUTS	Activated carbon		$6.50 \cdot 10^{-5}$	kg/kWh	(NETL, 2010b)
	Ammonia	$5.36 \cdot 10^{-3}$	$7.56 \cdot 10^{-3}$	kg/kWh	(NETL, 2010b)
	Chemicals[a]	$4.10 \cdot 10^{-4}$	$7.10 \cdot 10^{-4}$	US$/kWh	(NETL, 2010b)
	Coal transport	$3.61 \cdot 10^{-1}$	$5.07 \cdot 10^{-1}$	kg/kWh	(NETL, 2010b)
	Discharge process water	$5.02 \cdot 10^{-1}$	$1.08 \cdot 10^{0}$	kg/kWh	(NETL, 2010b)
	Limestone	$3.58 \cdot 10^{-2}$	$5.16 \cdot 10^{-2}$	kg/kWh	(NETL, 2010b)
	Monoethanolamine		$2.33 \cdot 10^{-3}$	kg/kWh	(Veltman et al., 2010)
	Sodium hydroxide		$5.42 \cdot 10^{-4}$	kg/kWh	(NETL, 2010b)
	Sulfuric acid		$5.18 \cdot 10^{-4}$	kg/kWh	(NETL, 2010b)
	Water[b]	$1.93 \cdot 10^{-3}$	$3.56 \cdot 10^{-3}$	m^3/kWh	(NETL, 2010b)
OUTPUTS	Ammonia	$2.00 \cdot 10^{-7}$	$2.90 \cdot 10^{-7}$	kg/kWh	NETL, 2010e
	Carbon dioxide	$8.56 \cdot 10^{-1}$	$1.16 \cdot 10^{-1}$	kg/kWh	(NETL, 2010b)
	Ash disposal	$3.50 \cdot 10^{-2}$	$4.91 \cdot 10^{-2}$	kg/kWh	(NETL, 2010b)
	Disposal of hazardous waste		$3.47 \cdot 10^{-3}$	kg/kWh	(Singh, 2011b)
	Carbon monoxide	$1.00 \cdot 10^{-4}$	$1.40 \cdot 10^{-4}$	kg/kWh	(NETL, 2010c)
	Lead	$5.90 \cdot 10^{-9}$	$8.40 \cdot 10^{-9}$	kg/kWh	(NETL, 2010c)
	Mercury	$4.54 \cdot 10^{-9}$	$5.53 \cdot 10^{-9}$	kg/kWh	(NETL, 2010c)
	Methane	$1.10 \cdot 10^{-5}$	$1.50 \cdot 10^{-5}$	kg/kWh	(NETL, 2010c)
	Monoethanolamine		$6.59 \cdot 10^{-5}$	kg/kWh	(Veltman, 2010)
	Nitrogen oxide	$2.78 \cdot 10^{-4}$	$3.39 \cdot 10^{-4}$	kg/kWh	(NETL, 2010e)
	Nitrous oxide	$1.60 \cdot 10^{-5}$	$2.30 \cdot 10^{-5}$	kg/kWh	(NETL, 2010e)
	Particulates	$5.20 \cdot 10^{-5}$	$6.30 \cdot 10^{-5}$	kg/kWh	(NETL, 2010e)
	Sulfur dioxide	$3.41 \cdot 10^{-4}$	$7.60 \cdot 10^{-6}$	kg/kWh	(NETL, 2010e)
	Sulfur hexafluoride	$2.60 \cdot 10^{-10}$	$2.60 \cdot 10^{-10}$	kg/kWh	(NETL, 2010e)
	VOC	$1.20 \cdot 10^{-5}$	$1.70 \cdot 10^{-5}$	kg/kWh	(NETL, 2010e)
	Waste heat	$6.18 \cdot 10^{0}$	$1.02 \cdot 10^{1}$	MJ/kWh	-
	Water-to-air	$1.93 \cdot 10^{0}$	$3.56 \cdot 10^{0}$	kg/kWh	(NETL, 2010b)

(Bouman et al., 2015)

a: The chemicals listed here consist of a non-specified mix of makeup and waste/water treatment chemicals and catalyst as accounted for in the operating costs of the plants (NETL, 2010b). The process 'manufacture of chemicals and chemical products' from the EXIOPOL background is used as a proxy.
b: This is water used directly from natural resources, and corresponds with the water consumption. An equivalent amount of water is emitted to air in the outputs (please note the difference in unit). Process water discharge (to river) is separately modelled using an ecoinvent process. Together, these two processes form the total raw water withdrawal.

TABLE 3.15

Inventory for the supercritical pulverized coal (SCPC) plant operation unit process

SCPC Plant operation		Without CCS	With CCS	Unit	Reference
INPUTS	Activated carbon		$5.98 \cdot 10^{-5}$	kg/kWh	(NETL, 2010b)
	Ammonia	$5.03 \cdot 10^{-3}$	$7.00 \cdot 10^{-3}$	kg/kWh	(NETL, 2010b)
	Chemicals	$3.90 \cdot 10^{-4}$	$6.50 \cdot 10^{-4}$	US$/kWh	(NETL, 2010b)
	Coal transport	$3.38 \cdot 10^{-1}$	$4.68 \cdot 10^{-1}$	kg/kWh	(NETL, 2010b)
	Discharge process water	$4.48 \cdot 10^{-1}$	$9.70 \cdot 10^{-1}$	kg/kWh	(NETL, 2010b)
	Limestone	$3.35 \cdot 10^{-2}$	$4.72 \cdot 10^{-2}$	kg/kWh	(NET,L 2010b)
	Monoethanolamine		$2.15 \cdot 10^{-3}$	kg/kWh	(Veltman, 2010)
	Sodium hydroxide		$9.98 \cdot 10^{-4}$	kg/kWh	(NETL, 2010b)
	Sulfuric acid		$4.76 \cdot 10^{-4}$	kg/kWh	(NETL, 2010b)
	Water[a]	$1.75 \cdot 10^{-3}$	$3.20 \cdot 10^{-3}$	m^3/kWh	(NETL, 2010b)
OUTPUTS	Ammonia	$2.56 \cdot 10^{-6}$	$1.95 \cdot 10^{-4}$	kg/kWh	NETL, 2010a; Koornneef, 2008
	Ash disposal	$3.27 \cdot 10^{-2}$	$4.53 \cdot 10^{-2}$	kg/kWh	(NETL, 2010b)
	Disposal of hazardous waste		$3.20 \cdot 10^{-3}$	kg/kWh	(Singh, 2011b)
	Carbon dioxide	$8.02 \cdot 10^{-1}$	$1.11 \cdot 10^{-1}$	kg/kWh	(NETL, 2010e)
	Carbon monoxide	$3.18 \cdot 10^{-7}$	$4.06 \cdot 10^{-7}$	kg/kWh	(NETL, 2010a)
	Lead	$4.79 \cdot 10^{-8}$	$4.79 \cdot 10^{-8}$	kg/kWh	(NETL, 2010a)
	Mercury	$4.27 \cdot 10^{-9}$	$5.16 \cdot 10^{-9}$	kg/kWh	(NETL, 2010b)
	Methane	$8.72 \cdot 10^{-9}$	$7.59 \cdot 10^{-7}$	kg/kWh	(NETL, 2010a)
	Monoethanolamine		$6.08 \cdot 10^{-5}$	kg/kWh	(Veltman, 2010)
	Nitrogen oxide	$2.61 \cdot 10^{-4}$	$3.16 \cdot 10^{-4}$	kg/kWh	(NETL, 2010b)
	Nitrous oxide	$2.43 \cdot 10^{-9}$	$3.66 \cdot 10^{-9}$	kg/kWh	(NETL, 2010a)
	Particulates	$4.90 \cdot 10^{-5}$	$5.90 \cdot 10^{-5}$	kg/kWh	(NETL, 2010b)
	Sulfur dioxide	$3.20 \cdot 10^{-4}$	$7.00 \cdot 10^{-6}$	kg/kWh	(NETL, 2010b)
	Sulfur hexafluoride	$3.53 \cdot 10^{-10}$	$3.53 \cdot 10^{-10}$	kg/kWh	(NETL, 2010a)
	VOC	$2.08 \cdot 10^{-8}$	$2.13 \cdot 10^{-8}$	kg/kWh	(NETL, 2010a)
	Waste heat	$5.56 \cdot 10^{0}$	$9.08 \cdot 10^{0}$	MJ/kWh	-
	Water-to-air	$1.75 \cdot 10^{0}$	$3.20 \cdot 10^{0}$	kg/kWh	(NETL, 2010b)

(Bouman et al., 2015)

a: This is water used directly from natural resources, and corresponds with the water consumption. An equivalent amount of water is emitted to air in the outputs (please note the difference in unit). Process water discharge (to river) is separately modelled using an ecoinvent process. Together, these two processes form the total raw water withdrawal.

3.7.2 SUB AND SUPERCRITICAL PULVERIZED COAL POWER PLANTS

The coal fired power plant without and with CCS is modelled assuming that the net amount of electricity produced is the same, i.e., a net production capacity of 550 MW (see Table 3.13 for key parameters). The respective net plant efficiencies are 36.8 per cent and 39.3 per cent for the EXPC and SCPC without CCS and 26.2 per cent and 28.4 per cent for the EXPC and SCPC with CCS.

Inputs to the plant operation unit processes include fuel, limestone for the flue gas desulfurization, water for cooling duties, ammonia for the selective catalytic reduction of NO_x emissions, MEA, caustic soda and activated carbon for the capture process. Ash disposal and discharge of process water is modelled using ecoinvent processes. Detailed emissions data for the operation of the subcritical coal fired power plant is not available in the NETL baseline report. Therefore, as a proxy the LCI data of a similar, but of somewhat lower production capacity, sub-critical power plant, is used (NETL, 2010e).The key emissions from the power plant without CCS are carbon dioxide, waste heat and water vapour, PM, SO_2, and NO_x. Note that although gypsum is an economic by-product of the flue gas desulfurization in the coal-fired power plant, a conservative approach is taken and no allocation towards gypsum is done. Consequentially, it is not represented in the LCI.

The power plant infrastructure is modelled using the detailed cost data of the NETL baseline report. Plant infrastructure, land use and financing costs are included and distributed over different sectors in the EXIOBASE input-output database. The CCS on-site infrastructure, such as the capture unit and CO_2 compressor, but excluding the CO_2 pipeline and well, is modelled similarly. CO_2 capture requires heat, electricity, MEA and inorganic chemicals. The electricity used by plant processes, such as CO_2 compression, is produced on-site. This electricity use is accounted in the efficiency penalty induced by CCS in the power plant. Table 3.14 and Table 3.15 show the inventory for the plant operation unit process for the EXPC and SCPC plants without and with CCS respectively.

In this study monoethanolamine (MEA) is used as the solvent in post-combustion capture. Due to the presence of oxygen, PM, and acid gases in the flue gas, MEA can degrade into ammonia and heat stable salts. In the NETL report an amine washer and advanced low temperature solvent reclaimer have been installed in the capture unit resulting in lower MEA loss levels reported compared to values reported in literature (Koornneef et al., 2008; Singh et al., 2011b; Veltman et al., 2010). However, in this report a conservative scenario is assumed by examining the potential impacts of the power plant when no further cleaner units are installed. Ammonia emissions due to MEA degradation are estimated using the equation below. Note that a higher MEA consumption rate of 2.15 g/kg CO_2 captured is chosen to reflect better the values reported in literature) (Koornneef et al., 2008):

$$NH_{3,emission} = \frac{MEA_{nom.loss} f_{oxidation} M_{NH3}}{M_{MEA}}$$

in which $MEA_{nom.loss}$ is the MEA nominal loss (64 per cent of MEA consumption) $f_{oxidation}$ is the oxidation factor (0.5) and MNH_3 and M_{MEA} are the molar masses of ammonia and MEA respectively (17 and 61 g/mol) (Koornneef et al., 2008).

3.7.3 INTEGRATED GASIFICATION COMBINED CYCLE

A LCI was made of an IGCC power plant without and with CCS. Net plant efficiencies are 42.1 per cent and 31.2 per cent, respectively. As noted before, the same coal transport process is used for the IGCC plants as for the previous coal power plant cases. Besides coal, the main inputs to plant operation are catalyst for the COS hydrolysis unit[24] (in the case of the power plant without CCS, modelled as 'chemical' from the EXIOBASE database) and the Claus Scott unit, and activated carbon for the removal of mercury. Sulfur is

24 In this until COS is converted to H_2S.

a by-product of the IGCC power plant. As with the gypsum production in the supercritical power plant, a conservative approach is taken and impacts are not allocated with respect to sulfur.

The power plant and carbon capture process of IGCC uses selexol (in the case of an IGCC without CCS, selexol is used in the acid gas removal unit). Since there is no ecoinvent unit process for selexol production, it is chosen to use the process for dimethyl ether as a proxy (Singh et al., 2011b). Detailed emissions are based on a previous LCA model of a similar IGCC plant (NETL, 2010c). Similar to the previous inventory, all plant infrastructure and CCS on-site infrastructure is modelled by distributing infrastructure costs (NETL, 2010b) over different sectors in the EXIOPOL input-output database. As in the other cases, the electricity used by plant processes is accounted for in the energy penalty due to CO_2 capture and compression.

TABLE 3.16

Inventory for the integrated gasification combined cycle (IGCC) plant operation process

IGCC Plant operation		Without CCS	With CCS	Unit	Reference
INPUTS	Activated carbon	$3.09 \cdot 10^{-6}$	$6.63 \cdot 10^{-6}$	kg/kWh	(NETL, 2010b)
	Chemicals	$6.70 \cdot 10^{-4}$	$5.60 \cdot 10^{-4}$	US$/kWh	(NETL, 2010b)
	Coal transport	$3.15 \cdot 10^{-1}$	$4.25 \cdot 10^{-1}$	kg/kWh	(NETL, 2010b)
	Discharge process water	$2.86 \cdot 10^{-1}$	$4.51 \cdot 10^{-1}$	kg/kWh	(NETL, 2010b)
	Selexol (dimethyl ether)		$7.81 \cdot 10^{-5}$		(NETL, 2010b)
	Water	$1.21 \cdot 10^{-3}$	$2.06 \cdot 10^{-3}$	m^3/kWh	(NETL, 2010b)
OUTPUTS	Carbon dioxide	$7.23 \cdot 10^{-1}$	$1.09 \cdot 10^{-1}$	kg/kWh	(NETL, 2010b)
	Disposal of catalyst	$2.85 \cdot 10^{-6}$	$6.12 \cdot 10^{-6}$	kg/kWh	(NETL, 2010b)
	Ash disposal	$3.15 \cdot 10^{-2}$	$4.25 \cdot 10^{-2}$	kg/kWh	(NETL, 2010c)
	Carbon monoxide	$3.75 \cdot 10^{-7}$	$4.30 \cdot 10^{-7}$	kg/kWh	(NETL, 2010d)
	Lead	$1.33 \cdot 10^{-8}$	$1.60 \cdot 10^{-8}$	kg/kWh	(NETL, 2010c)
	Mercury	$1.79 \cdot 10^{-9}$	$2.09 \cdot 10^{-9}$	kg/kWh	(NETL, 2010b)
	Methane	$1.03 \cdot 10^{-5}$	$1.18 \cdot 10^{-8}$	kg/kWh	(NETL, 2010c)
	Nitrogen oxide	$1.85 \cdot 10^{-4}$	$1.80 \cdot 10^{-4}$	kg/kWh	(NETL, 2010b)
	Nitrous oxide	$2.86 \cdot 10^{-9}$	$3.27 \cdot 10^{-9}$	kg/kWh	(NETL, 2010c)
	Particulates	$2.20 \cdot 10^{-5}$	$2.60 \cdot 10^{-5}$	kg/kWh	(NETL, 2010b)
	Sulfur dioxide	$1.30 \cdot 10^{-5}$	$8.00 \cdot 10^{-6}$	kg/kWh	(NETL, 2010b)
	Sulfur hexafluoride	$3.32 \cdot 10^{-10}$	$3.80 \cdot 10^{-10}$	kg/kWh	(NETL, 2010c)
	VOC	$2.46 \cdot 10^{-8}$	$2.81 \cdot 10^{-8}$	kg/kWh	(NETL, 2010c)
	Waste heat	$4.94 \cdot 10^{0}$	$6.66 \cdot 10^{0}$	MJ/kWh	-
	Water-to-air	$1.21 \cdot 10^{0}$	$2.06 \cdot 10^{0}$	kg/kWh	(NETL, 2010b)

Source: Bouman et al., 2015

Table 3.16 shows the inventory for the plant operation unit process for the plant without and with CCS and CO_2 operation unit processes for the IGCC plant.

3.7.4 NATURAL GAS COMBINED CYCLE

Unlike in the previous cases, the natural gas plant and its inputs are not scaled to keep the net output of the power plant equal, but the output is scaled according to plant size, i.e., the turbine/generator capacities are the same for both CCS cases, but due to the energy penalty of carbon capture the electricity produced by the plant with CCS is lower. The NGCC plant without CCS system has a net production of 555 MW, whereas the NGCC plant with CCS system has a net production of 474 MW. The corresponding efficiencies are respectively 50.2 per cent and 42.8 per cent based on HHV. We assume an offshore pipeline length of 1000 km between the natural gas extraction site and the power plant location. Consequentially, the transport requirement for 1 kg of gas is 1 ton-km and the transport requirement for 1 m^3 (at standard conditions) of natural gas is 0.731 ton-km. The ecoinvent process natural gas, at production, North America, with updated fugitive emissions (Burnham et al., 2011) is used as proxy for the natural gas extraction process.

Besides natural gas, the main plant inputs are ammonia for the selective catalytic reduction of NO_x emissions, process water for cooling duties and chemicals such as the catalyst of the SCR unit. Inputs to the CO_2 capture operation process are activated carbon and MEA. Table 3.17 shows the inventory for the plant operation unit process for the plant without and with CCS.

TABLE 3.17

Inventory for the natural gas combined cycle (NGCC) plant operation unit process

NGCC Plant operation		Value without CCS	Value with CCS	Unit	Reference
INPUTS	Activated carbon		$2.36 \cdot 10^{-6}$	kg/kWh	(NETL, 2010b)
	Ammonia	$4.13 \cdot 10^{-4}$	$4.83 \cdot 10^{-4}$	kg/kWh	(NETL, 2010b)
	Chemicals	$1.60 \cdot 10^{-4}$	$3.00 \cdot 10^{-4}$	US$/kWh	(NETL, 2010b)
	Discharge process water	$2.16 \cdot 10^{-1}$	$4.81 \cdot 10^{-1}$	kg/kWh	(NETL, 2010b)
	Monoethanolamine		$8.26 \cdot 10^{-4}$	kg/kWh	(Veltman, 2010)
	Natural gas	$1.87 \cdot 10^{-1}$	$2.19 \cdot 10^{-1}$	m3/kWh	(NETL, 2010b)
	Water	$7.46 \cdot 10^{-4}$	$1.43 \cdot 10^{-3}$	m^3/kWh	(NETL, 2010b)
OUTPUTS	Ammonia	$2.00 \cdot 10^{-5}$	$5.78 \cdot 10^{-4}$	kg/kWh	(NETL, 2010d; Koorneef, 2008
	Disposal of hazardous waste		1.23 ·10-3	kg/kWh	(Singh, 2011b)
	Carbon dioxide	$3.65 \cdot 10^{-1}$	$4.26 \cdot 10^{-2}$	kg/kWh	(NETL, 2010b)
	Carbon monoxide	$3.12 \cdot 10^{-7}$	$3.63 \cdot 10^{-7}$	kg/kWh	(NETL, 2010d)
	Lead	$2.44 \cdot 10^{-9}$	$2.44 \cdot 10^{-9}$	kg/kWh	(NETL, 2010d)
	Methane	$8.56 \cdot 10^{-9}$	$9.95 \cdot 10^{-9}$	kg/kWh	(NETL, 2010d)
	Monoethanolamine		$2.34 \cdot 10^{-5}$	kg/kWh	(Veltman, 2010)
	Nitrogen oxide	$3.23 \cdot 10^{-5}$	$3.76 \cdot 10^{-5}$	kg/kWh	(NETL, 2010d)
	Nitrous oxide	$2.38 \cdot 10^{-9}$	$2.77 \cdot 10^{-9}$	kg/kWh	(NETL, 2010d)
	Particulates	$2.83 \cdot 10^{-8}$	$3.29 \cdot 10^{-8}$	kg/kWh	(NETL, 2010d)
	Sulfur dioxide	$2.23 \cdot 10^{-9}$	$2.60 \cdot 10^{-9}$	kg/kWh	(NETL, 2010d)
	Sulfur hexafluoride	$3.47 \cdot 10^{-10}$	$4.03 \cdot 10^{-10}$	kg/kWh	(NETL, 2010d)
	VOC	$2.05 \cdot 10^{-8}$	$2.38 \cdot 10^{-8}$	kg/kWh	(NETL, 2010b)
	Waste heat	$3.59 \cdot 10^{0}$	$4.82 \cdot 10^{0}$	MJ/kWh	-
	Water-to-air	$7.46 \cdot 10^{-1}$	$1.43 \cdot 10^{0}$	kg/kWh	(NETL, 2010b)

Source: Bouman et al., 2015

3.7.5 TRANSPORT AND STORAGE OF CARBON DIOXIDE

3.7.5.1 Pipeline transport of carbon dioxide

CO_2 is captured at the plant and is subsequently transported to an underground aquifer by pipeline. CO_2 is transported in supercritical phase (P > 7.38 MPa, T > 304 K). In this report it is assumed a transport distance of 150 km. At this distance and with a pressure inlet of 15MPa intermediate CO_2 booster stations are not required. Following the approach by Singh et al. (2011b), the pipeline inventory data is based on the LCI of a high capacity offshore natural gas pipeline that is available in the ecoinvent database (Singh et al., 2011b). In order to develop a scaling factor between the CO_2 pipeline and the natural gas pipeline, the material volumes per km pipeline were compared. The natural gas pipeline in ecoinvent is made of carbon steel and has an internal diameter of 1000 mm and a steel thickness of 25 mm (Faist et al., 2007). Engineering calculations were used to estimate the internal diameter and of the CO_2 pipeline. The internal diameter of the CO_2 pipeline is dependent on the in- and outlet pressures of the CO_2 pipeline, the total mass flow, the density of the CO_2, the friction factor and the pipeline length. The friction factor in turn is dependent on the internal diameter. Thickness of the pipeline was calculated to be 11.5 mm for the EXPC, SCPC and IGCC cases and 8.5 mm for the NGCC case. The calculated scaling factor for the four different technologies is presented in Table 3.18.

TABLE 3.18

Pipeline parameters and scaling factor between ecoinvent unit processes

	EXPC	SCPC	IGCC	NGCC
Pressure inlet (MPa)	15.3	15.3	15.3	15.3
Pressure outlet (MPa)	9.9	10.7	12.3	11.2
Distance (km)	150	150	150	150
Mass flow (kg/s)	165.7	152.4	122.9	50.6
Internal Diameter calculated (m)	0.386	0.36	0.33	0.29
Wall thickness (m)	0.0115	0.0115	0.0115	0.0085
ΔP_{design} (Pa/m)	43	43	43	43
ΔP_{actual} (Pa/m)	36	30	20	27
Scaling factor	0.18	0.17	0.15	0.10

Source: Bouman et al., 2015.

TABLE 3.19

Carbon dioxide leakage rate

Case	CO_2 leakage rate (ton CO_2/km··year)
EXPC	3.31
SCPC	3.04
IGCC	2.32
NGCC	1.23

Source: Bouman et al., 2015.

Finally, it is assumed that there are fugitive emissions from the pipeline. These emissions were rescaled using data from Koornneef et al. (2008) who report a CO_2 leakage rate of 2.32 ton CO_2/km··year for a pipeline transporting 3.1 million tons CO_2/year (Koornneef et al., 2008). The CO_2 leakage rates used in this report are listed in Table 3.19.

3.7.5.2 Storage

Captured CO_2 is stored in a deep saline aquifer. For this study we have used the following specifications as specified in the NETL report. The saline formation lies at a depth of 1200 m, has a thickness of 161 m, and has a permeability of 22 mdarcy and a formation pressure of 8.4 MPa. It is assumed that the diameter of the injection pipe is of sufficient size, so that no booster compression is required at the well-head. The injection rate per well is 9,400 tons CO_2/day (NETL, 2010b). The storage well is modelled as an offshore drilling well from ecoinvent (Singh et al., 2011b). Table 3.20 lists the number of wells needed per technology based on the mass flow and well injection rate. Emissions associated with monitoring of the storage well are not included in the inventory.

TABLE 3.20

Carbon dioxide mass flows and number of wells required for carbon storage from specific power plants in the design assumed in this study

	EXPC	SCPC	IGCC	NGCC
CO_2 (kg/s)	166	152	123	51
Number of wells	2	2	1	1

3.8 MODELLING RESULTS

A life cycle impact assessment was performed for all inventories and all nine IEA regions (see for more information Chapter 2). In this section, the modelling results are presented compared to the Chinese regional electricity mix for the base year 2010. All results are presented on a functional unit basis, i.e., per kWh electricity produced. A comparison of the results for all technologies described in this report is presented in Chapter 10.

Total impacts for the inventories presented in the previous section are shown in Table 3.21. In Figure 3.29 to Figure 3.32 the impacts relative to the Chinese electricity mix are presented. In order to show the relative impact of the addition of CCS, the results for each technology with and without CCS are plotted in the same diagram. From the radar diagrams it can be seen that, while the carbon capture significantly reduces the impacts on climate change, the efficiency penalty and consequent increased resource use induce a mild increase in all other impact categories. A foreground contribution analysis is shown in Figure 3.33 to Figure 3.40. In presenting the results, a regrouping was made of the foreground processes presented in Figure 3.28: all on-site infrastructure is included in plant infrastructure, the CO_2 pipeline and well are grouped into (CO_2) transport and storage infrastructure and operation, carbon capture and electricity generation are included in plant operation, and raw material transport and extraction have remained as separate foreground systems.

The results presented fall within the range of results reported in literature and previously discussed in Chapter 3.6. With respect to the impacts relative to the 2010 Chinese electricity mix the results indicate that for most impact categories the impact is lower than those found for the average mix. This can be a result of the fact that the plants modelled in this chapter are based on state-of-the-art designs, with relatively high efficiencies compared to average power plants in the 2010 mix. For the subcritical and supercritical power plant, it can be seen that the addition of CCS almost completely counteracts the gains made by implementing a high-efficiency technology, effectively bringing back total impact (excl. climate change) to the impacts of the average mix. In the case of the coal-fired power plant, relatively high impacts can be observed for the freshwater ecotoxicity and eutrophication, which are related to the disposal processes of coal mining spoil and hard coal ash. The main contributor to ozone depletion for the natural gas power plants is the pipeline transport of natural gas. As China does not have an extensive natural gas pipeline infrastructure this explains the high relative value.

The contribution analysis shows that fuel extraction and power plant operation are the main drivers for the environmental impacts. This is reinforced by the energy penalty of carbon capture systems and the associated increase in fuel requirements.

TABLE 3.21

Life cycle impact assessment for all investigated technologies

Impact category	Unit	EXPC	EXPC w CCS	SCPC	SCPC w CCS	IGCC	IGCC w CCS	NGCC	NGCC w CCS
Agricultural land occupation	m^2a/MWh	12.6	18.0	11.8	16.6	10.9	14.7	0.388	0.533
Climate change	g CO_2-e/kWh	933	263	871	236	791	201	527	247
Fossil depletion	g oil-e/kWh	227	328	213	303	199	268	174	208
Freshwater ecotoxicity	g 1,4-DCB-e/kWh	8.15	12.8	7.58	11.7	6.72	9.24	6.3	8.11
Freshwater eutrophication	mg P-e/kWh	482	687	453	632	427	577	5.4	10.1
Human toxicity	g 1,4-DCB-e/kWh	111	172	104	158	91.1	125	88.0	112
Metal depletion	mg Fe-e/kWh	990	1990	929	1880	492	775	256	521
Natural land transformation	m_2/MWh	8.84	17.6	8.92	16.6	14.3	18.9	3.71	7.37
Ozone depletion	µg CFC-11e/kWh	2.94	5.05	2.75	4.68	0.960	1.33	20.0	23.7
Particulate matter formation	mg PM10-e/kWh	335	381	315	418	183	227	757	916
Photochemical oxidant formation	mg NMVOC/kWh	809	1160	762	1060	665	833	617	768
Terrestrial acidification	g SO_2e/kWh	1.10	1.23	1.05	1.61	0.720	0.927	3.78	4.68
Urban land occupation	m_2a/MWh	7.83	11.1	7.33	10.2	6.80	9.17	0.100	0.142
Water depletion	m_3/MWh	48.8	78.3	45.7	73.2	15.5	21.7	14.0	19.2

FIGURE 3.29

Environmental impacts for a subcritical pulverized coal power plant relative to the 2010 Chinese electricity mix of 2010

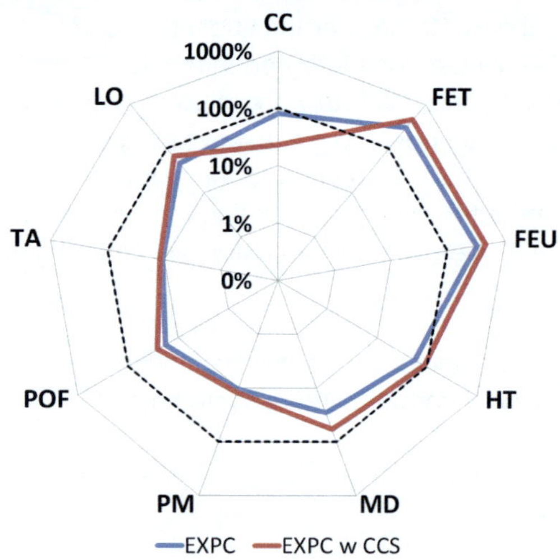

CCS: carbon dioxide capture and storage, CC: climate change, FET: freshwater ecotoxicity, FEU: freshwater eutrophication, HT: human toxicity, MD: metal depletion, PM: particulate matter, POF: photochemical oxidant formation, TA: terrestrial acidification, LO: land occupation.

FIGURE 3.30

Environmental impacts for a supercritical power plant relative to the 2010 Chinese electricity mix

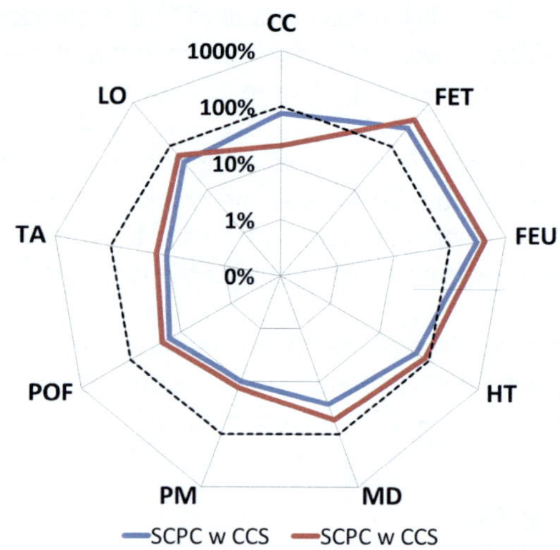

CCS: carbon dioxide capture and storage, CC: climate change, FET: freshwater ecotoxicity, FEU: freshwater eutrophication, HT: human toxicity, MD: metal depletion, PM: particulate matter, POF: photochemical oxidant formation, TA: terrestrial acidification, LO: land occupation.

FIGURE 3.31

Environmental impact of the integrated gasification combined cycle power plant with and without CCS relative to the 2010 Chinese electricity mix

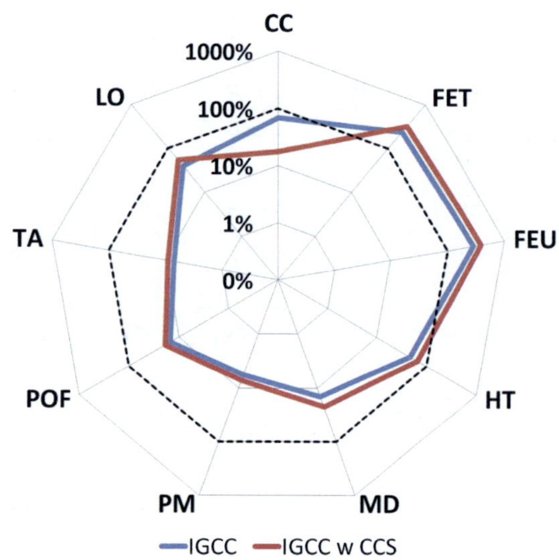

CC: climate change, FET: freshwater ecotoxicity, FEU: freshwater eutrophication, HT: human toxicity, MD: metal depletion, PM: particulate matter, POF: photochemical oxidant formation, TA: terrestrial acidification, LO: land occupation.

FIGURE 3.32

Environmental impacts of the natural gas combined cycle power plant with and without CCS relative to the 2010 Chinese electricity mix

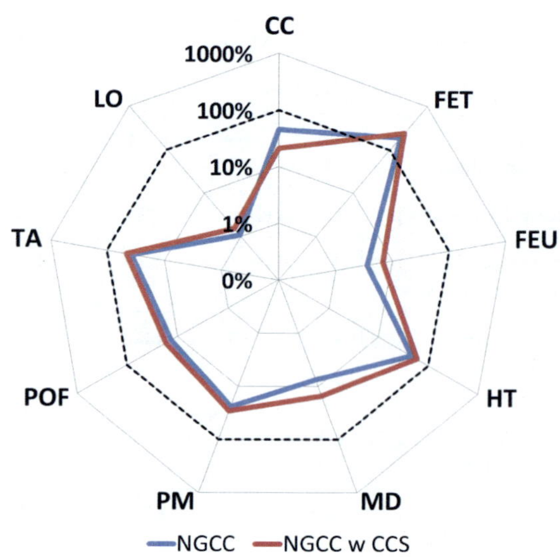

CC: climate change, FET: freshwater ecotoxicity, FEU: freshwater eutrophication, HT: human toxicity, MD: metal depletion, PM: particulate matter, POF: photochemical oxidant formation, TA: terrestrial acidification, LO: land occupation.

FIGURE 3.33

Contribution analysis for the subcritical coal fired power plant

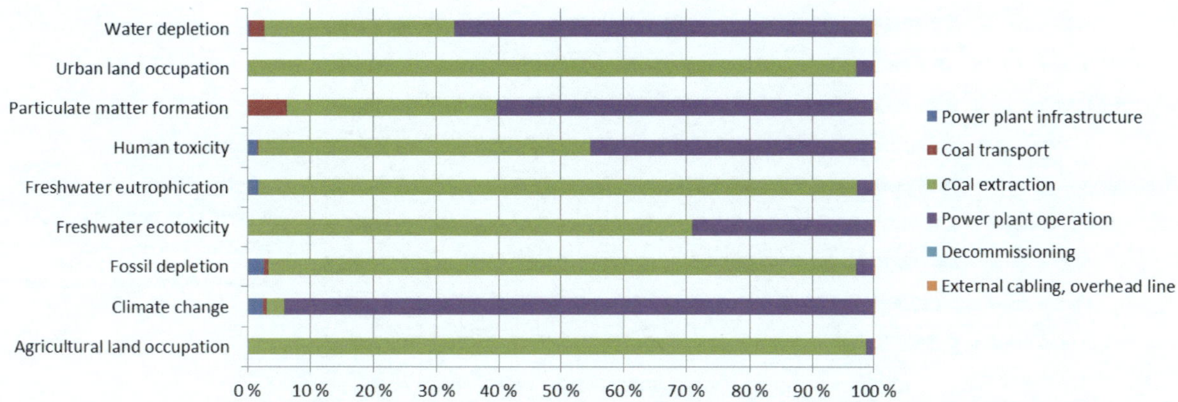

FIGURE 3.34

Contribution analysis for the subcritical coal fired power plant with carbon dioxide capture and storage

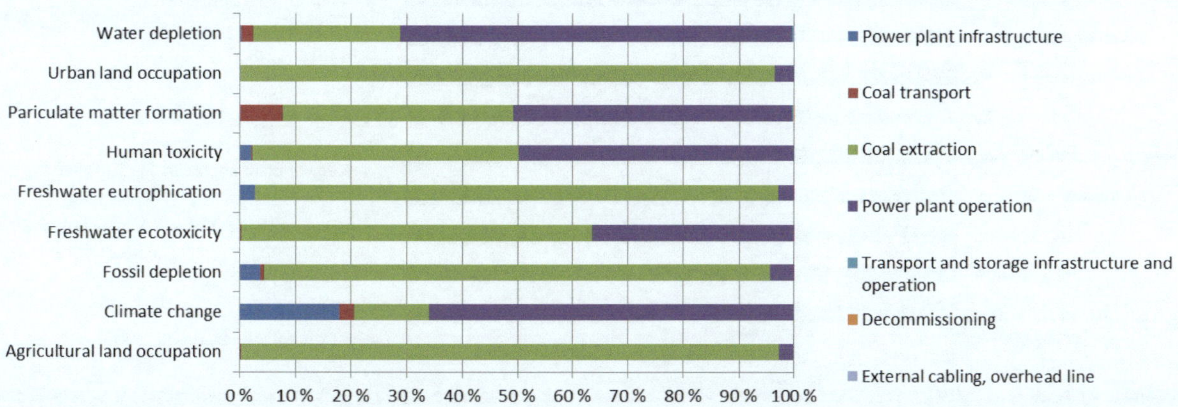

FIGURE 3.35

Contribution analysis for the supercritical coal fired power plant

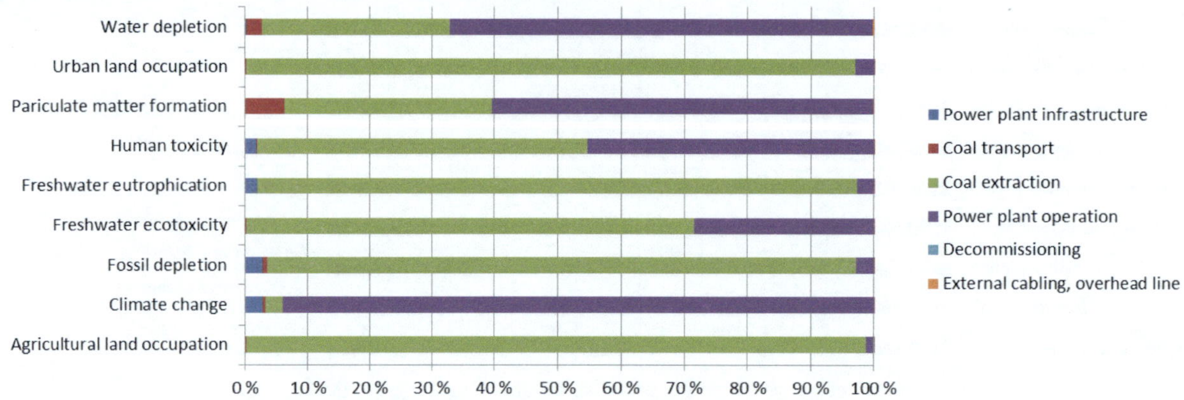

FIGURE 3.36

Contribution analysis for the supercritical coal fired power plant with carbon dioxide capture and storage

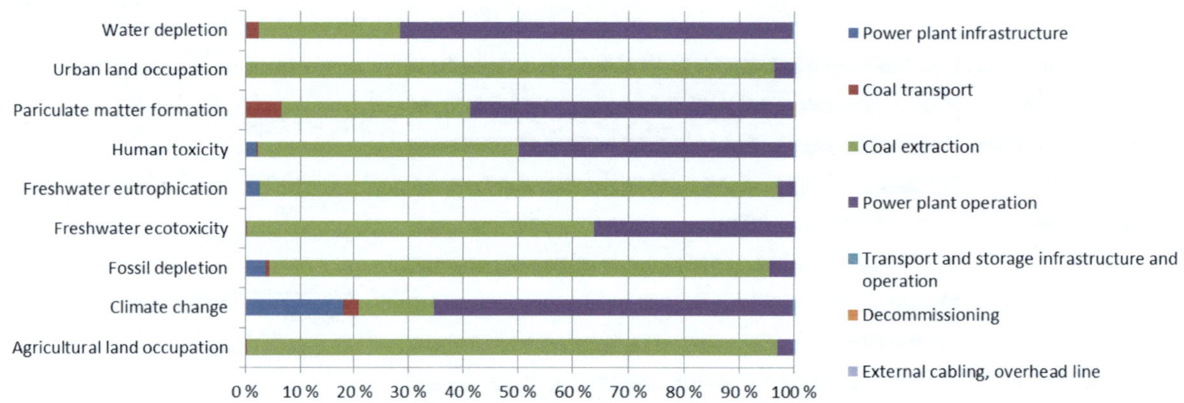

FIGURE 3.37

Contribution analysis for the integrated gasification power plant

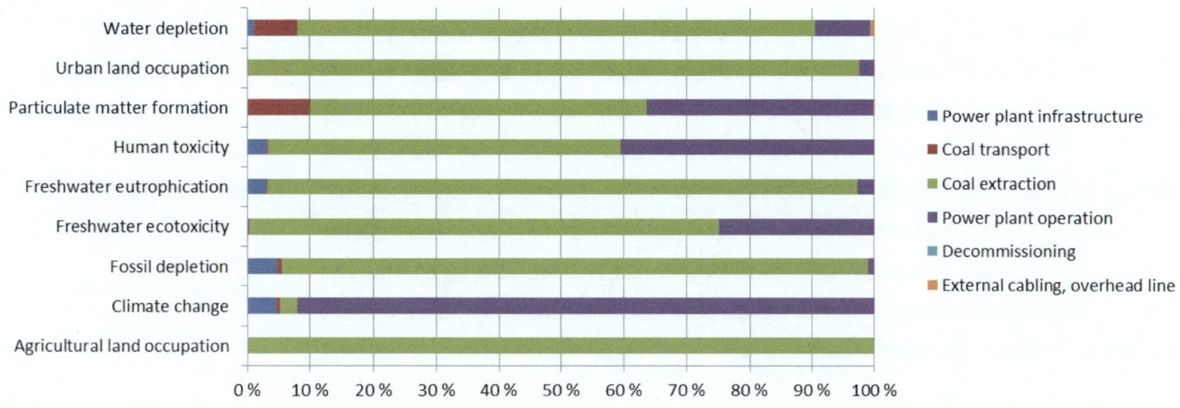

FIGURE 3.38

Contribution analysis for the integrated gasification power plant with carbon dioxide capture and storage

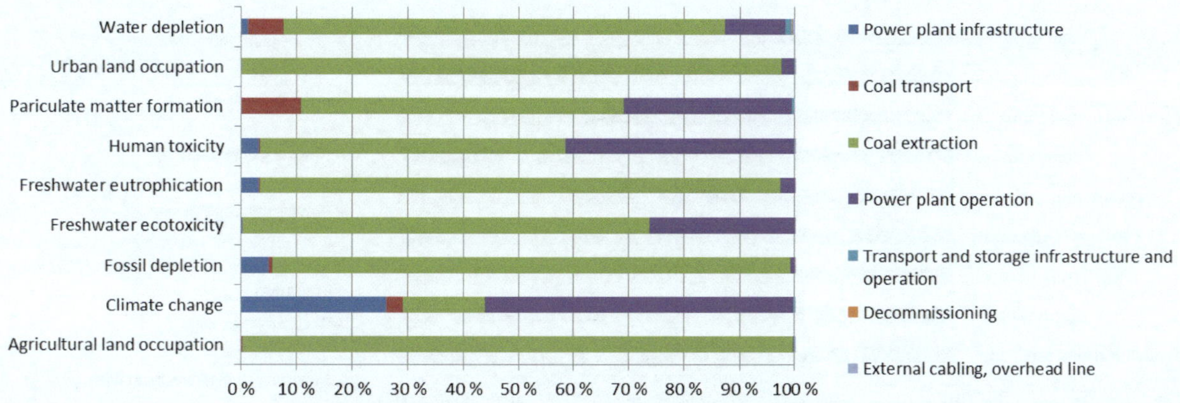

FIGURE 3.39

Contribution analysis for the natural gas fired power plant

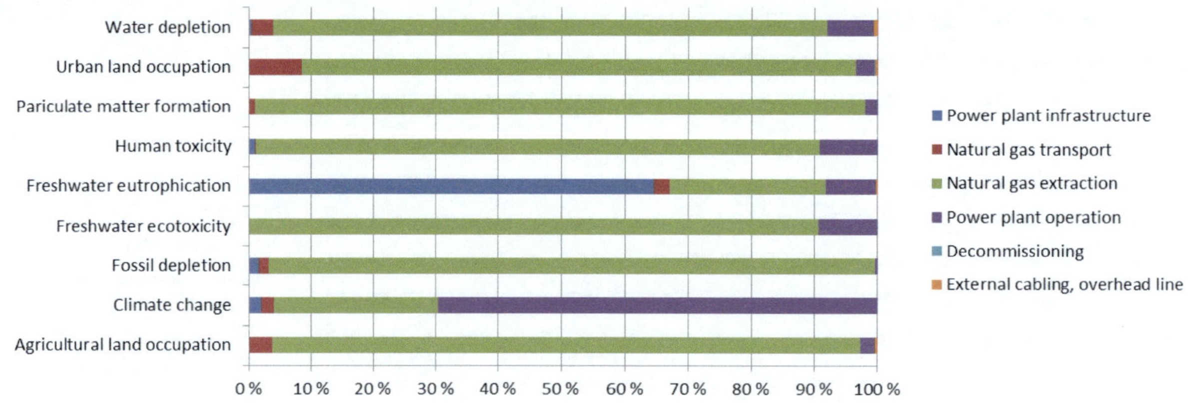

FIGURE 3.40

Contribution analysis for the natural gas fired power plant with carbon dioxide capture and storage

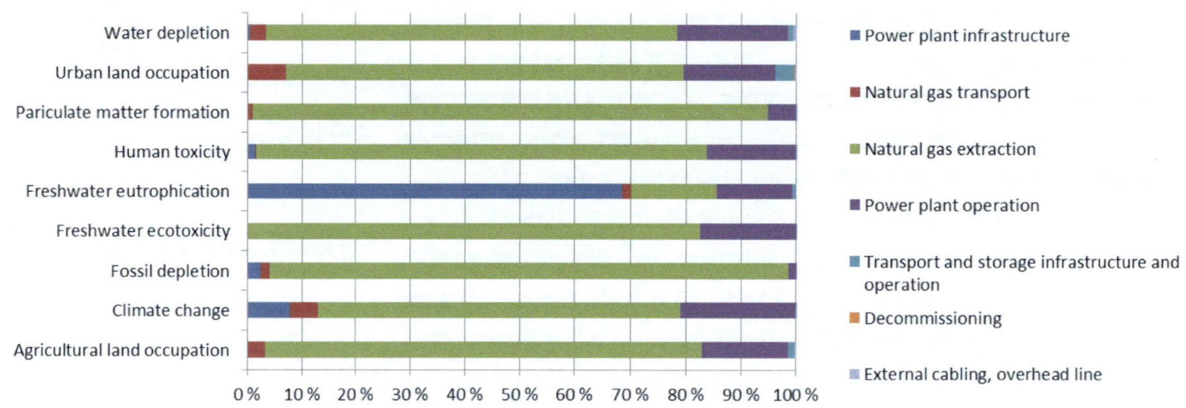

3.9 OVERALL CONCLUSIONS

Fossil fuels are currently the largest source of energy and their contribution is expected to remain significant in the coming decades. The literature overview presented in this chapter addresses the status and environmental impacts of conventional and unconventional fossil fuel extraction (e.g., oil sands, shale gas, coal-bed methane); examines current and mid-term options to generate heat and electricity from fossil fuels, and provides a systematic overview of the potential environmental impacts through their life cycle. Given the significant role of fossil fuel combustion on increasing CO_2 emissions in the atmosphere, technologies that can be used (as well as their potential environmental impact) to capture, transport and storage CO_2 from fossil fuel power plants are examined.

The findings confirm that decreasing the additional energy requirements induced by CCS is a key component in enhancing the environmental performance of fossil fuel fired power plants in terms of GWP while minimizing trade-offs in other environmental impact categories. Technologies that have lower energy penalties also show better performances for the different environmental categories. The chapter, however, also indicates that the environmental issues associated with the exploration, production, refining and distribution of fossil fuels are significant during the life cycle of power plants. Their role is further exacerbated by the deployment of CCS. This is due to the energy penalty induced by the capture process and the consequent need for additional fuel in order to produce the same amount of output. Remarkably, it is the upstream leakage of methane the main cause of the relative low performance – in terms of decreasing global warming potential- of natural gas power plants with CCS.

Results of the modelling work carried out for four types of fossil fuel power plants with and without CCS show that, for coal fired power plants without CCS, direct emissions account as the main contributor to GWP, PM formation and water depletion. Indirect emissions due to upstream and downstream processes make the largest contribution to fresh water eutrophication, ecotoxicity and fossil fuel depletion. For coal power plants with CCS, indirect emissions appear as the main contributor to GWP, acidification and human toxicity potential. In the case of natural gas fired power plants with and without CCS, the modelling results indicate that with exception of GWP, indirect emissions are the main contributor to the impacts in all environmental categories.

Although it was not possible to fully study their impact in the current study, information found so far indicates that the exploitation of unconventional fossil fuels will further augment the life cycle impacts of fossil fuel in the environment and, for power plants with CCS, could become the key issue to be addressed when optimizing the environmental performance of such chains.

A core assumption of the modelling work conducted in this chapter is that, for the CCS cases, the CO_2 would be permanently stored (i.e., no leakage). Leakage will not only have large implications on the effectiveness of the option –in terms of GWP- but could result on environmental impacts such as acidification, mobilization of heavy metals, contamination of underground water tables. Appropriate monitoring, verification and accounting programs are therefore fundamental requirements in CCS projects and are a key aspect to gain public acceptance, which is currently the main non-technical issue hindering the deployment of CCS technologies.

Engaging communities early in the life of CCS projects is essential, particularly because CCS remains a relatively new technology. Early dialogue with impacted communities will help keep such communities abreast of the proposed CCS project.

3.10 REFERENCES

ACAA. 2013. *Beneficial Use of Coal Combustion Products: An American Recycling Success Story.* Farmington Hills Michigan: American Coal Ash Association.

Acharaya, H. R., C. Henderson, M. Hope, H. Kommepalli, B. Moore, and H. Wang. 2011. *Cost Effective Recovery of Low-TDS Frac Flowback Water for Re-use*. United States Department of Energy.

Ackermann, T., G. Andersson, and L. Söder. 2001. Distributed generation: a definition. *Electric Power Systems Research* 57(3): 195-204.

ADB. 2009. *People's Republic of China: Recycling Waste Coal for Power Generation*. Asian Development Bank.

Aghalayam, P. 2010. Underground Coal Gasification: A Clean Coal Technology. In *Handbook of Combustion*: Wiley-VCH Verlag GmbH & Co. KGaA.

Aguilera, R. and T. Harding. 2008. State-of-the-art Tight Gas Sands Characterization and Production Technology. *Journal of Canadian Petroleum Engineering* 47(12): 37-41.

Akai, M., N. Nomura, H. Waku, and M. Inoue. 1997. Life-cycle analysis of a fossil-fuel power plant with CO_2 recovery and a sequestering system. *Energy* 22(2–3): 249-255.

Alley, B., A. Beebe, J. Rodgers Jr, and J. W. Castle. 2011. Chemical and physical characterization of produced waters from conventional and unconventional fossil fuel resources. *Chemosphere* 85(1): 74-82.

Allison, J. E. and J. Lents. 2002. Encouraging distributed generation of power that improves air quality: can we have our cake and eat it too? *Energy Policy* 30(9): 737-752.

Ally, J. and T. Pryor. 2007. Life-cycle assessment of diesel, natural gas and hydrogen fuel cell bus transportation systems. *Journal of Power Sources* 170(2): 401-411.

Alsalam, J. and S. Ragnauth. 2011. *Draft Global Anthropogenic Non-CO_2 Greenhouse Gas Emissions: 1990-2030*. Washington: US EPA.

Ambrose, W. A., E. C. Potter, and R. Briceno. 2008. An "Unconventional" Future for Natural Gas in the United States. *Geotimes*.

Ansolobehere, S., J. Beer, J. Deutch, A. D. Ellerman, J. Friedman, H. Herzog, H. Jacoby, P. Joskow, G. McRae, R. Lester, E. Moniz, and E. Steinfeld. 2007. *The Future of Coal: Options for a Carbon-Constrained World*. Cambridge, MA: Massachusetts Institute of Technology.

Arteconi, A., C. Brandoni, D. Evangelista, and F. Polonara. 2010. Life-cycle greenhouse gas analysis of LNG as a heavy vehicle fuel in Europe. *Applied Energy* 87(6): 2005-2013.

Ashworth, P. and C. Cormick. 2011. Enabling the Social Shaping of CCS Technology. In *Carbon Capture and Storage: Emerging Legal and Regulatory Issues*, edited by I. Havercroft, et al. Oxford, UK: Hart Publishing Ltd.

Ashworth, P., G. Paxton, and S. Carr-Cornish. 2011. Reflections on a process for developing public trust in energy technologies: Follow-up results of the Australian large group process. *Energy Procedia* 4(0): 6322-6329.

Ashworth, P., N. Boughen, M. Mayhew, and F. Millar. 2010a. From research to action: Now we have to move on CCS communication. *International Journal of Greenhouse Gas Control* 4(2): 426-433.

Ashworth, P., N. Boughen, M. Mayhew, and F. Millar. 2009a. An integrated roadmap of communication activities around carbon capture and storage in Australia and beyond. In *Greenhouse Gas Control Technologies 9*, edited by J. Gale, et al.

Ashworth, P., S. Carr-Cornish, N. Boughen, and K. Thambimuthu. 2009b. Engaging the public on Carbon Dioxide Capture and Storage: Does a large group process work? In *Greenhouse Gas Control Technologies 9*, edited by J. Gale, et al.

Ashworth, P., J. Bradbury, C. F. J. Feenstra, S. Greenberg, G. Hund, T. Mikunda, and S. Wade. 2010b. *Communication, project planning and management for carbon capture and storage projects: An international comparison*. Pullenvale, Australia: CSIRO.

Bachu, S. 2008. CO_2 storage in geological media: Role, means, status and barriers to deployment. *Progress in Energy and Combustion Science* 34(2): 254-273.

Bachu, S., D. Bonijoly, J. Bradshaw, R. Burruss, S. Holloway, N. P. Christensen, and O. M. Mathiassen. 2007. CO_2 storage capacity estimation: Methodology and gaps. *International Journal of Greenhouse Gas Control* 1(4): 430-443.

Battino, R. 1984. The Ostwald coefficient of gas solubility. *Fluid Phase Equilibria* 15(3): 231-240.

Bauer, C. and T. Heck. 2009. Carbon capture and storage: Life cycle assessment and external costs of future fossil power generation. Paper presented at 4th International Conference on Life Cycle Management, September 6-9, 2009, Cape Town, South Africa.

Bauer, C., T. Heck, R. Dones, O. Mayer-Spohn, and M. Blesl. 2008. *New Energy Externalities Developments for Sustainability Deliverable n°7.2 - RS 1a "Final report on technical data , costs , and life cycle inventories of advanced fossil power generation systems"*. Stuttgart, Germany: University of Stuttgart.

Benson, S. M., K. Bennaceur, P. Cook, J. Davison, H. de Coninck, K. Farhat, A. Ramirez, D. Simbeck, T. Surles, P. Verma, and I. Wright. 2012. Chapter 13 - Carbon Capture and Storage. In *Global Energy Assessment - Toward a Sustainable Future*. Cambridge University Press, Cambridge, UK and New York, NY, USA and the International Institute for Applied Systems Analysis, Laxenburg, Austria.

Bergerson, J. A. and D. W. Keith. 2010. The Truth About Dirty Oil: Is CCS the Answer? *Environmental Science & Technology* 44(16): 6010-6015.

Bergstein, B. 2008. Carbon nanotubes mimic asbestos in early study. 20 May, 2008.

Bouman, E. A., A. Ramirez, and E. G. Hertwich. 2015. Multiregional environmental comparison of fossil fuel power generation—Assessment of the contribution of fugitive emissions from conventional and unconventional fossil resources. *International Journal of Greenhouse Gas Control* 33(0): 1-9.

Bradbury, J. and S. Wade. 2010. *Case Study of the Carson CCS Project.* Washington, USA: Global CCS Institute.

Bradbury, J., I. Ray, T. Peterson, S. Wade, G. Wong-Parodi, and A. Feldpausch. 2009. The Role of Social Factors in Shaping Public Perceptions of CCS: Results of Multi-State Focus Group Interviews in the U.S. *Energy Procedia* 1(1): 4665-4672.

Brandt, A. R. 2012. Variability and Uncertainty in Life Cycle Assessment Models for Greenhouse Gas Emissions from Canadian Oil Sands Production. *Environmental Science & Technology* 46(2): 1253-1261.

Brouwer, J. 2006. Hybrid Gas Turbine Fuel Cell Systems. In *Gas Turbine Handbook.* United States Department of Energy.

Brunsting, S., B. van Bree, C. F. J. Feenstra, and M. Hekkenberg. 2011. *Public perceptions of low carbon energy technologies: Results from a Dutch large group workshop*. Amsterdam, Netherlands: Global CCS Institute.

Bugge, J., S. Kjær, and R. Blum. 2006. High-efficiency coal-fired power plants development and perspectives. *Energy* 31(10–11): 1437-1445.

Burnham, A., J. Han, C. E. Clark, M. Wang, J. B. Dunn, and I. Palou-Rivera. 2011. Life-Cycle Greenhouse Gas Emissions of Shale Gas, Natural Gas, Coal, and Petroleum. *Environmental Science & Technology* 46(2): 619-627.

Canova, A., G. Chicco, G. Genon, and P. Mancarella. 2008. Emission characterization and evaluation of natural gas-fueled cogeneration microturbines and internal combustion engines. *Energy Conversion and Management* 49(10): 2900-2909.

CAPP. 2011. *Crude Oil: Forecasts, Markets, and Pipelines*. Canadian Association of Petroleum Producers

Carpentieri, M., A. Corti, and L. Lombardi. 2005. Life cycle assessment (LCA) of an integrated biomass gasification combined cycle (IBGCC) with CO_2 removal. *Energy Conversion and Management* 46(11–12): 1790-1808.

Carter, K. M., J. A. Harper, K. W. Schmid, and J. Kostelnik. 2011. Unconventional natural gas resources in Pennsylvania: The backstory of the modern Marcellus Shale play. *Environmental Geosciences* 18(4): 217-257.

Cathles III, L. M., L. Brown, M. Taam, and A. Hunter. 2012. A commentary on "The greenhouse-gas footprint of natural gas in shale formations" by R.W. Howarth, R. Santoro, and Anthony Ingraffea. *Climatic Change* 113(2): 525-535.

Cathles, L. M. 2012. Assessing the greenhouse impact of natural gas. *Geochemistry, Geophysics, Geosystems* 13(6).

CGES. 2011. Canadian Oil Sands Output Expected to Rise by 150,000 bpd this Year. http://www.cges.co.uk/news/705-canadian-oil-sands-output-expected-to-rise-by-150000-bpd-this-year. Accessed June, 2011.

Chang, K. 2008. In study, researchers find nanotubes may pose health risks similar to asbestos. *The New York Times* May 21, 2008, section 22.

Charpentier, A. D., J. A. Bergerson, and H. L. MacLean. 2009. Understanding the Canadian oil sands industry's greenhouse gas emissions. *Environmental Research Letters* 4(1): 40051-400511.

Chen, R., Q.-S. Geng, and X.-B. SU. 2006. The Formation and Accumulation of Water Soluble Gas. *Journal of Henan Polytechnic University (Natural Science)*(3): 205-208.

Choi, Y.-S., S. Nesic, and D. Young. 2010. Effect of Impurities on the Corrosion Behavior of CO_2 Transmission Pipeline Steel in Supercritical CO_2-Water Environments. *Environmental Science and Technology* 44(23): 9233-9238.

Clark, C. and J. Veil. 2011. Produced Water Volume Estimates and Management Practices. *Society of Petroleum Engineers* 26(3): 234-239.

Colborn, T., C. Kwiatkowski, K. Schultz, and M. Bachran. 2011. Natural Gas Operations from a Public Health Perspective. *Human and Ecological Risk Assessment* 17(5): 1039-1056.

Corsten, M., A. Ramírez, L. Shen, J. Koornneef, and A. Faaij. 2013. Environmental impact assessment of CCS chains – Lessons learned and limitations from LCA literature. *International Journal of Greenhouse Gas Control* 13(0): 59-71.

Croiset, E., K. Thambimuthu, and A. Palmer. 2000. Coal combustion in O_2/CO_2 mixtures compared with air. *Canadian Journal of Chemical Engineering* 78(2): 402-407.

CRS. 2008. *North American Oil Sands: History of Development, Prospects for the Future*. Washington, USA: Congressional Research Service (CRS)

CRS. 2009. *Unconventional Gas Shales: Development, Technology, and Policy Issues*. Washington, USA: Congresstional Research Service (CRS).

Cui, Z., A. Aroonwilas, and A. Veawab. 2010. Simultaneous capture of mercury and CO_2 in amine-based CO_2 absorption process. *Industrial & Engineering Chemistry Research* 49: 12576-12586.

Curry, T., D. M. Reiner, S. Ansolabehere, and H. J. Herzog. 2005. How aware is the public of carbon capture and storage? In *Greenhouse Gas Control Technologies 7*, edited by E. S. R. W. K. F. G. Wilson and T. M. G. Thambimuthu. Oxford: Elsevier Science Ltd.

Damen, K., A. Faaij, and W. Turkenburg. 2006. Health, Safety and Environmental Risks of Underground Co2 Storage – Overview of Mechanisms and Current Knowledge. *Climatic Change* 74(1-3): 289-318.

Davies, E. G. R., P. Kyle, and J. A. Edmonds. 2013. An integrated assessment of global and regional water demands for electricity generation to 2095. *Advances in Water Resources* 52(0): 296-313.

Davis, S. D. and C. Frohlich. 1993. Did (or will) fluid injection cause earthquakes? - Criteria for a rational assessment. *Seismological Research Letters* 64(3-4): 207-224.

Davison, J. 2007. Performance and costs of power plants with capture and storage of CO_2. *Energy* 32(7): 1163-1176.

de Best-Waldhober, M., D. Daamen, and A. Faaij. 2009. Informed and uninformed public opinions on CO_2 capture and storage technologies in the Netherlands. *International Journal of Greenhouse Gas Control* 3(3): 322-332.

de Best-Waldhober, M., M. Paukovic, S. Brunsting, and D. Daamen. 2011. Awareness, knowledge, beliefs, and opinions regarding CCS of the Dutch general public before and after information. *Energy Procedia* 4: 6292-6299.

de Marliave, L. 2009. Case study: Communicating CCS and public dialogue – Demonstrating CCS in an onshore site in Europe: The current status of the Lacq integrated CCS project, in Ashworth. In *Case Study of the CO_2CRC Otway Project*, edited by P. S. Rodriguez and A. Miller. Canberra, Australia: Global CCS Institute.

Desbarats, J., P. Upham, H. Riesch, D. Reiner, S. Brunsting, M. De best Waldhober, E. Duetschke, C. Oltra, R. Sala, and C. McLachlan. 2010. *Review of the public participation practices for CCS and non-CCS projects in Europe*. Brussels: Institute for European Environmental Policy.

DNV. 2008. *Mapping of potential HSE issues related to large-scale capture, transport and storage of CO_2*. 2008-1993. Stavanger, Norway: Det Norske Veritas AS (DNV).

DNV. 2010. *Design and Operation of CO_2 pipelines*. J202. Det Norske Veritas AS (DNV).

Doctor, D. R., J. C. Molburg, and N. F. Brockmeier. 2001. Life-cycle analysis of a Shell gasification-based multi-product system with CO_2 recovery *Journal of Energy & Environmental Research* 1: 41-66.

Dones, R., C. Bauer, and A. Röder. 2007a. *Kohle*. Dübendorf, CH: Paul Scherrer Institut Villigen, Swiss Centre for Life Cycle Inventories.

Dones, R., C. Bauer, R. Bolliger, B. Burger, M. Faist Emmeneger, R. Frischknecht, T. Heck, N. Jungbluth, A. Röder, and M. Tuchsmid. 2007b. *Life Cycle Inventories of Energy Systems: Results for Current Systems in Switzerland and other UCTE Countries*. Dübendorf, CH: Paul Scherrer Institut Villigen, Swiss Centre for Life Cycle Inventories.

Dooley, J. J., P. Kyle, and E. G. R. Davies. 2013. Climate Mitigation's Impact On Global and Regional Electric Power Sector Water Use in the 21st Century. *Energy Procedia* 37(0): 2470-2478.

Duan, H. 2010. The public perspective of carbon capture and storage for CO_2 emission reductions in China. *Energy Policy* 38(9): 5281-5289.

Ecoba-Eurelectric. 2011. *Joint EURELECTRIC/ECOBA Briefing: The Classification of Coal Combustion Products under the Revised Waste Framework Directive (2008/98/EC).*

Ecoba. 2006. *Coal Combustion Products in Europe – Valuable Raw Materials for European Coal Combustion the Construction Industry*. Essen, DE: European Coal Combustion Products Association.

Einsiedel, E., A. Boyd, and J. Medlock. 2011. *Publics and Energy: Results from Calgary, Alberta (Canada) Workshop.* Alberta, CA: University of Calgary.

Entrekin, S., M. Evans-White, B. Johnson, and E. Hagenbuch. 2011. Rapid expansion of natural gas development poses a threat to surface waters. *Frontiers in Ecology and the Environment* 9(9): 503-511.

EPRI. 2002. *Water and Sustainability (Volume 3): US Water Consumption for Power Production-The Next Half Century*. 1006786. Palo Alto, CA: Electric Power Research Institute.

ERCB. 2008. *Alberta's energy reserves 2007 and supply/demand outlook 2008-2017*. Calgary, Alberta: Energy Resources Conservation Board.

Esposito, A. and S. M. Benson. 2011. Remediation of possible leakage from geologic CO_2 storage reservoirs into groundwater aquifers. *Energy Procedia* 4(0): 3216-3223.

Euroheat & Power. 2011. Combined Heat and Power.

Faist, E. M., T. Heck, and N. Jungbluth. 2007. *Final report ecoinvent No. 6-V*. Dübendorf, CH: Paul Scherrer Institut Villigen, Swiss Centre for Life Cycle Inventories.

Falcke, T. J., A. F. A. Hoadley, D. J. Brennan, and S. E. Sinclair. 2011. The sustainability of clean coal technology: IGCC with/without CCS. *Process Safety and Environmental Protection* 89(1): 41-52.

Feenstra, C. F. J., T. Mikunda, and S. Brunsting. 2010. *What happened in Barendrecht? Case Study on the planned onshore carbon dioxide storage in Barendrecht, the Netherlands*. Petten, The Netherlands: Energy Research Centre of the Netherlands.

Figueroa, J. D., T. Fout, S. Plasynski, H. McIlvried, and R. D. Srivastava. 2008. Advances in CO_2 capture technology—The U.S. Department of Energy's Carbon Sequestration Program. *International Journal of Greenhouse Gas Control* 2(1): 9-20.

Finkenrath, M. 2011. *Cost and performance of CO_2 capture from power generation. Working paper*. Paris, France: OECD/IEA.

Fischer, M., O. Schuller, S. Albrecht, and M. Faltenbacher. 2008. Exergy Efficiency as Enhancement of Energy Efficiency – an LCA Perspective. Paper presented at LCA8, 30 September - 2 October, 2008, Seattle, Washington.

Fleig, D., F. Normann, K. Andersson, F. Johnsson, and B. Leckner. 2009. The fate of sulfur during oxy-fuel combustion of lignite. Paper presented at Energy Procedia, 383-390

Friedmann, S. J., R. Upadhye, and F.-M. Kong. 2009. Prospects for underground coal gasification in carbon-constrained world. *Energy Procedia* 1(1): 4551-4557.

Gagnon, N., C. A. S. Hall, and L. Brinker. 2009. A preliminary investigation of energy return on energy investment for global oil and gas production. *Energies* 2(3): 490-503.

GCP. 2014. Global Carbon Project. http://www.globalcarbonproject.org/carbonbudget/14/hl-full.htm. Accessed 10 October 2014. [Archive accessed 31 May 2016: http://www.globalcarbonproject.org/carbonbudget/archive.htm#CB2014].

GEA. 2012. *Global Energy Assessment - Toward a Sustainable Future*. Cambridge University Press, Cambridge, UK and New York, NY, USA and the International Institute for Applied Systems Analysis, Laxenburg, Austria.

Global CCS Institute. 2011. *The global status of CCS: 2011*. Canberra, Australia: Global CCS Institute.

Goff, G. S. and G. T. Rochelle. 2004. Monoethanolamine Degradation: O_2 Mass Transfer Effects under CO_2 Capture Conditions. *Industrial & Engineering Chemistry Research* 43(20): 6400-6408.

Grace, J., T. Collett, F. Colwell, P. Englezos, E. Jones, R. Mansell, J. P. Meekison O.C., R. Ommer, M. Pooladi-Darvish, M. Riedel, J. Ripmeester, C. Shipp, and E. Willoughby. 2008. *Energy from Gas Hydrates: Assessing the Opportunities and Challenges for Canada*. Ottawa, Canada: Council of Canadian Academies.

Greb, S. F., C. F. Eble, D. C. Peters, and A. R. Papp. 2006. *Coal and the Environment*. Alexandria, VA: American Geological Institute.

Gregory, K. B., R. D. Vidic, and D. A. Dzombak. 2011. Water Management Challenges Associated with the Production of Shale Gas by Hydraulic Fracturing. *Elements*, June, 2011, 181-186.

Gue, E. H. 2010. Why Some Natural Gas Is Worth $7.28. *Investing Daily.*

Hangx, S. van der Linden, A. Marcelis, F. Bauer, A. 2013. The effect of CO_2 on the mechanical properties of the Captain Sandstone: Geological storage of CO_2 at the Goldeneye field (UK). *International Journal of Greenhouse Gas Control* 2013, 19, (0), 609-619.

Hangx, S. J. T. Spiers, C. J. Peach, C. J. 2010. Mechanical behavior of anhydrite caprock and implications for CO_2 sealing capacity. *Journal of Geophysical Research* 2010, 115, (B7), B07402.

Hansson, A. and M. Bryngelsson. 2009. Expert opinions on carbon dioxide capture and storage--A framing of uncertainties and possibilities. *Energy Policy* 37(6): 2273-2282.

Harrison, S. S. 1983. Evaluating System for Ground-Water Contamination Hazards Due to Gas-Well Drilling on the Glaciated Appalachian Plateau. *Ground Water* 21(6): 689-700.

Heemann, R., J. Ketzer, G. Hiromoto, and A. Scislewski. 2011. Assessment of the Geological Disposal of Carbon Dioxide and Radioactive Waste in Brazil, and Some Comparative Aspects of Their Disposal in Argentina. In *Geological Disposal of Carbon Dioxide and Radioactive Waste: A Comparative Assessment*, edited by F. L. Toth: Springer Netherlands.

Hertwich, E. G., M. Aaberg, B. Singh, and A. H. Strømman. 2008. Life-cycle Assessment of Carbon Dioxide Capture for Enhanced Oil Recovery. *Chinese Journal of Chemical Engineering* 16(3): 343-353.

Holditch, S. A. 1993. Completion Methods in Coal-Seam Reservoirs. *Journal of Petroleum Technology* 45(3): 270-276.

Holland, A. 2011. *Examination of Possibly Induced Seismicity from Hydraulic Fracturing in the Eola Field, Garvin County, Oklahoma*. Norman, Oklahoma: Oklahoma Geological Survey.

Hou, Q., H. Li, J. Fan, Y. Ju, T. Wang, X. Li, and Y. Wu. 2012. Structure and coalbed methane occurrence in tectonically deformed coals. *Science China Earth Sciences* 55(11): 1755-1763.

Howarth, R., R. Santoro, and A. Ingraffea. 2011. Methane and the greenhouse-gas footprint of natural gas from shale formations. *Climatic Change* 106(4): 679-690.

Hu, Y. Q., N. Kobayashi, and M. Hasatani. 2001. The reduction of recycled-NO_x in coal combustion with O_2/recycled flue gas under low recycling ratio. *Fuel* 80(13): 1851-1855.

Hultman, N., D. Rebois, M. Scholten, and C. Ramig. 2011. The greenhouse impact of unconventional gas for electricity generation. *Environmental Research Letters* 6(4): 40081-40089.

IEA. 2007. *Tracking Industrial Energy Efficiency and CO_2 Emissions*. Paris, France: OECD/IEA.

IEA. 2008. *Combined Heat and Power - Evaluating the benefits of greater global investment*. Paris, France: OECD/IEA.

IEA. 2009a. *Coal Mine Methane in Russia - Capturing the safety and environmental benefits*. Paris: International Energy Agency.

IEA. 2009b. *Technology Roadmap: Carbon Capture and Storage*. Paris, France: OECD/IEA.

IEA. 2010. *CO_2 Emissions from Fuel Combustion*. Paris, France: OECD/IEA.

IEA. 2012a. *Energy Technology Perspectives 2012: Pathways to a Clean Energy System*. Paris, France: OECD/IEA.

IEA. 2012b. *World Energy Outlook*. Paris, France OECD/IEA.

IEA. 2012c. *Technology Roadmap: High-Efficiency, Low-Emissions Coal-Fired Power Generation*. Paris, France: OECD/IEA.

IEA. 2013. *World Energy Outlook*. Paris, France OECD/IEA.

IEA GHG. 2006. *Environmental impact of solvent scrubbing of CO_2*. 2006/14. Cheltenham, UK: International Energy Agency Greenhouse Gas R&D Programme.

IEA GHG. 2007. *CO_2 capture from medium scale combustion installations*. Cheltenham, UK: International Energy Agency Greenhouse Gas R&D Programme.

IEA GHG. 2011. *Evaluation and analyses of water usage of power plants with CO_2 capture*. 2010/05. Cheltenham, UK: International Energy Agency Greenhouse Gas R&D Programme.

Ikeda, E., A. Lowe, C. Spero, and J. Stubington. 2007. *Technical performance of electric power generation systems. Selected flow sheets adapted to Australian Coal and conditions*. Vol. 1, *Technology assessment report 63. Cooperative Research Centre for Coal in Sustainable Development*. Pullenvale, Australia.

IPCC. 2006. *IPCC Reference Document on Best Available Technics for Large Combustion Plants*. Sevilla, Spain: European IPPC Bureau.

IPCC. 2005. *IPCC Special Report on Carbon Dioxide Capture and Storage. Prepared by Working Group III of the Intergovernmental Panel on Climate Change*. Cambridge, United Kingdom and New York, NY, USA: International Panel on Climate Change.

Itaoka, K., A. Saito, and M. Akai. 2005. Public acceptance of CO_2 capture and storage technology: A survey of public opinion to explore influential factors. In *Greenhouse Gas Control Technologies 7*, edited by E.S.Rubin, et al. Oxford: Elsevier Science Ltd.

Itaoka, K., Y. Okuda, A. Saito, and M. Akai. 2009. Influential information and factors for social acceptance of CCS: The 2nd round survey of public opinion in Japan. *Energy Procedia* 1(1): 4803-4810.

Jaramillo, P., W. M. Griffin, and H. S. Matthews. 2007. Comparative Life-Cycle Air Emissions of Coal, Domestic Natural Gas, LNG, and SNG for Electricity Generation. *Environmental Science & Technology* 41(17): 6290-6296.

Jiang, M., W. M. Griffin, C. Hendrickson, P. Jaramillo, J. VanBriesen, and A. Venkatesh. 2011. Life cycle greenhouse gas emissions of Marcellus shale gas. *Environmental Research Letters* 6(3): 40141-40149.

Johnson, M. R. and A. R. Coderre. 2011. An analysis of flaring and venting activity in the Alberta upstream oil and gas industry. *Journal of the Air & Waste Management Association* 61(2): 190-200.

Johnson, N., T. Gagnolet, R. Ralls, E. Zimmerman, B. Eichelberger, C. Tracey, G. Kreitler, S. Orndorff, J. Tomlinson, S. Bearer, and S. Sargent. 2010. *Pennsylvania Energy Impacts Assessment - Report 1: Marcellus Shale Natural Gas and Wind.* Arlington, VA: The Nature Conservancy.

Jordaan, S. M., D. W. Keith, and B. Steltfox. 2009. Quantifying land use of oil sands production: A life cycle perspective. *Environmental Research Letters* 4(2): 40041-400415.

Karacan, C. Ö., F. A. Ruiz, M. Cotè, and S. Phipps. 2011. Coal mine methane: A review of capture and utilization practices with benefits to mining safety and to greenhouse gas reduction. *International Journal of Coal Geology* 86(2–3): 121-156.

Kargbo, D. M., R. G. Wilhelm, and D. J. Campbell. 2010. Natural Gas Plays in the Marcellus Shale: Challenges and Potential Opportunities. *Environmental Science & Technology* 44(15): 5679-5684.

Kenny, J. F., N. L. Barber, S. S. Hutson, K. S. Linsey, J. K. Lovelace, and M. A. Maupin. 2009. *Estimated use of water in the United States in 2005*. Reston, Virginia: United States Geological Survey.

Kerr, R. A. 2010. Natural Gas From Shale Bursts Onto the Scene. *Science* 328(5986): 1624 -1626.

Ketzer, J. M., R. Iglesias, and S. Einloft. 2012. Reducing Greenhouse Gas Emissions with CO_2 Capture and Geological Storage. In *Handbook of Climate Change Mitigation*, edited by W.-Y. Chen, et al.: Springer US.

Khadse, A., M. Qayyumi, S. Mahajani, and P. Aghalayam. 2007. Underground coal gasification: A new clean coal utilization technique for India. *Energy* 32(11): 2061-2071.

Khoo, H. H. and R. B. H. Tan. 2006a. Life cycle evaluation of CO_2 recovery and mineral sequestration alternatives. *Environmental Progress* 25(3): 208-217.

Khoo, H. H. and R. B. H. Tan. 2006b. Life Cycle Investigation of CO_2 Recovery and Sequestration. *Environmental Science & Technology* 40(12): 4016-4024.

Kim, S. and B. Dale. 2005. Life Cycle Inventory Information of the United States Electricity System (11/17 pp). *The International Journal of Life Cycle Assessment* 10(4): 294-304.

Knoope, M. M. J., A. Ramírez, and A. P. C. Faaij. 2013. A state-of-the-art review of techno-economic models predicting the costs of CO_2 pipeline transport. *International Journal of Greenhouse Gas Control* 16(0): 241-270.

Koh, C. A., A. K. Sum, and E. D. Sloan. 2009. Gas hydrates: Unlocking the energy from icy cages. *Journal of Applied Physics* 106(6): 611011-06110114.

Koornneef, J., T. van Keulen, A. Faaij, and W. Turkenburg. 2008. Life cycle assessment of a pulverized coal power plant with post-combustion capture, transport and storage of CO_2. *International Journal of Greenhouse Gas Control* 2(4): 448-467.

Koornneef, J., A. Ramírez, W. Turkenburg, and A. Faaij. 2012. The environmental impact and risk assessment of CO_2 capture, transport and storage – An evaluation of the knowledge base. *Progress in Energy and Combustion Science* 38(1): 62-86.

Korre, A., Z. Nie, and S. Durucan. 2010. Life cycle modelling of fossil fuel power generation with post-combustion CO_2 capture. *International Journal of Greenhouse Gas Control* 4(2): 289-300.

Kulkarni, P. 2010. Arrival of IOCs and increasing legislative interest signal critical mass for Marcellus. *World Oil*, March, 2010, 77-85.

Kuramochi, T., W. Turkenburg, and A. Faaij. 2011. Competitiveness of CO_2 capture from an industrial solid oxide fuel cell combined heat and power system in the early stage of market introduction. *Fuel* 90(3): 958-973.

Lacombe, R. H. and J. Parsons. 2007. *Technologies, Markets and Challenges for Development of the Canadian Oil Sands Industry*. 07-006WP. Cambridge, Mass: Center for Energy and Environmental Policy Research.

Lee, J.-Y., T. C. Keener, and Y. J. Yang. 2009. Potential Flue Gas Impurities in Carbon Dioxide Streams Separated from Coal-Fired Power Plants. *Journal of the Air & Waste Management Association* 59(6): 725-732.

Liteanu, E. Spiers, C. J. de Bresser, J. H. P. 2013. The influence of water and supercritical CO_2 on the failure behavior of chalk. Tectonophysics 2013, 599, 157-169.

Liu, H. and Y. Shao. 2010. Predictions of the impurities in the CO_2 stream of an oxy-coal combustion plant. *Applied Energy* 87(10): 3162-3170.

Lombardi, L. 2003. Life cycle assessment comparison of technical solutions for CO_2 emissions reduction in power generation. *Energy Conversion and Management* 44(1): 93-108.

Lund, H., B. Mathiesen, P. Christensen, and J. Schmidt. 2010. Energy system analysis of marginal electricity supply in consequential LCA. *The International Journal of Life Cycle Assessment* 15(3): 260-271.

Markewitz, P., A. Schreiber, S. Vögele, and P. Zapp. 2009. Environmental impacts of a German CCS strategy. *Energy Procedia* 1(1): 3763-3770.

McGee, B. C. W. and C. W. McDonald. 2009. *ET-DSP Proof of Concept and Expanded Field Test Annual Performance Presentation.* Alberta, CA: Energy Resources Conservation Board.

McKinsey & Company. 2008. *Carbon capture and storage: assessing the economics*. McKinsey Climate Change Initiative.

Medlock III, K. B., A. M. Jaffe, and P. R. Hartley. 2011. *Shale gas and U.S. National Security*. Houston, Texas: James A. Baker III Institute for Public Policy - Rice University.

Miller, B. G. 2004. *Coal Energy Systems*. Burlington: Academic Press.

Modahl, I., C. Askham, K.-A. Lyng, and A. Brekke. 2012. Weighting of environmental trade-offs in CCS—an LCA case study of electricity from a fossil gas power plant with post-combustion CO_2 capture, transport and storage. *The International Journal of Life Cycle Assessment* 17(7): 932-943.

Molofsky, L. J., J. A. Connor, S. K. Farhat, A. S. Wylie, and T. Wagner. 2011. Methane in Pennsylvania water wells unrelated to Marcellus shale fracturing. *Oil & Gas* 109: 54-65.

Moore, T. A. 2012. Coalbed methane: A review. *International Journal of Coal Geology* 101(0): 36-81.

Murazeki, F. 2009. Natural gas, is it a fossil fuel anyway? Realizing a low carbon society through the advanced natural gas technologies. In *Future direction of Japanese gas industry in a low carbon society*. Tokyo, Japan.

Murphy, D. J. and C. A. S. Hall. 2010. Year in review—EROI or energy return on (energy) invested. *Annals of the New York Academy of Sciences* 1185(1): 102-118.

NEB. 2006. *Canada's Oil Sands: Opportunities and Challenges to 2015*. Calgary, Alberta: National Energy Board (NEB).

NEB. 2010. *Short-term Canadian Natural Gas Deliverability, 2010-2012*. Calgary, Alberta: National Energy Board (NEB).

NEEDS. 2009. *Deliverable no. 6.1-RS1a: External costs from emerging electricity generation technologies*. New Energy Externalities Developments for Sustainability.

NETL. 2007. *Cost and Performance Baseline for Fossil Energy Plants - Volume 1: Bituminous Coal and Natural Gas to Electricity. Final Report Revision 1*. DOE/NETL-2007-1281. Pittsburgh, PA: National Energy Technology Laboratory.

NETL. 2009. *Assessment of power plants that meet proposed greenhouse gas emission performance standards*. Pittsburgh, PA: National Energy Technology Laboratory.

NETL. 2010a. *Life Cycle Analysis: Supercritical Pulverized Coal (SCPC) Power Plant*. DOE/NETL-403-110609. Pittsburgh, PA: National Energy Technology Laboratory.

NETL. 2010b. *Cost and Performance Baseline for Fossil Energy Plants-Volume 1: Bituminous Coal and Natural Gas to Electricity-Final Report Revision 2*. Pittsburgh, PA: National Energy Technology Laboratory.

NETL. 2011a. Advanced Research: High Performance Materials. http://www.alrc.doe.gov/technologies/coalpower/advresearch/Ultrasupercritical.html. Accessed January, 2011.

NETL. 2011b. *Life Cycle Greenhouse Gas Inventory of Natural Gas Extraction, Delivery and Electricity Production*. DOE/NETL-2011/1522. National Energy Technology Laboratory.

NETL. 2013. *Best practices for risk analysis and simulation for geological storage of CO_2 - 2013 Revised version.* DOE/NETL-2013/1603. Pittsburgh, PA: National Energy Technology Laboratory.

NETL, National Energy Technology Laboratory. 2010c. *Life Cycle Analysis: Integrated Gasification Combined Cycle (IGCC) Power Plant*. DOE/NETL-403-110209. Pittsburgh PA:

NETL, National Energy Technology Laboratory. 2010d. *Life Cycle Analysis: Natural Gas Combined Cycle (NGCC) Power Plant*. DOE/NETL-403-110509Pittsburgh PA:

NETL, National Energy Technology Laboratory. 2010e. *Life Cycle Analysis: Existing Pulverized Coal (EXPC) Power Plant*. DOE/NETL-403-110809. Pittsburgh PA:

Nicholson, C. and R. L. Wesson. 1990. *Earthquake hazard associated with deep well injection: a report to the U.S. Environmental Protection Agency*, *U.S. Geological Survey Bulletin 1951*. Denver, CO: U.S. Government Printing Office.

Nie, Z. 2009. Life cycle modelling of carbon dioxide capture and geological storage in energy productionthesis, Imperial College London, London, UK.

NPCC. 2002. *Natural Gas Combined-cycle Gas Turbine Power Plants*. Northwest Power Planning Council (NPCC).

NRC. 2012. *Induced Seismicity Potential in Energy Technologies*. Washington, DC.

NYSDEC. 2011. *Revised Draft: Supplemental Generic Environmental Impact Statement On The Oil, Gas and Solution Mining Regulatory Program*. Albany, NY: New York State Department of Environmental Conservation (NYSDEC).

Odeh, N. A. and T. T. Cockerill. 2008. Life cycle GHG assessment of fossil fuel power plants with carbon capture and storage. *Energy Policy* 36(1): 367-380.

Odenberger, M. and F. Johnsson. 2010. Pathways for the European electricity supply system to 2050—The role of CCS to meet stringent CO_2 reduction targets. *International Journal of Greenhouse Gas Control* 4(2): 327-340.

Ogden, J. M. 2002. Hydrogen: The Fuel of the Future? *Physics Today* 55(4): 69-75.

OGP. 2010. *Environmental performance in the E&P industry: 2009 data*. 442. International Association of Oil & Gas Producers.

Okamura, T., M. Furukawa, and H. Ishitani. 2007. Future forecast for life-cycle greenhouse gas emissions of LNG and city gas 13A. *Applied Energy* 84(11): 1136-1149.

Okazaki, K. and T. Ando. 1997. NO_x reduction mechanism in coal combustion with recycled CO_2. *Energy* 23(2-3): 207–215.

Oltra, C., R. Sala, R. Solà, M. Di Masso, and G. Rowe. 2010. Lay perceptions of carbon capture and storage technology. *International Journal of Greenhouse Gas Control* 4(4): 698-706.

Osborn, S. G., A. Vengosh, N. R. Warner, and R. B. Jackson. 2011. Methane contamination of drinking water accompanying gas-well drilling and hydraulic fracturing. *Proceedings of the National Academy of Sciences* 108(20): 8172-8176.

Packham, R., Y. Cinar, and R. Moreby. 2011. Simulation of an enhanced gas recovery field trial for coal mine gas management. *International Journal of Coal Geology* 85(3–4): 247-256.

PADEP. 2004. *Methane Gas and Your Water Well*. Pennsylvania Department of Environmental Protection.

PADEP. 2008. *Erosion and Sediment Regulated*. Pennsylvania Department of Environmental Protection.

Palmer, I. D., S. W. Lambert, and J. L. Spiller. 1993. Coalbed methane well completions and stimulations. In *Hydrocarbons from Coal: AAPG Studies in Geology 38*, edited by B. E. Law and D. D. Rice.

PCAST. 2005. *The National Nanotechnology Initiative at Five Years: Assessment and Recommendations of the National Nanotechnology Advisory Panel.* President's Council of Advisors on Science and Technology.

Pehnt, M. 2008. Environmental impacts of distributed energy systems—The case of micro cogeneration. *Environmental Science & Policy* 11(1): 25-37.

Pehnt, M. and J. Henkel. 2009. Life cycle assessment of carbon dioxide capture and storage from lignite power plants. *International Journal of Greenhouse Gas Control* 3(1): 49-66.

Pétron, G., G. Frost, B. R. Miller, A. I. Hirsch, S. A. Montzka, A. Karion, M. Trainer, C. Sweeney, A. E. Andrews, L. Miller, J. Kofler, A. Bar-Ilan, E. J. Dlugokencky, L. Patrick, C. T. Moore, T. B. Ryerson, C. Siso, W. Kolodzey, P. M. Lang, T. Conway, P. Novelli, K. Masarie, B. Hall, D. Guenther, D. Kitzis, J. Miller, D. Welsh, D. Wolfe, W. Neff, and P. Tans. 2012. Hydrocarbon emissions characterization in the Colorado Front Range: A pilot study. *Journal of Geophysical Research: Atmospheres* 117(D4): D04304.

PGA. 2010. *The power of renewable: opportunities and challenges for China and the United States*. Washington, DC: Chinese Academy of Sciences; Chinese Academy of Engineering

Pluymakers, A. M. H. Samuelson, J. E. Niemeijer, A. R. Spiers, C. J., 2014. Effects of temperature and CO_2 on the frictional behavior of simulated anhydrite fault rock. *Journal of Geophysical Research: Solid Earth* 2014, 119, (12), 2014JB011575.

Rao, A. B. and E. S. Rubin. 2002. A Technical, Economic, and Environmental Assessment of Amine-Based CO_2 Capture Technology for Power Plant Greenhouse Gas Control. *Environmental Science & Technology* 36(20): 4467-4475.

Rath, B. B. and J. M. Marder. 2007. Powering the Future: Natural Gas - Challenges and Solutions. *Advanced Materials & Processes*, March, 2007, 27-29.

RECCS. 2008. *Ecological, economic and structural comparison of renewable energy technologies with carbon capture and storage – an integrated approach*. Berlin, Germany: Federal Ministry for the Environment, Nature Conservation and Nuclear Safety (BMU).

Reiner, D. 2010. The public(s) and CCS: An overview: FENCO workshop "CCS and public engagement": When science meets reality, Amsterdam, 19 May 2010.

Reiner, D. M., T. E. Curry, M. A. de Figueiredo, H. J. Herzog, S. D. Ansolabehere, K. Itaoka, F. Johnsson, and M. Odenberger. 2006. American Exceptionalism? Similarities and Differences in National Attitudes Toward Energy Policy and Global Warming. *Environmental Science & Technology* 40(7): 2093-2098.

Remme, U., N. Trudeau, D. Graczyk, and P. Taylor. 2011. *Technology development prospects for the Indian power sector*. Paris, France: OECD/IEA.

Réveillère, A. and J. Rohmer. 2011. Managing the risk of CO_2 leakage from deep saline aquifer reservoirs through the creation of a hydraulic barrier. *Energy Procedia* 4(0): 3187-3194.

Rogner, H.-H. 1997. an assessment of world hydrocarbon resources. *Annual Review of Energy and the Environment* 22(1): 217-262.

Rooney, R. C., S. E. Bayley, and D. W. Schindler. 2012. Oil sands mining and reclamation cause massive loss of peatland and stored carbon. *Proceedings of the National Academy of Sciences* 109(13): 4933-4937.

Rubin, E. S. 2008. CO_2 Capture and Transport. *Elements*, October, 2008, 311-317.

Rubin, E. S., C. Chen, and A. B. Rao. 2007. Cost and performance of fossil fuel power plants with CO_2 capture and storage. *Energy Policy* 35(9): 4444-4454.

Saghafi, A. 2012. A Tier 3 method to estimate fugitive gas emissions from surface coal mining. *International Journal of Coal Geology* 100(0): 14-25.

Samuelson, J. Spiers, C. J. 2012. Fault friction and slip stability not affected by CO_2 storage: Evidence from short-term laboratory experiments on North Sea reservoir sandstones and caprocks. International Journal of Greenhouse Gas Control 2012, 11S, S78-S90.

Saripalli, K. P., N. M. Mahasenan, and E. M. Cook. 2003. Risk and Hazard Assessment for Projects Involving the Geological Sequestration of CO_2. In *Greenhouse Gas Control Technologies - 6th International Conference*, edited by J. G. Kaya. Oxford: Pergamon.

Scheffknecht, G., L. Al-Makhadmeh, U. Schnell, and J. Maier. 2011. Oxy-fuel coal combustion—A review of the current state-of-the-art. *International Journal of Greenhouse Gas Control* 5, Supplement 1(0): S16-S35.

Schenk, C. J. and R. M. Pollastro. 2002. *Natural Gas Production in the United States: National Assessment of Oil and Gas Series*. U.S. Geological Survey.

Scheufele, D. and B. Lewenstein. 2005. The Public and Nanotechnology: How Citizens Make Sense of Emerging Technologies. *Journal of Nanoparticle Research* 7(6): 659-667.

Schreiber, A., P. Zapp, and W. Kuckshinrichs. 2009. Environmental assessment of German electricity generation from coal-fired power plants with amine-based carbon capture. *The International Journal of Life Cycle Assessment* 14(6): 547-559.

Seiersten, M. and K. O. Kongshaug. 2005. Chapter 16 - Materials Selection for Capture, Compression, Transport and Injection of CO_2. In *Carbon Dioxide Capture for Storage in Deep Geologic Formations*, edited by D. C. Thomas. Amsterdam: Elsevier Science.

Senapati, M. R. 2011. Fly ash from thermal power plants - Waste management and overview. *Current Science* 100(12): 1791-1794.

Serpa, J., J. Morbee, and E. Tzimas. 2011. *Technical and Economic Characteristics of a CO_2 Transmission Pipeline Infrastructure*. EUR 24731 EN Luxembourg: Institute for Energy.

Shackley, S., S. Mander, and A. Reiche. 2006. Public perceptions of underground coal gasification in the United Kingdom. *Energy Policy* 34(18): 3423-3433.

Shafirovich, E. and A. Varma. 2009. Underground coal gasification: A brief review of current status. *Industrial and Engineering Chemistry Research* 48(17): 7865-7875.

Sharp, J. D., M. K. Jaccard, and D. W. Keith. 2009. Anticipating public attitudes toward underground CO_2 storage. *International Journal of Greenhouse Gas Control* 3(5): 641-651.

Shindell, D. T., G. Faluvegi, D. M. Koch, G. A. Schmidt, N. Linger, and S. E. Bauer. 2009. Improved attribution of climate forcing to emissions. *Science* 326(5953): 716-718.

Singh, B., A. H. Strømman, and E. G. Hertwich. 2011a. Comparative impact assessment of CCS portfolio: Life cycle perspective. Paper presented at Energy Procedia.

Singh, B., A. H. Strømman, and E. Hertwich. 2011b. Life cycle assessment of natural gas combined cycle power plant with post-combustion carbon capture, transport and storage. *International Journal of Greenhouse Gas Control* 5(3): 457-466.

Skovholt, O. 1993. CO_2 transportation system. *Energy Conversion and Management* 34(9-11): 1095-1103.

Sloan, E. D. and C. A. Koh. 2008. *Clathrate Hydrates of Natural Gases, Third Edition*. Edited by T. Francis. Boca Raton, FL: CRC Press.

Smart, A. and A. Aspinall. 2009. *Water and the electricity generation industry. Implications of use*. Canberra, Australia: National Water Commission.

Song, Y., S. Liu, Q. Zhang, M. Tao, M. Zhao, and F. Hong. 2012. Coalbed methane genesis, occurrence and accumulation in China. *Petroleum Science* 9(3): 269-280.

Spath, P. and M. Mann. 2004. *Biomass power and conventional fossil systems with and without CO_2 sequestration – Comparing the energy balance, greenhouse gas emissions and economics*. NREL/TP-510-32575 Golden, Colorado: National Renewable Energy Laboratory.

Stephenson, T., J. E. Valle, and X. Riera-Palou. 2011. Modeling the relative GHG emissions of conventional and shale gas production. *Environmental Science and Technology* 45(24): 10757-10764.

Strachan, N., R. Hoefnagels, A. Ramírez, M. van den Broek, A. Fidje, K. Espegren, P. Seljom, M. Blesl, T. Kober, and P. E. Grohnheit. 2011. CCS in the North Sea region: A comparison on the cost-effectiveness of storing CO_2 in the Utsira formation at regional and national scales. *International Journal of Greenhouse Gas Control* 5(6): 1517-1532.

Stultz, S. C. and J. B. Kitto. 1992. *Steam, its generation and use, 40th edition*. Barberton, Ohio Babcock & Wilcox.

Su, S., J. Han, J. Wu, H. Li, R. Worrall, H. Guo, X. Sun, and W. Liu. 2011. Fugitive coal mine methane emissions at five mining areas in China. *Atmospheric Environment* 45(13): 2220-2232.

Sundkvist, S., Å. Klang, M. Sjödin, K. Wilhelmsen, K. Åsen, A. Tintinelli, S. McCahey, and H. Ye. 2004. AZEP gas turbine combined cycle power plants – thermal optimalisation and LCA analysis. Paper presented at Seventh International Conference on Greenhouse Gas Control Technologies, GHGT-7, Vancouver, Canada.

Svanes, E. 2008. *Evaluation of direct toxic impact of CO_2 capture chemicals in view of REACH regulation*. OR 09.08. Fredrikstad, Norway: Østfoldforskning AS.

Tabe, Y., S. Hirai, and K. Okazaki. 2000. Massive CO_2 clathrate hydrate growth at a high-polar-energy surface. *Journal of Crystal Growth* 220(1–2): 180-184.

Ter Mors, E., M. W. H. Weenig, N. Ellemers, and D. D. L. Daamen. 2010. Effective communication about complex environmental issues: Perceived quality of information about carbon dioxide capture and storage (CCS) depends on stakeholder collaboration. *Journal of Environmental Psychology* 30(4): 347-357.

Terwel, B. W., F. Harinck, N. Ellemers, and D. D. L. Daamen. 2009. Competence-Based and Integrity-Based Trust as Predictors of Acceptance of Carbon Dioxide Capture and Storage (CCS). *Risk Analysis* 29(8): 1129-1140.

Thitakamol, B., A. Veawab, and A. Aroonwilas. 2007. Environmental impacts of absorption-based CO_2 capture unit for post-combustion treatment of flue gas from coal-fired power plant. *International Journal of Greenhouse Gas Control* 1(3): 318-342.

Tokushige, K., K. Akimoto, and T. Tomoda. 2007. Public perceptions on the acceptance of geological storage of carbon dioxide and information influencing the acceptance. *International Journal of Greenhouse Gas Control* 1(1): 101-112.

Toman, M., A. E. Curtright, D. S. Ortiz, J. Darmstadter, and B. Shannon. 2008. *Unconventional Fossil-Based Fuels: Economic and Environmental Trade-Offs*. Arlington, VA: RAND Corporation.

Tzimas, E., A. Mercier, C.-C. Cormos, and S. D. Peteves. 2007. Trade-off in emissions of acid gas pollutants and of carbon dioxide in fossil fuel power plants with carbon capture. *Energy Policy* 35(8): 3991-3998.

UNECE. 2010. *Best Practice Guidance for Effective Methane Drainage and Use in Coal Mines*. New York and Geneva: United Nations Economic Commission for Europe.

UNEP. 2007. *Environmental management in oil and gas exploration and production. An overview of issues and management approaches*. UNEP IE/PAC Technical Report 37; E&P Forum Report 2.72/254. Paris, France.

URS Corporation. 2011. *Water-related issues associated with gas production in the Marcellus shale*. Fort Washington, PA URS Corporation.

US DOE. 2004a. *Fuel cell handbook*. Washington, DC: United States Department of Energy

US DOE. 2004b. *Coal Bed Methane Primer - New Source of Natural Gas–Environmental Implications*. Washington, DC: United States Department of Energy.

US DOE. 2009. *Modern Shale Gas Development in the United States: A Primer*. Washington, DC: United States Department of Energy.

US EIA. 1993. *Drilling sideways - A review of horizontal well technology and its domestic application*. Washington, DC: United States Energy Information Administration.

US EIA. 2011a. *Review of Emerging Resources: U.S. Shale Gas and Shale Oil Plays*. Washington, DC: United States Energy Information Administration.

US EIA. 2011b. International Energy Statistics: Petroleum, reserves. United States Energy Information Administration.

US EIA. 2011c. *Annual Energy Outlook 2011 with Projections to 2035*. Washington, DC: United States Energy Information Administration.

US EIA. 2013a. *Technically Recoverable Shale Oil and Shale Gas Resources: An Assessment of 137 Shale Formations in 41 Countries Outside the United States.* Washington, DC: United States Energy Information Administration.

US EIA. 2013b. *Natural Gas Annual 2013*. Washington, DC: United States Energy Information Administration.

US EPA. 2008. *An Assessment of the Environmental Implications of Oil and Gas Production: A Regional Case Study*. United States Environmental Protection Agency.

US EPA. 2010. *Available and emerging technologies for reducing greenhouse gas emissions from coal-fired generating units*. Research Triangle Park, North Carolina: United States Environmental Protection Agency.

US EPA. 2014. *Catalogue of CHP Technologies*. U.S. Environmental Protection Agency. Combined Heat and Power Partnership. Washington, DC: United States Environmental Protection Agency, March 2014.

USGS. 2000. *Coal-Bed Methane: Potential and Concerns*. U.S. Geological Survey.

van den Broek, M., A. Ramírez, H. Groenenberg, F. Neele, P. Viebahn, W. Turkenburg, and A. Faaij. 2010. Feasibility of storing CO_2 in the Utsira formation as part of a long term Dutch CCS strategy: An evaluation based on a GIS/MARKAL toolbox. *International Journal of Greenhouse Gas Control* 4(2): 351-366.

Vassolo, S. and P. Döll. 2005. Global-scale gridded estimates of thermoelectric power and manufacturing water use. *Water Resources Research* 41(4): W04010.

Veil, J. A. 2010. *Water Management Technologies Used by Marcellus Shale Gas Producers*. ANL/EVS/R-10/3. Washington, D.C.: Environmental Science Division, Argonne National Laboratory for the U.S. Department of Energy, Office of Fossil Energy.

Veltman, K., B. Singh, and E. G. Hertwich. 2010. Human and Environmental Impact Assessment of Postcombustion CO_2 Capture Focusing on Emissions from Amine-Based Scrubbing Solvents to Air. *Environmental Science & Technology* 44(4): 1496-1502.

Venkatesh, A., P. Jaramillo, W. M. Griffin, and H. S. Matthews. 2011. Uncertainty in life cycle greenhouse gas emissions from United States natural gas end-uses and its effects on policy. *Environmental Science and Technology* 45(19): 8182-8189.

Venkatesh, A., P. Jaramillo, W. M. Griffin, and H. S. Matthews. 2012. Uncertainty in life cycle greenhouse gas emissions from United States coal. *Energy and Fuels* 26(8): 4917-4923.

Viebahn, P., J. Nitsch, M. Fischedick, A. Esken, D. Schüwer, N. Supersberger, U. Zuberbühler, and O. Edenhofer. 2007. Comparison of carbon capture and storage with renewable energy technologies regarding structural, economic, and ecological aspects in Germany. *International Journal of Greenhouse Gas Control* 1(1): 121-133.

Voosen, P. 2010. Frightened, furious neighbours undermine German CO_2: Trapping power project. *The New* York Times, 7 April 2010.

Waku, H., I. Tamura, M. Inoue, and M. Akai. 1995. Life cycle analysis of fossil power plant with CO_2 recovery and sequestering system. *Energy Conversion and Management* 36(6–9): 877-880.

Wang, G., S. Liu, W. Su, W. Sun, D. Wang, H. Yuan, G. Xu, and C. Zou. 2008. Water soluble gas in deep carbonate reservoir, Sichuan basin, Southwest China *Journal of China University of Geosciences* 19(6): 636-644.

Waxman, H. A., E. J. Markey, and D. DeGette. 2011. *Chemicals Used in Hydraulic Fracturing*. Washington, D.C.: United States House of Representatives.

Weber, C. L. and C. Clavin. 2012. Life cycle carbon footprint of shale gas: Review of evidence and implications. *Environmental Science and Technology* 46(11): 5688-5695.

Weisser, D. 2007. A guide to life-cycle greenhouse gas (GHG) emissions from electric supply technologies. *Energy* 32(9): 1543-1559.

Weng, Z., G. M. Mudd, T. Martin, and C. A. Boyle. 2012. Pollutant loads from coal mining in Australia: Discerning trends from the National Pollutant Inventory (NPI). *Environmental Science and Policy* 19-20: 78-89.

Williams, H. F. L., D. L. Havens, K. E. Banks, and D. J. Wachal. 2008. Field-based monitoring of sediment runoff from natural gas well sites in Denton County, Texas, USA. *Environmental Geology* 55(7): 1463-1471.

WorleyParsons and Schlumberger. 2011. *Economic Assessment of Carbon Capture and Storage Technologies, 2011 Update*. Canberra, Australia Global CCS Institute.

Yang, X. and C. Lant. 2011. A geographical and temporal analysis of U.S. thermoelectric withdrawals in comparison to renewable water supplies. *Applied Geography* 31(3): 1154-1165.

Yeh, S. and E. S. Rubin. 2007. A centurial history of technological change and learning curves for pulverized coal-fired utility boilers. *Energy* 32(10): 1996-2005.

Yu, F., J. Chen, F. Sun, S. Zeng, and C. Wang. 2011. Trend of technology innovation in China's coal-fired electricity industry under resource and environmental constraints. *Energy Policy* 39(3): 1586-1599.

Zapp, P., A. Schreiber, J. Marx, M. Haines, J.-F. Hake, and J. Gale. 2012. Overall environmental impacts of CCS technologies—A life cycle approach. *International Journal of Greenhouse Gas Control* 8(0): 12-21.

ZEP. 2011. *The costs of CO_2 capture, transport and storage: Post-demonstration CCS in the EU*. European Technology Platform for Zero Emission Fossil Fuel Power Plants.

Zhai, H., E. S. Rubin, and P. L. Versteeg. 2011. Water use at pulverized coal power plants with postcombustion carbon capture and storage. *Environmental Science and Technology* 45(6): 2479-2485.

Zhang, Z. 1995. On the water soluble gas. *Natural Gas Geoscience* 6(5): 29-34.

Zhou, W., W. L. Chen, H. C. Deng, and Q. M. Zhou. 2011. Distribution, status and problems of world water-soluble gas resources. *Journal of Mineralogy and Petrology* 31(2): 73-78.

Zoback, M. D. Gorelick, S. M., 2012. Earthquake triggering and large-scale geologic storage of carbon dioxide. Proceedings of the National Academy of Sciences 2012, 109, (26), 10164-10168.

AMERICAN
150 TON

Chapter 4

Hydropower

Lead author: Edgar G. Hertwich
Contributing authors: Claudia Peña, Thomas Gibon, Mabel Vega

4.1 INTRODUCTION

Hydropower is currently the most important source of renewable electricity, supplying 3,288 TWh in 2009, which amounted to 6.1 per cent of the global primary energy supply. The amount of energy supplied from this source is currently increasing by approximately 3 per cent annually; the unexploited technical potential of hydropower is on the order of 10,000 to 15,000 TWh per year (Turkenburg et al., 2012). Important unexploited resources are concentrated in regions such as Africa and South America, where the initiation of hydropower projects has great potential for accelerating economic development, which may hence incentivise the development of these resources. Important drivers for hydropower deployment are energy security and climate protection. Hydropower can be inexpensive, easy to regulate, and offer black start capability and energy storage, although these benefits depend on the type and location of the facility. Hydropower plants tend to have longer lifetime than other power plants and are more likely to be refurbished than completely removed. At the same time, hydropower alters river flow patterns and leads to large changes in river landscapes and ecology. Hydropower projects have caused large resettlements and spearheaded development in some remote regions, with both positive and negative consequences. In terms of environmental concerns, freshwater ecosystem impacts associated with the dams, reservoirs and flow patterns, concerns about water quality, and biogenic greenhouse gas (bGHG) emissions are the largest consequences of hydropower projects.

Hydropower is often only one of several purposes for which dams, reservoirs, and associated channels are constructed (Table 4.1). Such facilities can additionally serve to store freshwater, enable river navigation and irrigation, control floods, and allow for new fisheries and tourism (World Commission on Dams, 2000). As this study focuses on the external benefits, risk, and environmental costs of electricity production to society at large, benefits such as navigation, river regulation, irrigation, flood control, and tourism may be counted as co-benefits of hydropower. On the other hand, these purposes are often important factors in the decision-making, and electricity may hence be viewed as a co-product of irrigation or navigation projects. The literature on multiple objective analysis connected to dam projects to a large degree addresses a trade-off between identified purposes such as electricity and irrigation, and various environmental and economic costs (Bai et al., 2013; Reed et al., 2013; Hurford et al., 2013; Chen et al., 2013; Ziv et al., 2012; Kuenzer et al., 2013).

Based on the records of the global reservoir and dams database, 16.7 million reservoirs larger than 100 m^2 were estimated to exist as of 2011 (Lehner et al., 2011). Nearly half of large rivers with a flow rate of more than 1,000 m^3s^{-1} are dammed (Lehner et al., 2011). Dams have increased the land area covered by surface water by 305,000 km^2. This area is approximately equivalent to 7 per cent of the area covered by naturally occurring surface water. Combined, these dams provide a storage capacity of 8,000 km^3. For comparison, the surface area of Lake Superior is 58,000 km^2 and has a volume of 12,100 km^3, while Lake Tanganyika covers 33,000 km^2 and has a volume of 18,900 km^3.

TABLE 4.1

Different types of hydropower plants, their purposes and specific social and environmental characteristics

Hydropower plant type	Energy and water management services	Main environmental and social characteristics (corresponding subsection)
All	Renewable electricity generation Increased water management options	Barrier for fish migration and navigation (4.2.2), and sediment transport (4.2.1.1, 4.2.3.4); Physical modification of riverbed and shorelines (4.2.1.4)
Run-of-river	Limited flexibility and increased variability in electricity generation output profile. Water quality (but no water quantity) management	Unchanged river flow when powerhouse in dam toe; when localized further downstream reduced flow between intake and powerhouse
Reservoir (Storage)	Storage capacity for energy and water; Flexible electricity generation output; Water quantity and quality management; groundwater stabilization; Water supply and flood management	Alteration of natural and human environment by impoundment and resulting in impacts on ecosystems and biodiversity (4.2.1.4, 4.2.4). Modification of volume and seasonal patterns of river flow (4.2.3.2, 4.2.3.3), changes in water temperature and quality (4.2.3.5), land use change-related GHG emissions (4.3)
Multipurpose	As for reservoir HPPs; Dependent on water consumption of other uses.	As for reservoir HPP; Possible water use conflicts; Driver for regional development (4.2.1.5)
Pumped storage	Storage capacity for energy and water; Net consumer of electricity due to pumping; No water management options.	Impacts confined to a small area; often operated outside the river basin as a separate system that only exchanges the water from a nearby river from time to time

Source: Kumar et al., 2011, Table 5.5

Contrasting perspectives on hydropower and its role in development and environmental protection have given rise to efforts to reconcile the different interests. The World Commission on Dams (WCD) was a high level effort to bring together representatives from different interests in order to study the effects of large dam development on communities, ecosystems and economic development (World Commission on Dams, 2000). The resulting report presents a comprehensive set of guidelines for dam development. Since then, the International Hydropower Association has continued working on sustainability issues, defining guidelines for the assessment of sustainability and the measurement of greenhouse gas (GHG) emissions arising from hydropower. Non-governmental organizations have continued campaigns opposing hydropower development and influencing its design to mitigate large impacts.

The environmental impacts of hydropower are very much project-specific, as they depend on the precise geographical conditions; climate, geology, ecosystems, settlement patterns, hydrological regimes, gradients and project size. A number of impacts can be avoided or reduced through the proper selection and design of projects. The strong dependency on local factors and design makes it difficult to provide a generalized conclusion regarding the impacts of hydropower. The largest environmental impacts from a macro perspective may arise from a few projects. Knowing the average impact says little about the merit of an individual project. With these limitations in mind, we try to gain an understanding of the environmental impacts and benefits of hydropower from a global perspective, noting the high variability among projects.

This chapter surveys the environmental issues related to hydropower at a similar level of detail to the IPCC Special Report on Renewable Energy (SRREN), which was a major resource in our work. This report goes a step further than the SRREN in the review and discussion of bGHG emissions and the accounting of bGHG in life cycle assessment (LCA). The climate benefit of hydropower is currently poorly understood because of the weak understanding of bGHG emissions (Fearnside and Pueyo, 2012) and the omission of the climate impacts from increased evaporation in hydropower reservoirs. SRREN emphasized the difference between different dam types, which differ in environmental impacts (Table 4.1). However, other power plant characteristics that have a greater influence on environmental impacts than plant type, so we treat hydropower plants in this study as a continuum rather than distinct classes. Such a treatment is in line with the literature on ecological impacts. In recent years, global assessments (Sathaye et al., 2011; GEA, 2012) have increasingly relied on LCA to evaluate GHG emissions and climate benefits of energy technologies. However, as the underlying LCA literature lacked a consistent treatment of bGHG emissions for hydropower, previous studies have not been particularly useful for assessing the climate benefit of hydropower.

4.2 ECOLOGICAL IMPACTS OF HYDROPOWER PLANTS

This section provides an overview of ecological impacts that can be caused by dams and reservoirs. As previously mentioned, the actual impacts depend very much on the specific project and biogeography. Dams can serve many purposes such as flood control, irrigation and navigation; hydropower generation is often only one of several purposes. The construction of hydropower plants modifies creeks and streams and often leads to the flooding of land areas; they are hence associated with changes in both aquatic and terrestrial habitat (Alho, 2011). The ecological consequences of such habitat change are site-specific and difficult to generalize. Habitat change leads to a change in species that populate these habitats, with potential consequences for larger regions, including the areas downstream of dams. The ecological impacts of hydropower dams are subject to controversy, and there are efforts to mitigate adverse consequences. Hydropower plants are massive civil engineering projects that may involve substantial earth movement, dam construction, tunnelling, and weir, pipe, turbine and electrical equipment installation. Although some environmental impacts associated with construction and machinery have been assessed with LCA, these assessments have gaps. A full review of the environmental impacts of hydropower was not possible as part of this work; we hence limit ourselves to a short description.

In the following section, we will first discuss the upstream impacts from the dam resulting from reservoir formation, those caused by the dam through the blocking of migration pathways, and downstream impacts resulting from changes in the flow regime and water properties. Finally, we address macroecological effects caused by hydropower projects over a larger region, as they are often placed several in series. In the subsequent section, we briefly address opportunities to mitigate these impacts.

4.2.1 RESERVOIR

The creation of a reservoir transforms terrestrial and riparian ecosystems into aquatic lake ecosystems. The inundation of land and embankments can affect both terrestrial and aquatic species. Shallow water habitat represents important breeding grounds and depends on the interaction of terrestrial and hydrologic processes. It is thus vulnerable to environmental change (Alho, 2011).

4.2.1.1 Sedimentation

Sediment carrying capacity is directly related to the current velocity and slop of the water body. As a result, the reduction of stream velocity leads to the sediment deposition in the reservoir. Kumar et al. (2011) point to earlier work which indicates that 0.5-1.0 per cent of the global freshwater storage capacity of reservoirs is lost annually as a result of sedimentation. The filling of reservoirs by sediments can raise the riverbed and increase flood risks, as was the case in the lower reaches of the Yellow River (Xu, 2002). The extent of sedimentation depends on the sediment flow of the drainage basin which again is a function of geological and climatic conditions. Human activities such as agriculture, mining, urbanization, river regulation, and infrastructure projects influence both the amount and composition of sediments.

4.2.1.2 Water quality

Sediments consist not only of minerals but also of organic matter, which contains nutrients. This nutrient input improves growing conditions for phytoplankton and algae, further increasing the organic content of the reservoir. Reduced turbulent mixing through reduced velocity and increased depth results in thermal stratification. The oxidation of organic matter in a stratified reservoir leads to the depletion of oxygen and may thus result in the formation of an anoxic zone (Kumar et al., 2011). Under anoxic conditions, the degradation of organic matter leads to the formation of CH_4. Water quality of the tributaries, especially the input of nutrients and organic matter, are important factors in determining the water quality and health of a reservoir.

4.2.1.3 Public health

Increased still water and poor water quality provide good habitats for disease vectors for malaria, river blindness, dengue or yellow fever, amongst others (Kumar et al., 2011; Ziegler et al., 2013). This is especially a problem in tropical and subtropical regions and in cases where dam construction or population displacement leads to higher concentrations of humans. In other cases, the deforestation associated with dam construction also contributes to the spread of disease vectors. For example, Vilela et al. (2011) investigate the spread of leishmaniasis as a result of the construction of the Luís Eduardo Magalhães Hydroelectric Plant in Brazil.

The creation of anoxic conditions can lead to the release of the mercury bound in soil and accumulated biomass, and the subsequent methylation of this mercury by sulfate and iron reducing bacteria (Driscoll et al., 2013). The resulting methylmercury can then enter aquatic food chains and lead to toxic effects in humans (Gump et al., 2012). Reservoirs also reduce the transport of mercury from other natural or anthropogenic sources, such as mining and fossil fuel power plants, to the oceans, thus leading to the mercury accumulation in freshwater bodies (Kumar et al., 2011).

Hydropower can also contributed to improved public health, in most cases through the development of the local economy that goes hand in hand with hydropower development. Dam operators sometimes finance public health programs to protect and improve the health of the local population.

4.2.1.4 Habitat change

Impoundment leads to a substantial change of habitat for fish and other aquatic species such as amphibians and crustaceans. Species adapted to fast flowing rivers are replaced by species adapted to lake-type environments. From an anthropocentric perspective, this can be positive or negative. Although the total biomass often increases with dam construction, it tends to favour species of lower the commercial value. Habitat created by dams can be ecologically valuable for birds, as these habitats replace wetland lost to agriculture in nearby areas (Kumar et al., 2011). However, from a biodiversity perspective, the reservoir is often of lower value than wetlands as it contains fewer ecological niches and often becomes the habitat of alien introduced species, which displace native species. Globally, there is a loss of riparian habitat and associated biodiversity.

4.2.1.5 Social impacts

The creation of a reservoir can also lead to the displacement of populations (Bao, 2010; Heming et al., 2001; Nakayama et al., 1999) and the flooding of cultural heritage sites (Kumar et al., 2011). Scudder (Scudder, 2002, 2005) surveyed 50 cases of dam construction involving the resettlement of a total of 1.5 million people. About half of the affected population was classified as tribal or indigenous, and the majority consisted of smallholder farmers. Scudder found that in 82 per cent of the investigated cases, resettlement lead to a deterioration of living conditions for the affected population; living conditions improved in only 7 per cent of cases. The results are complicated by the fact that in many cases, the resettlement process had not been completed at the time of the survey. The affected populations were found to suffer from unemployment and landlessness, implying a loss of livelihood resulting from the construction project. The survey found that displaced native populations were unable to compete with migrants attracted by the construction project. This was an important contributing factor to the overall negative outcome.

4.2.2 DAM

Dams present large, physical barriers to passage up and down rivers. This obstruction leads to habitat fragmentation, decrease of in-stream habitat and blockage of migrating fish (Finer and Jenkins 2012; Renöfält et al., 2010; Wollebæk et al., 2011b; Ziv et al., 2012; Sheaves et al., 2008; McLellan et al., 2008; Saunders et al., 1991).

4.2.2.1 Obstruction of fish migration

Dams obstruct the migration of migratory fish species, thus interfering with their life cycles. Blocking migrating species is a serious problem caused by damming the river. Many fish populations have been, or are expected to become, extinct because of dam blocking. Diadromous fish that live in salt water and spawn in freshwater or vice versa are, in many cases, entirely unable to reach their spawning grounds. Salmon and shad have become locally extinct due to dam construction at several sites (Mann and Plummer, 2000; Larinier, 2001; Thorstad et al., 2008). The dams in Elwha river have obstructed the upstream migration of *salmonidae* to over 90 per cent of the watershed for over 90 years in Washington State (Pess et al., 2008). In some cases, these impacts can be mitigated through fish ladders, e.g., for salmon.

4.2.2.2 Habitat fragmentation

Dams also isolate local fish, insects and larval clam populations (Wollebæk et al., 2011b). Reduced genetic exchange between populations can lead to decreased survivability. Biological interactions also play a part. For example, reductions in insects and larval clam populations, which serve as food for organisms higher up on the food chain, can have indirect effects on fish populations (Finer and Jenkins 2012). Dams can also compromise the dispersal of seeds (Nilsson and Berggren 2000).

4.2.3 DOWNSTREAM IMPACTS

Dams affect the natural fluctuations in water flow. Although reservoirs prevent seasonal flooding, this can reduce the deposition of nutrients on flood plains and affect species and ecosystems that are dependent on regular flooding (Kunz et al., 2011). We discuss below some of the concerns that can arise from hydropower projects.

4.2.3.1 Volume and timing of water release

Such changes affect many downstream species and habitats. Fish and amphibians require specific conditions on banks and in flood pools to spawn and rear. These conditions may be affected by the timing, volume, ramping and pulsing of water flow from the dam, leading to reproductive failure (Yarnell et al., 2012; Young et al., 2011; Guo et al., 2011; Arias et al., 2014). For example, Yarnell et al., investigate the effect of seasonal water pulses of regulated waterways in Northern California on the reproduction of foothill yellow-legged frog (*Rana boylii*) populations and find that there is a disconnect between suitable sites that protect egg masses and tadpole habitats. In addition, the timing of pulses e.g., spring floods can be the cue for species to begin migrating, and a hydropower driven modification can disturb these signals (Young et al., 2011). In extreme cases, hydropower reservoir operation can cause rivers to temporarily run dry (Fu et al., 2008). Regulating the river flow to avoid or mitigate spring floods and other seasonal flow variations may be included in the purpose of the dam and can have both positive and negative effects on local wildlife and economy (Arias et al., 2014).

4.2.3.2 Flood plains

Dam construction reduces seasonal flooding and associated nutrient deposition, affecting the extent and fertility of flood plains (Zeilhofer and de Moura 2009). Flood plains tend to have a high biodiversity and high productivity. On the Zambezi river, the completion of the Itezhi-Tezhi Reservoir in 1978 has led to a reduction of nitrogen and phosphorus transport to the floodplains of the Kafue flood plains by 50 per cent and 60 per cent, respectively (Kunz et al., 2011). The regulation of flow can hence have an important impact on downstream terrestrial habitat.

4.2.3.3 Reduced sediment flow

Sedimentation in reservoirs reduces the sediment load in rivers, changes their morphology, and can lead to a deepening of rivers, a reduction of water tables, and the erosion of river deltas, subsequently affecting these downstream ecosystems. Sediment starvation attributed to retention by dams can alter the substrate composition downstream, which is important for spawning and rearing habitat formation. In coastal areas, the erosion caused by waves is no longer counteracted by deposition of sediment; the WCD reports that the coastline of Togo and Benin has decreased by 10-15 meters per year after the Akosombo Dam on the Volta River was completed (World Commission on Dams, 2000). For the Nile River, the Aswan High Dam has stopped the flow of sediment, resulting in a significant erosion of the riverbed and banks and a retreat of its estuary. As a result of lowering the river bed by 2-3 m, irrigation intakes were left dry and bridges undermined (Kumar et al., 2011).

There are indications that erosion may also contribute to floodplain fertility (Arias et al., 2014). Downstream, changes in flows of freshwater and in nutrient levels can influence the estuarine habitats where many marine fish come to spawn. Lowered nutrient levels can result in lowered overall productivity from a diminished primary food source, i.e. less primary production, as occurred with the Aswan High Dam in Egypt. Furthermore, increases in salinity resulting from reduced freshwater flows can allow marine predators to invade, lowering recruitment rates (WCD, 2000).

A reduction of the sediment flow can have a substantial impact on marine ecosystems through reduced input of silica and other nutrients, which affect algal ecology (Ittekkot et al., 2000). The reduction of freshwater input to estuaries also affects the composition of fish found in these habitats (Vorwerk et al., 2008).

4.2.3.4 Changes in water quality of downstream waterways

Stratification effects

A reservoir affects a number of variables, including the water temperature through thermal stratification and the content of dissolved gases through the hydrostatic pressure and creation of potentially anoxic conditions. Hydropower plants sometimes draw water from deeper layers of the reservoir, where the temperature and dissolved gas content can be substantially different from the natural conditions of the river. Lower water temperatures in a river can have an impact on sensitive native fish species and life history processes of invertebrates. In the long term, susceptible species may be eliminated altogether from the downstream habitat. Coldwater releases have been found to delay spawning by up to 30 days in some fish species (Sherman et al., 2007; Miles and West, 2011). However, human activities generally tend to increase water temperatures, e.g., by cooling systems for thermal power plants (Hester and Doyle, 2011), and dams are sometimes used to reduce water heating effects and thus maintain more natural water temperatures.

Dissolved gas supersaturation

Spills over dams may cause supersaturation of waters downstream, which influence the physiological processes in aquatic fauna. For example, supersaturated waters absorbed by fish during respiration cause the formation of gas bubbles in the bloodstream (Johnson et al., 2007; Li et al., 2009). This is called gas bubble disease. The physiological effects are similar to decompression-induced supersaturation occurring when divers emerge too quickly from deep dives (Beyer et al., 1976). Gas bubble disease damages the fish's tissue. If extensive, it can even lead to the fish's death (Weitkamp et al., 2003).

4.2.4 MACROECOLOGICAL IMPACTS

In the previous sections, we listed a range of individual impacts that can occur as a result of the construction of hydropower stations with the associated infrastructure of reservoirs and dams. Many of these impacts are strongly influenced by other anthropogenic activities, such as erosion resulting from agriculture, forestry activity and infrastructure, water pollution, or the regulation of waterways for navigation and flood protection.

Dams can have a substantial influence on biodiversity and ecosystems. The extent of these consequences can be identified only when looking at the macro level, i.e., at entire river basins. Some impacts only become apparent at this level because dam construction affects migratory species and because dams are often built in series rather than in isolation (Dudgeon, 2000, 2011; Van Looy et al., 2014; Carrara et al., 2014). The effects from the different dams interact with each other and other human development effects, and the total impact can only be understood taking into consideration the interaction between all of these factors (Xu, 2013). The impact is strongest where dam construction induces development in previously undeveloped areas and therefore necessitates road construction, deforestation, and the construction of settlements (Finer and Jenkins, 2012).

4.2.5 MITIGATION OF IMPACTS OF HYDROELECTRIC DAMS

The ecological and health impacts of hydroelectric dams can be reduced in a number of ways (Liu et al., 2013). Mitigation initiatives can be categorised by their goal, or effect: measures to ensure the continued migration of fish, controlled flooding to simulate conditions in natural river habitats, upstream water quality improvements and erosion control, mitigation measures related to sediment transport, and the compensation of habitat loss through the construction of new shallow-water habitat. Some measures, such as the maintenance of a minimum "environmental flow" can fulfil several of these services at once. The success of mitigation measures must be monitored and verified.

4.2.5.1 Measures to allow fish migration

A number of measures have been developed to allow migratory fish to pass dams. Upstream passage is ensured through gateways such as fish ladders (Wollebæk et al., 2011b, 2011a), while downstream passage turbines for run-of-the-river plants have been developed to allow fish to pass through the turbine on the way downstream (Deng et al., 2010). A range of other, sometimes species-specific devices is under development (Hassinger 2011). The overall success of such mitigation strategies, however, has been questioned (Brown et al., 2013).

4.2.5.2 Environmental flow

As emphasized above, the water flow in rivers is an important parameter defining the habitat of species. Many hydropower dams alter the flow regime, reduce floods and shift the timing of water flow variations. It has been found that in many cases, adjusting the operation of hydropower dams can substantially reduce ecological impacts while having only a small impact on power production (Guo et al., 2011; Esselman and Opperman, 2010). Such "environmental flow" regimes include a minimum flow requirement and the simulation of seasonal floods to allow for sediment transport and trigger life cycle processes of specific species (McCartney et al., 2009; Kang et al., 2010; Poff and Matthews 2013). Issues of flow management also include avoiding undesirable pulses to meet peak demand, for example, in order to avoid the stranding of fish amongst other consequences (Young et al., 2011).

4.2.5.3 Habitat enhancement and offsets

It is possible to design reservoirs such that they offer more habitat for endemic aquatic species, or to construct or enhance adequate habitat in nearby areas, such as tributaries, dead arms etc. The focus is often on shallow water and wetland habitats that may otherwise be lost due to reservoir construction. The objective is to offer a diversity of habitats to ensure a diversity of species (Wen et al., 2008).

Existing literature pays significant attention to the development of mitigation measures, but few encompass a systematic, comparative study of the effectiveness of such measures. The focus is often on individual species, but sometimes, appropriate attention is given to landscape level issues relating to interactions between ecosystems. The multitude of relevant effects and species concerned requires comprehensive knowledge for optimal design and operation of dams and power plants. Awareness and competence issues or a lack of data often leads to inadequate project design or inappropriate environmental flow management (Renöfält et al., 2010; Esselman and Opperman, 2010). Minimizing the ecological impacts of hydropower projects requires further knowledge about both design and operational issues and their influence on biodiversity and threatened freshwater species. Available guidelines are process-oriented and require adequate attention and competent execution. There is a need for more research, for the appropriate training of responsible personnel, and for follow-up control and evaluation of the measures taken to ensure a learning cycle.

4.2.6 SUMMARY OF ECOLOGICAL IMPACTS

In this section, we briefly reviewed ecological impacts of hydropower. We have not found a general, systematic basis for a summary evaluation and synthesis of what type of project causes which impacts, or the success of potential mitigation measures. The hydropower industry points to environmental benefits of flow regulation, but we have found few peer-reviewed studies documenting such benefits. Where summary evaluations of individual projects or regions have been undertaken, the net ecological impacts of hydropower tend to be negative (Fu et al., 2014; Reed et al., 2013; Bai et al., 2013).

Some people argue that that ecological impacts of hydropower can be reduced by pursuing small hydropower projects rather than large ones (Abbasi and Abbasi, 2011). Current evidence, however, suggests that small dams may have disproportionately large impacts on ecosystems (Kibler and Tullos, 2013; Lehner et al., 2011; Kareiva, 2012). In a study of the multiple effects of hydropower stations on the Nu River in China, Kibler and Tullos (2013) find that smaller dams impact longer stretches of the river channel and have larger impacts on the diversity of habitats, hydrological regimes, water quality, and areas designated as biodiversity and conservation priority. Larger dams have a larger influence on flooded land areas and sediment transport. On the Mekong River, the completion of dams on the tributaries would have larger impacts on fish productivity and biodiversity than constructing dams on the main river (Ziv et al., 2012). This review indicates that there are still substantial gaps in understanding both the impacts of hydropower plants and the effectiveness of mitigation measures.

4.3 BIOGENIC GREENHOUSE GAS EMISSIONS ASSOCIATED WITH HYDROELECTRIC DAMS

4.3.1 INTRODUCTION

Biogenic GHG emissions of hydropower plants are related to bacterial digestion of organic matter, which produces carbon dioxide (CO_2), methane (CH_4) and dinitrogen oxide (N_2O). Biogenic CO_2 and CH_4 are produced by the mineralization of biomass or detritus, organic carbon matter in soil or sediments. The principal concern from a climate perspective is methane formation, as organic carbon would eventually have oxidized to CO_2 regardless of dam construction, but methane has a stronger climate forcing effect. The organic carbon comes from the flooding of biomass and soil when the reservoir is filled (land use change), transport of upstream biomass to the reservoir by rivers, or growth occurring within the reservoir (Demarty and Bastien, 2011; Tremblay et al., 2005b). Nitrous oxide forms as part of the denitrification of nitrogen bound in organic matter or through partial nitrate reduction. There has been relatively little research on N_2O emissions. While Demarty and Bastien (2011) suggest that emissions of N_2O are relatively minor in boreal reservoirs, more research is required to evaluate their importance in other regions. Dams may also increase evaporation from land surfaces, thus increasing the latent heat flux to the atmosphere, which has a potential climate effect. Hydropower reservoirs also affect albedo. These effects have not been addressed by studies we have reviewed and are not considered further here.

The aim of this section is review the issue of bGHG emissions and recommend how these could be addressed in LCA. The work focuses on questions as they are posed by international assessments like those conducted for the IPCC, where a broader insight into the environmental impacts of a technology is required, and not decisions about an individual project. A review of how bGHG emissions have been addressed in existing LCA case studies has been conducted for the IPCC SRREN (Sathaye et al., 2011). The section reviews the scientific literature on bGHG emissions from dams and interprets the insights provided by this literature from the perspective of assessing the life cycle impacts of energy systems.

In SRREN, the IPCC has thoroughly reviewed the environmental aspects of energy technologies, relying also on LCA and ecological studies (Kumar et al., 2011; Sathaye et al., 2011). It did, however, fail to systematically account for biogenic emissions. While these are discussed both for bioenergy and hydropower, they are left out of the comparison charts presented in the summary. The hydropower chapter in SRREN (Kumar et al., 2011) presents a detailed discussion of mechanisms for biogenic emissions, but the emissions rates are provided per reservoir area, and are not related to the power generated. Its review of hydropower LCAs identified 27 estimates of life cycle GHG emissions from 11 distinct references. Sixteen estimates from seven references included bGHG emissions, and only three estimates from two references include emissions from the decommissioning phase. The assessment combines LCAs that consider and ignore bGHG emissions, and it is unclear whether the available cases are representative. Similarly, a recent review of hydropower LCAs (Raadal et al., 2011) does not assess the importance of biogenic emissions. Also, the concept of gross versus net emissions used in the environmental science literature is not explained, which may potentially lead to misunderstandings.

In this section, we first present an overview of the role of rivers in the global carbon cycle and identify the mechanisms by which dams interfere with this carbon cycle and thus affect the concentration of GHGs in the atmosphere. Second, we discuss the mechanisms and pattern of biogenic CH_4 and CO_2 emissions and their measurement. Third, we review reported emissions from reservoirs and the discussion surrounding these emissions. Fourth, we provide a tentative estimate of global emissions per unit electricity generated. Finally, we discuss the need for further assessments and give recommendations for how to conduct such assessments.

4.3.2 ORIGIN OF INCREASED METHANE PRODUCTION

4.3.2.1 Rivers in the global carbon cycle

While often neglected in relevant studies, rivers and lakes have an important role in the global carbon cycle. Freshwater is both the recipient of organic matter from soil and terrestrial biomass and a medium for further biomass growth, which fixes atmospheric CO_2 (Cole et al., 2007; Tranvik et al., 2009). Rivers transport carbon to the ocean in both organic and inorganic forms, and freshwater returns part of the carbon to the atmosphere (Figure 4.1). In an initial assessment of the carbon flows through freshwater, Cole et al. (2007) estimate that freshwater bodies receive 1,900 million tons carbon/year from the terrestrial landscape. Of these, 230 million tons are buried in sediments, 750 million tons or more are released to the atmosphere, and 900 million tons are delivered to the ocean. There is, however, significant uncertainty in these estimates; Tranvik et al. (2009) estimate 2.9 billion tons carbon/year of input to freshwaters, sedimentation of 0.6 billion tons and atmospheric emissions of 1.4 billion tons. Measurements from Asia may potentially reveal that freshwater plays a larger role in the carbon cycle than previously believed (Huang et al., 2012). On the other hand, models for the nutrient export of rivers (Mayorga et al., 2010) indicate carbon flows in line with Cole et al., For comparison, the net primary production of terrestrial plants accounts for approximately 60 billion tons carbon/year, while total fossil fuel emissions account for 7.7 billion tons carbon/year. The carbon flow in rivers consists of dissolved inorganic carbon (DIC), that is, CO_2, carbonic acid and its dissociated forms, approximately 0.3 billion tons carbon/year, dissolved organic carbon (DOC), like humic acids, approximately 0.2 billion tons carbon/year, and particulate organic carbon, that is, dead plant matter, corresponding to 0.1-0.4 billion tons carbon/year. Groundwater flow directly to estuaries delivers about 0.2 billion tons carbon/year (Cole et al., 2007; Mayorga et al., 2010). The ultimate fate of the organic carbon transported by rivers to the coastal or open ocean is not yet well investigated. Much of the organic carbon is mineralized, but 10-20 per cent of the particulate organic carbon reaching the ocean floor will be buried with the sediment and thus escape mineralization (Burdige, 2007). Globally, burial in marine sediments removes only 0.5 per cent of the organic carbon formed each year, but it is assumed to remove 9-17 per cent of the terrestrial organic carbon reaching the oceans (Burdige, 2007). In addition, carbon mineralized in the deep ocean is removed from the carbon cycle for thousands of years.

FIGURE 4.1

Schematic showing the carbon cycling in freshwater bodies

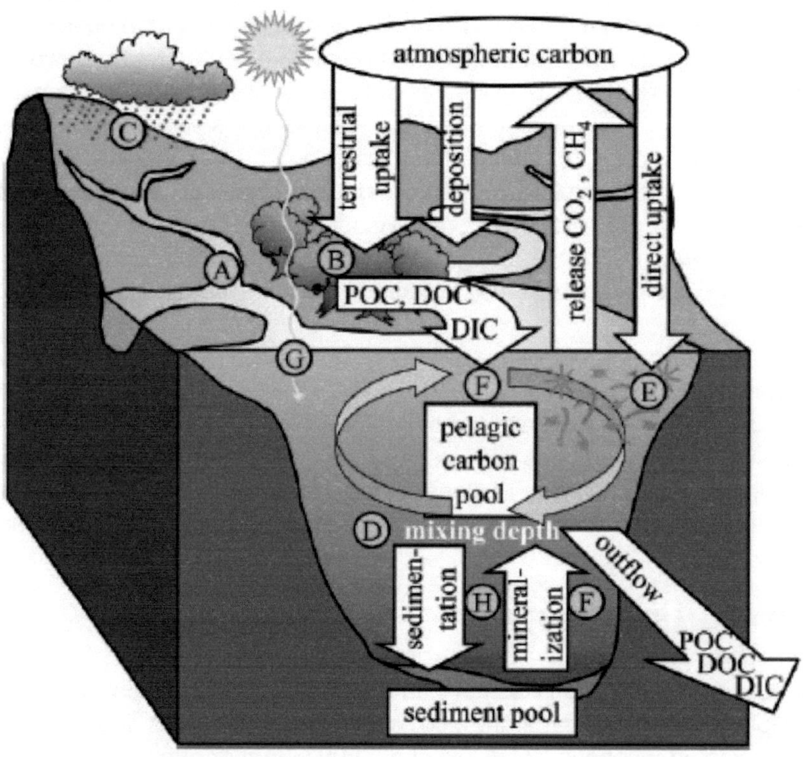

Source: Tranvik et al., 2009

4.3.2.2 Freshwater and the methane balance

Global methane concentrations have increased by almost 150 per cent since the onset of the industrial revolution and contribute 20 per cent to the increased radiative forcing from GHGs. Lakes and rivers play an important role in the methane balance of the atmosphere. Methane constitutes around 4 per cent of the carbon released from lakes, according to Bastviken et al. (2011). Global methane emissions estimates have large ranges because both freshwater area and emissions rates are uncertain. In a review of the methane balance literature, Kirschke et al. (2013) estimate the contribution from wetlands to be on the order of 200 million tons CH_4/year and specify the freshwater component as 40 million tons CH_4/year. Bastviken et al. (2011) estimate emissions from freshwater lakes and river as 100 million tons CH_4/year. By comparison, the total natural and anthropogenic emissions are on the order of 600 million tons CH_4/year. Two methods are available to estimate methane emissions. Bottom-up methods measure emission rates at selected sites and scale the results according to the area. Top-down methods measure concentration gradients and rely on inverse modelling of atmospheric processes to specify emissions sources required to produce the measured concentrations. The measurement of carbon isotope ratios is used to assign emissions to source categories (Kirschke et al., 2013).

Humans interfere with the natural methane balance in several ways. Methane emissions have increased due to increased populations of ruminants such as cows and sheep producing methane via enteric fermentation, rice paddies, leakage from fossil fuel systems, increased biogenic carbon input to freshwaters through soil erosion and eutrophication.

Hydropower dams interfere with the carbon cycle in several ways, potentially changing the carbon cycle, the storage of carbon in sediments or the deep ocean, and the form in which carbon is returned to the atmosphere, i.e. as CO_2 or CH_4. The following processes are relevant.

1. Dams turn land or wetland surface into reservoirs, often flooding plants and soils, and thereby submerging organic carbon. Labile carbon is then slowly released over a period ranging from several years to decades, either as CO_2 or CH_4, depending on climate and reservoir characteristics, as well as the amount and character of the organic carbon (Tremblay et al., 2005a). Dams also turn river surface into reservoir surface, increasing the fraction of carbon released as methane rather than CO_2 due to the common formation of an anoxic bottom layer in the reservoir.

2. Dams interfere with the river transport of organic matter and nutrients to the oceans, leading to sedimentation or decay of some of this organic matter (Maeck et al., 2013), but also affecting downstream biomass production (see Chapter 1.2.3). Dams reduce the flow of particulate organic carbon to the ocean and hence its potential removal to the deep ocean. Reservoirs often have anoxic conditions, leading to the anaerobic digestion of organic carbon to CH_4. Dams also lead to the build-up of organic matter, which can be beneficial if it permanently increases the carbon stored in sediments or can be detrimental if dam removal or dredging leads to methane emissions (Pacca, 2007).

3. Reservoirs provide an opportunity for freshwater biomass growth. These conditions can lead to a net absorption of carbon, which is subsequently either captured in the reservoir's sediment or transported downstream (Chanudet et al., 2011).

To understand the impact of hydropower dams on GHG concentrations, it is hence important to understand the fate of carbon in the area converted to a reservoir both before and after damming of a river. Measurements of emissions after the completion of the dam measure what literature calls "gross emissions", which is the total flux of carbon to or from the surface (UNESCO/IHA, 2010). Note that these gross emissions constitute a net effect of emissions and absorption of carbon. To obtain what the literature refers to as "net emissions", one has to subtract the emissions that occurred before the dam was built. These emissions are often estimated based on literature studies (Demarty and Bastien, 2011). Pre-flooding field measurements have been conducted only by the Eastman 1 reservoir in Quebec, Canada (Demarty et al., 2011; Teodoru et al., 2012). In other cases, the pre-flooding emissions are estimated based on typical emissions factors for the land cover before the flooding, or are simply neglected. Note that emissions before the flooding may be positive as is the case when a wetland area is flooded, negative when a forest or agricultural area is flooded or close to zero.

4.3.3 BIOGENIC METHANE EMISSIONS AND THEIR MEASUREMENT[1]

Understanding the net climate effect of hydropower dams is far from trivial. In this section, we describe emission mechanisms and discuss recent studies. There is a lively debate in literature and disagreements on important issues (Cullenward and Victor, 2006; Fearnside, 1996; Rosa et al., 2006; Fearnside and Pueyo, 2012). We provide our own assessment of the literature. The focus of the discussion will be on tropical and equatorial regions, where measurements show large variations. However, similar variations have also been identified in recent work on non-tropical reservoirs (Sobek et al., 2012).

Organic matter occurs in freshwater bodies either through the growth of plants or from the land surface in the form of plant matter or soil organic matter. GHG emissions from freshwater are connected to the degradation of this organic matter. The organic matter will be mineralized to CO_2 by aerobic bacteria and to methane by anaerobic bacteria, that is, under anoxic conditions. Sediments also contain organic matter, so not all organic matter necessarily decays within the normal lifetime of a dam.

1 This section is largely based on Hertwich (2013). Copyright American Chemical Society, reprinted with permission.

FIGURE 4.2

Possible pathways for biogenic methane and carbon dioxide emissions from reservoir hydropower stations

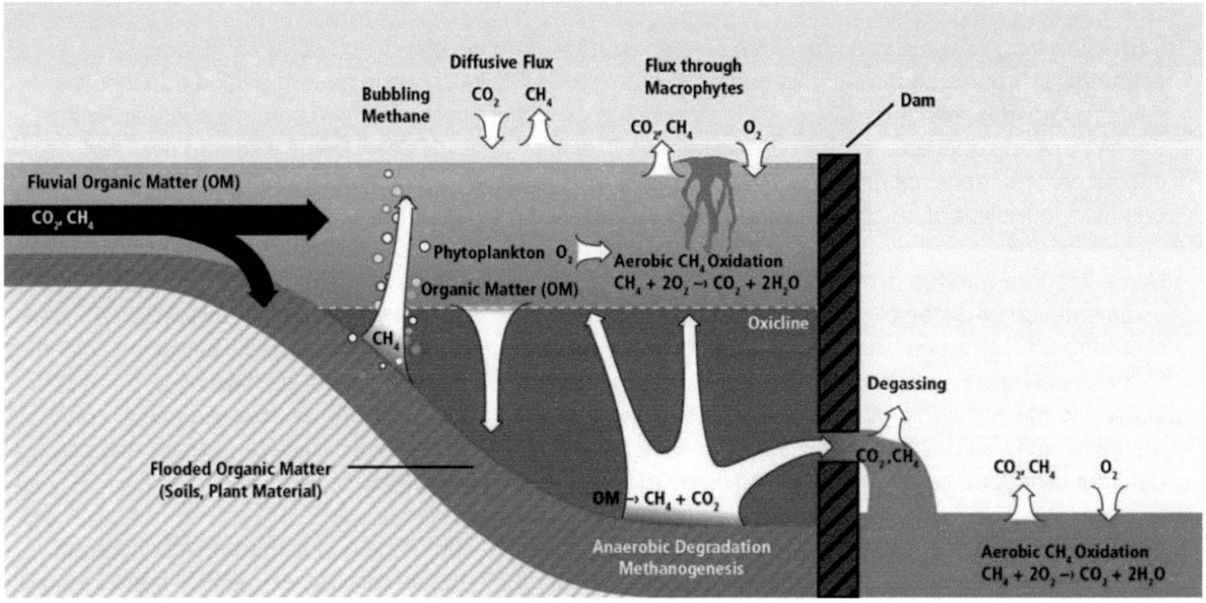

Source: Kumar et al., 2011

FIGURE 4.3

Emissions of methane over time from the Petit-Saut Reservoir in French Guyana

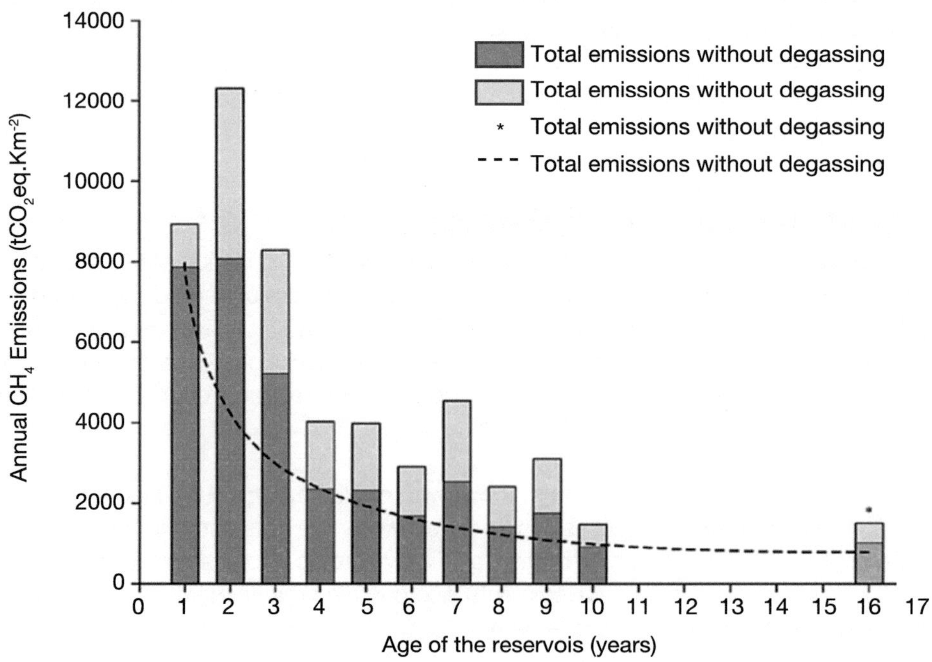

Source: Demarty and Bastien, 2011. The decreasing emissions reflect the degradation of initially present labile biomass and soil organic carbon. The stabilization of emissions is due to the depletion of that initial reservoir; the remaining emissions are mostly from organic carbon transported to the reservoir by tributaries or from plants growing in the reservoir.

Empirical studies need to address a number of emission sources (Figure 4.2).

1. Diffusion of CO_2, CH_4 and NO_2 across the air-water interface: The gas flux depends on a number of variables such as wind speed, rainfall, temperature, relative gas concentrations in air and water. Diffusive emissions can be directly measured using surface floating chambers (Abril et al., 2005; Guérin et al., 2006) or derived from boundary layer models (Vachon and Prairie 2013; Schilder et al., 2013).

2. Bubble emissions: Methane produced through anaerobic digestion in sediments leads to bubbling. Temperature and hydrostatic pressure affect the bubbling rate. Bubbles come in bursts and not as a steady flow, with uneven bubbling events estimated contain a significant proportion of the total amount of methane released (Eugster et al., 2011; Delsontro et al., 2011). Gas transport can also be mediated by aquatic plants, macrophytes (Kumar et al., 2011). Methane bubbles are usually measured using funnels (Tremblay et al., 2005b), eddy covariance (Schubert et al., 2012), echosounders (Delsontro et al., 2011), or floating chambers (Bastviken et al., 2010).

3. Downstream emissions: Water in hydropower plants is often drawn from some depth in the reservoir. At this depth, methane and CO_2 concentrations are higher than the saturation vapour pressure at the surface. Part of the methane is released directly at the hydropower plant after the water has passed through the turbines. This is called degassing. Another part of the methane is released from supersaturated water through diffusion or bubbling some distance from the dam (Guérin et al., 2006; Kemenes et al., 2007). Downstream emissions are often neglected or underestimated (Demarty and Bastien, 2011; Kemenes et al., 2011; Fearnside and Pueyo, 2012).

Emissions vary due to weather and seasonal effects (Eugster et al., 2011; Cheng et al., 2012). A proper assessment must address these variations through measurements that extend over seasons (Demarty and Bastien, 2011), ideally supplemented with modelling exercises that build a mechanistic understanding of the processes involved (Tremblay et al., 2005b; Delsontro et al., 2010). Emissions also change over the lifetime of the reservoir. It is widely acknowledged that emissions are highest in early years and decrease as the initially flooded biomass decays. This feature is nicely illustrated by measurements at the Petit Saut reservoir in French Guyana (Figure 4.3), where emissions decreased by a factor of three and methane emissions by a factor of almost five from the average of the first three years to measurements ten years later (Abril et al., 2005). Further measurements indicate that a reservoir can reach a steady state with constant emissions (Abril et al., 2005), but recharge, e.g., through storms and floods, may also occur.

4.3.4 EXAMPLES OF HYDROPOWER DAM EMISSIONS

A review of measurements presented by IPCC SRREN (Kumar et al., 2011) suggested that the gross emissions of both CH_4 and CO_2 may be temperature dependent and thus potentially higher in tropical regions than in temperate and boreal regions. This temperature dependence is also suggested by seasonal comparisons across existing reservoirs (Delsontro et al., 2010). However, the temperature dependence is not strongly supported by the statistical analysis conducted later. The rate of biomass growth/input serves as an alternative explanatory factor but co-varies with temperature, creating a challenge for the statistical analysis.

Kemenes et al. (2007) provides a detailed account of the Balbina reservoir in the Brazilian Amazon, which was built in 1987, has an average area of 1,770 km^2 and an installed hydroelectric capacity of 250 MW. Methane emissions were measured from the reservoir, below the reservoir, and concentrations of methane flowing through the turbines and floodgates throughout most of 2005. The diffusive emissions of the reservoir varied from 5 to 343 with an average value of 47 mg C m^{-2} day^{-1}. The flux through the dam showed a similar range of variation. The measurements were integrated to determine total flows. Emissions from the reservoir amounted to 34,000 tons carbon/year, degassing at the turbine outflow is also 34,000 tons carbon/year, while diffusive emissions within 30 km downstream from the plant are 5,000 tons carbon/year. Kemenes et al. (2011) document measurements of CO_2 emissions at the same power plant. These emissions are 2.4 million tons carbon/year from the reservoir and 0.08 million tons carbon/year downstream. The measurements show that while CO_2 emissions mostly occur directly from the reservoir, for methane emissions downstream of the power plant are equally important as those upstream.

Balbina is in many ways an extreme case, producing only ca. 1,100 GWh per year from a large reservoir. Per kWh, gross emissions of CH_4 are 2.2 kg CO_2 equivalent and of CO_2 are 8.5 kg, amounting to ten times the emissions intensity of coal power. Similar measurements were also conducted for the Petit Saut power station in French Guyana, which produces 560 GWh/year (Demarty and Bastien, 2011; Abril et al., 2005; Guérin et al., 2008) and releases 370,000 tons CO_2 and 35,000 tons CH_4 (Abril et al., 2005), equivalent to 0.65 and 1.55 kg CO_2e/kWh.

A counter-example is provided by the Nam Ngum Reservoir in Laos, for which a net absorption of CO_2 during the measurement period was reported, in addition to small methane fluxes and no downstream degassing (Chanudet et al., 2011). The data are not expressed relative to electricity production. For the upstream Nam Leuk Reservoir in the study, the emissions are on the order of 0.05-0.1 kg CO_2-eq/kWh. The carbon budget presented points to a significant accumulation of carbon in the sediments.

The measurements at Balbina and Petit Saut are the only measurements in literature that include downstream emissions of CH_4 dissolved in the water that streams across the dam either through the turbine or the weir. For the Brazilian Tucurui (Fearnside, 2002) and Curua Una (Fearnside, 2005) reservoirs, CH_4emissions have also been estimated to exceed 1 kg CO_2-eq/kWh, even though there is some controversy surrounding these estimates. Demarty and Bastien review a large range of measurements and estimates of methane emission from reservoirs in tropical regions, indicating that emissions can have a large range, from only 2 g to 4,000 g CO_2-eq/kWh (Demarty and Bastien, 2011). Important determinants include the amount of vegetation that is inundated, the inflow of organic carbon, and the size of the reservoir relative to the electricity generated. In general, large reservoir areas lead to high methane emissions (dos Santos et al., 2006; Demarty and Bastien, 2011). Not all power plants have such high emissions. Currently, analysis to evaluate the methane emissions from the Three Gorges Dam in China is ongoing (Chen et al., 2011; Yang et al., 2012).

4.3.5 EMISSION FACTOR ESTIMATION

To compare hydropower to other electricity supply technologies, we would like to understand its emissions per unit electricity generated. Based on approximately 150 available measurements of the flux of CO_2 and CH_4 from reservoirs, Barros et al. (2011) estimate the total global emissions from reservoirs. The strategy was to use emissions measurements per surface area to derive a regression equation. This equation was then used to estimate average emission rates for different climate zones, which were then multiplied by the reservoir area to obtain estimates for total emissions.

Hertwich (2013) supplemented data from Barros et al. (2011) with information regarding electricity generation from various internet sources and the potential net primary productivity of the area (Haberl et al., 2007) as additional variables and reanalysed the data set. Figure 4.4 shows the frequency distribution of measurements of CH_4 emissions per kWh electricity distributed. The measurements have an unweighted mean of 54 g CH_4/kWh, a geometric mean of 0.61 g CH_4/kWh and a geometric standard deviation of 46. The weighted mean is 3.5 g CH_4/kWh. The highest emissions rate is from the Lokka power plant in Finland (Huttunen et al., 2002), which sits partly on peat and produces only 0.5 GWh/year, but regulates the flow of a river with a series of power stations further downstream. The water released from the reservoir produces in total about 365 GWh/year, which we used as power generation from the reservoir in the further analysis. Across our sample, power station capacity factor, latitude, NPP, land use (m^2y/kWh) and the measurement of bubbling emissions show certain correlations with CH_4 emissions. Given the lognormal distribution seen in Figure 4.4, one may desire to predict the logarithm of the emissions factor; while this would not preserve the mean, the estimate is more likely to be closer to the actual value for the individual power plant than an estimate based on a linear regression. The best prediction of the emission factor (g CH_4/kWh) is offered by including the logarithms of land use and NPP, as well as age as a linear variable (adjusted r^2=0.78). Regression coefficients and statistics are shown in Table 4.2. Figure 4.5 relates emission factors to explanatory variables and indicates the predicted value of the regression equations. Of the different dependent variables, land use (r^2=0.64) offers the high predictive value, as Figure 4.5 shows. If viewed on a linear scale, the predictive value for CH_4 is much less, but for CO_2, it is high.

TABLE 4.2

Regression coefficient and statistical parameters for the regression of per-kWh emissions of carbon dioxide (CO_2) and methane (CH_4) as a function of land use, age and potential net primary production (NPP)

	CO_2 emissions per kWh	CH_4 emissions per kWh
B_{Const}	0.8 (-1.8 - 3.8)	0.18 (-0.5 - 0.87)
$B_{Land\ use}$	0.97 (0.84-1.11)	1.26 (1.07-1.44)
B_{Age}	-0.006 (-0.011- -0.0009)	-0.014 (-0.022- -0.006)
B_{NPP}	0.737 (-0.16 - 1.64)	0.0017 (0 - 0.0024)
r^2	0.69	0.79
F	77.5	93
p	<0.001	<0.001

The regression equation is $\log(E) = const + B_{land\ use} * \log(Land\ use) + B_{Age}*Age + B_{NPP}*\log(NPP)$.
Reprinted with permission from (from Hertwich 2013).

4.3.6 A GLOBAL EMISSIONS ESTIMATE

Based on available measured CO_2 and CH_4 flux data from reservoirs, Barros et al. (2011) estimate the total global emissions from reservoirs. The strategy was to use emissions measurements per surface area to derive a regression equation. This equation was then used to estimate average emission rates for different climate zones, which were multiplied by the reservoir area to obtain estimates for total emissions. The study neglects downstream emissions, and some measurements do not include bubbling. In addition, the regression equation is applied to the logarithm of the flux estimates, an operation that is not mean-preserving and systematically underestimates the importance of high measurements; it is hence not suitable to determine the global total emissions, which will strongly depend on the influence of power stations with high emissions.

Given that reservoir area is strongly correlated with methane emissions, global emissions from hydropower reservoirs are best predicted based on the reservoir area. Given the lack of information on location and age of all dams, we use climate zones as a proxy. The resulting estimate of total CH_4 emissions from reservoirs (Table 4.3) is 2.5 times higher than the estimate provided by Barros et al. (2011) and similar that of Maeck et al. (2013). Correcting for bubbling emissions in studies where these were ignored would raise methane emissions estimate further to 10 million tons carbon/year. Relating the estimates in Table 4.2 to a total hydropower production of 3,288 TWh in 2009, the average direct emissions from hydropower dams corresponds to 3 g CH_4/kWh, which is close to our sample average.

TABLE 4.3

Estimate of carbon dioxide (CO_2) and methane (CH_4) emissions from global reservoirs

Climate zone	Area	CO_2	CH_4	CO_2	CH_4
	10^3 km^2	kg/m^2/year	g/m^2/year	10^6 tons C/year	10^6 tons C/year
Boreal	80	0.97	40	21	2.4
Temperate	130	0.42	7.2	15	0.7
Tropical	120	1.2	46	50	4.1
Total	330	-	-	76	7.3

The uncertainty is estimated to be a multiplicative factor of 2. Reprinted with permission from Hertwich (2013). Copyright 2013 American Chemical Society.

FIGURE 4.4

Distribution of methane (CH_4) emissions rates per kWh of electricity produced across the sample of hydropower stations

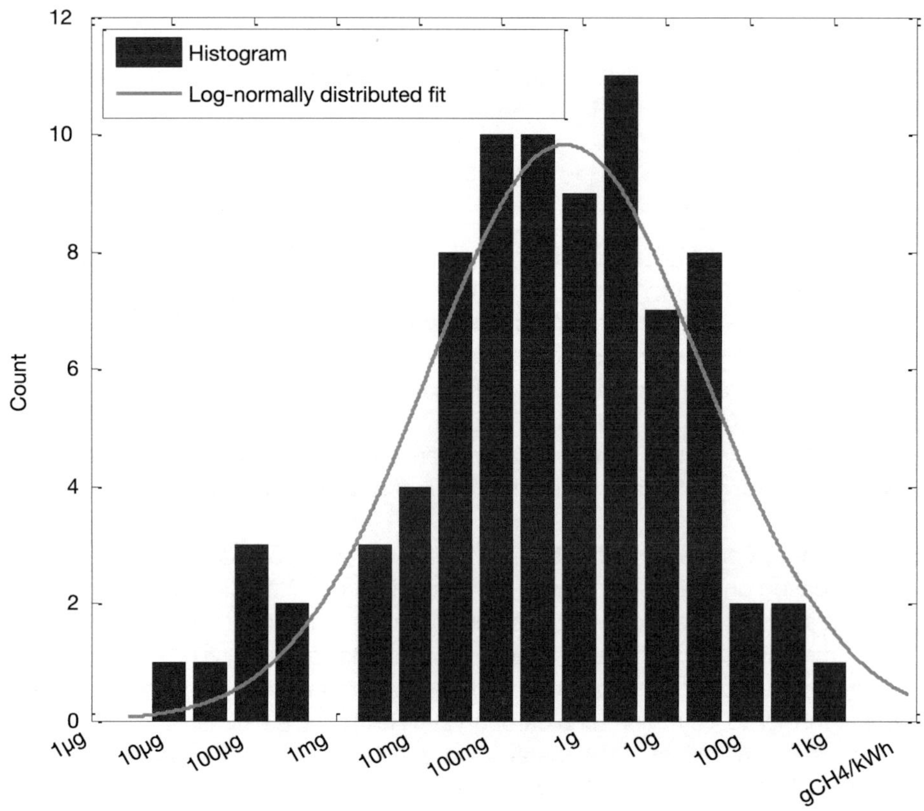

Reprinted with permission from Hertwich (2013). Copyright 2013 American Chemical Society.

FIGURE 4.5

Methane emissions per unit electricity produced from hydropower dams as related to the reservoir area (a), age of the reservoir (b) and potential net primary productivity NPP0 (c)

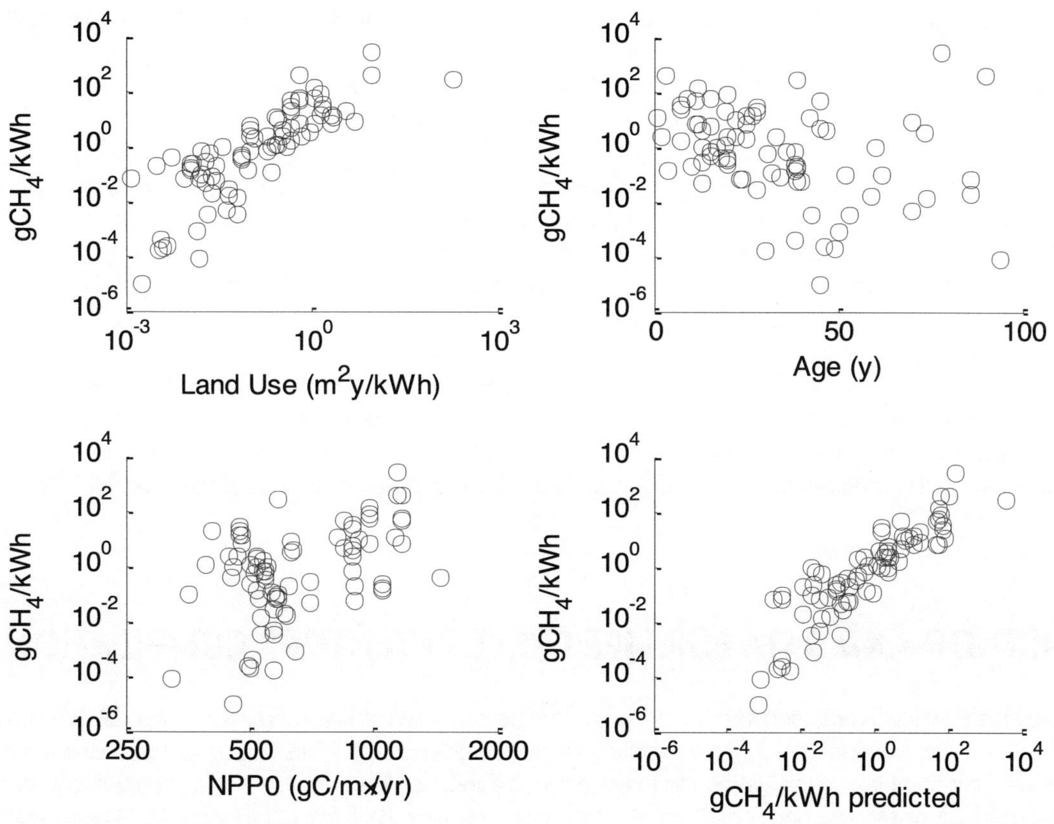

Panel (d) shows measured versus predicted values based on a trivariate regression. The figure indicates that land use is the most important explanatory variable.
Reprinted with permission from Hertwich (2013). Copyright 2013 American Chemical Society.

We emphasize that there is a large degree of uncertainty associated with the estimate presented in Table 4.3, including the actual area of reservoirs. The uncertainty is not least related to the high variability of emissions rates from individual reservoirs across time and space and the differences found across reservoirs, in combination with the low fraction of reservoirs sampled. Further, in some cases, measurements often address gross but not net emissions and they systematically neglect important sources of CH_4 emission, including downstream emissions and bubbling (Fearnside and Pueyo, 2012). Further, there may be a selection bias. On the one hand, the reservoirs that were investigated are likely to be those where a suspicion of high emissions was given. On the other hand, the sample had a poor coverage of Asia, where high emissions have been recorded (Li and Lu, 2012).

What is important for the future development of hydropower is to take notice that some hydropower plants have very high direct emissions of GHGs and others have very low emissions. The high-emissions plants raise the average to significant levels. There may, however, be an opportunity to capture some of the methane being generated and exploit it as an energy source (Ramos et al., 2009); such opportunities should be urgently investigated. However, for future projects, it is important that the potential bGHG generation in reservoirs is appropriately considered and that only those projects with low life cycle emissions are developed.

An issue that has not been considered much in the literature and that has been neglected here is the fate of the carbon that is accumulated in the reservoirs. The accumulation of sediments slowly reduces reservoir volume and reservoirs have to be dredged or eventually abandoned. These processes could lead to potentially significant emissions (Pacca, 2007).

4.4 METHOD AND DATA FOR LIFE CYCLE INVENTORY COMPILATION

The life cycle inventory (LCI) data for hydropower came from several case studies of reservoir hydroelectric plants located in Chile, where rivers are mostly fed by melting ice from highlands (CDEC-SIC, 2011). The cases are from a large hydroelectric complex on the Baker and Pascua river basins, between latitude 47° and 49°S (Patagonia), involving five reservoir plants and one pass-through, with a total installed capacity of around 2.76 GW. The cradle-to-gate approach includes the construction process, building materials, machinery, electricity generators, transportation, and other upstream processes. Plant operation, maintenance, mechanical component replacement and end-of-life activities were not included here. Data were obtained from primary sources and official environmental reports (Centrales Hidroeléctricas de Aysen S.A., 2008).

The inventory data did not show a direct correlation between land occupation and installed power due to differences in hydrology. Additionally, dam size and height, and specific building material requirements proved to be highly dependent on local topographical features. Transport required for the plants hence also differed significantly.

Others relevant factors not included in this assessment are biodiversity impacts through habitat change and the obstruction of migration patterns, changes in the amount and composition of sediment swept down the river basin, social displacement, and impacts from decommissioning. Such impacts have been considered as individual research topics for specific cases (Bunn and Arthington, 2002; Chanudet et al., 2011; Pacca, 2007; Rosa et al., 2004) or recommended more generically (Ribeiro and da Silva, 2010), but there is no systematic approach for addressing these issues in LCA.

The design of hydropower plants and hence their life cycle impacts depend significantly on local factors. Indeed, in addition to the basin topographical features mentioned above, consideration has to be given to the organic matter content of the water system. In this respect, it must be mentioned that cold high mountain Andean rivers exhibit a supersaturated oxygen level, neutral pH, low temperature, conductivity and negligible organic matter content (Centrales Hidroeléctricas de Aysen S.A., 2008). Thus, CH_4 generation due to anaerobic digestion should be much less significant than reported values for tropical latitudes, where the high organic matter load constitutes a considerable carbon source for biological processes, as shown in the CH_4 emissions reported in the literature.

4.5 MODELLING RESULTS

The results presented hereafter describe the environmental profile of the two selected types of dams for nine categories. In the next section, a contribution analysis is also shown for three impact categories: climate change, metal depletion, and land use.

FIGURE 4.6

Environmental impacts for a 660 MW hydroelectric plant (Baker 1) relative to the Latin American electricity mix of 2010

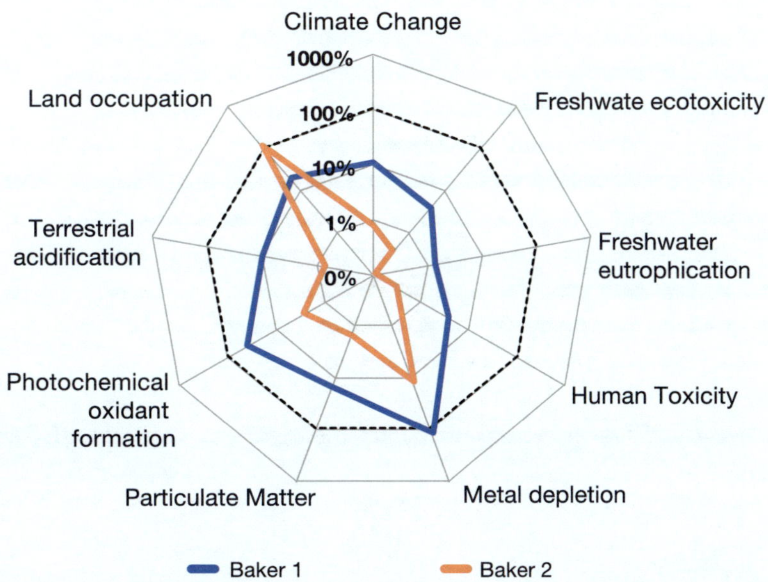

Panel (d) shows measured versus predicted values based on a trivariate regression. The figure indicates that land use is the most important explanatory variable. Reprinted with permission from Hertwich (2013). Copyright 2013 American Chemical Society.

As shown on Figure 4.6, when compared to the 2010 Latin American background electricity mix, the Baker 1 (660 MW capacity) and Baker 2 (360 MW) plants show a worse or equal impact for mineral depletion and land occupation, respectively. The land occupation of the reservoir per unit power generated varies widely across hydropower plants, ranging from 1-10000 m^2 per MWh/a of electricity generated (Figure 4.5). The land occupation of the two investigated power plants (7-46 m^2 per MWh/a) is below the average of hydropower and Latin America has some hydropower plants with especially large reservoirs and a relatively large share of hydropower in its mix, so that the comparison above is in part to other hydropower plants. Coal power plants have a similar land occupation, mostly due to mining (Chapter 3). Building the two plants necessitated the transportation of cement over thousands of kilometres, by truck, which, in turn, increased the need for roads in the remote area chosen for the project. Mineral depletion results from the use of iron and steel for the turbine, the grid connection and infrastructure. For Baker 1, a significant amount of transportation is required, which leads to a high consumption of iron and steel. The impact changes little over the 40-year time period because of the large contribution of direct impacts such as emissions from transportation and direct land use. Furthermore, hydropower is a mature technology and therefore not expected to achieve significant technological improvements over the studied period. We assume no increase in efficiency over time.

Figure 4.7 and Figure 4.8 show the contribution of each life cycle process to the set of environmental impacts introduced in Figure 4.6. In both cases, transportation contributes chiefly (i.e. more than 90% for Baker 1, and more than 55 per cent for Baker 2) to most impacts, whereas the area directly occupied by the reservoir contributes to most of the impact on land occupation.

FIGURE 4.7

Process contribution to a set of environmental impacts of a 660 MW dam hydroelectric plant (Baker 1), regional background Latin America, 2010

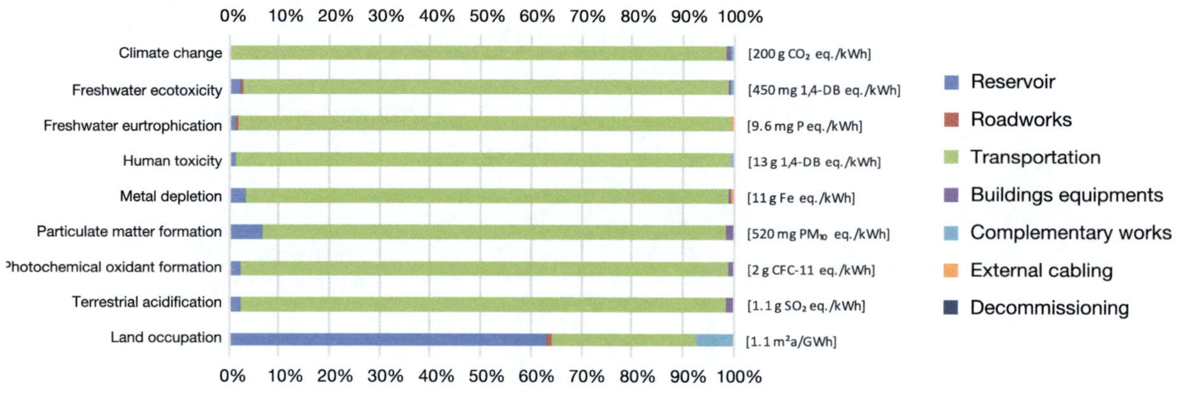

FIGURE 4.8

Process contribution to a set of environmental impacts of a 360 MW dam hydroelectric plant (Baker 2), regional background Latin America, 2010

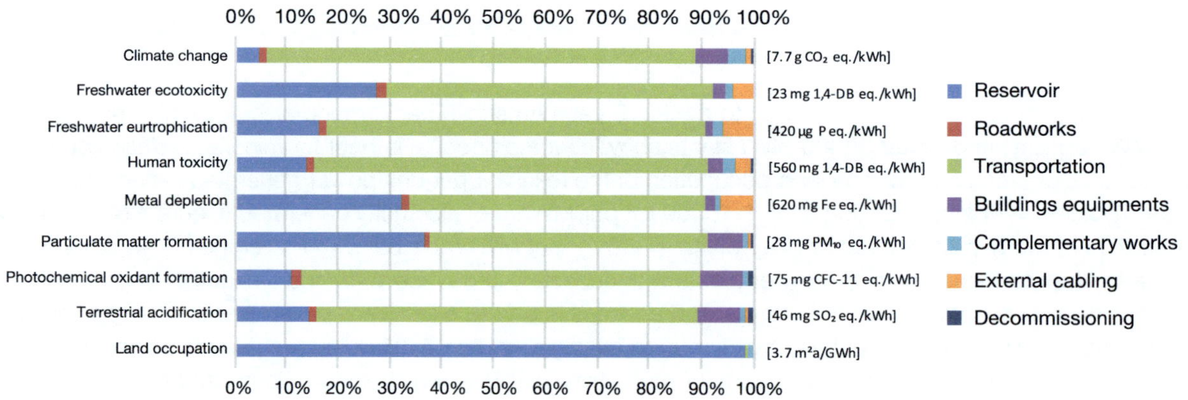

FIGURE 4.9

Contribution to climate change of a 660 MW dam hydroelectric plant (Baker 1), regional background Latin America, in g CO_2 e/kWh

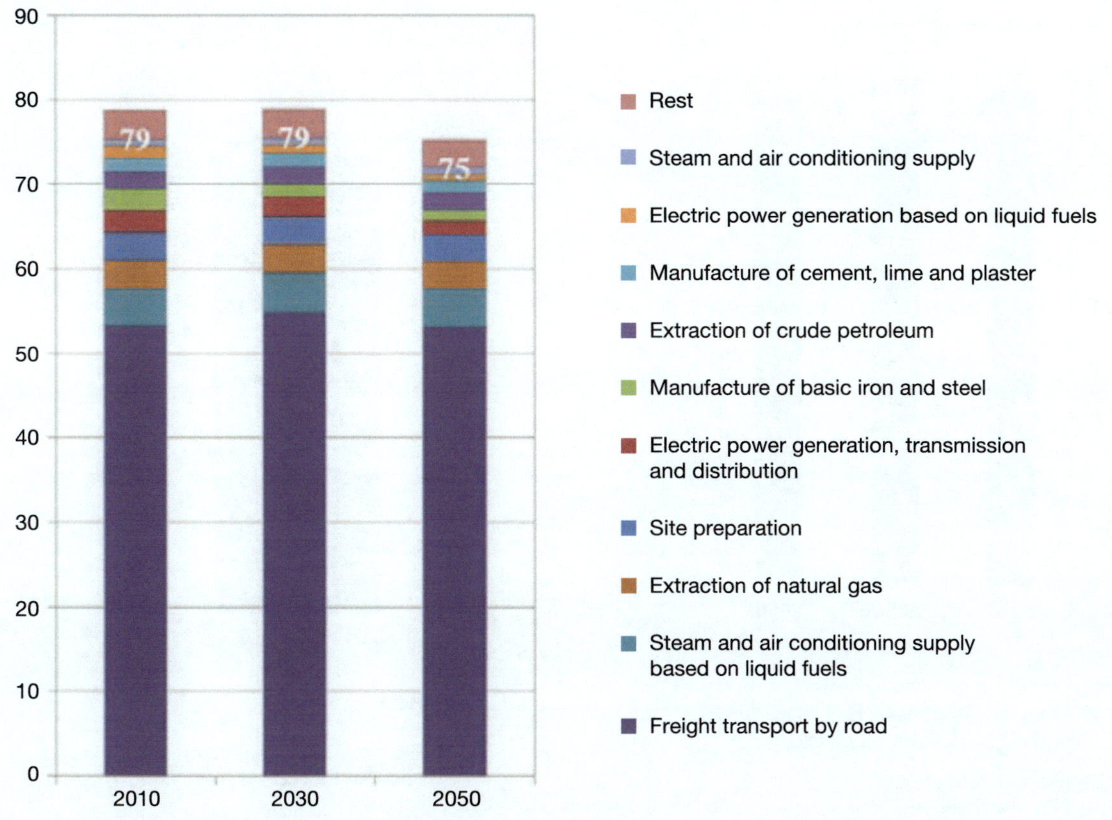

Figure 4.9 shows how various processes contribute to the total GHG emissions per kWh of electricity from the Baker 1 plant. The emissions are at the very upper end of the hydropower LCAs reviewed by the IPCC (Kumar et al., 2011). Roughly two thirds of life cycle emissions originate from the long-distance transportation of materials by road. It should be noted that this high contribution are specific to this plant and cannot be generalized to other hydroelectric plants. However, it indicates the importance of accounting for transportation when assessing remote hydropower projects. Most other background processes are energy related, while the foreground process of site preparation emits GHGs directly.

FIGURE 4.10

Contribution to mineral depletion of a 660 MW dam hydroelectric plant (Baker 1), regional background Latin America, in g Fe eq./kWh

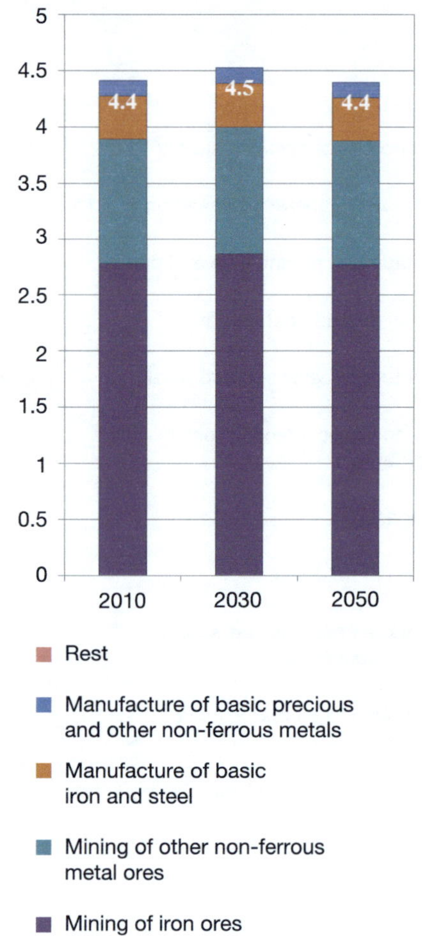

FIGURE 4.11

Contribution to land occupation of a 660 MW dam hydroelectric plant (Baker 1), regional background Latin America, in m²·a/MWh

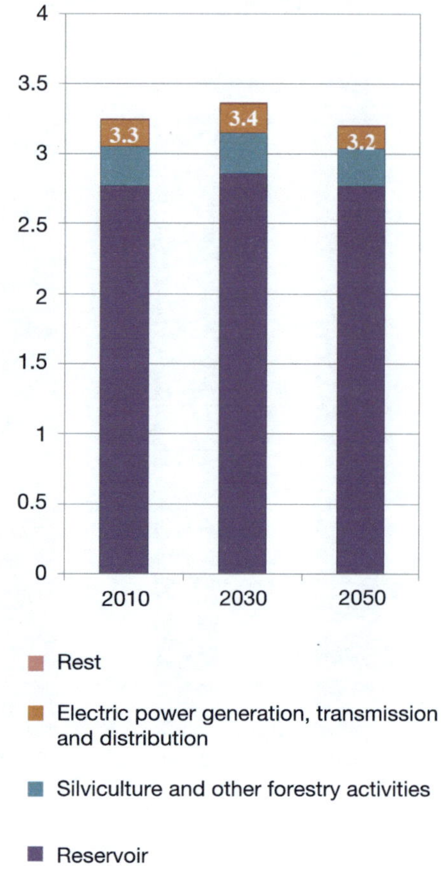

Figure 4.10 shows that more than half of the mineral depletion potential is caused by the extraction of iron ores. The extraction of non-ferrous metals is the second largest contributor, owing to the mining of copper, nickel and manganese, namely for the building of the grid connection and machinery. Other contributing processes are insignificant to mineral depletion.

Metal depletion shows no change through the years. The current model does not account for potential future changes in metal availability, and no technological change is assumed.

Figure 4.11 illustrates that the reservoir contributes by far the most to land occupation. As far as indirect land use is concerned, forestry activity for the construction of underground coal mines comes second, but contributes less than 10 per cent of the total impact, while electricity generation, transmission and distribution represents 5-6 per cent of the impact.

FIGURE 4.12

Contribution to climate change of a 360 MW dam hydroelectric plant (Baker 2), regional background Latin America, in g CO_2 eq./kWh

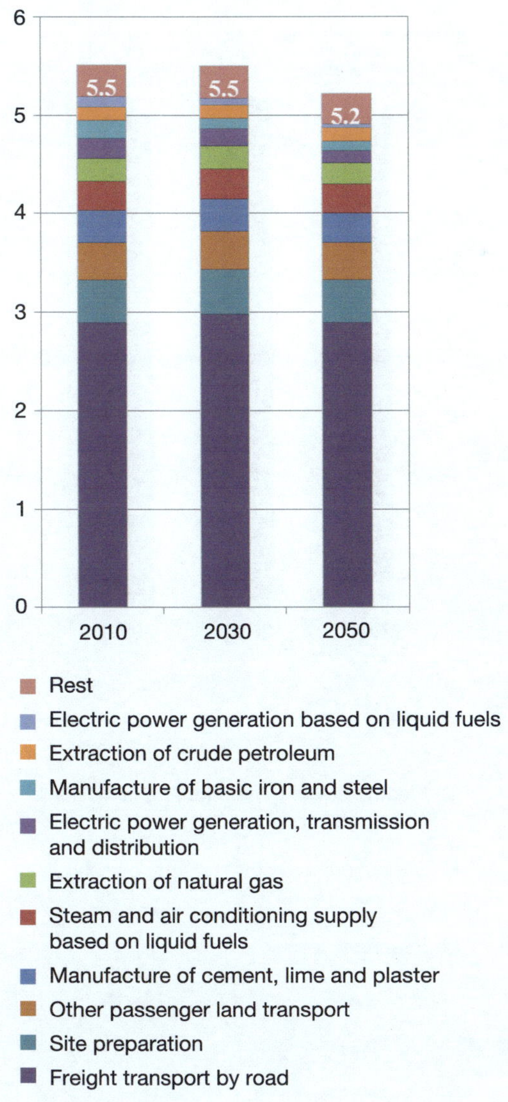

FIGURE 4.13

Contribution to metal depletion of a 360 MW dam hydroelectric plant (Baker 2), regional background Latin America, in g Fe eq./kWh

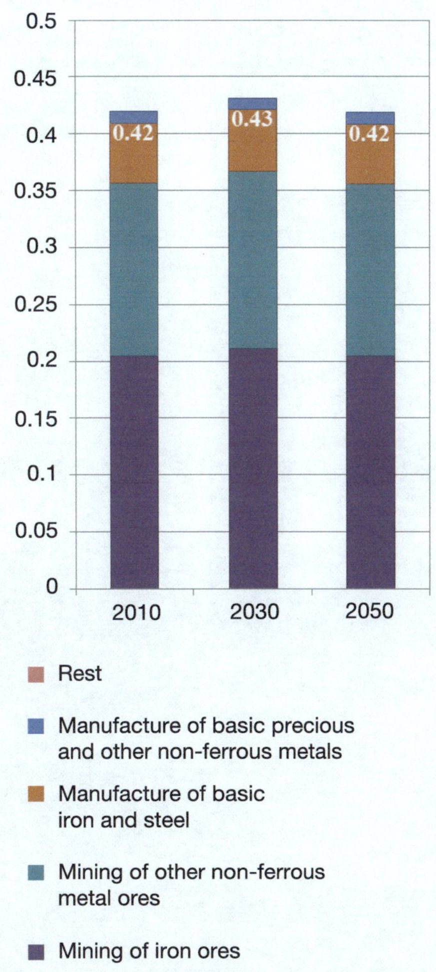

Figure 4.12 shows the breakdown of GHG emissions for the Baker 2 plant. The total impact is significantly lower than Baker 1 due to reduced material transport distances. Transport still represents half of the total impact, while the site preparation, other land transportation and the production of cement contribute equally, with less than 10 per cent each.

Processes contributing to mineral, or metal, depletion are shown in Figure 4.13. The extraction of iron ore and production of steel contribute to 60 per cent of the impact, while other non-ferrous metals, principally manganese, copper and nickel, contribute to the rest.

FIGURE 4.14

Contribution to land occupation of a 360 MW dam hydroelectric plant (Baker 2), regional background Latin America, in m²·a/MWh

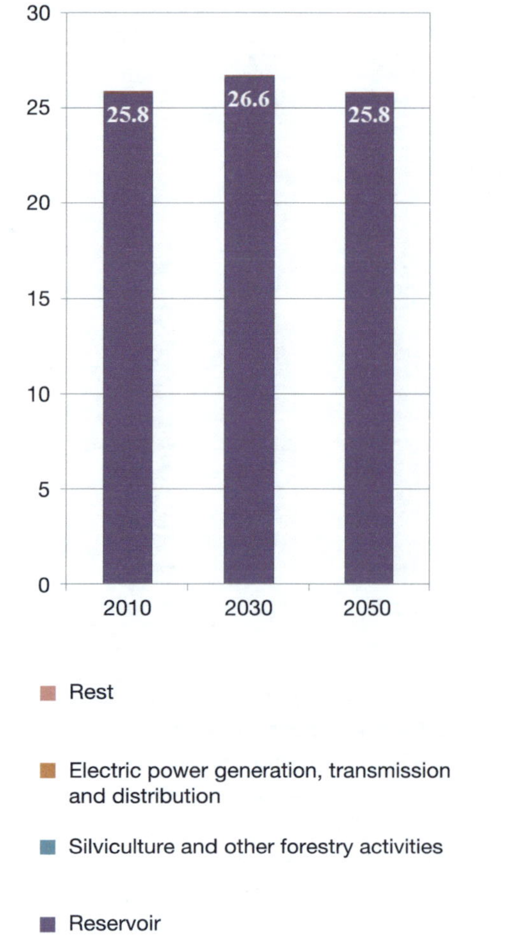

Figure 4.14 shows how land use is distributed across background processes. The direct land occupation of the dam and reservoir themselves dominates the impact on land use. With 26-27 m²·a/MWh, land use occupation is high in comparison to the average impact of electricity mix or the Baker 1 plant.

4.6 CONCLUSIONS

As discussed in Chapter 4.2, variability and site-specificity are critical parameters that influence environmental impacts from hydropower plants. Although they share the same geographical location, the two hydropower plant projects selected for this report have significantly different impact assessment results in many impact categories. The project location can alter the impact results by one order of magnitude not only from the life cycle perspective, but also for direct emissions, as described in Chapter 4.3. The other particularity of hydropower in comparison to other energy technologies is the way per-kWh impacts are unaffected by the changes in the background of processes and sectors modelled in this study. This uniformity over time stems from two factors. The first is the assumption that no further improvements in hydropower technology will be made, as it is a mature technology. The second is the relatively small demand of electricity; electricity contributes little to the total impacts and therefore the improvement of the electricity mix over time has little effect on total impacts. Further, we do not consider improvements in freight transport or construction machinery in our work, as these activities contribute much less to the other energy technologies studied, but may have a significant role in hydropower. The lack of improvement in the technology and the relatively low importance of manufacturing contrasts with other, less developed renewable energy technologies such as wind and solar power.

The impacts of the hydroelectric plants are very site-dependent. Even land occupation is dependent on the characteristics and design features of the hydroelectric production system, despite being the least site-dependent and hence most predictable impact related to hydropower investigated in this study. The study shows that results depend strongly on location-specific conditions such as temperature, material transport distances, flow and fall height of waterfalls, among others, and the processes design, even for plants located in the same region of a country. Such conditions need to be analysed in a life cycle perspective and on a case-to-case basis.

4.7 REFERENCES

Abbasi, T. and S. A. Abbasi. 2011. Small hydro and the environmental implications of its extensive utilization. *Renewable and Sustainable Energy Reviews* 15(4): 2134-2143.

Abril, G., F. Guérin, S. Richard, R. Delmas, C. Galy-Lacaux, P. Gosse, A. Tremblay, L. Varfalvy, M. A. Dos Santos, and B. Matvienko. 2005. Carbon dioxide and methane emissions and the carbon budget of a 10-year old tropical reservoir (Petit Saut, French Guiana). *Global Biogeochemical Cycles* 19(4).

Alho, C. J. R. 2011. Environmental effects of hydropower reservoirs on wild mammals and freshwater turtles in Amazonia: A review. *Oecologia Australis* 15(3): 593-604.

Arias, M. E., T. A. Cochrane, M. Kummu, H. Lauri, G. W. Holtgrieve, J. Koponen, and T. Piman. 2014. Impacts of hydropower and climate change on drivers of ecological productivity of Southeast Asia's most important wetland. *Ecological Modelling* 272: 252-263.

Bai, Y., H. Zheng, Z. Ouyang, C. Zhuang, and B. Jiang. 2013. Modeling hydrological ecosystem services and tradeoffs: A case study in Baiyangdian watershed, China. *Environmental Earth Sciences* 70(2): 709-718.

Bao, G. 2010. Study on the ecological impacts of hydropower resettlement in the Nujiang area. *Shuili Fadian Xuebao/Journal of Hydroelectric Engineering* 29(5): 120-124.

Barros, N., J. J. Cole, L. J. Tranvik, Y. T. Prairie, D. Bastviken, V. L. M. Huszar, P. Del Giorgio, and F. Roland. 2011. Carbon emission from hydroelectric reservoirs linked to reservoir age and latitude. *Nature Geoscience* 4(9): 593-596.

Bastviken, D., L. J. Tranvik, J. A. Downing, P. M. Crill, and A. Enrich-Prast. 2011. Freshwater methane emissions offset the continental carbon sink. *Science* 331(6013): 50.

Bastviken, D., A. L. Santoro, H. Marotta, L. Q. Pinho, D. F. Calheiros, P. Crill, and A. Enrich-Prast. 2010. Methane emissions from Pantanal, South America, during the low water season: Toward more comprehensive sampling. *Environmental Science and Technology* 44(14): 5450-5455.

Beyer, D. L., B. G. D'Aoust, and L. S. Smith. 1976. Decompression induced bubble formation in salmonids: Comparison to gas bubble disease. *Undersea Biomedical Research* 3(4): 321-338.

Brown, J. J., K. E. Limburg, J. R. Waldman, K. Stephenson, E. P. Glenn, F. Juanes, and A. Jordaan. 2013. Fish and hydropower on the U.S. Atlantic coast: Failed fisheries policies from half-way technologies. *Conservation Letters* 6(4): 280-286.

Bunn, S. E. and A. H. Arthington. 2002. Basic Principles and Ecological Consequences of Altered Flow Regimes for Aquatic Biodiversity. *Environmental Management* 30(4): 492-507.

Burdige, D. J. 2007. Preservation of Organic Matter in Marine Sediments: Controls, Mechanisms, and an Imbalance in Sediment Organic Carbon Budgets? *Chemical Reviews* 107(2): 467-485.

Carrara, F., A. Rinaldo, A. Giometto, and F. Altermatt. 2014. Complex interaction of dendritic connectivity and hierarchical patch size on biodiversity in river-like landscapes. *American Naturalist* 183(1): 13-25.

CDEC-SIC. 2011. *Annual Report Statistic and Operation*. Santiago, Chile: Center for Economic Load Dispatch of Central Interconnected System.

Centrales Hidroeléctricas de Aysen S.A. 2008. *Environmental impact study "Aysen Hydroelectric Project"*.

Chanudet, V., S. Descloux, A. Harby, H. Sundt, B. H. Hansen, O. Brakstad, D. Serça, and F. Guerin. 2011. Gross CO 2 and CH 4 emissions from the Nam Ngum and Nam Leuk sub-tropical reservoirs in Lao PDR. *Science of the Total Environment* 409(24): 5382-5391.

Chen, H., X. Yuan, Z. Chen, Y. Wu, X. Liu, D. Zhu, N. Wu, Q. Zhu, C. Peng, and W. Li. 2011. Methane emissions from the surface of the Three Gorges Reservoir. *Journal of Geophysical Research D: Atmospheres* 116(21).

Chen, J., S. Guo, Y. Li, P. Liu, and Y. Zhou. 2013. Joint Operation and Dynamic Control of Flood Limiting Water Levels for Cascade Reservoirs. *Water Resources Management* 27(3): 749-763.

Cheng, B. H., Q. J. Hao, and C. S. Jiang. 2012. Research progress on the emission of greenhouse gases from reservoir and its influence factors. *Wetland Science* 10(1): 121-128.

Cole, J. J., Y. T. Prairie, N. F. Caraco, W. H. McDowell, L. J. Tranvik, R. G. Striegl, C. M. Duarte, P. Kortelainen, J. A. Downing, J. J. Middelburg, and J. Melack. 2007. Plumbing the global carbon cycle: Integrating inland waters into the terrestrial carbon budget. *Ecosystems* 10(1): 171-184.

Cullenward, D. and D. G. Victor. 2006. The dam debate and its discontents. *Climatic Change* 75(1-2): 81-86.

Delsontro, T., D. F. McGinnis, S. Sobek, I. Ostrovsky, and B. Wehrli. 2010. Extreme methane emissions from a Swiss hydropower reservoir: Contribution from bubbling sediments. *Environmental Science and Technology* 44(7): 2419-2425.

Delsontro, T., M. J. Kunz, T. Kempter, A. Wüest, B. Wehrli, and D. B. Senn. 2011. Spatial heterogeneity of methane ebullition in a large tropical reservoir. *Environmental Science and Technology* 45(23): 9866-9873.

Demarty, M. and J. Bastien. 2011. GHG emissions from hydroelectric reservoirs in tropical and equatorial regions: Review of 20 years of CH_4 emission measurements. *Energy Policy* 39(7): 4197-4206.

Demarty, M., J. Bastien, and A. Tremblay. 2011. Annual follow-up of gross diffusive carbon dioxide and methane emissions from a boreal reservoir and two nearby lakes in Québec, Canada. *Biogeosciences* 8(1): 41-53.

Deng, Z., T. J. Carlson, J. P. Duncan, M. C. Richmond, and D. D. Dauble. 2010. Use of an autonomous sensor to evaluate the biological performance of the advanced turbine at Wanapum Dam. *Journal of Renewable and Sustainable Energy* 2(5): Article number 053104.

dos Santos, M. A., L. P. Rosa, B. Sikar, E. Sikar, and E. O. dos Santos. 2006. Gross greenhouse gas fluxes from hydro-power reservoir compared to thermo-power plants. *Energy Policy* 34(4): 481-488.

Driscoll, C. T., R. P. Mason, H. M. Chan, D. J. Jacob, and N. Pirrone. 2013. Mercury as a Global Pollutant: Sources, Pathways, and Effects. *Environmental Science & Technology* 47(10): 4967-4983.

Dudgeon, D. 2000. Large-scale hydrological changes in tropical Asia: Prospects for riverine biodiversity. *BioScience* 50(9): 793-806.

Dudgeon, D. 2011. Asian river fishes in the Anthropocene: Threats and conservation challenges in an era of rapid environmental change. *Journal of Fish Biology* 79(6): 1487-1524.

Esselman, P. C. and J. J. Opperman. 2010. Overcoming information limitations for the prescription of an environmental flow regime for a Central American river. *Ecology and Society* 15(1).

Eugster, W., T. Delsontro, and S. Sobek. 2011. Eddy covariance flux measurements confirm extreme CH_4 emissions from a Swiss hydropower reservoir and resolve their short-term variability. *Biogeosciences* 8(9): 2815-2831.

Fearnside, P. M. 1996. Hydroelectric dams in Brazilian Amazonia: Response to Rosa, Schaeffer & dos Santos. *Environmental Conservation* 23(2): 105-108.

Fearnside, P. M. 2002. Greenhouse gas emissions from a hydroelectric reservoir (Brazil's Tucuruídam) and the energy policy implications. *Water, Air, and Soil Pollution* 133(1-4): 69-96.

Fearnside, P. M. 2005. Do hydroelectric dams mitigate global warming? The case of Brazil's Curuá-Una Dam. *Mitigation and Adaptation Strategies for Global Change* 10(4): 675-691.

Fearnside, P. M. and S. Pueyo. 2012. Greenhouse-gas emissions from tropical dams. *Nature Climate Change* 2(6): 382-384.

Finer, M. and C. N. Jenkins. 2012. Proliferation of hydroelectric dams in the Andean Amazon and implications for Andes-Amazon connectivity. *Plos One* 7(4).

Fu, B., Y. K. Wang, P. Xu, K. Yan, and M. Li. 2014. Value of ecosystem hydropower service and its impact on the payment for ecosystem services. *Science of the Total Environment* 472: 338-346.

Fu, X. C., N. C. Wu, S. C. Zhou, W. X. Jiang, F. Q. Li, and Q. H. Cai. 2008. Impacts of a small hydropower plant on macroinvertebrate habitat and an initial estimate for ecological water requirement of Xiangxi River. *Shengtai Xuebao/ Acta Ecologica Sinica* 28(5): 1942-1948.

GEA. 2012. *Global Energy Assessment - Toward a Sustainable Future*. Cambridge University Press, Cambridge, UK and New York, NY, USA and the International Institute for Applied Systems Analysis, Laxenburg, Austria.

Guérin, F., G. Abril, A. de Junet, and M. P. Bonnet. 2008. Anaerobic decomposition of tropical soils and plant material: Implication for the CO_2 and CH_4 budget of the Petit Saut Reservoir. *Applied Geochemistry* 23(8): 2272-2283.

Guérin, F., G. Abril, S. Richard, B. Burban, C. Reynouard, P. Seyler, and R. Delmas. 2006. Methane and carbon dioxide emissions from tropical reservoirs: Significance of downstream rivers. *Geophysical Research Letters* 33(21).

Gump, B. B., J. A. MacKenzie, A. K. Dumas, C. D. Palmer, P. J. Parsons, Z. M. Segu, Y. S. Mechref, and K. G. Bendinskas. 2012. Fish consumption, low-level mercury, lipids, and inflammatory markers in children. *Environmental Research* 112: 204-211.

Guo, W. X., H. X. Wang, J. X. Xu, and Z. Q. Xia. 2011. Ecological operation for Three Gorges Reservoir. *Water Science and Engineering* 4(2): 143-156.

Haberl, H., K. H. Erb, F. Krausmann, V. Gaube, A. Bondeau, C. Plutzar, S. Gingrich, W. Lucht, and M. Fischer-Kowalski. 2007. Quantifying and mapping the human appropriation of net primary production in Earth's terrestrial ecosystems. *Proceedings of the National Academy of Sciences of the United States of America* 104(31): 12942-12945.

Hassinger, R. 2011. New developments for the ecological enhancement of hydropower sites [Neue Entwicklungen zur gewässerökologischen Optimierung von Wasserkraftstandorten]. *WasserWirtschaft* 101(7-8): 61-65.

Heming, L., P. Waley, and P. Rees. 2001. Reservoir resettlement in China: past experience and the Three Gorges Dam. *The Geographical Journal* 167(3): 195-212.

Hertwich, E. G. 2013. Addressing Biogenic Greenhouse Gas Emissions from Hydropower in LCA. *Environmental Science & Technology* 47(17): 9604-9611.

Hester, E. T. and M. W. Doyle. 2011. Human impacts to river temperature and their effects on biological processes: A quantitative synthesis. *Journal of the American Water Resources Association* 47(3): 571-587.

Huang, T. H., Y. H. Fu, P. Y. Pan, and C. T. A. Chen. 2012. Fluvial carbon fluxes in tropical rivers. *Current Opinion in Environmental Sustainability* 4(2): 162-169.

Hurford, A. P., I. Huskova, and J. J. Harou. 2013. Using many-objective trade-off analysis to help dams promote economic development, protect the poor and enhance ecological health. *Environmental Science and Policy*.

Huttunen, J. T., T. S. Väisänen, S. K. Hellsten, M. Heikkinen, H. Nykänen, H. Jungner, A. Niskanen, M. O. Virtanen, O. V. Lindqvist, O. S. Nenonen, and P. J. Martikainen. 2002. Fluxes of CH_4, CO_2, and N_2O in hydroelectric reservoirs Lokka and Porttipahta in the northern boreal zone in Finland. *Global Biogeochemical Cycles* 16(1): 3-1-3-17.

Ittekkot, V., C. Humborg, and P. Schäfer. 2000. Hydrological Alterations and Marine Biogeochemistry: A Silicate Issue? *BioScience* 50(9): 776-782.

Johnson, E. L., T. S. Clabough, C. A. Peery, D. H. Bennett, T. C. Bjornn, C. C. Caudill, and M. C. Richmond. 2007. Estimating adult Chinook salmon exposure to dissolved gas supersaturation downstream of hydroelectric dams using telemetry and hydrodynamic models. *River Research and Applications* 23(9): 963-978.

Kang, L., Y. Y. Huang, Z. X. Yang, and X. M. Zhang. 2010. Reservoir ecological operation model and its application. *Shuili Xuebao/Journal of Hydraulic Engineering* 41(2): 134-141.

Kareiva, P. M. 2012. Dam choices: Analyses for multiple needs. *Proceedings of the National Academy of Sciences of the United States of America* 109(15): 5553-5554.

Kemenes, A., B. R. Forsberg, and J. M. Melack. 2007. Methane release below a tropical hydroelectric dam. *Geophysical Research Letters* 34(12).

Kemenes, A., B. R. Forsberg, and J. M. Melack. 2011. CO_2 emissions from a tropical hydroelectric reservoir (Balbina, Brazil). *Journal of Geophysical Research G: Biogeosciences* 116(3).

Kibler, K. M. and D. D. Tullos. 2013. Cumulative biophysical impact of small and large hydropower development in Nu River, China. *Water Resources Research* 49(6): 3104-3118.

Kirschke, S., P. Bousquet, P. Ciais, M. Saunois, P. Bergamaschi, D. Bergmann, L. Bruhwiler, P. Cameron-Smith, J. G. Canadell, S. Castaldi, F. Chevallier, E. J. Dlugokencky, L. Feng, A. Fraser, M. Heimann, E. L. Hodson, S. Houweling, B. Josse, J.-F. Lamarque, C. L. Quéré, V. Naik, P. I. Palmer, I. Pison, D. Plummer, B. Poulter, B. Ringeval, M. Santini, M. Schmidt, D. T. Shindell, R. Spahni, S. A. Strode, K. Sudo, S. Szopa, G. R. van der Werf, A. Voulgarakis, M. v. Weele, J. E. Williams, and G. Zeng. 2013. Three decades of methane sources and sinks: Budgets and variations. *Nature Geoscience* 6(10): 813-823.

Kuenzer, C., I. Campbell, M. Roch, P. Leinenkugel, V. Q. Tuan, and S. Dech. 2013. Understanding the impact of hydropower developments in the context of upstream-downstream relations in the Mekong River basin. *Sustainability Science* 8(4): 565-584.

Kumar, A., T. Schei, A. Ahenkorah, R. C. Rodriguez, J.-M. Devernay, M. Freitas, D. Hall, Å. Killingtveit, and Z. Liu. 2011. Hydropower. In *IPCC Special Report on Renewable Energy Sources and Climate Change Mitigation*, edited by O. Edenhofer, et al., Cambridge, United Kingdom and New York, NY, USA: Cambridge University Press.

Kunz, M. J., A. Wüest, B. Wehrli, J. Landert, and D. B. Senn. 2011. Impact of a large tropical reservoir on riverine transport of sediment, carbon, and nutrients to downstream wetlands. *Water Resources Research* 47(12).

Larinier, M. 2001. Environmental issues, dams and fish migration. *FAO Fisheries technical paper*: 45-90.

Lehner, B., C. R. Liermann, C. Revenga, C. Vörösmarty, B. Fekete, P. Crouzet, P. Döll, M. Endejan, K. Frenken, J. Magome, C. Nilsson, J. C. Robertson, R. Rödel, N. Sindorf, and D. Wisser. 2011. High-resolution mapping of the world's reservoirs and dams for sustainable river-flow management. *Frontiers in Ecology and the Environment* 9(9): 494-502.

Li, R., J. Li, K. F. Li, Y. Deng, and J. J. Feng. 2009. Prediction for supersaturated total dissolved gas in high-dam hydropower projects. *Science in China, Series E: Technological Sciences* 52(12): 3661-3667.

Li, S. and X. X. Lu. 2012. Uncertainties of carbon emission from hydroelectric reservoirs. *Natural Hazards* 62(3): 1343-1345.

Liu, J., J. Zuo, Z. Sun, G. Zillante, and X. Chen. 2013. Sustainability in hydropower development: A case study. *Renewable and Sustainable Energy Reviews* 19: 230-237.

Maeck, A., T. DelSontro, D. F. McGinnis, H. Fischer, S. Flury, M. Schmidt, P. Fietzek, and A. Lorke. 2013. Sediment Trapping by Dams Creates Methane Emission Hot Spots. *Environmental Science & Technology*.

Mann, C. C. and M. L. Plummer. 2000. Can science rescue salmon? *Science* 289(5480): 716.

Mayorga, E., S. P. Seitzinger, J. A. Harrison, E. Dumont, A. H. W. Beusen, A. F. Bouwman, B. M. Fekete, C. Kroeze, and G. Van Drecht. 2010. Global Nutrient Export from WaterSheds 2 (NEWS 2): Model development and implementation. *Environmental Modelling and Software* 25(7): 837-853.

McCartney, M. P., A. Shiferaw, and Y. Seleshi. 2009. Estimating environmental flow requirements downstream of the Chara Chara weir on the Blue Nile River. *Hydrological Processes* 23(26): 3751-3758.

McLellan, H. J., S. G. Hayes, and A. T. Scholz. 2008. Effects of reservoir operations on hatchery coastal rainbow trout in Lake Roosevelt, Washington. *North American Journal of Fisheries Management* 28(4): 1201-1213.

Miles, N. G. and R. J. West. 2011. The use of an aeration system to prevent thermal stratification of a freshwater impoundment and its effect on downstream fish assemblages. *Journal of Fish Biology* 78(3): 945-952.

Nakayama, M., B. Gunawan, T. Yoshida, and T. Asaeda. 1999. Resettlement issues of Cirata Dam project: A post-project review. *International Journal of Water Resources Development* 15(4): 443-458.

Nilsson, C. and K. Berggren. 2000. Alterations of riparian ecosystems caused by river regulation. *BioScience* 50(9): 783-792.

Pacca, S. 2007. Impacts from decommissioning of hydroelectric dams: A life cycle perspective. *Climatic Change* 84(3-4): 281-294.

Pess, G. R., M. L. McHenry, T. J. Beechie, and J. Davies. 2008. Biological impacts of the Elwha River dams and potential salmonid responses to dam removal. *Northwest Science* 82(sp1): 72-90.

Poff, N. L. and J. H. Matthews. 2013. Environmental flows in the Anthropocence: Past progress and future prospects. *Current Opinion in Environmental Sustainability* 5(6): 667-675.

Raadal, H. L., L. Gagnon, I. S. Modahl, and O. J. Hanssen. 2011. Life cycle greenhouse gas (GHG) emissions from the generation of wind and hydro power. *Renewable and Sustainable Energy Reviews* 15(7): 3417-3422.

Ramos, F. M., L. A. W. Bambace, I. B. T. Lima, R. R. Rosa, E. A. Mazzi, and P. M. Fearnside. 2009. Methane stocks in tropical hydropower reservoirs as a potential energy source. *Climatic Change* 93(1-2): 1-13.

Reed, P. M., D. Hadka, J. D. Herman, J. R. Kasprzyk, and J. B. Kollat. 2013. Evolutionary multiobjective optimization in water resources: The past, present, and future. *Advances in Water Resources* 51: 438-456.

Renöfält, B. M., R. Jansson, and C. Nilsson. 2010. Effects of hydropower generation and opportunities for environmental flow management in Swedish riverine ecosystems. *Freshwater Biology* 55(1): 49-67.

Ribeiro, F. d. M. and G. A. da Silva. 2010. Life-cycle inventory for hydroelectric generation: A Brazilian case study. *Journal of Cleaner Production* 18(1): 44-54.

Rosa, L. P., M. A. Dos Santos, B. Matvienko, E. O. Dos Santos, and E. Sikar. 2004. Greenhouse gas emissions from hydroelectric reservoirs in tropical regions. *Climatic Change* 66(1-2): 9-21.

Rosa, L. P., M. A. D. Santos, B. Matvienko, E. Sikar, and E. O. D. Santos. 2006. Scientific errors in the Fearnside comments on greenhouse gas emissions (GHG) from hydroelectric dams and response to his political claiming. *Climatic Change* 75(1-2): 91-102.

Sathaye, J., O. Lucon, A. Rahman, J. Christensen, F. Denton, J. Fujino, G. Heath, S. Kadner, M. Mirza, H. Rudnick, A. Schlaepfer, and A. Shmakin. 2011. Renewable Energy in the Context of Sustainable Energy. In *IPCC Special Report on Renewable Energy Sources and Climate Change Mitigation*, edited by O. Edenhofer, et al., Cambridge, United Kingdom and New York, NY, USA: Cambridge University Press.

Saunders, D. A., R. J. Hobbs, and C. R. Margules. 1991. Biological Consequences of Ecosystem Fragmentation: A Review. *Conservation Biology* 5(1): 18-32.

Schilder, J., D. Bastviken, M. Van Hardenbroek, P. Kankaala, P. Rinta, T. Stötter, and O. Heiri. 2013. Spatial heterogeneity and lake morphology affect diffusive greenhouse gas emission estimates of lakes. *Geophysical Research Letters* 40(21): 5752-5756.

Schubert, C. J., T. Diem, and W. Eugster. 2012. Methane emissions from a small wind shielded lake determined by eddy covariance, flux chambers, anchored funnels, and boundary model calculations: A comparison. *Environmental Science and Technology* 46(8): 4515-4522.

Scudder, T. 2002. A Comparative Survey of Dam-induced Resettlement in 50 Cases.

Scudder, T. 2005. *The Future of Large Dams: Dealing with Social, Environmental, Institutional and Political Costs*. London: Earthscan.

Sheaves, M., N. H. Duc, and N. X. Khoa. 2008. Ecological attributes of a tropical river basin vulnerable to the impacts of clustered hydropower developments. *Marine and Freshwater Research* 59(11): 971-986.

Sherman, B., C. R. Todd, J. D. Koehn, and T. Ryan. 2007. Modelling the impact and potential mitigation of cold water pollution on Murray cod populations downstream of Hume Dam, Australia. *River Research and Applications* 23(4): 377-389.

Sobek, S., T. Delsontro, N. Wongfun, and B. Wehrli. 2012. Extreme organic carbon burial fuels intense methane bubbling in a temperate reservoir. *Geophysical Research Letters* 39(1).

Teodoru, C. R., J. Bastien, M. C. Bonneville, P. A. Del Giorgio, M. Demarty, M. Garneau, J. F. Hélie, L. Pelletier, Y. T. Prairie, N. T. Roulet, I. B. Strachan, and A. Tremblay. 2012. The net carbon footprint of a newly created boreal hydroelectric reservoir. *Global Biogeochemical Cycles* 26(2).

Thorstad, E., F. Økland, K. Aarestrup, and T. Heggberget. 2008. Factors affecting the within-river spawning migration of Atlantic salmon, with emphasis on human impacts. *Reviews in Fish Biology and Fisheries* 18(4): 345-371.

Tranvik, L. J., J. A. Downing, J. B. Cotner, S. A. Loiselle, R. G. Striegl, T. J. Ballatore, P. Dillon, K. Finlay, K. Fortino, L. B. Knoll, P. L. Kortelainen, T. Kutser, S. Larsen, I. Laurion, D. M. Leech, S. Leigh McCallister, D. M. McKnight, J. M. Melack, E. Overholt, J. A. Porter, Y. Prairie, W. H. Renwick, F. Roland, B. S. Sherman, D. W. Schindler, S. Sobek, A. Tremblay, M. J. Vanni, A. M. Verschoor, E. Von Wachenfeldt, and G. A. Weyhenmeyer. 2009. Lakes and reservoirs as regulators of carbon cycling and climate. *Limnology and Oceanography* 54(6 PART 2): 2298-2314.

Tremblay, A., J. Therrien, B. Hamlin, E. Wichmann, and L. J. Ledrew. 2005a. GHG emissions from boreal reservoirs and natural aquatic ecosystems. *Greenhouse Gas Emissions - Fluxes and Processes: Hydroelectric Reservoirs and Natural Environments*: 209-232.

Tremblay, A., L. Varfalvy, C. Roehm, and M. Garneau, eds. 2005b. *Greenhouse Gas Emissions - Fluxes and Processes Hydroelectric Reservoirs and Natural Environments*. Berlin Heidelberg New York: Springer.

Turkenburg, W. C., D. J. Arent, R. Bertani, A. Faaij, M. Hand, W. Krewitt, E. D. Larson, J. Lund, M. Mehos, T. Merrigan, C. Mitchell, J. R. Moreira, W. Sinke, V. Sonntag-O'Brien, B. Thresher, W. van Sark, E. Usher, and E. Usher. 2012. Chapter 11 - Renewable Energy. In *Global Energy Assessment - Toward a Sustainable Future*. Cambridge University Press, Cambridge, UK and New York, NY, USA and the International Institute for Applied Systems Analysis, Laxenburg, Austria.

Vachon, D. and Y. T. Prairie. 2013. The ecosystem size and shape dependence of gas transfer velocity versus wind speed relationships in lakes. *Canadian Journal of Fisheries and Aquatic Sciences* 70(12): 1757-1764.

Van Looy, K., T. Tormos, and Y. Souchon. 2014. Disentangling dam impacts in river networks. *Ecological Indicators* 37(PART A): 10-20.

Vilela, M. L., C. G. Azevedo, B. M. Carvalho, and E. F. Rangel. 2011. Phlebotomine fauna (Diptera: Psychodidae) and putative vectors of leishmaniases in impacted area by Hydroelectric Plant, State of Tocantins, Brazil. *Plos One* 6(12).

Vorwerk, P. D., P. W. Froneman, A. W. Paterson, and A. K. Whitfield. 2008. Fish community response to increased river flow in the Kariega Estuary, a freshwater-deprived, permanently open southern African system. *African Journal of Aquatic Science* 33(3): 189-200.

Weitkamp, D. E., R. D. Sullivan, T. Swant, and J. DosSantos. 2003. Gas bubble disease in resident fish of the lower Clark Fork River. *Transactions of the American Fisheries Society* 132(5): 865-876.

Wen, M. X., S. L. Liu, B. S. Cui, and M. Yang. 2008. Impacts of hydroelectric project construction on nature reserve and assessment. *Shengtai Xuebao/ Acta Ecologica Sinica* 28(4): 1663-1671.

Wollebæk, J., J. Heggenes, and K. H. Røed. 2011a. Variations on a theme: Diversification of cuticular hydrocarbons in a clade of cactophilic Drosophila. *BMC Evolutionary Biology* 11(1).

Wollebæk, J., J. Heggenes, and K. H. Røed. 2011b. Population connectivity: dam migration mitigations and contemporary site fidelity in arctic char. *BMC Evolutionary Biology* 11: 207.

World Commission on Dams. 2000. *Dams and Development: A New Framework for Decision-Making*. London: Earthscan.

Xu, J. 2002. River sedimentation and channel adjustment of the lower Yellow River as influenced by low discharges and seasonal channel dry-ups. *Geomorphology* 43(1-2): 151-164.

Xu, J. 2013. Complex Response of Channel Fill-Scour Behavior to Reservoir Construction: An Example of the Upper Yellow River, China. *River Research and Applications* 29(5): 593-607.

Yang, L., F. Lu, X. Wang, X. Duan, W. Song, B. Sun, S. Chen, Q. Zhang, P. Hou, F. Zheng, Y. Zhang, X. Zhou, Y. Zhou, and Z. Ouyang. 2012. Surface methane emissions from different land use types during various water levels in three major drawdown areas of the Three Gorges Reservoir. *Journal of Geophysical Research D: Atmospheres* 117(10).

Yarnell, S. M., A. J. Lind, and J. F. Mount. 2012. Dynamic flow modelling of riverine amphibian habitat with application to regulated flow management. *River Research and Applications* 28(2): 177-191.

Young, P. S., J. J. Cech Jr, and L. C. Thompson. 2011. Hydropower-related pulsed-flow impacts on stream fishes: A brief review, conceptual model, knowledge gaps, and research needs. *Reviews in Fish Biology and Fisheries* 21(4): 713-731.

Zeilhofer, P. and R. M. de Moura. 2009. Hydrological changes in the northern Pantanal caused by the Manso Dam: Impact analysis and suggestions for mitigation. *Ecological Engineering* 35(1): 105-117.

Ziegler, A. D., T. N. Petney, C. Grundy-Warr, R. H. Andrews, I. G. Baird, R. J. Wasson, and P. Sithithaworn. 2013. Dams and Disease Triggers on the Lower Mekong River. *PLoS Neglected Tropical Diseases* 7(6).

Ziv, G., E. Baran, S. Nam, I. Rodríguez-Iturbe, and S. A. Levin. 2012. Trading-off fish biodiversity, food security, and hydropower in the Mekong River Basin. *Proceedings of the National Academy of Sciences of the United States of America* 109(15): 5609-5614.

©Håkan Dahlström/Flickr

Chapter 5

Wind power

Lead author: Anders Arvesen
Contributing authors: Wang Zhongying, Yuan Jingting, Liu Mingliang, Edgar G. Hertwich

5.1 INTRODUCTION

Electric power generation by wind turbines is the second largest contributor, next to hydro power, to world renewable electricity generation (IEA, 2013). Furthermore, wind power is commonly regarded as a key technology in addressing some of the greatest environmental and resource concerns of today, namely anthropogenic climate change and other negative consequences from air pollution, and energy security. Among other factors, a strong growth in today's markets and prospects of exploiting vast and as yet untapped resources, contribute to the anticipation that wind power will play a significant role in shifting energy markets away from fossil-based power generation towards renewables in coming decades (GWEC, 2011; Wiser et al., 2011b). Wind power likewise features prominently in the current body of global climate change mitigation scenarios produced by energy-economy models (IEA, 2010a, 2013; Krey and Clarke, 2011).

Although wind power is driven by the kinetic energy in air streams, which is a renewable energy flux, from a life cycle perspective wind power causes non-renewable resource demands and harmful emissions. These environmental and resource pressures can be quantified and assessed by methods of life cycle assessment (LCA). Other environmental concerns associated with wind power include local effects on ecosystems, particularly bird and bat fatalities, perceived negative effects on the visual quality of landscapes, and adverse effects on human health due to noise.

The aim of this chapter is to help decision makers to understand better the potential environmental impacts and resource requirements of wind power. We first present a review of existing wind power LCA literature in Chapter 5.2. Next, land use, issues connected with rare earth elements use, impacts on ecosystems and impacts on humans are treated in turn in Chapters 5.3 through 5.6. Chapter 5.7 presents the LCA of wind power conducted for this report. Chapter 5.8 concludes this chapter. Chapter 10 of this report complements the current chapter by presenting a comparative LCA of important technologies for electricity generation, including wind power.

The remainder of this introduction provides a brief outline of wind power technology basics (Chapter 5.1.1), and a brief discussion of resource potentials and market developments (Chapter 5.1.2).

5.1.1 TECHNOLOGY BASICS

The function of wind turbines is to convert the kinetic energy of moving air to electrical energy. Mechanical energy is intermediate between the kinetic and electrical energy forms. To be more specific, incoming air flow activates rotation of rotor blades, and the blades spin a low-speed shaft. In the most common wind turbine configuration, a gearbox transfers rotating power to a high-speed shaft, which again spins an electricity-producing generator. In direct-drive configurations, the rotor and generator are mounted on the same shaft. While a great number of variations in wind turbine designs exist, the market is dominated, broadly by the horizontal axis and three-bladed rotor design, with the rotor situated upwind of the tower and nacelle. Figure 5.1 shows the breakdown of a horizontal-axis wind turbine with gearbox into main components.

FIGURE 5.1

Breakdown of main components for a horizontal-axis wind turbine with a gearbox

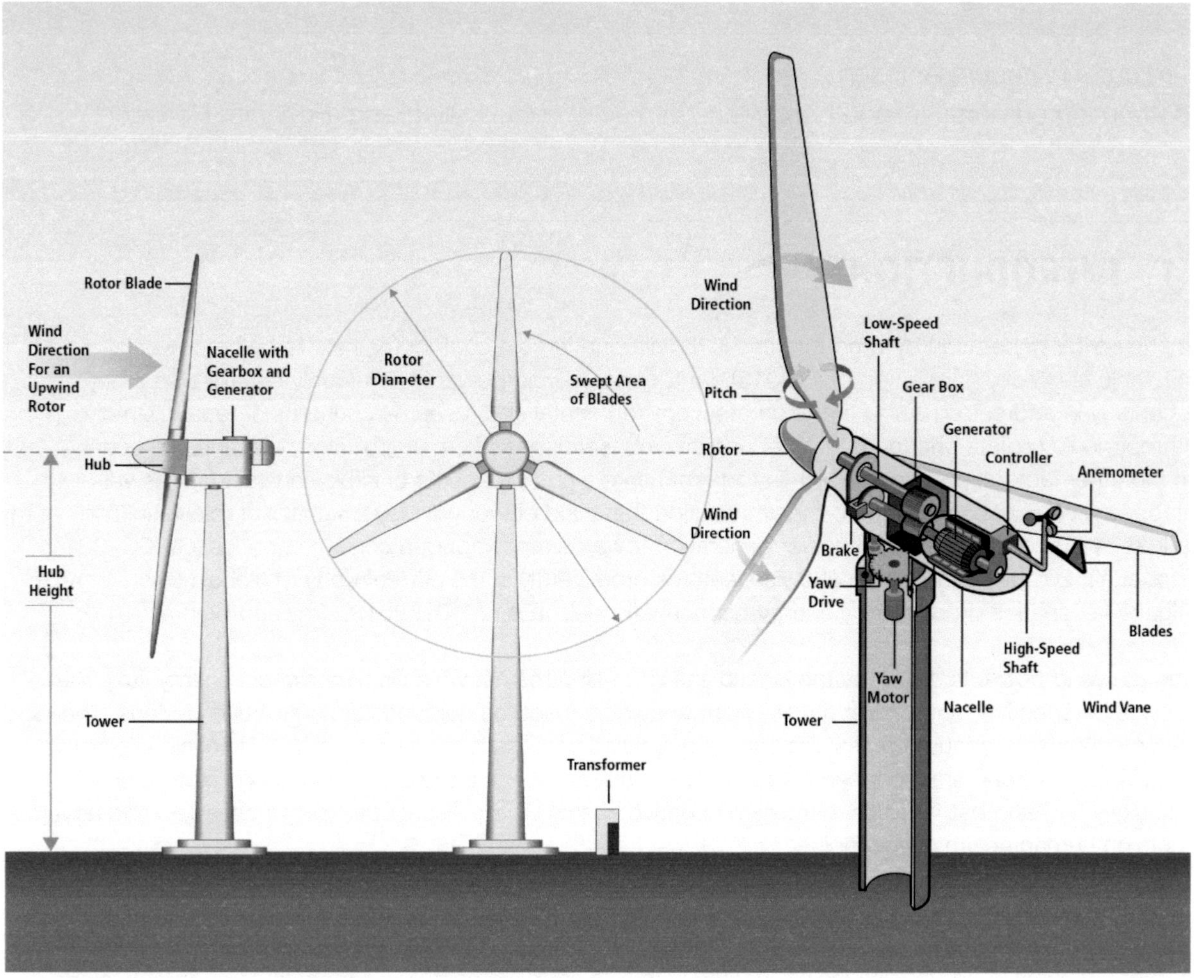

Source: Wiser et al., 2011b, Figure 7.5

Wind turbines may be situated onshore or offshore. The global share of offshore wind power to total wind power capacity was about two per cent in 2012. The offshore share is generally expected to grow in the future owing to, among other factors, many excellent wind resource sites and fewer problems with land use competition and public opposition offshore.

The average wind turbine size has grown substantially and continually since the first turbines with nominal power ratings of hundreds of kilowatts were installed in the 1980s. In 2001, the average size of newly installed wind turbines was 0.85 MW; in 2012, that figure was 1.8 MW. For offshore turbines larger units are preferred: the average size of turbines installed in 2012 was 3.8 MW. Globally, direct-drive wind turbines represent 19.5 per cent of new wind turbine installations in 2012 (Navigant, 2013). Some, but not all, gearless (direct-drive) configurations use rare earth permanent magnets. Reportedly, direct drive configurations with rare earth magnets represent about 5 per cent of wind turbine markets outside China and 25 per cent of those in China (US DOE, 2011).

FIGURE 5.2

Global cumulative installed capacity by region for 2006-2013

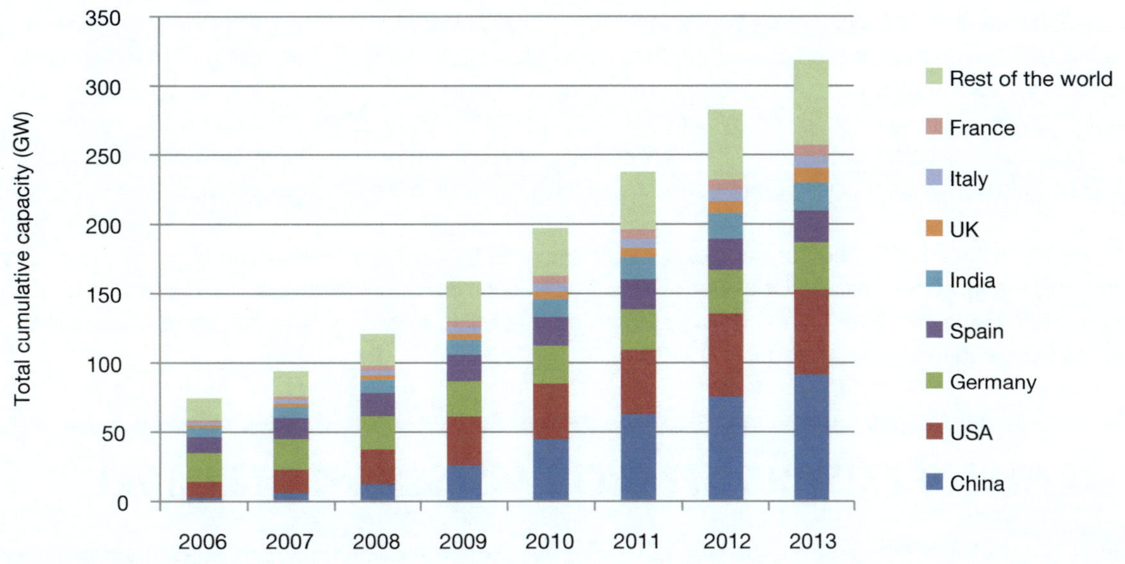

Source: Based on data from GWEC (2006-2013, 2014).

5.1.2 RESOURCE POTENTIALS AND MARKET DEVELOPMENTS

The availability and quality of wind resources are affected by many natural factors, particularly climate, terrain, and land and sea conditions. Wind resources are generally geographically dispersed, though not uniformly distributed. Wind resources are also variable in time over different scales, ranging from sub-hourly to seasonal or inter-annual. Nevertheless, owing to smoothing out of variability over wider geographical areas, wind resources are consistently strong in some regions, both geographically and in time. The portions of wind energy resources that can be harvested, furthermore, depend on characteristics of prevailing technologies and practices.

While the exact magnitude of wind energy output that can be obtained with contemporary technology remains a subject of debate, it is clear that the number is large in relation to current levels globally and for all major regions of the world, and by a wide margin sufficient to support future expansions as described in climate change mitigation scenarios (Wiser et al., 2011b; Rogner et al., 2012; Archer and Jacobson, 2013). Wind power deployment may also be subject to socio-economic and environmental constraints, some of which are treated Chapter 5.3 through 5.6. See Wiser et al. (2011b) for a more comprehensive discussion). Grid infrastructure constraints are discussed in Chapter 9.

In 2012, the global wind power generation was 434 TWh, corresponding to roughly 2 per cent of total global electricity generation (IEA, 2013). As is evident from Figure 5.2, the world's fleet of wind farms is growing rapidly. Global installed wind power capacity by the end of 2013 was 318 GW, up by 12 per cent from 2012. This represents a significant growth in one year, but the rate is lower than the average annual growth in global installed capacity of about 21 per cent over the past ten years. The slowdown in growth in 2013 is attributable to a marked drop in new capacity additions in the US (GWEC, 2013), coinciding with cuts in federal incentives that were expected for 2013 (Wiser and Bolinger, 2013).

To date, global wind power deployment has largely been concentrated in European, Asian (in particular Chinese) and US markets (Figure 5.2). At the end of 2013, the top five countries in terms of cumulative installed capacity (China, US, Germany, Spain, India) together comprised 72 per cent of the global installed capacity. At the same time, there is a clear trend of wind power becoming more geographically widespread. For example, Canada reached 7.8 GW cumulative installed capacity in 2013, and Mexico, Australia, Brazil, Japan all have 2-3 GW installed capacity each (GWEC, 2014). Wind energy deployment in Africa is currently slow, despite some areas being well endowed with wind resources (Wiser et al., 2011b; GWEC 2013, 2014). The cumulative global installed capacity in offshore locations was 5.4 GW in 2012, or roughly 2 per cent of all wind power capacity. Of this, more than 90 per cent was in Europe, and more than half was in the UK alone (GWEC, 2013).

By one evaluation (IEA, 2012), the current global expansion rate for onshore wind power is roughly consistent with achieving the target of limiting global warming to 2°C above pre-industrial levels, but expansion of offshore wind power is too slow. GWEC (2013) also notes that offshore wind power deployment is currently progressing slower than was expected.

5.2 REVIEW OF EXISTING LIFE CYCLE ASSESSMENT LITERATURE

Surveying LCA studies published from the year 2000 on, this section provides quantitative syntheses of the current state of knowledge about the potential life cycle environmental impacts of wind electricity. The primary objectives are to contribute to a better understanding of these potential impacts, to summarize results, and to identify not only areas that are well understood, but also weaknesses and gaps in knowledge that future research should address.

Several literature reviews of wind power LCAs are already available. Lenzen and Munksgaard (2002) survey 72 energy and CO_2 analyses of wind power systems published between 1977 and 2001. Kubiszewski and colleagues (2010) and Raadal and colleagues (2011) extend the work of Lenzen and Munksgaard (2002), adding additional analyses, focusing on energy demand and greenhouse gas (GHG) emissions, respectively. In another review in the IPCC Special Report on Renewable Energy Sources and Climate Change (Sathaye et al., 2011; Wiser et al., 2011b), 126 estimates from 49 studies are surveyed. The present LCA review aims to build on and supplement the previous assessments that have been made, providing new surveys and analyses of results as well as qualitative discussions. In particular, we attempt to make the following original contributions: *i)* taking a broader view of environmental impacts, focusing not only on cumulative energy demand and GHG emissions (Kubiszewski et al., 2010; Lenzen and Munksgaard, 2002; Raadal et al., 2011), but on a wider set of impact categories assessed in the LCA literature; *ii)* discussing important aspects that are not sufficiently treated in previous LCA reviews, including capacity factor assumptions, modeling of recycling benefits, techniques for calculating life cycle inventories (LCIs) (process-LCA or hybrid LCA), and static versus future-oriented LCA; *iii)* critically assessing the scope and quality of existing studies, identifying important knowledge gaps; and, *iv)* proposing directions that future research may take in order to gain a more complete and solid understanding of the environmental implications of wind power.

5.2.1 LIFE CYCLE ASSESSMENT: CONCEPTUAL BASIS AND CALCULATION TECHNIQUES

LCA is a method to explore how the delivery of or demand for a specific product or service (e.g., the delivery of one unit of electricity from wind) initiates processes that may cause environmental impacts. Through a systematic mapping of operations and associated environmental pressures along a product's life cycle, LCA strives to give a complete picture of the environmental burdens caused by one product.

Two approaches to quantifying LCI are in use. In conventional LCA methodology, henceforth referred to as process-LCA, a bottom-up approach is taken to define and describe operations in physical terms. This approach enables the use of data that are specific for the operations under consideration, meaning that results can potentially be generated at high levels of detail and accuracy. The disadvantage of such an approach is that there is a need to apply cut-off criteria to exclude operations that are not expected to make significant contributions. It is known, however, that the sum of the excluded contributions is significant (Lenzen, 2000; Majeau-Bettez et al., 2011). The second approach, environmentally extended input-output analysis (EEIOA), is a top-down technique in which inventories are quantified using monetary data at the level of economic sectors. As EEIOA does not require cut-offs to be made, it does not have the same problem with truncation as process-LCA. However, EEIOA operates at a high aggregation level; the sector resolution in EEIOA is generally too coarse for making LCAs of specific products. Hybrid methods – where process-LCA is used to model important operations, and EEIOA is used to model operations that would otherwise be omitted – can potentially exploit advantages of both approaches, but such hybrid methods are more challenging to employ (Lenzen, 2000; Majeau-Bettez et al., 2011; Suh et al., 2004). Also, depending on the method of hybridization and quality of data (Suh et al., 2004), most hybrid models may offer limited support for tracking physical flows through input-output supply chains, and hence transparency may be limited and it may be difficult to test the robustness of results empirically.

LCA results may be presented as inventories of individual stressors, or as environmental impact category indicators at 'midpoint' or 'endpoint' levels of aggregation. Midpoint indicators allow for environmental effects of several individual stressors to be assimilated into a single impact category. For example, a midpoint indicator for climate change impact combines the effects of emissions of individual GHGs such as carbon dioxide, methane and nitrous oxide. Endpoint indicators measure impact potentials by endpoints in the effect chain; human health, ecosystem health and natural resources are typically regarded as three such endpoints, but sometimes even a single indicator of environmental damage is used (Finnveden et al., 2009; Pennington et al., 2004).

5.2.2 LITERATURE SURVEY

In surveying published wind power LCA research, priority was given to cover publications in peer-reviewed journals, and, for the most part, studies were identified through searches in common scientific databases. However, other appropriate types of publications (such as environmental reports released by manufacturers and documentation of LCA databases) that have been accessed by the authors were also included. The LCA survey presented here differs from that of past reviews in that studies published prior to 2000 are excluded. The primary reason for this is the recent strong developments in wind power technologies, LCA methodologies and databases, and background economy characteristics in previous decades. Furthermore, the set of studies reviewed was judged to be large enough to provide interesting insights.

An overview of the reviewed LCA studies on wind power systems is provided in Table 5.1. Of the 42 reviewed studies (Table 5.1), 34 were selected for quantitative analysis. In general, the following guidelines were followed in constructing the set of results used for quantitative analysis: *i)* Only original LCA research was included; *ii)* studies of integrated wind power generation and energy storage systems were excluded in the cases where the contribution from the actual wind power system could not be extracted from the inventories presented; *iii)* for studies presenting a number of results that apply to different systems (e.g., onshore and offshore wind farms, or differently sized turbines), all reported results were included; and, *iv)* for studies presenting a number of results for one specific system, but with differing methods or assumptions (e.g., different capacity factors, or different approaches to modelling benefits of recycling), the default, or reference, scenario was surveyed if such a scenario was defined. Conversely, if a default scenario was not defined, an average of reported values was surveyed. Table S1 in the supplementary information of Arvesen and Hertwich (2012a) provides the raw data for the quantitative analysis in terms of system characteristics, and emission and impact indicator results.

Finally, we note that the set of studies and estimates included in the quantitative analysis is not a random sample. The identification of studies did not follow a formal, randomized procedure, and the identified studies are sometimes not independent, as they use common sets of assumptions or data. Also, some subjective choices such as regarding how multiple scenario results from single studies should be inventoried were involved in establishing inventories of results and may to some extent have influenced quantitative analysis results.

TABLE 5.1

Overview of assumptions, methodological characteristics and impact categories addressed in the reviewed LCA studies of wind power systems

First author	Year	Site	Size	Lifetime (years)	Credits	Geographical scope	Temporal scope	Method	Impact categories
Lenzen	2012	-	-	-		Global	2009-2100	-	C
Kabir	2012	Ons	S M	25	x	Canada		Pro	CC E A PO
Guezuraga	2012	Ons	L	20		Germany Denmark China		Pro	CC E
Arvesen	2011	Ons Off	L	20 (Ons) 25 (Off)		Europe/Global	2007-2050	Hyb	C CC A PO N
Chen	2011	Ons	L	20	x	China		IOA	CC E
Vestas	2011	Ons	L	20	x	Denmark		Pro	CC E R A P ET PO N W
Wagner	2011	Off	L	20		Germany		Pro	CC E A HT PO N
Wiedmann	2011	Off	L	20		UK		Hyb	C CC
Zhong	2011	Ons	M	20	(x)	Europe		Pro	CC E R A O HT ET L
Gonç. da Silva	2010	-	-	20		Brazil	20 years	-	E
Vattenfall	2010	Mix	L	20		Northern-Europe		Pro	C CC A O P PO N W τ
Crawford	2009	Ons	M L	20		Australia		Hyb	CC E
Fleck	2009	Ons	S	20		Canada		Pro	CC
Fthenakis	2009	Ons	-	30		Denmark		Pro	L
Martínez	2009	Ons	L	20	(x)	Spain		Pro	CC E R A O HT ET PO N L s
Rule	2009	Ons	L	-	x	New Zealand	100 years	Pro	C E
Tremeac	2009	Ons	S L	20	(x)	France		Pro	CC E h e r
Weinzettel	2009	Off	L	20	(x)	Norway		Pro	CC E R A HT ET PO N
Ardente	2008	Ons	M	20		Italy		Pro	C CC E A O P PO N W τ
Lee	2008	Ons	-	20		Taiwan		Pro	C E
NEEDS	2008	Off	L	20*		Denmark	2005-2050	Pro	C P L α τ
Pehnt	2008	Off	L	-		Germany	2005-2020	Pro	C

References: Lenzen and Schaeffer (2012); Kabir et al. (2012); Guezuraga et al. (2012); Arvesen and Hertwich (2011) (corrigendum: Arvesen and Hertwich (2012b)); Chen et al. (2011); Vestas (2011); Wagner et al. (2011); Wiedmann et al. (2011); Zhong et al. (2011); Gonçalves da Silva (2010); Vattenfall (2010); Crawford (2009); Fleck and Huot (2009); Fthenakis and Kim (2009); Martínez et al. (2009b), Martínez et al. (2009a) (related: Martínez et al. (2010)); Rule et al. (2009); Tremeac and Meunier (2009); Weinzettel et al. (2009); Ardente et al. (2008); Lee and Tzeng (2008); NEEDS (2008); Pehnt et al. (2008); ecoinvent (2007) (related: Burger and Bauer (2007), Jungbluth et al. (2005)); Celik et al. (2007); Pehnt (2006); Rankine (2006); Vestas (2006b); (Vestas, 2006a); White (2006); Hondo (2005); Khan et al. (2005); Elsam (2004); Lenzen and Wachsmann (2004); Wagner and Pick (2004); Chataignere and Le Boulch (2003); Properzi and Herk-Hansen (2003); Kemmoku et al. (2002); Pacca and Horvath (2002); Voorspools et al. (2000); Schleisner (2000).

Table 5.1 Overview of assumptions, methodological characteristics and impact categories addressed in the reviewed LCA studies of wind power systems (continued)

First author	Year	Site	Size	Lifetime (years)	Credits	Geographical scope	Temporal scope	Method	Impact categories
ecoinvent	2007	Ons Off	S M L	20		Switzerland/ Europe		Pro	C CC E R A O HT P ET PO N W L h e r s
Celik	2007	Ons	S	25		Turkey		Pro	C CC E
Pehnt	2006	Ons Off	L	-		Germany	2010	Pro	C CC E A P N α τ
Rankin	2006	Ons	S	20	(x)	UK		Pro	C E
Vestas	2006	Ons Off	L	20 (Ons) 20* (Off)	x	Denmark		Pro	C CC E A O HT ET PO N W
Vestas	2006	Ons	L	20	x	Denmark		Pro	C CC E A O HT ET PO N W
White	2006	Ons	M	20-30		US		Pro	C E
Hondo	2005	Ons	M	30		Japan		Pro	CC
Khan	2005	Ons	M	20		Canada		Pro	CC E
Elsam	2004	Ons Off	L	20 (Ons) 20* (Off)	x	Denmark		Pro	C CC E A O HT N W α
Lenzen	2004	Ons	M	-		Germany Brazil		Hyb	C E
Wagner	2004	Ons	M L	20		Germany		Pro	E
Chataignere	2003	Ons Off	M L	20		Europe		Pro	C α
Properzi	2003	Off	L	20*		Denmark		Pro	CC A HT ET PO W
Kemmoku	2002	Ons	M	15		Japan		Pro	C E
Pacca	2002	Ons	M	20		US	40 years	Pro	CC
Voorspools	2002	Ons Off	M	20		Denmark		Pro	C E α s
Schleisner	2002	Ons	M	20		Belgium		Pro IOA	CC E

Key :
Site: Ons = Onshore; Off = Offshore. Size: S = Small (< 100 kW); M = Medium (100 kW-1 MW); L = Large (> 1 MW). Lifetime: '*' means that longer lifetimes are assumed for some system components. Credits: 'x' means that the system is credited with indicator values that are perceived to be avoided through recycling at the end-of-life; A blank means that the system is not credited; '(x)' means that the system is credited, but results without credits are also made available. Geographical scope: Reference location of the system under study. Temporal scope: A blank means static assessment under assumptions of present technologies; Non-blank entries indicate future-oriented assessments. Method: Pro = Process-LCA; Hyb = Hybrid LCA; IOA = Environmentally extended input-output analysis. Impact categories: C = CO_2 emissions; CC = Climate change/global warming; E = Cumulative/total energy demand; R = Resource requirements, abiotic depletion, material intensity (different indicators exist but are here assimilated in one category); A = Acidification; O = Stratospheric ozone depletion; HT = Human toxicity; P = Particulate matter formation, particles/dust; ET = Ecotoxicity (different indicators exist but are here assimilated in one category); PO = Photochemical oxidation/ozone creation (smog); N = Nutrient enrichment, eutrophication; W = Solid waste generation; L = Land use, land transformation; h = human health endpoint impact indicator; e = natural environment endpoint impact indicator; r = natural resources endpoint impact indicator; s = single score endpoint indicator; α = non-toxic air emissions that provide additional information; τ = toxic emissions that provide additional information ('additional' with regards to the impact categories 1-13 that are accounted in this table column). Numbers are underlined if results are presented in generic units only (e.g., 'eco-points', 'person-equivalents').

5.2.3 SCOPE, ASSUMPTIONS AND METHODOLOGIES

The LCA literature covers the whole spectrum of available wind turbine sizes, from units sized for the generation of hundreds of watts (Fleck and Huot, 2009; Tremeac and Meunier, 2009) to multi-megawatt turbines in onshore and offshore locations. As is evident from Table 5.1, analyses of wind farms operating on land comprise the vast majority of studies and there is a predominance of analyses using European countries as their reference locations. A fair number of analyses (13) of ocean-based systems were also identified, however. With some exceptions, LCAs of offshore wind power study bottom-fixed wind turbines in relatively shallow waters. Two of the studies analyse a hypothetical wind farm consisting of floating units (Weinzettel et al., 2009) and an operational wind farm at a water depth of 30 m (Wagner et al., 2011). Only one study assesses a wind turbine with a gearless drive train configuration (Guezuraga et al., 2012), and no studies consider the environmental impacts associated with rare earth element-based permanent magnets that are used in most direct drive designs.

Manufacturing of the actual wind turbines is the only life cycle stage that is common to all analyses. In addition, all assessments based on wind turbines with capacities of hundreds of kilowatts and more include the manufacturing of foundations, and the majority model electrical connections such as internal cables within wind farm, external cabling and sometimes transformer stations that are needed to connect a wind farm to an existing grid. Most studies also take into consideration, albeit inconsistently, transportation activities and other operations associated with the installation, operation and maintenance of the system. A number of assessments (Celik et al., 2007; Fleck and Huot, 2009; Kemmoku et al., 2002; Khan et al., 2005; Pehnt et al., 2008) address integrated systems where wind energy converters are supplemented with other power generation technologies and/or technologies for energy storage.

The manner in which the end-of-life phase is modelled varies. Some studies make assumptions to model transport and disposal of waste, while others omit this part. End-of-life is unique among the life cycle phases in that it may *reduce* impact indicator values; negative contributions occur when analysts deduct indicator values that are perceived to be avoided when, after the operating lifetime, system components are recycled or incinerated to produce valuable outputs. In this way, the system is credited for returning usable resources such as recyclable steel to the technosphere. In the LCA literature, this is referred to as substitution by system expansion or the avoided burden method (Finnveden et al., 2009). LCA studies that employ the avoided burden method are in the minority, but nevertheless represent a significant share of the studies assessed (Table 5.1).

LCAs of wind power generally assume lifetimes of 20 years, for onshore and offshore wind farms alike (Table 5.1). Figure 5.3 displays capacity factor assumptions by region as a function of power rating. Three general trends may be observed from Figure 5.3, in overall terms consistent with general knowledge and expectations (Boccard, 2009; Wiser and Bolinger, 2013): *i*) performance in terms of capacity factor increases with wind turbine nominal capacity; *ii*) offshore wind farms exhibit greater energy capture than onshore farms; and *iii*) for a given power rating, sites in North America tend to show higher capacity factors than European sites. Across all regions, the assumed capacity factor mean value (± standard deviation) is 18 per cent (±5.4 per cent) for onshore wind turbines with nameplate capacity below 100 kW, 22 per cent (±5.1 per cent) for onshore with capacity 100 kW - 1 MW, 31 per cent (±8.1 per cent) for onshore with capacity greater than 1 MW, and 43 per cent (±8.3 per cent) for offshore (Table S2 in the supplementary information of Arvesen and Hertwich (2012a).

LCA studies of wind power have historically focused on energy demand and GHG emissions (Lenzen and Munksgaard, 2002), and so has the recent literature (Table 5.1). Smaller sample sizes are hence available for other pollutants (Figure 5.4). Estimates of climate change indicator values, measured in units of CO_2-equivalents (CO_2-eq.), consist of contributions from CO_2, CH_4 and N_2O, but in some cases, (e.g. *ecoinvent*, 2007; Wiedmann et al., 2011) fluorinated GHGs (SF_6, HFC, PFC) are also taken into account. Of the studies cited in Table 5.1, more than half include impact categories other than energy and GHGs. In general, the environmental stressors that are most thoroughly covered are air pollutants associated with production and combustion of fossil energy carriers: CO_2, CH_4, CO, NH_3, NMVOC, N_2O, NO_x, particulates and SO_2.

FIGURE 5.3

Capacity factor by location as a function of power rating

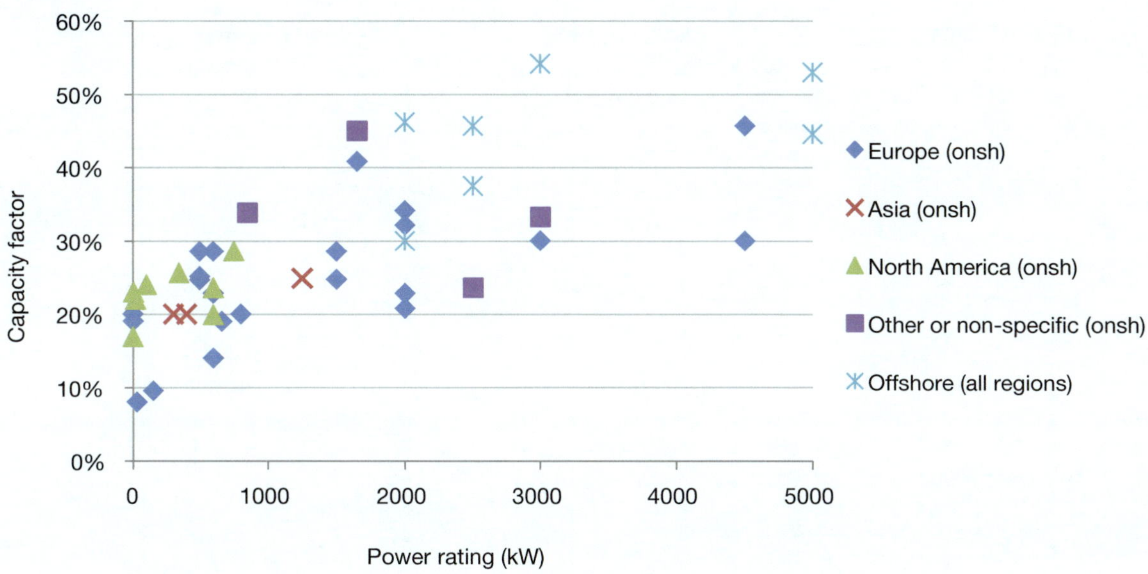

Such a set of pollutants facilitates meaningful impact assessments in the climate change, acidification, eutrophication and photochemical oxidation (smog) formation categories. In comparison to fossil fuel-related air emissions, other kinds of pollution have received little attention; only eight of the studies cited in Table 5.1 quantify characterized toxicity indicator results. Beyond fossil energy carriers, resource requirements and non-renewable resource depletion are scarcely addressed in detail. A handful of studies (Martínez et al., 2009a; Vestas, 2011; Weinzettel et al., 2009; Zhong et al., 2011) address non-renewable resource depletion; others (Vattenfall, 2010; Vestas, 2006b) present LCIs for individual mineral resources without applying any impact assessment. One publication (Fthenakis and Kim, 2009) examines in some detail direct and indirect land use of power generation technologies, including wind. Some studies quantify life cycle water use, but water use is generally not highlighted or discussed in detail. Fthenakis and Kim (2010) review previous studies and evaluate life cycle use of water in electricity supply by different technologies.

As is evident from Table 5.1, process-LCA studies dominate the wind power LCA literature, and few studies employ hybrid LCA methodologies. As a final point regarding methodology, we note that different kinds of future-oriented LCAs of wind energy have started to emerge, but are yet to gain widespread adoption (cf. 'temporal scope' column in Table 5.1).

5.2.4 STRESSOR AND IMPACT INDICATOR RESULTS

Figure 5.4 presents literature survey results with respect to total emissions and impact indicator values, and the numbers of estimates and studies that were surveyed. The energy indicator value in Figure 5.4 refers to the energy intensity, i.e., the ratio between total life cycle energy demand and electricity generated over the lifetime. For onshore and offshore wind power, respectively, the mean energy intensity value is 0.063 (±0.061 standard deviation on either side of the mean) and 0.055 (±0.037) kWh/kWh; mean GHG emissions are 18 (±13) and 15 (±9.5) g CO_2 eq./kWh; and mean CO_2 emissions 16 (±14) and 12 (±6.9) g/kWh. These relatively large standard deviations, and the broad ranges that can be observed for all categories displayed in Figure 5.4, illustrate that

FIGURE 5.4

Stressor and impact indicator results by 5 wind power system categories and 11 impact categories

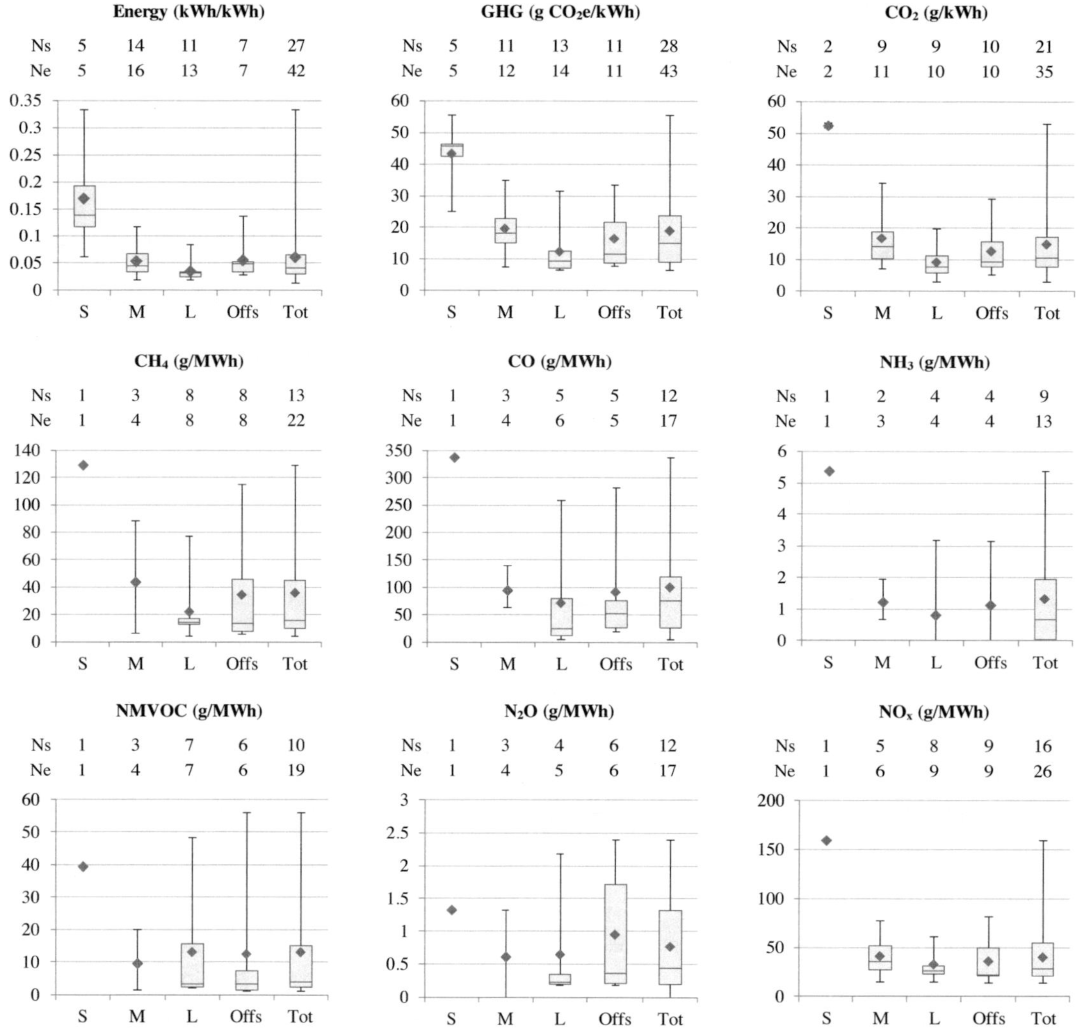

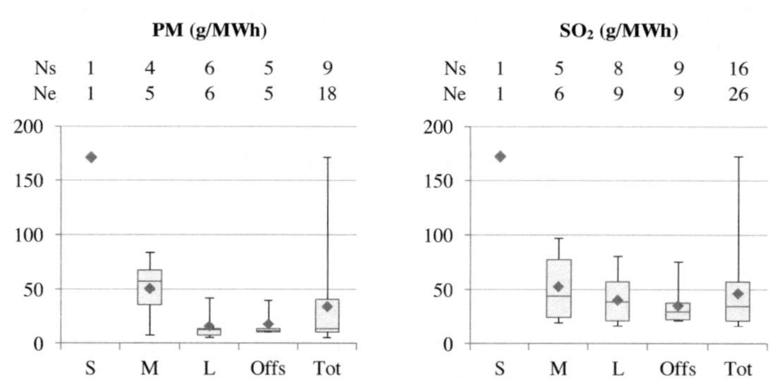

Box: range from first to third quartile; Horizontal bar within box: median value; Blue diamond: mean value; Upper and lower fences (whiskers): minimum and maximum values. 'S' means small wind turbine (< 100 kW) at onshore site; 'M' means medium wind turbine (100 kW - 1 MW) at onshore site; 'L' means large wind turbine (> 1 MW) at onshore site; 'Offs' means offshore wind power (any wind turbine size); 'Tot' denotes total sample. Ns = Number of studies; Ne = Number of estimates. Ne > Ns if more than one estimate was surveyed from one study.

In some cases, the total sample size slightly exceeds the sum of the shown sub-sample sizes; this is because estimates for wind farm portfolios were not assigned a specific wind power system type, but were included in the total sample. If Ns < 5, interquartile ranges (boxes) are not shown; If Ne = 1, the one value is shown as a blue diamond. Energy indicator value refers to the energy intensity, i.e. the ratio between total life cycle energy demand and electricity generated over the lifetime. Chemical formulas and abbreviations: GHG = greenhouse gases; CO_2 = carbon dioxide; CH_4 = methane; CO = carbon monoxide; NH3 = ammonia; NMVOC = non-methane volatile organic compounds; N_2O = nitrous oxide; NO_x = mono-nitrogen oxides; SO_2 = sulfur oxides. Numerical results in tabulated form are available in the supplementary information of Arvesen and Hertwich (2012).

results vary considerably. For example, reported energy intensity values across all wind power system categories form an interval of 0.014-0.333 kWh/kWh. If analyses of wind turbines with nameplate capacity less than 100 kW are excluded, however, the interval narrows to 0.014-0.137 kWh/kWh, thereby exemplifying a general pattern that the highest indicator values are observed for small wind turbine sizes (< 100 kW). It should be noted that the average size of newly installed wind turbines in 2012 was 1.85 MW (Navigant, 2013), and in Figure 5.4 it is primarily the categories representing large and offshore wind turbine units that are of importance for overall wind power capacity expansion.

Offshore wind power systems show comparable or slightly higher emissions than onshore systems consisting of large wind turbines (Figure 5.4), despite the systematically higher wind capacity factors assumed for offshore systems (Figure 5.3). This is due to the higher resource requirements of wind power systems located offshore. Another observation that can be made from Figure 5.4 is a tendency for estimates to concentrate in the lower part of the observed intervals (note for example that the mean values lie systematically above median values).

Releases of individual toxic substances in the life cycle of wind power systems are reported in some cases, but synthesizing these findings is difficult due to differences in which chemicals are reported and a lack of transparency on calculation methods and assumptions. Table 5.2 compares human toxicity and freshwater and terrestrial ecotoxicity indicator results from five studies. Marine aquatic eco-toxicity is not included due to weaknesses in current impact assessment methods (Pettersen and Hertwich, 2008), and because two of the cited studies (Wagner et al., 2011; Weinzettel et al., 2009) do not address this impact category at all. One of the publications (Vestas, 2011) cited in Table 5.2 report results that are up to three orders of magnitude smaller than those from the other studies. The reason for this discrepancy is unknown, but could possibly be a consequence of different impact characterization methods. Comparisons of toxicity of wind power and fossil fuel-based power are inconclusive. Two studies of wind power find, respectively, that a wind farm scores two to six times worse in toxicity impact categories than a natural gas combined cycle plant (Weinzettel et al., 2009), and that wind electricity is slightly worse than the average German electricity with respect to human toxicity (Wagner et al., 2011). A different picture is presented by other publications, whose findings suggest that wind power grossly outperforms European (Vestas, 2006b) and Spanish (Martínez et al., 2009a) average electricity mixes with respect to human toxicity. Again, it is possible that differences in impact assessment methods lead to different results, but this needs to be investigated further.

TABLE 5.2

Overview toxicity indicator results by three impact categories, as quantified by five studies

Citation	Wind turbine size, site	Stated impact characterization method	Results (g 1,4-DCBe/kWh)		
			HT	FET	TET
Vestas (2011)	1.85 MW, onshore	USETox (2008)	0.83	0.03	0.03
Wagner et al. (2011)	5 MW, offshore	-	69	-	-
Martínez et al. (2009a)	2 MW, onshore	CML (2000)	16	2.8	0.16
Weinzettel et al. (2009)	5 MW, offshore	CML (2000)	83	12	0.23
ecoinvent (2007)	800 kW, onshore	CML (2001)	54	10	0.16
ecoinvent (2007)	2 MW, offshore	CML (2001)	53	10	0.18

HT = human toxicity. FET = freshwater ecotoxicity. TET = terrestrial ecotoxicity. DCBe = 1,4-dichlorobenzene equivalents. See Chapter 2 for general descriptions of human toxicity and ecotoxicity impact indicators. Freshwater ecotoxicity measures toxicity to living organisms in freshwater ecosystems, while terrestrial ecotoxicity measures toxicity to organisms in terrestrial ecosystems.

Looking at the relative contribution from different life cycle stages to total energy use and climate change indicator, manufacturing of components dominates, and is sometimes of the order of 90 per cent of total impact indicator values (Figure 5.5. See also discussion in previous LCA reviews (Lenzen and Munksgaard, 2002; Raadal et al., 2011). Figure 5.5 compares breakdowns of energy use and GHG emissions by components and life cycle stages. It should be noted that ambiguity exists in the categories shown in Figure 5.5; for example, some studies separate transportation as an individual category, while other studies subsume transportation activities within other categories. Nevertheless, it is clear from Figure 5.5 that for onshore wind power systems, the wind turbine is the single most important component with regards to energy use and GHG emissions, followed by the substructure. The tower may hold a share of 30-70 per cent of total wind turbine indicator values. For offshore wind farms, the substructure becomes relatively more important.

Generally, emissions associated with transportation are found to be of minor importance or altogether negligible, though they are sometimes relatively more important for NMVOC and NO_x emissions. The results of Tremeac and Meunier (2009) (not included in Figure 5.5) stand out with large relative contributions from transportation; 34 per cent of GHG emissions are due to transportation). This observation could possibly be related to the choice of concrete as tower material in Tremeac and Meunier (2009), as opposed to lighter tubular steel towers modelled in most other studies. Emissions of heavy metals due to materials production is the primary cause of toxicity indicator results (Vestas, 2011; Wagner et al., 2011; Weinzettel et al., 2009).

FIGURE 5.5

Breakdown of energy intensity (EI) or GHG emission intensity by main components or life cycle stages according to 8 onshore and 3 offshore estimates

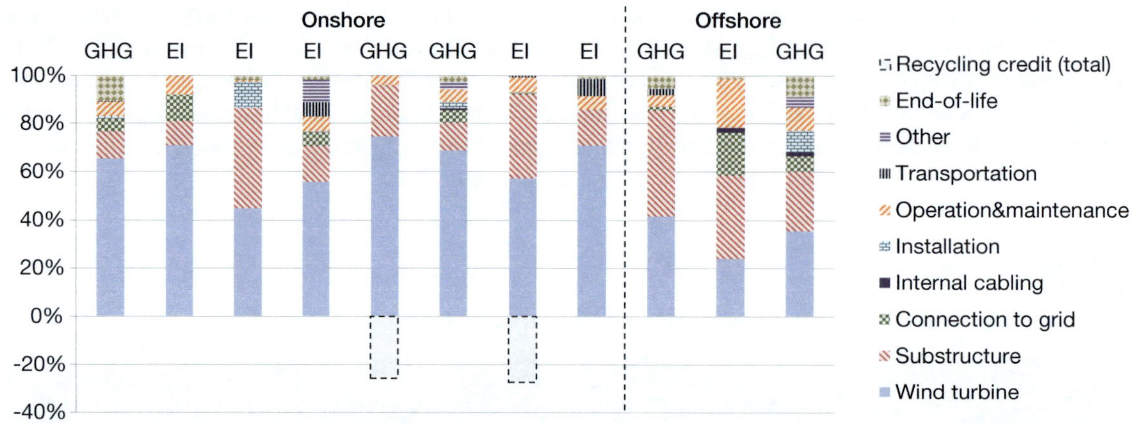

References for the 8 onshore and 3 offshore estimates in the figure are, from left to right, Chataignere and Le Boulch (2003), Wagner and Pick (2004), Lee and Tzeng (2008), Ardente et al. (2008), Martínez et al. (2009a), Arvesen and Hertwich (2011, 2012b), Chen et al. (2011), Guezuraga et al. (2012), Chataignere and Le Boulch (2003), Wagner et al. (2011), Arvesen and Hertwich (2011, 2012b). In some cases interpretation of results and/or reading off charts in the cited publications was necessary. Shown positive indicator shares from Chen et al. (2011) and Martínez et al. (2009a) do not include recycling credits.

If the avoided burden method is applied, the end-of-life phase typically yields considerable emissions reductions. Recycling credits approximately halve the energy or GHG emissions embodied in the wind turbine, and lower total indicator values by 26-27 per cent, in Chen et al. (2011) and Martínez et al. (2009a) (Figure 5.5). In another study, recycling credits lead to around 20 per cent and 40 per cent reductions in GHG emissions for a 4.5 MW and 250 W wind turbine, respectively (Tremeac and Meunier, 2009). In total, the end-of-life phase contributes -19 per cent to GHG emissions in Weinzettel et al. (2009).

5.2.5 EFFECTS OF WIND TURBINE SIZE AND METHOD FOR LIFE CYCLE INVENTORY

Previous reviews of LCA wind power studies maintain economies of scale in the life cycle environmental impacts of wind power systems. Lenzen and Munksgaard (2002) report that a 1 MW wind turbine appears to require only one third of the life cycle energy per unit output needed for a 1 kW sized unit. Kubiszewski et al. (2010) and Raadal et al. (2011) show evidence of energy use and GHG emissions decreasing with increasing wind turbine size, but in the former case, it remains unanswered to what extent the trend continues when moving into the MW size spectrum, and in both cases it appears that the practice of surveying old and arguably outdated analyses dating to the late 1970s on a par with recent analyses obscures the picture. With the results of the present survey, Figure 5.6 depicts GHG emissions with increasing wind turbine nameplate capacity. The figure confirms the presence of strong economies of scale for power ratings up to 1 MW or so, but a downward trend is not readily discernible for larger turbine sizes.

FIGURE 5.6

Total GHG or CO_2 emissions (in g CO_2-eq./kWh) as a function of wind turbine power rating by 4 combinations of methods (hybrid LCA and EEIOA versus process-LCA) and sites (onshore versus offshore)

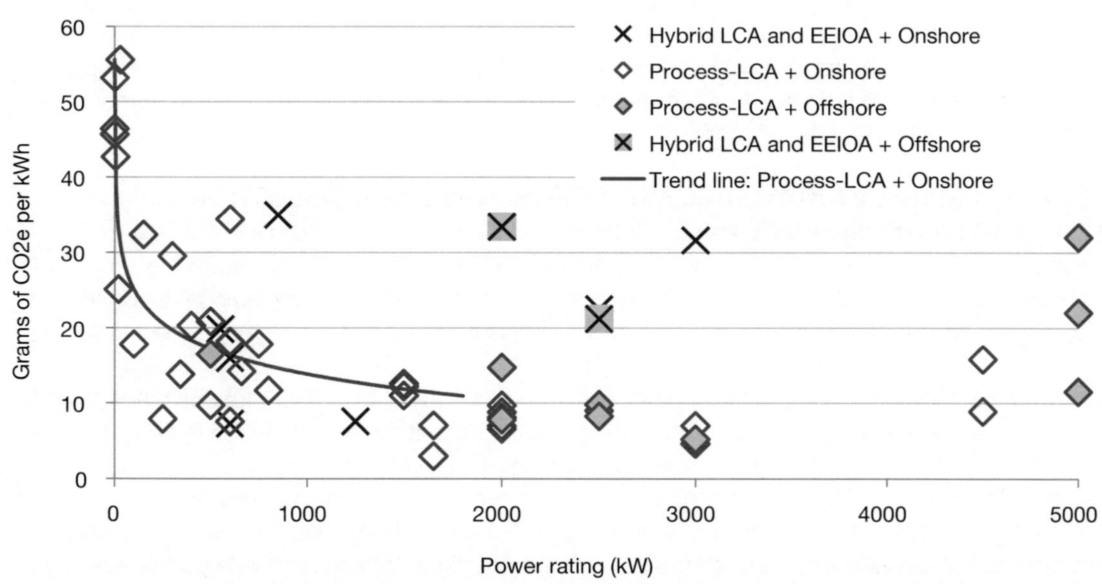

When available, total GHG emissions estimates were included in the figure. If GHG emissions estimates were not available, CO_2 emissions estimates were included. For estimates for offshore wind farms, markers are filled with grey. The trend line represents the sample characterized by a process-LCA calculation technique and onshore site only, and is cut at power rating 1800 kW. Trend line equation: $y = -4.9\ln(x) + 47.7$, with $x \in [0,1800]$. $R^2 = 0.68$

Theory and empirical evidence from the broader LCA literature predict that hybrid LCA and EEIO-based assessments give systematically higher impacts (Chapter 5.2.1). Some wind power hybrid LCA studies show large contributions from economic input-output sectors, for example, 23-26 g CO_2e/kWh from input-output sectors (corresponding to 74 per cent of totals) in Crawford (2009), and 19 g CO_2e/kWh from input-output sectors (57 per cent of total) in Wiedmann et al. (2011). The overall picture of Figure 5.6 is mixed, however, with even a few hybrid LCA or EEIOA results appearing in the lower range of results. The sample size of studies representing hybrid LCA in Figure 5.6 is too small to admit a robust assessment comparing results of hybrid LCA and process-LCA.

5.2.6 FUTURE-ORIENTED ASSESSMENTS

In a forward-looking study for Germany, Pehnt et al. (2008) couple LICI with a stochastic electricity market model to study the life cycle CO_2 emissions of wind power, grid expansion, energy storage by means of compression of air, and energy balancing requirements, in an integrated framework. Results for the year 2020 show only negligible emissions from storage and grid upgrades, but a relatively large emission penalty of 18-70 g CO_2/kWh arising from the balancing of variable wind electricity by fossil-fuelled power stations. A global scenario-based LCA is presented by Arvesen and Hertwich (2011, 2012b), who estimate around 3 Gt CO_2e caused by the expansion and operation of wind power plants in the 2007-2050 period to supply 22 per cent of worldwide electricity in 2050. The same study includes an integrated life cycle modelling of cumulative avoided emissions; results suggest that emissions avoided by wind power grossly exceed emissions caused by wind power. Lenzen and Schaeffer (2012) analyse caused and avoided climate change impacts of eight energy technologies towards 2100, with the primary objective to illustrate differences between emissions and temperature-based indicators for climate change mitigation potential; the authors argue that indicators of avoided temperature are more relevant for decision-making than avoided emissions. In yet another study, Gonçalves da Silva, (2010) proposes a mathematical framework for simulating the time dynamics in net and gross energy balances of renewable energy technology deployments; the computed results are favourable for wind power. Finally, the report on offshore wind technology in the NEEDS project (NEEDS, 2008) makes assumptions on design changes and economies of scale in wind electricity technologies to establish LCI for future offshore wind power systems. For all of the scenario assessments cited above, there are important simplifying assumptions, and thus careful interpretations of results are required. Indeed, the authors of the original publication also emphasize this point.

5.2.7 CURRENT STATE OF KNOWLEDGE AND RESEARCH NEEDS: A DISCUSSION

5.2.7.1 Capacity factor and lifetime assumptions

The strong influence of assumed capacity factors and lifetimes on results is obvious, as emissions per unit of electricity (in units of grams of CO_2 per kWh, or similar) are inversely proportional to the amount of electricity generated over the lifetime; this is analogous to calculations of generation costs in units of Euro per kWh, or similar.

With respect to capacity factor, one interesting comparison to make is that of assumptions made in LCAs (Figure 5.3) versus real-world experiences. The average realized capacity factor in EU15 in 2003-2007 is reported at 20.8 per cent, with country-level averages ranging from a low 18.3 per cent (Germany) to a high 26.1 per cent (UK) (Boccard, 2009). These real-world performances are significantly lower than the overall picture emerging from the assumed values shown in Figure 5.3 for onshore wind turbines in the range of 1 MW and above, but are relatively more consistent with assumptions for smaller turbine sizes. Real capacity factors in the US are generally higher, averaging at 30.3 per cent in the period 2000-2005 and 32.1 per cent in 2006-2012, according to Wiser and Bolinger (2013). However, there is a seemingly conflicting report of average capacity factor value for the US of 26 per cent in Boccard (2009). Wiser and Bolinger (2013) find that recent technology developments have contributed to higher capacity factors in the US, but at the same time the average wind resource quality for new projects has declined, and hence the average capacity factor for projects completed after 2005 has remained fairly constant. Average capacity factors for China are reported at 16-17 per cent in Yang et al. (2012) and 23 per cent in Cyranoski (2009). Data points representing North America and Asia in Figure 5.3 are, however, mostly at the lower end of the turbine size spectrum and therefore do not provide a good basis for comparison.

For the offshore case, LCA studies often assume capacity factors above 40 per cent and even 50 per cent (Figure 5.3), which also seems somewhat optimistic in comparison to currently available empirical data. One study (Lemming et al., 2009) concludes from a survey that "a typical offshore installation has an utilization time of 3,000 hours or more" (i.e., capacity factor 34 per cent or more) while, based on experiences from early Danish and Dutch wind farms, Feng et al. (2010) generally expect a 35 per cent capacity factor value for UK offshore wind farms, but find that the real average value for UK round 1 offshore wind farms is 29.5 per cent. Finally, we note that Lemming et al. (2009) propose a constant 37.5 per cent average capacity factor for offshore wind power to be used in scenario analysis towards 2050, while more optimistic figures are derived in Arvesen and Hertwich (2011) from IEA data (IEA, 2010a), leading to an average offshore load of 43 per cent in 2050.

Based on the above information, there appears to be a general tendency of wind power LCAs to assume higher capacity factors than current averages from real-world experiences. At the same time, it needs to be emphasized that many LCAs make assumptions that are specific to one technology or wind farm site and as such are not intended to be representative for overall trends.

Unlike capacity factors, real-world empirical evidence on the lifetimes of modern wind turbines is lacking; assumptions on lifetimes thus need to be guided by wind turbine specialists' evaluations and the design lifetime set by manufacturers. While LCAs typically assume lifetimes of 20 years for onshore and offshore systems alike (Table 5.1), Blanco (2009) finds that current economic assessments of wind power generally set lifetimes to 20 years onshore and 25-30 years offshore. On this basis, it appears that assumptions regarding lifetimes in the LCA literature are less favourable for offshore wind power, compared with equivalent assumptions underlying economic assessments.

5.2.7.2 Impact category coverage

The current survey shows that life cycle energy demand and GHG emissions of wind power are extensively covered in the existing literature. Also, there is a fairly large selection of studies quantifying air pollutants typically associated with the burning of fossil fuels (e.g., NO_x, SO_2) and the related impact categories, namely acidification, eutrophication, photochemical oxidant formation, and to a lesser extent, particulate matter formation. In the view of the authors, given the material intensive nature of wind power compared to fossil alternatives (Kleijn et al., 2011), and that toxic releases to the environment are known to originate from materials production (Vestas, 2011; Wagner et al., 2011; Weinzettel et al., 2009), the most serious gap in knowledge is the inadequate understanding of toxic emissions generated in the life cycle of wind power systems. From the viewpoint of the LCA practitioner, assessing toxic effects may be difficult because emissions data on toxic substances are missing or are incomplete, and because current impact assessment methods for toxicity produce contradictory results and hence lack robustness (Finnveden et al., 2009; Hauschild, 2005). The neglect (or incomplete modelling) of toxicity is not a problem specific to wind power LCAs, however, but applies to the LCA literature in general. For marine ecological impacts from emissions to water, robust impact assessment methods are in the early stage of development. In general, unresolved issues may be exemplified by contradictory results for toxic effects of long-term metal releases, as discussed in Pettersen and Hertwich (2008), and effects of particle emissions in (Veltman et al., 2011). This is unfortunate for LCA research on offshore wind power, as there are operations associated with offshore wind power taking place in ocean waters that must be modelled.

Another significant gap in knowledge is that represented by a lack of comprehensive evaluations of non-renewable, or abiotic, resource demands. As with toxicity, this is a much debated impact category in the LCA community for which there is no consensus on impact assessment methods (Finnveden et al., 2009). In the broader literature, concerns have been raised about future shortage of supply of neodymium and dysprosium, two metals belonging to the group of rare-earth elements that are increasingly used to manufacture direct-drive wind turbines (Chapter 9.3). Sustainability assessments of wind power must also adequately consider site-specific impacts, such as visual impacts, habitat change, and bird and bat collisions (Chapters 9.2 and 5.1). There is, however, little tradition for including such impact categories in LCA, and they are more frequently assessed using other impact assessment methods such as cost-benefit analysis (e.g., Snyder and Kaiser, 2009).

5.2.7.3 Life cycle phases: research coverage, research agreement and quality of knowledge

Table 5.3 summarizes our overall assessment of the current knowledge about potential environmental burdens associated with four life cycle phases of wind power systems. Extended life cycles such as those considering network integration or re-powering of systems are discussed in Chapter 9.1.3.5.

TABLE 5.3

Summary of authors' overall judgments regarding research coverage, research agreement and quality of knowledge for four life cycle phases (production of components, transportation and on-site construction, operation and maintenance, and end-of-life)

Life cycle phase	Coverage	Agreement	Quality	Remarks
Production of components	******	****	****	Complete coverage. Uncertainty about emissions embodied in materials. Detailed material compositions are often unknown. Toxic emissions from manufacturing are poorly understood; issues of mineral resource pressures are not well understood. Rare earth magnets are not considered. Studies assuming European energy systems dominate. Few studies of very large wind turbines and offshore wind turbines in deep waters and/or far from shore.
Transportation to site, on-site construction	****	***	***	Coverage is variable. Onshore: not important according to most studies. Offshore: possibly important; modeling appears simplistic; NO_x from fuel oil-burning may be significant. Few studies of wind turbines in deep waters and/or far from shore.
Operation and maintenance	****	***	***	Coverage is variable. Offshore transportation and on-site activities: modeling appears simplistic; NO_x from fuel oil-burning may be significant. Empirical basis for assumptions about replacement of parts seems to be lacking. Few studies of wind turbines in deep waters and/or far from shore.
End-of-life	***	****	**	Scarcely assessed in detail. Future waste handling practices for rotor blades are unknown. Assessments using the avoided burden method are often lacking in transparency and may be inconsistent.

The table shows authors' overall judgments regarding research coverage of life cycle phases in existing studies (the number of asterisks indicates the degree to which we find studies include the life cycle phase in their scope), the degree to which research results are in agreement (low number of asterisks indicates research do not agree, and that the reasons for the disagreements are hard to establish; more asterisks indicates higher level of agreement, or that reasons for disagreements are well understood), and quality of knowledge (the number of asterisks indicates the degree to which we judge current knowledge to be sound and transparent). The latter indicator (quality of knowledge) depends on the research coverage and agreement, but also our (qualitative) evaluations of level of uncertainty and transparency.

Production of components

Production of system components forms a natural part of any wind power LCA. Discrepancies concerning values for embodied energy and emissions in materials contribute to differences in impact indicator results. In some cases, values for the emissions embodied in materials are meant to be different across studies; this may be, for example, because the assumed energy mix in production is different. See, for example Guezuraga et al. (2012); Lenzen and Wachsmann (2004). In other cases, discrepancies may be due to different LCA databases being utilized, or arise as a result of different assumptions being made regarding material types such as steel alloys in the face of uncertainty of the exact types of materials used in each component.

A general problem is that component material compositions detailing exact material types are typically not available, and furthermore, that LCA databases provide LCIs for only a limited selection of generic materials. This creates uncertainty, which we illustrate here using a simple calculation exercise. In one study (Burger and Bauer, 2007), ferrous metal content in the 800 kW onshore wind turbine, excluding the foundation, consists of 7 per cent cast iron, 78 per cent low-alloy steel and 15 per cent high-alloy (chromium) steel, while in a second study (Martínez et al., 2009a) for a 2 MW onshore wind turbine, the corresponding shares are 16 per cent cast iron and 84 per cent reinforcing steel. Both studies utilize the ecoinvent LCA database to model materials manufacturing; we find the relevant GHG emission intensities in ecoinvent are 1.48, 1.45, 1.72 and 4.50 kg CO_2e/kg for cast iron, reinforcing steel, low-alloy steel and chromium steel, respectively (ecoinvent, 2007). Hypothetically, assuming a 2 MW wind turbine contains 200 tonnes of ferrous metals, has a lifetime of 20 years and 25 per cent capacity factor, these values correspond to either 4.9 or 3.3 g CO_2e/kWh, depending on whether the material composition factors of Burger and Bauer (2007) or Martínez et al. (2009a) are adopted. This exemplifies how material modelling choices which are often not justified and are scarcely discussed in LCA studies may significantly influence total impact indicator values. Another potentially important, poorly understood factor is the composite materials used in the rotor blades and nacelle.

Direct-drive wind turbines with rare earth magnets represent a growing segment of the wind power market. While there is widespread concern about environmental damage associated with the production of rare earth magnets (Chaper 9.3), this concern is not addressed in wind power LCA literature or, as far as we know, other LCA literature. This leaves a noticeable gap in the account of environmental impacts of wind power, as it remains unclear and untested to what extent findings for wind turbine designs with gearbox are representative of that for rare earth magnet-based designs.

Transportation, on-site construction, and operation and maintenance

The overall picture emerging from the current LCA literature is that emissions associated with transportation and on-site construction are small or negligible (Chapter 5.1.1.1.1). While this conclusion appears to be fairly well documented with respect to the energy use and GHG emissions for onshore wind farms, one could question to what extent it is valid for offshore projects with more complicated installation procedures than onshore units. This is perhaps especially the case for NO_x emissions originating mainly from transportation and construction activities. NO_x emissions are the main contributor to marine eutrophication and photochemical oxidant formation impact indicator values for the offshore wind farm modelled by Arvesen and Hertwich (2011). The same argument may apply to transportation and construction activities associated with maintenance. To our understanding, existing LCAs of offshore wind farms rely on rather simplistic and theoretical calculations for modelling on-site operations, and consistency with real-world conditions has not yet been demonstrated.

LCA studies either neglect replacement of parts (e.g., Burger and Bauer, 2007; Lee and Tzeng, 2008) or variably assume that certain shares of components must be replaced. For example, Crawford (2009) assumes 50 per cent gearbox replacement during lifetime, Ardente et al. (2008) one blade and 15 per cent generator replacement, and Weinzettel et al. (2009) 5 per cent complete wind turbine replacement. One study develops a high-maintenance scenario in which one generator, one gearbox and one set of blades requires replacement (Martínez et al., 2010). While these replacement assumptions are not uniform across studies, one can discern that gearboxes, generators and rotor blades are expected to be most susceptible to failure and replacement. An empirical basis for assumptions about replacement seems to be lacking, however (a similar point is made by Martínez et al. (2010)). One central question is how the assumed replacement rates relate to past experiences from operational wind farms; another question is how to extrapolate information from past experiences to modern wind turbines and more immature application areas such as wind farms in marine environments. In our judgment, these questions are not adequately addressed in the LCA literature.

End-of-life

Since LCAs typically assume the bulk of materials contained in wind power systems will either remain *in situ* or be recycled to be returned to usage as raw materials, waste disposal is generally not an important contributor to emissions. Excluding 'new' life cycles that are created when materials are recycled is common practice in LCA. Compare this with the cut-off allocation principle in open-loop recycling; see, for example, Shen et al. (2010).

There is considerable uncertainty surrounding the fate of fibre-reinforced plastic materials used in the rotor blades: Recycling fibre-reinforced plastic composites represents a technological challenge for which little practical experience exists (Correia et al., 2011; Larsen, 2009). While there is a consensus that the traditional practice of landfilling reinforced plastics is unsatisfactory, and regulatory measures to phase out landfilling of these materials are coming into place (Andersen et al., 2007; Larsen, 2009), the question of which waste treatment strategies are viable and should be chosen remains open (Correia et al., 2011). There are concerns about toxic emissions occurring when cutting the blades, which may be necessary to facilitate transport (Andersen et al., 2007), from waste treatment if the materials are landfilled (Zhong et al., 2011), and from flue gas and ashes if the materials are incinerated (Correia et al., 2011; Larsen, 2009). Future LCA research may have to address waste handling of rotor blades in order to ensure environmentally sound end-of-life phase for wind turbines.

A significant number of studies credit the system with perceived emissions reductions from end-of-life recycling (avoided burden method; Table 5.1). However, applications of the avoided burden method sometimes use inappropriate methodologies and are generally lacking in transparency. The root of the problems appears to be that it is not widely recognized that the two issues of 1) including recycled content as input materials in the production phase and 2) crediting the system with prevented environmental burdens at the end-of-life cannot be viewed independently. The share of secondary inputs in the production phase should always be zero for the materials for which avoided burden is calculated; otherwise one would use one perspective to model benefits of recycling in the production phase, and a different, inconsistent perspective to model benefits of recycling at the end-of-life, effectively double-counting the benefits of recycling. The crux of the issue is that analysts must decide whether benefits of recycling should belong to systems that *use* recycled materials (as is the implicit assumption if secondary materials are used as inputs in an LCA), or those that make available recyclable materials, as is the assumption if avoided burden method is applied, and not mix these two perspectives.

We are aware of one study (Weinzettel et al., 2009) that uses the avoided burden method appropriately, assuming no secondary resources as inputs in production when the avoided burden method is applied. (Another study (Vestas, 2011) in which the avoided burden method is used also assumes only virgin resource inputs in production, but the stated reason is lack of data on recycled content, and the assumption is inappropriately described as "very conservative".) One apparently inconsistent assessment is Martínez et al. (2009a), where materials containing significant amounts of recycled content (i.e., cast iron, reinforcing steel and copper ecoinvent processes (Classen et al., 2007)) are stated to be used in the production phase, while simultaneously, recycling credits are given for avoided production. Other LCAs use the avoided burden method while not specifying that only virgin resources are used in production.

5.2.7.4 Method for life cycle inventory and system boundary issues

Lenzen and Munksgaard (2002) recommended that future wind power LCA research employs hybrid LCA methodologies "in order to achieve system completeness while dispensing with the problem of selecting of a boundary for the production system". However, the current survey demonstrates that hybrid LCA studies on wind power are still relatively scarce; confirming that despite its acknowledged advantages, hybrid techniques have not yet become standard practice in LCA (Majeau-Bettez et al., 2011). Hybrid LCA is more challenging to conduct and requires additional data, which may be an explanation for its lack of use. Also, it is interesting to note that Wiedmann et al. (2011) employ two hybrid LCA calculation techniques separately, and find that while the total emission estimates obtained by the two techniques are comparable, there are considerable differences in the relative contribution from input-output sectors. This points to yet unresolved issues with IO-based LCA calculation techniques.

Notwithstanding the data and methodological challenges of hybrid methods, hybrid LCA is the only technique that can offer both process-level detail and a complete coverage of the entire product system. While there is no consensus in the LCA community on how to measure the truncation bias of process-LCA, in all explorations into this issue surveyed by Majeau-Bettez et al. (2011), it is found that process-LCA fails to account for 30 per cent or more of total indicator values. This predicates that the employment of hybrid LCA methodologies should be a goal of future LCA research on wind power; and that if hybrid techniques on the other hand are not applied, the problem of cut-off errors should at the least be recognized in the discussion and conclusions; this is not the case in existing literature.

5.2.7.5 Aspects of scale, temporal evolutions, and network integration

In recent years, LCA practitioners have remarked on the insufficiency of static, unit-based analyses for evaluating implications of future wind energy developments (Arvesen and Hertwich, 2011; Gonçalves da Silva, 2010; Pehnt, 2006; Pehnt et al., 2008). One shortcoming of existing research is the general failure to address the magnitudes of aggregated impacts: a transition away from conventional and towards lower-carbon energy systems in coming decades, as envisaged by contemporary climate change mitigation scenarios (IEA, 2010a; Krey and Clarke, 2011), will in itself cause harmful emissions. Due to the sheer scale of the transition, total emissions and resource use brought about by 'clean' energy technologies may be significant in the overall global system, even if unit-based assessments (i.e., assessments where indicator values are measured per kWh) indicate low impacts. In the literature, climate change mitigation scenario analyses explore energy transitions at the economy-wide level, but do not consider emissions arising from building and operating non-fossil power plants. Conversely, LCAs of power generation predominantly have a purely micro-level focus. The integration of these two perspectives could potentially provide valuable new insights on the economy-wide effects of large-scale energy transitions. Ideally, such scenario calculations incorporate some projections of future technological changes, as discussed below.

Inventories for wind power systems are not static, but change over time as new technological configurations are adopted, changes in background economies are experienced and economies of scale play a role in wind projects. Projections of impacts of research and scientific developments on future technological designs, based on technology forecasting studies or learning curve studies (Cohen et al., 2008; NEEDS, 2008), may provide LCA analysts with a basis for modelling future inventory changes, as demonstrated by Viebahn et al. (2011) for concentrating solar power. Besides changes in wind power technology configurations, impact indicator values are influenced by the characteristics of background economies through relatively clean or dirty manufacturing; indeed, it is the current economies' preoccupation with fossil fuels which is the very reason why electricity from wind is not CO_2-free. The importance of background energy system characteristics is illustrated by the results of Lenzen and (Wachsmann, 2004), where the embodied CO_2 is a factor of five lower for a wind turbine produced in Brazil compared to one produced in Germany; the difference stems entirely from the higher portion of renewable sources, namely hydro and biomass, in Brazil's energy supply. It is not just the energy mix as such that is important, however, but also the energy efficiency. The environmental impacts of metals supply change due to combined effects of technological advances in mining and manufacturing, the stringency and implementation of environmental regulations or voluntary practices in environmental management at the enterprise level, changes in the portion of secondary to primary materials used, and reduction in the ore grades (Mudd, 2010; Norgate and Jahanshahi, 2010; Yellishetty et al., 2011). Future research may address the effects of such changes through scenario analyses.

The final type of scaling or temporal aspect discussed here relates to the variable and partially unpredictable nature of wind power. Higher shares of intermittent electricity supply, such as electricity from wind, increase the overall costs of short-term balancing in the system (that is, matching electricity supply with demand over seconds to days), reduce overall peak-load system adequacy (because the contribution of a wind power plant to peak-load capacity adequacy is smaller than for conventional technologies), and may require upgrades in the electricity transmission infrastructure to admit transfer of electric power to the load centres. In the available literature, life cycle emissions of wind power and emission penalties due to the variability of wind power (e.g., Gross and Heptonstall, 2008; Luickx et al., 2010; Oswald et al., 2008) are generally analysed separately and lead to separate evaluations of emissions connected with wind power deployment. In a sense, these two areas of research form two independent departures from the notion that wind power is 'emissions-free', both aiming to provide a more complete picture. The potential exists to combine the assessments in these two research fields, as exemplified by the study by Pehnt et al. (2008) discussed in Chapter 9.1.2. This would indeed be congruent with the often-stated goal of LCA to provide holistic assessments, but on the other hand it involves substantial methodological and data challenges. In any case, when interpreting results of current LCA studies it is important to bear in mind the failure of LCA research to account for emission penalties due to intermittency. A more detailed discussion of grid balancing and storage is included in Chapter 9.

5.2.8 FINAL REMARKS AND RECOMMENDATIONS

Despite the considerable variability in results, and the limitations of current knowledge that have been mentioned, we conclude that existing LCA research provides many insights into and gives a fairly good overall understanding of the life cycle environmental impacts of wind power in terms of cumulative fossil energy demand and associated pollution. Discrepancies between studies can likely be explained by a combination of actual differences in the systems studied (e.g., small versus large wind turbines), key assumptions (e.g., capacity factor and lifetime), data inconsistencies (e.g., emission intensities of materials), and differences in methodologies and approaches (e.g., process-LCA or hybrid IO-LCA, accounting of recycling benefits). Previous LCA reviews (Lenzen and Munksgaard, 2002; Raadal et al., 2011; Wiser et al., 2011b) have duly noted that the large gap between low and high values limit the usefulness of results to decision-makers, and that compliance with some standardized sets of methods and assumptions in future analyses would be advantageous.

The problems of confusion and uncertainty due to variability in results, and incomprehensibility due to the complex networks of operations that are studied and many assumptions that are made, need to be given due attention. One measure that can be taken to alleviate these problems and which conforms to the guiding principle that LCAs should be transparent (ISO, 2006), is to make process-level inventory input data available together with LCA publications. Such a step would increase the transparency as to how results are obtained, and may help to give clarity on why results differ across studies, and allow for proper meta-analyses of wind power LCAs (Price and Kendall, 2012). Furthermore, making inventory input data at the level of unit processes available can contribute to a cumulative build-up of knowledge, rather than having efforts going into repetitions of sometimes cumbersome data collection processes.

This review has shown that to date, the largest research efforts have been devoted to studying typical onshore wind turbines or wind farms in European locations, placing the most emphasis on the production life cycle stage. Future research may focus attention on system types or life cycle phases for which research is still relatively scarce or robust assessments are lacking. This may include:

- Systems that are produced and operated under conditions of regions other than Europe.
- Direct drive wind turbines using rare earth permanent magnets.
- Installation and operation and maintenance phases, in particular for offshore systems.
- Large wind turbines (> 3 MW), and offshore systems in deep waters and/or far from shore.

Wind power LCAs have traditionally had their domain in assessing potential environmental pressures caused by one small reference unit (1 kWh of electricity), have primarily focused on fossil energy-related emissions, and have predominantly employed a process-LCA methodology. Such assessments have proved valuable in the past and are likely to continue to play a role in future research. At the same time, given the sizeable number of published studies that are similar with regards to goal and scope, one could expect that research had made further strides in analyses with different or broader scopes, or more sophisticated methodologies. In this respect, we call for future research efforts to be directed into:

- The employment of hybrid LCA methodologies.
- Broadening the scope with regards to environmental impacts, as far as available impact assessment methods allow it. In particular, we call for more detailed explorations of toxicity and mineral resource depletion.
- Exploring technology evolution through scenario analyses, addressing, for example, the scale of environmental burdens at regional or global levels, changes in LCIs as key technologies or background economies change, or emission penalties due to intermittency.

In all cases, future studies should avoid inconsistent modelling of recycling benefits.

5.3 LAND USE

Wind power plants require land or water areas to utilize the kinetic energy of moving air. Some horizontal area is used directly by infrastructure, notably turbines with foundations and access roads. This direct area cannot simultaneously be used for other purposes or combined with terrestrial wildlife. A much larger area encompassing all units of the power plant is used only by virtue of the presence of discontinuous infrastructure within that area, as the spacing between infrastructure elements can be natural habitat or used for agriculture, ranching or other purposes. An even larger area may be seen as being impacted by a wind farm if one takes into consideration potential indirect effects on wildlife, such as on migratory birds or bats or, from a visual amenity point of view, degradation of the quality of landscapes. In the literature, some studies use total area of wind power plants to assess the land use impacts of wind electricity and competing technologies (Fthenakis and Kim, 2009; McDonald et al., 2009; Scheidel and Sorman, 2012). Other sources use the direct area only (Burger and Bauer, 2007; Hirschberg et al., 2006) or emphasize the multifunctional nature of the total area (Jacobson, 2009; Jacobson and Delucchi, 2011).

Denholm et al. (2009) synthesize information on land requirements for 172 wind power projects in the United States. The survey sample includes only large wind power plants (> 20 MW) constructed or proposed after the year 2000. Table 5.4 shows survey results from Denholm et al. (2009) in terms of direct area (that is the land footprint of infrastructure, staging) and total area for wind power projects. Direct area is further divided into permanent and temporary direct area use, where the latter accrues mostly from unpaved access roads that are used during installation but not maintained during the operation phase Permanent roads comprise 79 per cent of total direct area, while turbine area contributes only 10 per cent. It should be noted that both direct and total area figures differ markedly across projects; for total area use, the standard deviation is 224 km^2/GW (Denholm et al., 2009). Denholm et al. (2009) identify some cases of overestimation in the underlying data for total area, but the overall significance of such overestimation appears unclear.

In other literature, estimates of total area are often based on theoretical considerations of the amount of land needed to convert wind energy to electricity or assuming a certain spacing between units to optimize energy extraction, for example, Smil (2003), MacKay (2009) and Jacobson (2009). Assumed values for onshore wind power vary considerably in existing literature: 0.5-1.5 W/m^2 values area assumed in Scheidel and Sorman (2012), 2 W/m^2 in MacKay (2009), 4-5 W/m^2 in McDonald et al. (2009), 5 W/m^2 in US DOE (2008) and 11 W/m^2 in Jacobson (2009).

TABLE 5.4

Overview of direct (permanent and temporary) and total area use of large wind power plants (> 20 MW) in the United States (Denholm et al., 2009)

	Direct area		Total area
	Permanent	Temporary	
Number of projects[a]	93	52	161
Capacity (MW)[b]	13897	8984	25438
Reported area (km^2)[b]	37.6	61.4	8779
Average area (km^2/GW)[c]	2.7	6.8	345
Land use per unit energy (m^2a/MWh)*d*	0.8-1.5	2.0-3.9	98-197
Average power density (W/m^2)[c] (total area only)			2.9

[a] The number of projects from which land use data are collected.
[b] Summed over projects.
[c] Calculated here from area and capacity numbers.
[d] For a capacity factor of 20-40 per cent

5.4 RARE EARTH ELEMENTS USE

In recent years, the rapidly growing use of rare earth elements and potential future disruptions in their supply have emerged as subjects of considerable academic, political and public attention (Du and Graedel, 2011; EC, 2010; Kleijn and van der Voet, 2010; Prins et al., 2011; US DOE, 2011). The rare earth elements neodymium and dysprosium are used to manufacture high-performance, compact permanent magnets used in certain direct drive wind turbines. Rare earth elements are of varying abundance in the earth's crust. The less abundant rare earth elements are those that form the heavier part of the group (atomic numbers > 64), dysprosium being one of them. Neodymium, on the other hand are, is one of the lighter and relatively less scarce rare earths. Furthermore, rare earth elements are commonly regarded as valuable or critical and their future supply as uncertain due to their unique properties, recent considerable increases in global use, and an almost complete Chinese dominance of current production and uncertainty about future exports which may be problematic for other regions (Erdmann and Graedel, 2011; JRC-IET, 2011; US DOE, 2011). Additionally, environmental damage resulting from rare earth mining is often quoted as a reason for concern.

Typically, a neodymium-iron-boron (NdFeB) alloy is used to manufacture the permanent magnets used in direct drive windmills. Dysprosium is frequently used as an additive to the NdFeB magnetic alloy to maintain its usable temperature range and enhance its resistance to demagnetisation with temperature. Praseodymium and terbium are occasionally used too, but their use in permanent magnets is of little significance compared to neodymium and dysprosium (Hoenderdaal et al., 2013; US DOE, 2011). Current permanent magnet-based direct-drive designs contain about 600 kg of magnet per MW capacity, of which roughly 30 per cent is neodymium and 4 per cent dysprosium (Hoenderdaal et al., 2013). Although detailed information does not appear to be available, it is clear that direct-drive configurations with NdFeB magnets, represent a small share of current wind turbine markets; a share of 5 per cent outside China and 25 per cent in China is reported (US DOE, 2011). The trend is towards increasing market share for turbines using NdFeB magnets however, and several major wind turbine manufacturers are pursuing the development of gearless NdFeB designs. Benefits of such designs in the form of reduced nacelle weight and maintenance requirements are considered to be particularly valuable for offshore installations and large wind turbines.

A combination of limited production capacity, local environmental costs and geopolitical factors will likely place some limit on the market uptake of wind turbines containing rare earths. In the literature, Kleijn and van der Voet (2010) find that neodymium constraints are likely to limit the fraction of wind turbines with NdFeB magnets in future markets. Such constraints can contribute to a less than optimal performance of wind power in the future, but are not expected by Kleijn and van der Voet (2010) to be a major impediment to wind power deployment. Jacobson and Delucchi (2011) contend that, as generator designs not involving rare earth elements exist or can be developed, constrained availability of rare earths is not likely to noticeably affect the overall cost of wind power deployment. Hoenderdaal et al. (2013) warn that a global dysprosium shortage is in sight in the short term towards 2020, but the opening of new mines and efficient recycling may contribute to remedy the situation in the long term. US DOE (2011) assigns a supply risk indicator value of 3 to neodymium and 4 to dysprosium, on a scale from 1 to 4 where 4 means high supply risk, in both the short- (toward 2015) and medium-term (2015-2025). Stocks of rare earth elements in in-use NdFeB magnets as of 2007 have been found to exceed the 2007 extraction rate by a factor of four, indicating some opportunity for recycling to alleviate issues with primary resource availability (Du and Graedel, 2011). Several publications have focused on the need for and development of recycling strategies for managing dysprosium shortage (Alonso et al., 2012; Rademaker et al., 2013), but recent research indicates that the development of NdFeB magnets using less or no dysprosium to be more promising (Liu et al., 2012; Brown et al., 2014; Seo and Morimoto, 2014).

5.5 IMPACTS ON ECOSYSTEMS

Wind power projects have the potential to affect ecosystems in a number of ways. This section aims to give a brief overview of such potential impacts, with an emphasis on direct bird and bat mortality (Chapters 9.4.1 and 9.4.2), as these are two widespread concerns for which a good deal of research has been carried out. For more detailed discussions on the environmental impacts of wind power on ecosystems, see, for example, Inger et al. (2009), Wilson et al. (2010), Wiser et al. (2011a), or, Lovich and Ennen (2013).

5.5.1 BIRD COLLISION FATALITIES

Research on potential effects of wind turbines on bird migration or bird populations has been undertaken in a number of countries, including the United States, Canada, Ireland, Germany, Spain, Denmark, and China (Desholm et al., 2006; Desholm and Kahlert, 2005; Winkelman, 1989; Xu et al., 2010). Birds can be directly affected by wind power projects through collision or interaction with wind turbine blades or tower. They can be indirectly affected through roosting or feeding site disturbance due to wind turbine operation or maintenance activities, or through habitat damage that occurs as a consequence of construction activities (EWEA, 2002; CWS, 2005c; EPHC, 2009; Langston and Pullan, 2003; Ruddock and Whitfield, 2007; Wiser et al., 2011a).

A survey by the US National Research Council found that overall bird mortality estimates range from 0.95 to 12/MW/yr (NRC, 2007); in Northeast China, the bird mortality in Liaoning wind farm is estimated at about 0.55 (Shan et al., 2010). Wind turbines pose a collision risk for birds that are resident to the wind farm areas and for migrant birds when roosting and foraging in wind farm areas (Desholm, 2005; Sun et al., 2007). Low-flying birds such as songbirds are vulnerable by collision to wind turbines (NRC, 2007; Barrios and Rodriguez, 2004; Kuvlesky et al., 2007; Smallwood and Thelander, 2008; Wiser et al., 2011a). Although songbird species dominate bird collision observations in terms of fatality numbers, the most serious concerns often relate to raptor fatalities, as raptor populations are more often small and vulnerable and perhaps already subject to conservation concerns (Wiser et al., 2011a: May et al., 2009). As offshore wind energy projects have become increasingly more widespread, concerns have also been raised about seabirds specifically (Garthe and Hüppop, 2004; Xu et al., 2010). Studies demonstrate that some marine birds modify their flight behaviour to avoid offshore wind farms, but very little evidence is available on collision fatality rates at offshore sites (Wilson et al., 2010).

In general, fatality rates recorded at wind farms vary widely and depend on many factors, including species type, region, season, turbine design and site characteristics. Further, there seems to be a considerable lack of agreement in the literature on appropriate estimation methods and results for total numbers of birds killed by wind turbines (Sovacool, 2009; Willis et al., 2009; Loss et al., 2013; Smallwood, 2013; Huso and Erickson, 2013; Smallwood et al., 2013), with estimated total number of annual bird fatalities in the US ranging from twenty thousand (Sovacool, 2012) to hundreds of thousands (Smallwood, 2013; Loss et al., 2013). More research is needed to understand the magnitude and importance of cumulative effects, and because much remains to be explained about how different factors affect bird collision rates (Dong Energy, 2006; NRC, 2007; De, 2004; Desholm and Kahlert, 2005; Drewitt and Langston, 2006; Everaert and Stienen, 2007; Kuvlesky et al., 2007; Shan et al., 2010; Wiser et al., 2011a).

In comparison to bird mortality due to collisions with other man-made structures such as buildings, transmission lines and communications towers, and also compared to the number of birds killed by domestic or feral cats, the number of bird fatalities caused by wind power plants appears very low (NRC, 2007; Erickson et al., 2005; Sovacool, 2009, 2012; Wiser et al., 2011a). Moreover, it is worthwhile to note that all energy supply options may negatively impact bird populations through collisions or habitat modifications, or by causing emissions of GHGs or other types of pollution (NABCI, 2010c; Lilley and Firestone, 2008; Sovacool, 2009, 2012). At the same time, aggregate comparisons where fatality rates are summed across species and sites will not reflect any concerns that are specific to specific bird species or local populations, such as concerns about wind farms causing eagle deaths (Smallwood and Karas, 2009; Drewitt and Langston, 2006; May et al., 2012). In addition, observed historical fatality rates may not reflect impacts caused by future large-scale deployments.

Opportunities for reducing bird collision mortality exist. These opportunities include optimized local planning, such as keeping clear of bird migration corridors, avoiding construction near bird protection zones, and painting patterns or different colours on wind turbine parts, all of which have demonstrated some level of success (Cui et al., 2008; Shan et al., 2010; Wiser et al., 2011a).

5.5.2 BAT COLLISION AND BAROTRAUMA FATALITIES

Bats, like birds, are flying animals that may be injured or killed by colliding with wind turbines or other human-made structures. In some respects, certain species of bats may be more vulnerable to damage induced by wind turbines than birds. One reason for this is that bats, in search of roosts or for other reasons, may be attracted to wind turbines (NRC, 2007; Horn et al., 2008; Cryan and Barclay, 2009). Another reason is that, besides collisions, there is some indication that bats are susceptible to injuries (barotrauma) when passing through low air-pressure regions close to moving wind turbine blade tips (Baerwald et al., 2008). Further research is warranted, however, to clarify the causal mechanisms underlying bat mortality at wind power facilities, and how these mechanisms may be influenced by factors such as weather, site and turbine characteristics and turbine operation (Arnett et al., 2008; Kunz et al., 2007b; Wiser et al., 2011a; Smallwood, 2013).

Bat fatalities at wind power plants have been researched less than bird fatalities, and data exists for fewer countries (NRC, 2007; Cryan and Barclay, 2009; Dürr and Bach, 2004; Kunz et al., 2007b; Wiser et al., 2011a). Very high fatality rates for bats of certain species are reported for some wind power facilities. For example, Arnett et al. (2005) estimate that in the order of 2 000 bats were killed by 63 onshore turbines during six weeks at two sites in the eastern US (Arnett et al., 2005). Recorded bat fatality rates at various wind power plants vary widely, however, and some investigations show very low fatality rates (NRC, 2007; Arnett et al., 2008). Surveying bat fatality estimates reported in literature, the US National Research Council finds rates of fatality ranging from 0.8 to 41.1 bats per MW per year (NRC, 2007). Arnett et al. (2008) find fatality rates between 0.2 and 53.3 bats per MW per year. Based on evidence from North America, the bats that are at most risk of damage by wind turbines appear to be migratory species that use trees as roosts (Kunz et al., 2007b; Cryan and Barclay, 2009), while evidence from Europe shows that migratory and non-migratory bat fatality numbers are comparable (Rydell et al., 2010). Evidence suggests that the number of bat fatalities per turbine increases with increasing turbine tower height (Barclay et al., 2007; Rydell et al., 2010).

There are concerns that wind power has already become a significant mortality factor for some bat species in North America, and that the future expansion of wind power represents a potential serious threat to bat populations (Kunz et al., 2007b; Boyles et al., 2011; Willis et al., 2010; Smallwood, 2013). Current knowledge on cumulative impacts of wind energy deployment on bat populations remains uncertain, however.

Though many aspects of bat fatality patterns at wind power facilities remain not well understood, mitigation measures have been proposed and some have been found to be effective (Wiser et al., 2011a; Arnett et al., 2011). For example, it is clear that bat fatality rates are disproportionally high on nights with low wind speed (below 6 metres per second) (Arnett et al., 2008), and study findings studies suggest that curtailing the operation of wind turbines during low wind situations may reduce the number of bat fatalities to half or even less, with only marginal loss in power output (Arnett et al., 2011; Baerwald et al., 2009).

5.5.3 OTHER IMPACTS ON ECOSYSTEMS

Besides causing bird and bat mortality through collisions or exposure to air pressure changes, wind power facilities may impact ecosystems in a number of ways. Compared to direct bird and bat mortality, more indirect effects on wildlife, and effects on flora and fauna other than birds and bats have received less attention in peer-reviewed scientific literature (Northrup and Wittemyer, 2013; Lovich and Ennen, 2013). Negative impacts on terrestrial wildlife can potentially result from habitat fragmentation or land conversion to install wind turbines, access roads and transmission lines, due to noise during construction or operational stages affecting animal behaviour, due to shadow flicker from moving rotor blades, or for other reasons (Kuvlesky et al., 2007; Wiser et al., 2011a; Lovich and Ennen, 2013). The fact that negative impacts can occur does not imply that they have to occur or be important, however. Many wind farms are installed in areas that have already been transformed by human activities, and sometimes the wind farm area is simultaneously used for other purposes such as for growing agricultural crops or for pasture. In such cases, negative impacts on ecosystems can generally be expected to be much smaller than in cases where wind farms are built in undisturbed forest areas, as is the example used by Wiser et al. (2011a).

Impacts of offshore wind farms on marine wildlife may be positive or negative. Positive effects can be produced when bottom-mounted substructures or floating structures act as artificial reefs or fish aggregation devices, thus creating new habitat (Wilson and Elliott, 2009). Negative impacts, on the other hand, can be caused by noise or vibration generated during construction or operational phases, or from electromagnetic fields generated by submarine power cables (Inger et al., 2009; Wilson et al., 2010; Lovich and Ennen, 2013). Underwater noise effects are a particular concern for marine mammals, as this group of animals is dependent on auditory means of communicating, and they make use of sound (echolocation) to forage for food and to navigate (Wilson et al., 2010; Simmonds and Brown, 2010; Bailey et al., 2010). Some authors emphasize pile driving operations during the construction phase as a particularly problematic activity with respect to noise generation (Snyder and Kaiser, 2009; Tougaard et al., 2009; Wilson et al., 2010; Simmonds and Brown, 2010). It is clear that some types of impacts can be mitigated through good siting practices and avoiding particularly sensitive areas (Wilson et al., 2010; Punt et al., 2009).

In addition to the aspects noted above, large wind farms have the potential to affect local climate (e.g., Roy and Traiteur, 2010; Fiedler and Bukovsky, 2011; Zhou et al., 2012), with unknown, if any, consequences for ecosystems. In general, environmental impacts of wind power on ecosystems are highly project-specific, and the evidence-base is, in many areas of study, still fragmented and sparse. This makes it difficult to reach generalized conclusions.

5.6 IMPACT ON HUMANS

Besides ecological damage, there are concerns about wind power negatively affecting the scenic qualities of landscapes, qualities that may be highly valued by humans. Such concerns are discussed in brief in Chapter 9.5.1. Chapter 9.5.2 addresses possible impacts of wind power on human health. The discussions in Chapters 9.5.1 and 9.5.2 are largely based on Wiser et al. (2011a).

5.6.1 VISUAL INTRUSION IN LANDSCAPES

As Wiser et al. (2011a) notes, large wind turbines tend to be dominant elements in landscapes or seascapes, because they are tall and often intentionally sited at high elevations and in areas where there are few other visually dominating elements. Also, as wind power deployment continues and uncontroversial sites are increasingly becoming scarce, developers may need to increasingly turn towards areas that are valued for their scenic attributes (Wiser et al., 2011a). At the same time, it is also true that wind farms are often built on land that has already been impacted by land clearing, and they can coexist with agricultural uses such as crops or grazing. According to Wiser et al., (2011a), visual intrusion in landscapes is generally one of the primary concerns of communities considering wind power projects, of people living in the vicinity of existing projects, and of institutions involved in wind power planning. Visual impacts are highly subjective and difficult to quantify. They are also highly site-specific.

While recognizing that concerns about visual amenity cannot be eliminated entirely, Wiser et al. (2011a) remark that many authorities require that visual impacts are assessed in the siting process. Wiser et al., also point to a number of mitigation measures proposed in literature; this includes using similar type turbines, painting turbines in light colours, choosing few large wind turbine units as opposed to many small ones, and making sure that blades move in the same direction. One question which arises is to what degree the adjusted siting processes or other measures to mitigate visual impacts will lead to reduced productivity.

5.6.2 IMPACTS ON HUMAN HEALTH

Wind turbines can generate audible sound and sub-audible sound (i.e., infrasound, sound with frequency below the nominal limit of human hearing) during operation. Two literature reviews (Wiser et al., 2011a; MDEP, 2012) find that available evidence do not indicate that audible noise from wind turbines can cause health effects directly, but residents living near wind power plants can experience annoyance over wind turbine sound, and this again can cause sleep disruption and affect well-being in those residents. Besides the noise itself, such annoyance depends on the visibility of the turbine and attitude to the wind power project (Wiser et al., 2011a; MDEP, 2012). As regards sub-audible sound from turbines, an array of studies and government reports cited in Wiser et al. (2011a), and one government report published after (MDEP, 2012), find that to date there is insufficient evidence to support claims that such sound can cause health damage. On the other hand, some experts also caution that possible effects of low-frequency vibration on humans are not well understood, and note the need for further study (NRC, 2007).

Another concern relates to moving shadows from wind turbines in some places reaching residential areas. This can be a significant nuisance for some residents. No clear association between exposure to shadow flicker and health damage appears to have been established in literature (MDEP, 2012). Further, during particular weather conditions, ice may form on turbine blades and subsequently fall or be thrown off during operation. This represents a potential safety hazard, and appropriate precautions should be taken to guard against physical harm (MDEP, 2012). Finally, safety hazards to workers are present in the wind power industry, as in any industry. In recent years, several incidents with severe and fatal injuries have been reported in China and received considerable attention (Yu, 2011; HAWS, 2011). Most incidents occur due to failure to comply with safety procedures. A comparison of risk assessment results for different energy technologies in Sathaye et al. (2011) indicates that wind power exhibits relatively low fatal accident risk, however.

5.7 LIFE CYCLE ASSESSMENT CONDUCTED FOR THIS REPORT

This section presents the LCA of wind power conducted for this report. The section is divided into two subsections. The first provides accounts of data and assumptions for the wind power systems studied. The second presents results of the assessment of wind power, as well as a comparison to corresponding results obtained for the global mix of electricity sources in 2010. Besides these presentations dealing with wind power specifically, the general characteristics of the LCA model used for all technology assessments in this report are described in Chapter 2, and a detailed comparison of LCA results for a portfolio of power generation technologies, including wind power, are presented in Chapter 10.

5.7.1 DATA AND ASSUMPTIONS

This section describes the inventory data that underlie the investigations of global environmental pressures and impacts of wind power in this report. Assumptions and data are to a large extent adopted from Arvesen and Hertwich (2011) and Arvesen et al. (2013), with original sources cited therein. The wind power technology descriptions cover onshore and offshore systems described in terms of their general characteristics in Table 5.5. The offshore system is further divided into two subcategories depending on whether foundations are made of steel or concrete. We assume the onshore and offshore systems are representative of overall land- and ocean-based developments until 2050. Future increases in average wind load factors are assumed in the IEA scenarios and are adopted here (Table 5.5). Chapter 2 gives a general description of methods and data for the prospective LCAs carried out for this report.

TABLE 5.5

Key data for conceptual onshore and offshore wind farms

	Onshore	Offshore
Nominal capacity wind farm	150 MW	350 MW
Nominal capacity wind turbine	2.5 MW	5 MW
Lifetime	20 years	25 years
Average load (year 2010)	23.6%	37.5%
Average load (year 2030)	27.5%	42.2%
Average load (year 2050)	28.9%	43.0%
Internal cabling, length	48 km	63 km
Number of transformer stations	1	2
Grid connection length, submarine		50 km
Grid connection length, underground	15 km	10 km
Grid connection length, overhead	15 km	10 km
Land use	0.4 km^2	
Foundations made of steel, share		50%
Foundations made of concrete, share	100%	50%

Source: Adapted from Arvesen and Hertwich (2011) and Arvesen et al. (2013).

TABLE 5.6

Breakdowns of components and materials for onshore and offshore wind turbines and foundations

Component	Subcomponent	Material	Quantity	
			Onshore	Offshore
Rotor	Blades	Glass-reinforced plastics	21 t	53 t
	Hub with nose cone	Cast iron	13 t	35 t
		Low-alloy steel	7.4 t	21 t
		Glass-reinforced plastics	0.50 t	1.4 t
Nacelle	Generator	Aluminium	0.10 t	0.34 t
		Copper	3.1 t	10 t
		Electrical steel	7.0 t	23 t
	Gearbox	Aluminium	0.25 t	0.83 t
		Cast iron	12 t	41 t
		High-alloy steel	12 t	41 t
	Housing	Glass-reinforced plastics	3.1 t	10 t
	Main frame	Cast iron	11 t	35 t
		Low-alloy steel	5.9 t	19 t
	Main shaft	High-alloy steel	8.1 t	27 t
		Low-alloy steel	1.4 t	4.8 t
	Transformer	Aluminium	0.08 t	0.26 t
		Copper	2.4 t	7.8 t
		Electrical steel	5.3 t	18 t
Tower	Tubular steel	Low-alloy steel	200 t	350 t
	Tower internals	Aluminium	2.6 t	2.6 t
		Copper	1.3 t	1.3 t
Foundation (concrete)	Ballast	Gravel	-	5200 t
	Concrete	Concrete	410 m^3	1300 m^3
	Reinforcement	Reinforcement steel	35 t	560 t
Foundation (steel)	Steel structure	Low-alloy steel	-	600 t
	Corrosion protection	Aluminium anode	-	5 t

Each onshore unit has a nominal capacity 2.5 MW and each offshore unit 5 MW. Offshore foundations are made of either concrete or steel.

Source: Adapted from Jonkman (2009), Arvesen and Hertwich (2011), and Arvesen et al. (2013).

TABLE 5.7

Internal and external cabling total weight and material composition

	Total weight (t/km)	Material composition (%)						Notes, source
		Al	Cu	Pb	St	PE	PP	
Internal, underground	1.57	42	21	—	—	22	15	Al conductor. *Source*: Adapted from Vestas (2006a).
Internal, submarine	23.8	—	22	26	41	6.2	4.2	Three-core 33 kV Cu. Average, mix of cable dimensions used in one wind farm. *Source*: Arvesen et al. (2013).
External, underground	22.5	35	8.8	—	—	36	20	Set of three single-core 132 kV 1000 mm² Al cables. One set is needed to connect the onshore wind farm, two sets for offshore. *Source*: Adapted from ABB (2010), Vestas (2006a).
External, submarine	65.2	—	27	25	35	8.7	4.9	Three-core 132 kV 3x630 mm² Cu. Two cables are needed (offshore wind farm). *Source*: Adapted from Birkeland (2011).

Al: Aluminium; Cu: copper; Pb: lead; St: steel, galvanized; PE: polyethelene; PP: polyproylene.

Area requirements of wind farms are included only in terms of the permanent area used directly by infrastructure, excluding spacing area between infrastructure elements and temporary land use (see also related discussions in chapter 9.2). We assume that land use of onshore projects amounts to 2.7 km²/GW, which is the average permanent direct area use of wind power plants surveyed in Denholm et al. (2009) (see Chapter 9.2). Seabed or water surface area for offshore projects is excluded.

Table 5.6 shows the breakdowns of components and materials assumed for onshore and offshore wind turbines and foundations. The total weight of rotor, hub, nacelle and tubular tower components correspond with a conceptual wind turbine modelled in Arvesen and Hertwich (2011) (onshore case) and a reference offshore wind turbine defined in Jonkman et al. (2009) (offshore case). Foundation weights are adopted from Arvesen and Hertwich (2011) and Arvesen et al. (2013).

Internal underground (onshore) or submarine (offshore) cables connect the wind turbines to a transformer station; external underground or submarine cables or aerial lines serve as transmission links to an existing grid. Assumed total internal cabling and grid connection lengths are shown in Table 5.5 and material compositions for underground and submarine cables in Table 5.7, which also provides data sources. Further, we assume material requirements for overhead lines as in Jorge et al. (2012) and for onshore and offshore transformer stations respectively as in Arvesen and Hertwich (2011) and Arvesen et al. (2013). We also take into account some additional materials processing such as wire drawing, based on data gathered in previous work (Arvesen et al., 2013; Arvesen and Hertwich, 2011) and from manufacturer reports.

The installation stage comprises transport of components to site and on-site construction. For the onshore case, we adopt the physical inventories of Arvesen and Hertwich (2011) for the installation phase and assume decommissioning impacts correspond to 10 per cent of installation impacts. For the offshore wind farm, the inventories for installation and decommissioning are identical to the physical inventory data in Arvesen et al. (2013); this includes the direct emission factors for construction ships used in Arvesen et al. (2013). For the operations stage, we model transport and on-site activities based on physical inventories in Arvesen and Hertwich (2011) (onshore case) and Arvesen et al. (2013) (offshore). Additionally, we include large and small replacement parts at annual replacement rates as in Arvesen et al. (2013).

5.7.2 RESULTS AND DISCUSSION

TABLE 5.8

Impact on climate change (g CO_2-eq./kWh) for onshore and offshore wind power by nine regions (year 2010)

	CN	IN	EU	NA	PAC	EIT	LA	AS	AME
Onshore	14.5	13.5	9.2	12.3	12.1	11.8	10.7	12.7	12.7
Offshore, steel foundation	16.4	15.6	11.1	14.3	14.2	13.9	12.7	14.8	14.8
Offshore, gravity-based foundation	16.3	15.5	10.8	14.1	14.1	13.7	12.5	14.7	14.7

CN: China; IN: India; EU: OECD Europe; NA: OECD North America; PAC; OECD Pacific; EIT: Economies in transition; LA: Latin America; AS: Other Developing Asia; AME: Africa and Middle East.

Table 5.8 shows the GHG emission intensity of onshore and offshore wind power calculated for each region, assuming year 2010 technology. Across all regions, the emission intensities for onshore wind electricity range between 9.2 and 14 g CO_2-eq./kWh and for offshore wind 11 and 16 g CO_2-eq./kWh. The lowest emission intensity is seen for the Europe region. Offshore systems are more material and energy-demanding than land-based systems, but on the other hand benefit from more favourable wind load factor and lifetime assumptions in our analysis; thus total impact scores are similar for the onshore and offshore cases. Differences between onshore and offshore wind power projects appear when comparing relative contributions of components and activities, however, as is shown later in Figure 5.7. The overall GHG emission intensity range of 9-16 g CO_2-eq./kWh is roughly comparable to the median of observed results in previous LCAs, when looking only at studies assuming megawatt-sized wind turbines (see Chapter 5.2.4).

FIGURE 5.7

LCA results for onshore and offshore wind power systems by main components (year 2010; Europe region; steel foundations assumed for the offshore wind farm)

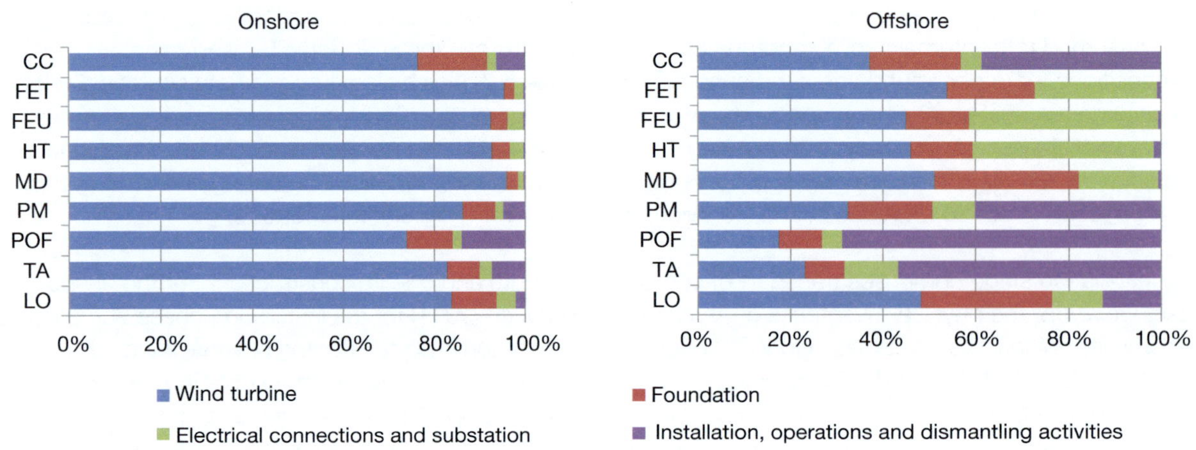

CC: climate change; FET: freshwater ecotoxicity; FEU: freshwater eutrophication; HT: human toxicity; MD: metal depletion; PM: particulate matter; POF: photochemical oxidant formation; TA: terrestrial acidification; LO: land occupation.

While the total impact scores for the three wind power systems are similar, the relative contributions of components differ between land- and ocean-based systems. This is illustrated by Figure 5.7, where it can be seen that the wind turbine is responsible for the bulk of the impacts for onshore wind parks, but is a less dominant contributor for the offshore case. Installation, operations and decommissioning activities by marine vessels contribute significantly to the impact of offshore wind power. See Arvesen et al. (2013) for a detailed discussion on the importance of ships in LCAs of offshore wind power. The contribution of the electrical connections is also larger than for the onshore system. For wind farms situated offshore, electrical connections give disproportionally high contributions for toxicity and eutrophication impact categories. These large toxicity and eutrophication impacts are largely attributable to a high copper content of submarine cables and electrical equipment, and long-term leakages of toxic and eutrophying substances from tailings and overburden material deposits in connection with copper mining.

As is evident from Figure 5.8, manufacture of base materials iron, steel, and plastic, is a main contributor to climate change impacts in 2010. This contribution is reduced in 2030 and 2050 owing to more de-carbonized electricity production and cleaner manufacturing processes. The contributions from cement production and steam and air conditioning supply decrease as well, albeit less considerably. Similarly as for the onshore wind farm, production of iron, steel, plastics, and cement are responsible for significant shares of the total impact on climate change for the offshore wind farms (Figure 5.9 and Figure 5.10). The most salient difference when comparing the onshore (Figure 5.8) and offshore cases (Figure 5.9 and Figure 5.10) is the significant portion of total emissions caused by ships in the offshore cases.

FIGURE 5.8

Impact on climate change (g CO_2-eq./kWh) for conventional onshore wind power (EU region) in 2010-2050

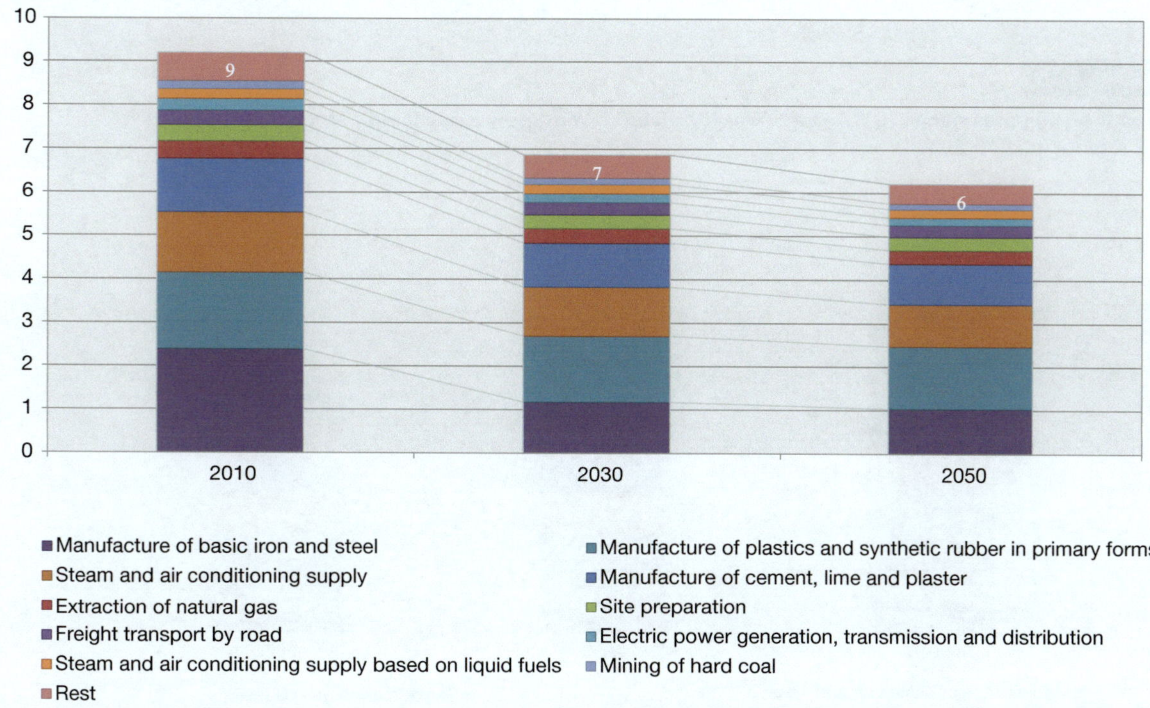

FIGURE 5.9

Impact on climate change (g CO_2 eq./kWh) for offshore wind power, assuming gravity-based substructures (EU region) in 2010-2050

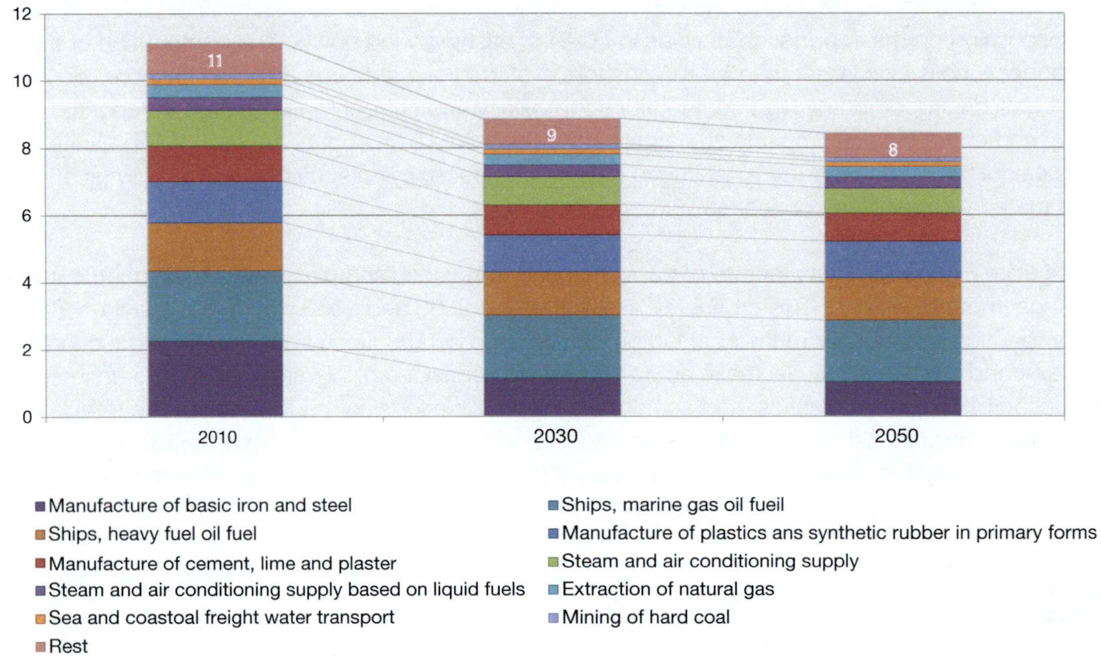

FIGURE 5.10

Impact on climate change (g CO_2-eq./kWh) for offshore wind power, assuming steel substructures (EU region) in 2010-2050

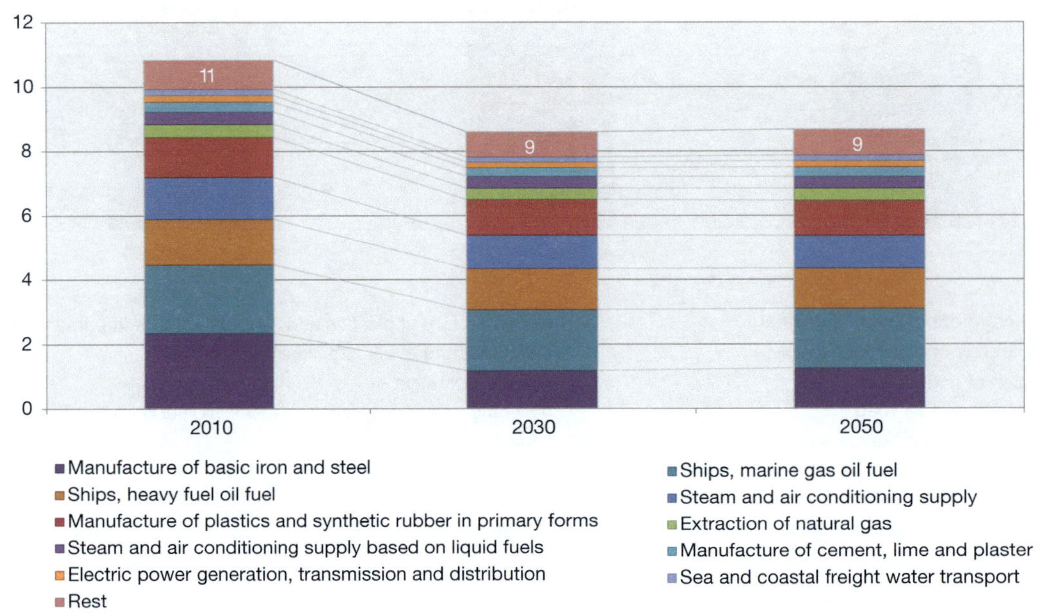

Figure 5.11 shows the climate change impact results for the onshore wind farm by nine regions and the years 2010, 2030 and 2050. According to these results, emission intensities may be reduced by a third or more from 2010 to 2050, illustrating that significant improvements in onshore wind power's carbon footprint can be achieved with cleaner background economies and increased wind load factors.

FIGURE 5.11

Impact on climate change (g CO_2-eq./kWh) for conventional onshore wind power by nine regions in 2010-2050

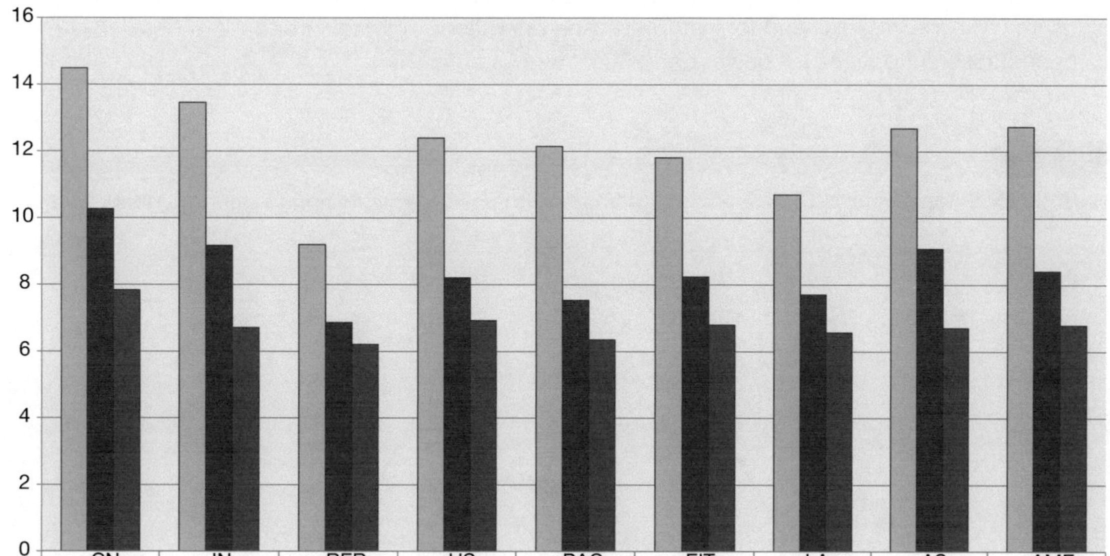

	CN	IN	RER	US	PAC	EIT	LA	AS	AME
2010	14	13	9	12	12	12	11	13	13
2030	10	9	7	8	8	8	8	9	8
2050	8	7	6	7	6	7	7	7	7

CN: China; IN: India; EU: OECD Europe; NA: OECD North America; PAC; OECD Pacific; EIT: Economies in transition; LA: Latin America; AS: Other Developing Asia; AME: Africa and Middle East.

Figure 5.12 compares the estimated total life cycle impact potentials of wind power to that of the global mix of electricity sources in 2010. As is evident from the figure, onshore and offshore wind power score 1-2 orders of magnitude better than the global electricity mix for all the assessed impact categories except metal depletion. Copper, iron and the steel alloying elements manganese and nickel are the main contributors to metal depletion impacts in these results, which do not reflect concerns about potential shortages of rare earth elements. Human toxicity and freshwater ecosystem toxicity impacts of wind power appear small but not negligible in comparison to that of the global electricity mix. For wind power, toxic effects in humans and ecosystems are largely the result of leakages from disposed copper and iron mine tailings and overburden material. Total impact scores for onshore and offshore wind facilities are similar, though the offshore system exhibits worse performance in acidification and photochemical oxidants and particulate formation. The higher acidifying and photochemical oxidant and particulate matter formation impacts for offshore are attributable to emissions from marine vessels used for installation and operations activities. For a more detailed comparative assessment of different options for power generation, see Chapter 10.

FIGURE 5.12

LCA results for Europe, year 2010 onshore and offshore wind power systems normalized to global electricity mix

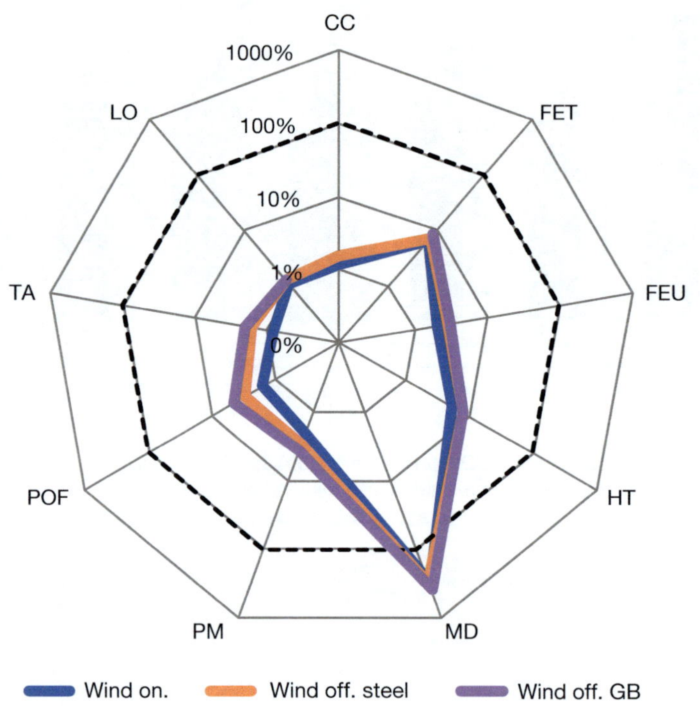

CC: climate change; FET: freshwater ecotoxicity; FEU: freshwater eutrophication; HT: human toxicity; MD: metal depletion; PM: particulate matter formation; POF: photochemical oxidant formation; TA: terrestrial acidification; LO: land occupation. Wind on. = Wind onshore. Wind off. steel = Wind offshore, steel foundation. Wind off. GB = Wind offshore, gravity-based foundation.

The LCA results for land use impacts do not take into account the spacing between installations, such as wind turbines and roads, in a wind farm. This is because the spacing between installations can be natural habitat or used for agriculture or other purposes. At the same time, we also recognize that excluding the spacing area from the analysis could understate the land use impacts of wind power, as it can be argued that the land in between windmills is not unaffected but may in some ways be degraded. If an entire wind park area is considered, land use would be at least two orders of magnitude higher and yield results exceeding the global electricity mix in Figure 5.12. Further, it should be emphasized that the results for metal depletion do not reflect any concerns about shortages of rare earth elements.

Further remarks concerning simplifications for the life cycle inventory

We assume the generic onshore and offshore wind technology descriptions satisfactorily represent future developments toward 2050 when increases in wind load factors are taken into account. This is a major simplification and overlooks the introduction of different or new material solutions such as relatively increased or reduced use of glass, carbon or natural fibre reinforcement in rotor blades, or towers made of concrete. These simplifications also disregard possible implementation of different design types (e.g., foundation types offshore or floating wind power plants, drive train configurations), impacts of changing site characteristics and availability (e.g., taller towers or offshore developments in deeper waters or farther from shore as suitable good wind sites become increasingly scarce), and scaling effects as wind turbine units or wind farms become ever larger. At the same time, the current inventory data set represents modern, large wind turbines and wind power plants and should, with the exception of rare earth elements, cover the spectrum of important material types involved in component manufacturing in coming decades. We do not anticipate radically different technologies becoming widespread before 2050.

The representations of wind turbines employed in this work concern conventional drive train layouts where the rotor shaft is connected to an electric generator via a gearbox. In some configurations, however, the rotor and generator are mounted on the same shaft. These configurations represent more 19.5 per cent of the added capacity in 2012 (Navigant, 2013). Such direct drive, gearless designs may use electromagnets or rare earth permanent magnets; the former option has thus far led to heavier nacelles while the latter may offer overall weight reductions. Different types of hybrid solutions also exist. The utilization of rare earth elements in wind turbines has come under increasing scrutiny in recent years (Chapter 9.3). Besides resource availability issues associated with rare earth elements, environmental damage resulting from mining of rare earths is often cited as a reason for concern. Existing LCI databases and impact assessment methods do not support investigations of such resource and environmental concerns, however. Recently, studies have started to look at production of rare earths in a life cycle framework (Talens Peiró and Villalba Méndez, 2013; Sprecher et al., 2014).

We assume relatively large wind turbine and wind farm sizes in order for the data to be suitable for prospective analysis, but the current data do not explicitly incorporate or allow for studying effects of scale. Positive effects of scale in terms of lower material requirements per unit of rated power may generally be expected for tower and foundation, whereas only small effects or negative effects may occur for rotor and nacelle, particularly for multi-megawatt units (Caduff et al., 2012; EWEA, 2009; Lenzen and Munksgaard, 2002). Literature survey results in this study show evidence of strong positive effects of scale in terms of less GHG emissions per unit of electricity delivered in the lower end of the turbine size spectrum, but show no clear evidence for such effects for megawatt-sized units (FigURE 5.6). A recent analysis (Caduff et al., 2012) coupling published LCIs from ten sources with statistics on installed wind power capacity suggests that a doubling of total installed capacity reduces the GHG emission intensity of new capacity with 14 per cent. These effects are attributable to both scaling and learning effects over time (Caduff et al., 2012).

The submarine grid connection in the offshore wind farm model consists of a pair of high-voltage AC (HVAC) cables running from offshore substations to shore, hence not capturing a shift towards increased use of high-voltage direct current (HVDC) transmission in the future. HVDC becomes relevant for connections running over longer distances than those assumed here, but at the same time, HVDC cables are lighter than their AC counterparts for a given transmission capacity. We expect the composition of materials used in subsea HVDC power cables (Birkeland, 2011) to be similar to that assumed for HVAC in this work. HVDC systems require power converters at AC/DC interfaces both offshore, where internal cables link with export cable, and onshore, where export cable link with power grid (Nes, 2012).

Several studies in the literature suggest that more comprehensive use of input-output analysis to tackle the issue of cut-off errors would give higher impact results (Chapter 9.1.1). This applies not only to wind power, however, but also the other technologies addressed in this report. It should also be noted that the LCA does not consider emissions resulting from grid reinforcement or extension, energy storage and system balancing.

5.8 OVERALL CONCLUSIONS

Wind energy is a renewable source of electricity, and over the last decades it has steadily become more important.. Its deployment is also becoming increasingly more geographically widespread, albeit with most of the development seen so far taking place in Asian, North American and European countries. Most of the current installed wind power capacity is onshore (98 per cent in 2012), but the offshore segment is growing and is expected to grow also in the future. Wind power technology is characterized by an increasing size of power plants and technical improvements resulting in increasing capacity factors (more energy harvested) and lower cost. Novel technologies aim at increasing the reliability and further reducing costs. While evidence on the exact magnitude of wind energy output that can be obtained with contemporary technology remains inconclusive, it is clear that the number is by a wide margin sufficient to support future expansions as described in climate change mitigation scenarios.

The survey of results from past LCA studies of wind power indicate that when using megawatt-sized wind turbines – which is the turbine size spectrum dominating overall wind power capacity expansion – electricity can be generated with about 7-33 g CO_2-eq. emitted per kWh of electricity. The life cycle GHG emissions for wind power estimated in this report amount to 9-16 g CO_2-eq./kWh (year 2010 results), thus falling in the lower part of the observed interval from previous assessments. These results indicate large potential climate benefits of wind power displacing fossil fuel-based power. Further, an explicit comparison of LCA results for wind power versus the global mix of electricity sources shows that wind power outperforms the global electricity mix by 1-2 orders of magnitude for eight out of nine impact categories. The one exception is metal depletion, for which wind power exhibits larger impacts than the global electricity mix. Copper and iron requirements constitute the two main causes of metal depletion for wind power, according to the results. The LCA results for land use of wind power presented in this chapter do not take into account the spacing between installations, because this land can be natural habitat or used for agriculture or other purposes. If an entire wind park area is considered, land use would be at least two orders of magnitude higher and exceed the land use of the global electricity mix.

Beyond the LCA results for wind power presented in this chapter, a detailed comparison of LCA results for a portfolio of power generation technologies, including wind power, is presented in Chapter 10. It should also be noted that emissions due to grid, storage or balancing requirements to accommodate variable renewables in electricity networks lie outside the scope of the LCAs presented in this report; see Chapter 9 for a discussion of such issues.

Beyond the types of environmental and resource concerns addressed in the LCA, a range of site-specific impacts receive public interest and should be adequately considered in decision-making processes. One concern is visual intrusion and noise intrusion in landscapes, seascapes or communities. Such impacts can reduce the human-perceived quality of landscapes or cause annoyance or nuisance in humans in other ways, but are highly subjective, site-specific and difficult to quantify. Another concern is bird and bat mortality due to collision or interaction with wind turbines. There are concerns that wind power has become a significant mortality factor for certain bat species in North America, although much remains unknown about bat mortality from wind turbines. Overall bird mortality due to wind power appears low when compared with mortality caused by other man-made structures such as buildings and power lines, but wind turbines may tend to kill different types of birds than, for example, buildings. The most serious concerns about bird mortality due to wind power often seem to relate to raptors. At least to some degree and for some types of impacts on ecosystems and humans, adverse effects can be avoided or mitigated through proper siting practices.

Yet another issue not captured by LCAs is unreliable or uncertain supply of rare earth elements, in particular dysprosium, and environmental damage resulting from rare earth element mining and processing. Wind turbines containing rare earth elements comprise a small but increasing proportion of current markets. Moving towards low-dysprosium magnets is seen as an effective strategy to counter dysprosium shortage.

The LCA results presented in this chapter and later in Chapter 10 show that wind power can deliver substantial reductions in GHG emissions and many other types of pollutant emissions. At the same time, environmental benefits of wind power can only arise if and when worse alternatives, such conventional fossil fuel-based power, is avoided, and this may not happen automatically. In practice the environmental performance of wind power and the effectiveness of wind power deployment as a strategy to mitigate climate change will depend on to what degree fossil fuel combustion is eliminated. This again depends on the effectiveness and stringency of climate policy.

5.9 REFERENCES

ABB. 2010. XLPE land cable systems. User's guide. www.abb.com.

Alves-Pereira, M., Branco, C. 2007. In-home wind turbine noise is conducive to vibroacoustic disease, Presented at the Second International Meeting on Wind, Lyon, France.

Alonso, E., Sherman, M., Wallington, T.J., Everson, M.P., Field, F.R., Roth, R., Kirchain. 2012. Evaluating rare earth element availability: A case with revovutionary demand from clean technologies. Environmental Science & Technology. 46(6): 3406-2414.

Andersen, P.D., Borup, M., Krogh, T. 2007. Managing long-term environmental aspects of wind turbines: A prospective case study. International Journal of Technology, Policy and Management 7, 339-354.

Archer, C. L. and M. Z. Jacobson. 2013. Geographical and seasonal variability of the global "practical" wind resources. *Applied Geography* 45: 119-130.

Ardente, F., Beccali, M., Cellura, M., Lo Brano, V. 2008. Energy performances and life cycle assessment of an Italian wind farm. Renewable and Sustainable Energy Reviews 12, 200-217.

Arnett, E.B., Brown, W.K., Erickson, W.P., Fiedler, J.K., Hamilton, B.L., Henry, T.H., Jain, A., Johnson, G.D., Kerns, J., Koford, R.R. 2008. Patterns of bat fatalities at wind energy facilities in North America, Journal of Wildlife Management, pp. 61-78.

Arnett, E.B., Erickson, W.P., Kerns, J., Horn, J. 2005. Relationships between Bats and Wind Turbines in Pennsylvania and West Virginia: An Assessment of Fatality Search Protocols, Patterns of Fatality, and Behavioral Interactions with Wind Turbines, Journal of Wildlife Management.

Arnett, E.B., Huso, M.M.P., Schirmacher, M.R., Hayes, J.P. 2011. Altering turbine speed reduces bat mortality at wind-energy facilities, Frontiers in Ecology and the Environment, pp. 209-214.

Arnett, E.B., Schirmacher, M., Huso, M.M.P., Hayes, J.P. 2009. Effectiveness of Changing Wind Turbine Cut-in Speed to Reduce Bat Fatalities at Wind Facilities, Austin, Texas, USA.

Arvesen, A., Birkeland, C., Hertwich, E.G. 2013. The importance of ships and spare parts in LCAs of offshore power. Environmental Science & Technology 47, 2948-2956.

Arvesen, A., Hertwich, E.G. 2011. Environmental implications of large-scale adoption of wind power: A scenario-based life cycle assessment. Environmental Research Letters 6, 045102.

Arvesen, A., Hertwich, E.G. 2012a. Assessing the life cycle environmental impacts of wind power: A review of present knowledge and research needs. Renewable and Sustainable Energy Reviews 16, 5994-6006.

Arvesen, A., Hertwich, E.G. 2012b. Corrigendum: Environmental implications of large-scale adoption of wind power: A scenario-based life cycle assessment. Environmental Research Letters 3, 039501.

Baerwald, E.F., Edworthy, J., Holder, M., Barclay, R.M.R. 2009. A large-scale mitigation experiment to reduce bat fatalities at wind energy facilities, Journal of Wildlife Management, pp. 1077-1082.

Baerwald, E.F., Genevieve, D.A., Klug, B.J., Barclay, R.M.R. 2008. Barotrauma is a significant cause of bat fatalities at wind turbines. Current Biology, pp. 695–696.

Baidya, R.S., Traiteur, J.J. 2010. Impacts of wind farms on surface air temperatures, Proceedings of the National Academy of Sciences, pp. 17899-17904.

Barclay, R.M.R., Baerwald, E.F., Gruver, J.C. 2007. Variation in bat and bird fatalities at wind energy facilities: Assessing the effects of rotor size and tower height. Canadian Journal of Zoology, 85(3), 381-387.

Barrios, L., Rodriguez, A. 2004. Behavioral and environmental correlates of soaring bird mortality at onshore wind turbines, Journal of Applied Ecology, pp. 72-81.

Berkhout, V., Faulstich, S., Görg, P., Kühn, P., Linke, K., Lyding, P., Pfaffel, S., Rafik, K., Rohrig, K., Rothkegel, R., Stark, E. 2013. Windenergie Report Deutschland 2012. Fraunhofer Institut für Windenergie und Energiesystemtechnik, http://windmonitor.iwes.fraunhofer.de/bilder/upload/Windenergie_Report_Deutschland_2012.pdf. Accessed 28 June 2013.

Birkeland, C. 2011. Assessing the life cycle environmental impacts of offshore wind power generation and power transmission in the North Sea. MSc Department of Energy and Process Engineering. Norwegian University of Science and Technology, Trondheim, Norway.

Blanco, M.I. 2009. The economics of wind energy. Renewable and Sustainable Energy Reviews 13, 1372-1382.

Boccard, N. 2009. Capacity factor of wind power realized values vs. estimates. Energy Policy 37, 2679-2688.

Boyles, J.G., Cryan, G.F., McCracken, G.F., Kunz, T.H. 2011. The Economic Importance of Bats in Agriculture. Science 332 (6025), 41-42.

Burger, B., Bauer, C. 2007. Windkraft, in: Dones, R.E.e.a. (Ed.), Sachbilanzen von Energiesystemen: Grundlagen für den ökologischen Vergleich von Energiesystemen in Ökobilanzen für die Schweiz. Final report ecoinvent No. 6-XIII. Paul Scherrer Institut Villigen, Swiss Centre for Life Cycle Inventories, Dübendorf, CH.

Caduff, M., Huijbregts, M.A.J., Althaus, H.-J., Koehler, A., Hellweg, S. 2012. Wind Power Electricity: The Bigger the Turbine, The Greener the Electricity? Environmental Science & Technology 46, 4725-4733.

Celik, A.N., Muneer, T., Clarke, P. 2007. An investigation into micro wind energy systems for their utilization in urban areas and their life cycle assessment. Proceedings of the Institution of Mechanical Engineers, Part A: Journal of Power and Energy 221, 1107-1117.

Chataignere, A., Le Boulch, D. 2003. Wind turbine (WT) systems. Final report. ECLIPSE project: Environmental and Ecological Life Cycle Inventories for present and future Power Systems in Europe.

Chen, G.Q., Yang, Q., Zhao, Y.H. 2011. Renewability of wind power in China: A case study of nonrenewable energy cost and greenhouse gas emission by a plant in Guangxi. Renewable and Sustainable Energy Reviews 15, 2322-2329.

Classen, M., Althaus, H.-J., Blaser, S., Tuchschmid, M., Jungbluth, N., Doka, G., Faist Emmenegger, M., Scharnhorts, W. 2007. Life Cycle Inventories of Metals. Final report ecoinvent data v2.0, N° 10, Dübendorf, Switzerland.

Cohen, J., Schweizer, T., Laxson, A., Butterfield, S., Schreck, S., Fingersh, L., Veers, P., Ashwill, T., 2008. Technology Improvement Opportunities for Low Wind Speed Turbines and Implications for Cost Energy Reductions. National Renewable Energy Laboratory (NREL).

Correia, J.R., Almeida, N.M., Figueira, J.R. 2011. Recycling of FRP composites: Reusing fine GFRP waste in concrete mixtures. Journal of Cleaner Production 19, 1745-1753.

Crawford, R.H. 2009. Life cycle energy and greenhouse emissions analysis of wind turbines and the effect of size on energy yield. Renewable and Sustainable Energy Reviews 13, 2653-2660.

Cryan, P.M., Barclay, R.M.R. 2009. Causes of bat fatalities at wind turbines: Hypotheses and predictions, Journal of Mammalogy, pp. 1330-1340.

Cui, H., Yang, Q., Zhang, S. 2008. Analyses of bird-turbine collision factors and protectin measurements, Environmental Science Survey, pp. 52-56.

CWS. 2005c. Wind Turbines and Birds - A Background Review for Environmental Assessment. Canadian Wildlife Service (CWS).

Cyranoski, D. 2009. Renewable energy: Beijing's windy bet. Nature 457, 372-374.

De, L. 2004. The effects of a wind farm on birds in a migration point: The Strait of Gibraltar, Biodiversity and Conservation, pp. 395-407.

Denholm, P., Hand, M., Jackson, M., Ong, S. 2009. Land-use requirements of modern wind power plants in the United States. National Renewable Energy Laboratory (NREL), http://www.nrel.gov/docs/fy09osti/45834.pdf. Accessed 20 December 2012.

Desholm, M. 2005. Preliminary Investigations of bird-turbine collisions at Nysted offshore wind farm and final quality control of thermal animal detection system (TADS), Ronde, Denmark.

Desholm, M., Fox, A., Beasley, P., et al. 2006. Remote techniques for counting and estimating the number of bird-wind turbine collisions at sea: A review, Ibis, pp. 148 (Suppl.)176-189.

Desholm, M., Kahlert, J. 2005. Avian collision risk at an offshore wind farm, Biology Letter, pp. 296-298.

DONG Energy. 2006. Danish Offshore Wind: Key Environmental Issues. DONG Energy, Vattenfall, the Danish Energy Authority and the Danish Forest and Nature. Copenhagen, Denmark, p. 142.

Drewitt, A.L., Langston, R.W. 2006. Assessing the impacts of wind farms on birds, Ibis, pp. 29-42.

Du, X., Graedel, T.E. 2011. Global Rare Earth In-Use Stocks in NdFeB Permanent Magnets. Journal of Industrial Ecology 15, 836-843.

Dürr, T., Bach, L. 2004. Bat deaths and wind turbines – a review of current knowledge, and of the information available in the database for Germany, Bremer Beiträge für Naturkunde und Naturschutz, pp. 253-264.

EC. 2010. Critical raw materials for the EU. Report of the Ad-hoc Working Group on defining critical raw materials. European Commission. Enterprise and Industry, http://ec.europa.eu/enterprise/policies/raw-materials/files/docs/report_en.pdf. Accessed 24 June 2013.

ecoinvent. 2007. Life cycle inventory database v2.1. Swiss Centre for Life Cycle Inventories.

Elsam. 2004. Life cycle assessment of offshore and onshore sited wind farms. Elsam Engineering A/S.

EPHC. 2009. National Wind Farm Development Guidelines Public Consultation Draft—October 2009. Environment Protection and Heritage Council (EPHC). NEPC.

Erdmann, L., Graedel, T.E. 2011. Criticality of Non-Fuel Minerals: A Review of Major Approaches and Analyses. Environmental Science & Technology 45, 7620-7630.

Erickson, W.P., Johnson, G.D., Young, D.P.J. 2005. A Summary and Comparison of Bird Mortality from Anthropogenic Causes with an Emphasis on Collisions, Washington,DC, USA, p. 14

Everaert, J., Stienen, E.W.M. 2007. Impact of wind turbines on birds in Zeebrugge (Belgium), Biodiversity and Conservation, pp. 3345-3359.

EWEA. 2002. European best practice guidelines for wind energy development. European Wind Energy Association (EWEA).

EWEA. 2009. Wind Energy - The Facts. Part I. Technology. European Wind Energy Association (EWEA). http://www.wind-energy-the-facts.org/documents/download/Chapter1.pdf. Accessed 13 March 2013.

Fastl, H., Zwicker, E. 2007. Psychoacoustics: Facts and Models, 3rd ed.Springer, Berlin and Heidelberg, Germany and New York, NY, USA, p. 428.

Feng, Y., Tavner, P.J., Long, H. 2010. Early experiences with UK round 1 offshore wind farms. Proceedings of the ICE - Energy 163, 167-181.

Finnveden, G., Hauschild, M.Z., Ekvall, T., Guinée, J., Heijungs, R., Hellweg, S., Koehler, A., Pennington, D., Suh, S. 2009. Recent developments in Life Cycle Assessment. Journal of Environmental Management 91, 1-21.

Fleck, B., Huot, M. 2009. Comparative life-cycle assessment of a small wind turbine for residential off-grid use. Renewable Energy 34, 2688-2696.

Frandsen, S.T., Barthelmie, R.J., Pryor, S.C., Rathmann, O., Larsen, S., Højstrup, J., Thøgersen, M. 2006. Analytical modelling of wind speed deficit in large offshore wind farms, Wind Energy, pp. 39-53.

Frandsen, S.T., Jørgensen, H.E., Barthelmie, R.J., Rathmann, O., Badger, J., Hansen, K., Ott, S., Rethore, P., Larsen, S.E., Jensen, L.E. 2009. The making of a second-generation wind farm efficiency model complex, Wind Energy, pp. 445-458.

Fthenakis, V., Kim, H.C. 2009. Land use and electricity generation: A life-cycle analysis. Renewable and Sustainable Energy Reviews 13, 1465-1474.

Fthenakis, V., Kim, H.C. 2010. Life-cycle uses of water in U.S. electricity generation. Renewable and Sustainable Energy Reviews 14, 2039-2048.

Garthe, S., Hüppop, O. 2004. Scaling possible adverse effects of marine wind farms on seabirds: Developing and applying a vulnerability index, Journal of Applied Ecology, pp. 724-734.

Gonçalves da Silva, C. 2010. Renewable energies: Choosing the best options. Energy 35, 3179-3193.

Gross, R., Heptonstall, P. 2008. The costs and impacts of intermittency: An ongoing debate: "East is East, and West is West, and never the twain shall meet." Energy Policy 36, 4005-4007.

Guezuraga, B., Zauner, R., Pölz, W. 2012. Life cycle assessment of two different 2 MW class wind turbines. Renewable Energy 37, 37-44.

GWEC. 2006-2013. Global wind report. Annual market update, Various editions (2006-2013). Global Wind Energy Council (GWEC). Accessed at www.gwec.net 15 February 2014.

GWEC. 2014. Global wind statistics 2013. Global Wind Energy Council (GWEC). http://www.gwec.net/wp-content/uploads/2014/02/GWEC-PRstats-2013_EN.pdf. Accessed 17 February 2014.

Hauschild, M.Z. 2005. Assessing Environmental Impacts in a Life-Cycle Perspective. Environmental Science & Technology 39, 81A-88A.

HAWS. 2011. Hebei Administrator of Work Safety. "1.5" electric shock accident at Zhangjiakou wind farm. Hebei Administrator of Work Safety (HAWS). 18 April 2011.

Hirschberg, S., Dones, R., Heck, T., Burgherr, P., Schenler, W., Bauer, C. 2006. Strengths and weaknesses of current energy chains in a sustainable development perspective. International Journal for Nuclear Power 51, 447-457.

Hoenderdaal, S., Tercero Espinoza, L., Marscheider-Weidemann, F., Graus, W. 2013. Can a dysprosium shortage threaten green energy technologies? Energy 49, 344-355.

Hondo, H. 2005. Life cycle GHG emission analysis of power generation systems: Japanese case. Energy 30, 2042-2056.

Horn, J.W., Arnett, E.B., Kunz, T.H. 2008. Behavioral responses of bats to operating wind turbine. Journal of Wildlife Management 72, 123-132.

IEA. 2010a. Energy technology perspectives 2010. Scenarios & strategies to 2050. International Energy Agency (IEA), Paris.

IEA. 2012. Energy technology perspectives 2012. Pathways to a clean energy system. International Energy Agency (IEA), Paris.

IEA. 2010b. World energy outlook: 2010. International Energy Agency (IEA), Paris.

IEA. 2013. World energy outlook: 2013. International Energy Agency (IEA), Paris.

ISO. 2006. Environmental management - Life cycle assessment - Principles and framework (ISO 14040: 2006). International Organization for Standardization.

Jacobson, M.Z. 2009. Review of solutions to global warming, air pollution and energy security. Energy and Environmental Science 2, 26.

Jacobson, M.Z., Delucchi, M.A. 2011. Providing all global energy with wind, water, and solar power, Part I: Technologies, energy resources, quantities and areas of infrastructure, and materials. Energy Policy 39, 1154-1169.

Jonkman, J.M., Butterfield, S., Musial, W., Scott, G. 2009. Definition of a 5 MW reference wind turbine for offshore system development. National Renewable Energy Laboratory. Technical report NREL/TP-500-38060.

Jorge, R., Hawkins, T., Hertwich, E. 2012. Life cycle assessment of electricity transmission and distribution—Part 1: power lines and cables. The International Journal of Life Cycle Assessment 17, 9-15.

JRC-IET. 2011. Critical metals in strategic energy technologies. Assessing rare metals as supply-chain bottlenecks in low-carbon energy technologies. European Commission. Joint Research Centre (JRC). Institute for Energy and Transport (IET), http://publications.jrc.ec.europa.eu/repository/bitstream/111111111/22726/1/reqno_jrc65592_critical%20metals%20in%20strategic%20energy%20technologies%20%28online%29.pdf. Accessed 24 June 2013.

Jungbluth, N., Bauer, C., Dones, R., Frischknecht, R. 2005. Life Cycle Assessment for Emerging Technologies: Case Studies for Photovoltaic and Wind Power. The International Journal of Life Cycle Assessment 10, 24-34.

Kabir, M.R., Rooke, B., Dassanayake, G.D.M., Fleck, B.A. 2012. Comparative life cycle energy, emission, and economic analysis of 100 kW nameplate wind power generation. Renewable Energy 37, 133-141.

Keen, J. 2008. Neighbors at odds over noise from wind turbines, USA TODAY.

Kemmoku, Y., Ishikawa, K., Nakagawa, S., Kawamoto, T., Sakakibara, T. 2002. Life cycle CO_2 emissions of a photovoltaic/wind/diesel generating system. Electrical Engineering in Japan 138, 14-23.

Khan, F.I., Hawboldt, K., Iqbal, M.T. 2005. Life Cycle Analysis of wind-fuel cell integrated system. Renewable Energy 30, 157-177.

Kleijn, R., van der Voet, E. 2010. Resource constraints in a hydrogen economy based on renewable energy sources: An exploration. Renewable and Sustainable Energy Reviews 14, 2784-2795.

Kleijn, R., van der Voet, E., Kramer, G.J., van Oers, L., van der Giesen, C. 2011. Metal requirements of low-carbon power generation. Energy 36, 5640-5648.

Krey, V., Clarke, L. 2011. Role of renewable energy in climate mitigation: a synthesis of recent scenarios. Climate Policy 11, 1131-1158.

Kubiszewski, I., Cleveland, C.J., Endres, P.K. 2010. Meta-analysis of net energy return for wind power systems. Renewable Energy 35, 218-225.

Kunz, T.H., Arnett, E.B., Cooper, B.M., Erickson, W.P., Lark, R.P., Mabee, T., Morrison, M.L., Strickland, M.D., Szewczak, J.M. 2007a. Assessing impacts of wind-energy development on nocturnally active birds and bats: A guidance document, Journal of Wildlife Management, pp. 2449-2486.

Kunz, T.H., Arnett, E.B., Erickson, W.P., Hoar, A.R., Johnson, G.D., Larkin, R.P., Strickland, M.D., Thresher, R.W., Tuttle, M.D. 2007b. Ecological impacts of wind energy development on bats: Questions, research needs, and hypotheses, Frontiers in Ecology and the Environment, pp. 315-324.

Kuvlesky, W.P., Brennan, L.A., Morrison, M.L., Boydston, K.K., Ballard, B.M., Bryant, F.C. 2007. Wind energy development and wildlife conservation:challenges and opportunities, Journal of Wildlife Management, pp. 2487-2498.

Langston, R.H.W., Pullan, J.D. 2003. Wind farms and birds: An analysis of the effects of wind farms on birds, and guidance on environmental assessment criteria and site selection issues. Report T-PVS/Inf (2003).

Larsen, K. 2009. Recycling wind. Reinforced Plastics 53, 20-23, 25.

Lee, Y., Tzeng, Y. 2008. Development and life-cycle inventory analysis of wind energy in Taiwan. Journal of Energy Engineering 134, 53-57.

Lemming, J., Morthorst, P.E., Clausen, N.-E., Jensen, P.H. 2009. Contribution to the Chapter on Wind Power Energy Technology Perspectives 2008. Risø National Laboratory for Sustainable Energy.

Lenzen, M. 2000. Errors in Conventional and Input-Output-based Life-Cycle Inventories. Journal of Industrial Ecology 4, 127-148.

Lenzen, M., Munksgaard, J. 2002. Energy and CO_2 life-cycle analyses of wind turbines - Review and applications. Renewable Energy 26, 339-362.

Lenzen, M., Schaeffer, R. 2012. Historical and potential future contributions of power technologies to global warming. Climatic Change 112, 601-632.

Lenzen, M., Wachsmann, U. 2004. Wind turbines in Brazil and Germany: An example of geographical variability in life-cycle assessment. Applied Energy 77, 119-130.

Liu, Z., Qian, D., Zeng, D. 2012. Reducing Dy content by Y substitution in nanocomposite NdFeB alloys with enhanced magnetic properties and thermal stability. Magnets, IEEE Transactions on Magnetics 48(11): 2797-2799.

Lilley, M.B., Firestone, J. 2008. Wind power, wildlife, and the Migratory Bird Treaty Act: A way forward, Environmental Law, pp. 1167-1214.

Loss, S.R., Will, T., Marra, P.P. 2013. Estimates of bird collision mortality at wind facilities in the contigous United States. Biological Conservation 168, 201-209.

Luickx, P.J., Delarue, E.D., D'Haeseleer, W.D. 2010. Impact of large amounts of wind power on the operation of an electricity generation system: Belgian case study. Renewable and Sustainable Energy Reviews 14, 2019-2028.

MacKay, D.J.C. 2009. Sustainable energy: Without the hot air. UIT, Cambridge.

Majeau-Bettez, G., Strømman, A.H., Hertwich, E.G. 2011. Evaluation of Process- and Input–Output-based Life Cycle Inventory Data with Regard to Truncation and Aggregation Issues. Environmental Science & Technology 45, 10170-10177.

Martínez, E., Jiménez, E., Blanco, J., Sanz, F. 2010. LCA sensitivity analysis of a multi-megawatt wind turbine. Applied Energy 87, 2293-2303.

Martínez, E., Sanz, F., Pellegrini, S., Jiménez, E., Blanco, J. 2009a. Life-cycle assessment of a 2-MW rated power wind turbine: CML method. International Journal of Life Cycle Assessment 14, 52-63.

Martínez, E., Sanz, F., Pellegrini, S., Jiménez, E., Blanco, J. 2009b. Life cycle assessment of a multi-megawatt wind turbine. Renewable Energy 34, 667-673.

May, R., Bevanger, K., van Dijk, J., Petrin, Z., Brende, H. 2012. Renewable energy respecting nature. A synthesis of knowledge on environmental impacts of renewable energy financed by the Research Council of Norway. Norwegian Institute for Nature Research. www.nina.no/archive/nina/PppBasePdf/rapport/2012/874.pdf.Accessed 11 April 2013.

McCunney, R.J., Meyer, J. 2007. Occupational exposure to noise, Environmental and Occupational Medicine. 4th ed. W.N. Rom (ed.). Lippincott Williams and Wilkins, Baltimore, MD, pp. 1295-1238.

McDonald, R.I., Fargione, J., Kiesecker, J., Miller, W.M., Powell, J. 2009. Energy Sprawl or Energy Efficiency: Climate Policy Impacts on Natural Habitat for the United States of America. PLoS ONE 4, e6802.

MDEP. 2012. Wind turbine health impact study: Report of independent expert panel. Prepared for: Massachusetts Department of Environmental Protection (MDEP), Massachusetts Department of Public Health. http://www.mass.gov/eea/docs/dep/energy/wind/turbine-impact-study.pdf. Accessed 14 March 2014.

Mudd, G.M. 2010. The Environmental sustainability of mining in Australia: Key mega-trends and looming constraints. Resources Policy 35, 98-115.

NABCI, 2010c. The State of the Birds: 2010 Report on Climate Change. North American Bird Conservation Initiative (NABCI), US Committee, US Department of the Interior. Washington, DC, USA.

Navigant. 2013. World market update 2012. Executive summary. http://www.navigantresearch.com/wp-assets/uploads/2013/03/WWMU-13-Executive-Summary.pdf. Accessed 17 February 2014.

NEEDS. 2008. Life cycle approaches to assess emerging energy technologies. Final report on offshore wind technology.

Nes, R.N. 2012. Life cycle assessment of an offshore electricity grid interconnecting Northern Europe. MSc Department of Energy and Process Engineering. Norwegian University of Science and Technology, Trondheim, Norway.

Norgate, T., Jahanshahi, S. 2010. Low grade ores - Smelt, leach or concentrate? Minerals Engineering 23, 65-73.

NRC. 2007. Environmental Impacts of Wind-Energy Projects. National Research Council (NRC). The National Academy Press, Washington, DC, USA, p. 394.

Oswald, J., Raine, M., Ashraf-Ball, H. 2008. Will British weather provide reliable electricity? Energy Policy 36, 3212-3225.

Pacca, S., Horvath, A. 2002. Greenhouse Gas Emissions from Building and Operating Electric Power Plants in the Upper Colorado River Basin. Environmental Science & Technology 36, 3194-3200.

Pedersen, E., van den Berg, F., Bakker, R., Bouma, J. 2010. Can road traffic mask sound from wind turbines? Response to wind turbine sound at different levels of road traffic sound, Energy Policy, pp. 2520-2527.

Pedersen, E., Waye, K.P. 2007. Wind turbine noise, annoyance and self-reported health and well-being in different living environments, Occupational and Environmental Medicine, pp. 480-486.

Pedersen, E., Waye, K.P. 2008. Wind turbines – low level noise sources interfering with restoration?, Environmental Research Letters, pp. 1-5.

Pehnt, M. 2006. Dynamic life cycle assessment (LCA) of renewable energy technologies. Renewable Energy 31, 55-71.

Pehnt, M., Oeser, M., Swider, D.J. 2008. Consequential environmental system analysis of expected offshore wind electricity production in Germany. Energy 33, 747-759.

Pennington, D.W., Potting, J., Finnveden, G., Lindeijer, E., Jolliet, O., Rydberg, T., Rebitzer, G. 2004. Life cycle assessment Part 2: Current impact assessment practice. Environment International 30, 721-739.

Pettersen, J., Hertwich, E.G. 2008. Critical review: Life-cycle inventory procedures for long-term release of metals. Environmental Science & Technology 42, 4639-4647.

Price, L., Kendall, A. 2012. Wind Power as a Case Study. Improving Life Cycle Assessment Reporting to Better Enable Meta-analyses. Journal of Industrial Ecology 16, S22-S27.

Prins, A.G., Slingerland, S., Manders, T., Lucas, P., Hilderink, H., Kok, M. 2011. Scarcity in a sea of plenty? Global resource scarcities and policies in the European Union and the Netherlands. PBL Netherlands Environmental Assessment Agency. Available from www.pbl.nl/en. Accessed 26 February 2013.

Properzi, S., Herk-Hansen, H. 2003. Life cycle assessment of a 150 MW offshore wind turbine farm at Nysted/ Roedsand, Denmark. International Journal of Environment and Sustainable Development 1, 113-121.

Prospathopoulos, J.M., Voutsinas, S.G. 2005. Noise propagation issues in wind energy applications, Journal of Solar Energy Engineering, pp. 234-241.

Punt, M.J., Groeneveld, R.A., van Ierland, E.C., Stel, J.H. 2009. Spatial planning of offshore wind farms: A windfall to marine environmental protection?, Ecological Economics, pp. 93-103.

Raadal, H.L., Gagnon, L., Modahl, I.S., Hanssen, O.J. 2011. Life cycle greenhouse gas (GHG) emissions from the generation of wind and hydro power. Renewable and Sustainable Energy Reviews 15, 3417-3422.

Rademaker, J.H., Kleijn, R., Yang, Y. 2013. Recycling as a strategy against rare earth element criticality: A systemic evaluation of the potential yield of NdFeB magnet recycling. Environmental Science & Technology 47(18): 10129:10136.

Rankine, R. 2006. Energy and carbon audit of a rooftop wind turbine. Proceedings of the Institution of Mechanical Engineers Part A, Journal of Power and Energy 220, 643-654.

RenewableUK. 2010. Noise from Wind Turbines - The Facts, Renewable UK. http://www.bwea.com/ref/noise.html 4 2010.

Rogner, H.-H., R. F. Aguilera, C. Archer, R. Bertani, S. C. Bhattacharya, M. B. Dusseault, L. Gagnon, H. Haberl, M. Hoogwijk, A. Johnson, M. L. Rogner, H. Wagner and V. Yakushev. 2012. Chapter 7 - Energy Resources and Potentials. In *Global Energy Assessment - Toward a Sustainable Future*, Cambridge University Press, Cambridge, UK and New York, NY, USA and the International Institute for Applied Systems Analysis, Laxenburg, Austria, pp. 423-512.

Ruddock, M., Whitfield, D.P. 2007. A Review of Disturbance Distances in Selected Bird Species. A report from Natural Research (Projects) Ltd to Scottish Natural Heritage.

Rule, B.M., Worth, Z.J., Boyle, C.A. 2009. Comparison of Life Cycle Carbon Dioxide Emissions and Embodied Energy in Four Renewable Electricity Generation Technologies in New Zealand. Environmental Science & Technology 43, 6406-6413.

Rydell, J., Bach, L., Dubourg-Savage, M-J., Green, M., Rodrigues, L., Hedenström, A. 2010. Bat mortality at wind turbines in Northwestern Europe. Acta Chiropterologica, 12(2), 261-274.

Sathaye, J., Lucon, O., Rahman, A., Christensen, J., Denton, F., Fujino, J., Heath, G., Kadner, S., Mirza, M., Rudnick, H., Schlaepfer, A., Shmakin, A. 2011. Renewable Energy in the Context of Sustainable Energy. In IPCC Special Report on Renewable Energy Sources and Climate Change Mitigation [O. Edenhofer, R. Pichs-Madruga, Y. Sokona, K. Seyboth, P. Matschoss, S. Kadner, T. Zwickel, P. Eickemeier, G. Hansen, S. Schlömer, C. von Stechow (eds)].

Scheidel, A., Sorman, A.H. 2012. Energy transitions and the global land rush: Ultimate drivers and persistent consequences. Global Environmental Change 22, 588-595.

Schleisner, L. 2000. Life cycle assessment of a wind farm and related externalities. Renewable Energy 20, 279-288.

Shan, X., Chen, X., et al. 2010. Impacts on birds of wind farm constructions at Liaoning and protection measurements, Science Association Forum.

Shen, L., Worrell, E., Patel, M.K. 2010. Open-loop recycling: A LCA case study of PET bottle-to-fibre recycling. Resources, Conservation and Recycling 55, 34-52.

Simmonds, M.P., Brown, V.C. 2010. Is there a conflict between cetacean conservation and marine renewable-energy developments? Wildlife Research 37, 688-694.

Smallwood, K.S., Karas, B. 2009. Avian and bat fatality rates at old-generation and repowered wind turbines in California. The Journal of Wildlife Management 73(7), 1062-1071.

Smallwood, K.S., Thelander, C. 2008. Bird mortality in the Altamont Pass Wind Resource Area, California, Journal of Wildlife Management, pp. 215-223.

Smallwood, K.S. 2013. Comparing bird and bat fatality-rate estimates among North American wind-energy projects. Wildlife Society Bulletin, 37(1), 19-33.

Smil, V. 2003. Energy at the crossroads: Global perspectives and uncertainties. MIT Press, Cambridge, Mass.

Snyder, B., Kaiser, M.J. 2009. Ecological and economic cost-benefit analysis of offshore wind energy. Renewable Energy 34, 1567-1578.

Sprecher, B., Xiao, A., Walton, J., Speight, J., Harris, R. Kleijn, R., Visser, G., Kramer, G.J. 2014. Life cycle inventory of the production of rare earths and the subsequent produciton of NdFeB rare earth permanent magnets. Environmental Science & Technology 48(7): 3951:3958

Sovacool, B.K. 2009. Contextualizing avian mortality: A preliminary appraisal of bird and bat fatalities from wind, fossil-fuel, and nuclear electricity, Energy Policy, pp. 2241-2248.

Sovacool, B.K. 2012. The avian benefits of wind energy: A 2009 update, Renewable Energy, pp. 1-6.

Suh, S., Lenzen, M., Treloar, G.J., Hondo, H., Horvath, A., Huppes, G., Jolliet, O., Klann, U., Krewitt, W., Moriguchi, Y., Munksgaard, J., Norris, G. 2004. System Boundary Selection in Life-Cycle Inventories Using Hybrid Approaches. Environmental Science and Technology 38, 657-664.

Sun, J., Qian, Y., Xu, W., et al. 2007. Impacts on birds of Dafeng windfarm at Jiangsu, Anhui Agriculture Science, pp. 9920-9922.

Talens Peiró, L., Villalba Méndez, G. 2013. Material and energy requirement for rare earth production. JOM 65(10): 1327:1340.

Tougaard, J., Carstensen, J., Teilmann, J., Skov, H., Rasmussen, P. 2009. Pile driving zone of responsiveness extends beyond 20 km for harbor porpoises, The Journal of the Acoustical Society of America, pp. 11-14.

Tremeac, B., Meunier, F. 2009. Life cycle analysis of 4.5 MW and 250 W wind turbines. Renewable and Sustainable Energy Reviews 13, 2104-2110.

US DOE. 2008. 20% Wind Energy by 2030. Increasing Wind Energy's Contribution to U.S. Electricity Supply. U.S. Department of Energy.

US DOE. 2011. Critical Materials Strategy. U.S. Department of Energy.

van den Berg, G. 2004. Effects of the wind profile at night on wind turbine sound, Journal of Sound and Vibration, pp. 955-970.

van den Berg, G.P. 2005. The beat is getting stronger: The effect of atmospheric stability on low frequency modulated sound of wind turbines, Noise Notes, pp. 15-40.

van den Berg, G.P. 2008. Wind turbine power and sound in relation to atmospheric stability, Wind Energy, pp. 151-169.

Vattenfall. 2010. Vattenfall wind power. Certified environmental product declaration of electricity from Vattenfall's wind farms.

Veltman, K., Huijbregts, M.A.J., Rye, H., Hertwich, E.G. 2011. Including impacts of particulate emissions on marine ecosystems in life cycle assessment: The case of offshore oil and gas production. Integrated Environmental Assessment and Management 7, 678-686.

Vestas. 2006a. Life cycle assessment of electricity produced from onshore sited wind power plants based on Vestas V82-1.65 MW turbines.

Vestas. 2006b. Life cycle assessment of offshore and onshore sited wind power plants based on Vestas V90-3.0 MW turbines, 2nd ed. Vestas Wind Systems A/S, Randers, Denmark.

Vestas. 2011. Life cycle assessment of electricity production from a V112 turbine wind plant.

Viebahn, P., Lechon, Y., Trieb, F. 2011. The potential role of concentrated solar power (CSP) in Africa and Europe—A dynamic assessment of technology development, cost development and life cycle inventories until 2050. Energy Policy 39, 4420-4430.

Voorspools, K.R., Brouwers, E.A., D'Haeseleer, W.D. 2000. Energy content and indirect greenhouse gas emissions embedded in 'emission-free' power plants: results for the Low Countries. Applied Energy 67, 307-330.

Wagner, H.-J., Baack, C., Eickelkamp, T., Epe, A., Lohmann, J., Troy, S. 2011. Life cycle assessment of the offshore wind farm Alpha Ventus. Energy 36, 2459-2464.

Wagner, H.-J., Pick, E. 2004. Energy yield ratio and cumulative energy demand for wind energy converters. Energy 29, 2289-2295.

Weinzettel, J., Reenaas, M., Solli, C., Hertwich, E.G. 2009. Life cycle assessment of a floating offshore wind turbine. Renewable Energy 34, 742-747.

White, S. 2006. Net Energy Payback and CO_2 Emissions from Three Midwestern Wind Farms: An Update. Natural Resources Research 15, 271-281.

Wiedmann, T.O., Suh, S., Feng, K., Lenzen, M., Acquaye, A., Scott, K., Barrett, J.R. 2011. Application of hybrid life cycle approaches to emerging energy technologies – The case of wind power in the UK. Environmental Science & Technology 45, 5900-5907.

Willis, C.K.R, Barclay, R.M.R., Boyles, R., Brigham, R.M., Brack Jr., V., Waldien, D.L., Reichard, J. 2010. Bats are not birds and other problems with Sovacool's (2009) analysis of animal fatalities due to electricity generation, Energy Policy, 38(4). 2067-2069.

Wilson, J.C., Elliott, M. 2009. The habitat-creation potential of offshore wind farms, Wind Energy, pp. 203-212.

Winkelman, J. 1989. birds and the wind park near Urk: Collisions victims and disturbance of ducks, geese and swans. RIN report.

Wiser, R., Bolinger, M. 2013. 2012 Wind technologies market report. US Department of Energy (DOE). http://www1.eere.energy.gov/wind/pdfs/2012_wind_technologies_market_report.pdf. Accessed 8 February 2014.

Wiser, R., Yang, Z., Hand, M., Hohmeyer, O., Infield, D., Jensen, P. 2011a. Wind Energy. In IPCC Special Report on Renewable Energy Sources and Climate Change Mitigation [O.Edenhofer, R. Pichs-Madruga, Y.Sokona, K.Seyboth, P.Matschoss, S.Kadner, T. Zwickel, P. Eickemeier, G. Hansen, S.Schlomer, C. Von Stechow (eds)].

Wiser, R., Yang, Z., Hand, M., Hohmeyer, O., Infield, D., Jensen, P.H., Nikolaev, V., O'Malley, M., Sinden, G., Zervos, A. 2011b. Wind Energy. In IPCC Special Report on Renewable Energy Sources and Climate Change Mitigation [O. Edenhofer, R. Pichs-Madruga, Y. Sokona, K. Seyboth, P. Matschoss, S. Kadner, T. Zwickel, P. Eickemeier, G. Hansen, S. Schlömer, C. von Stechow (eds)].

Xu, X., Zheng, Y., Yang, L., Lv, S. 2010. Influence of wind power field on birds in Yancheng National Rare Waterfowls Nature Reserve of Jiangsu, Chinese Journal of Ecology, pp. 560-565.

Yang, M., Patiño-Echeverri, D., Yang, F. 2012. Wind power generation in China: Understanding the mismatch between capacity and generation. Renewable Energy 41, 145-151.

Yellishetty, M., Mudd, G.M., Ranjith, P.G. 2011. The steel industry, abiotic resource depletion and life cycle assessment: A real or perceived issue? Journal of Cleaner Production 19, 78-90.

Yu, C. 2011. Five people died in the wind farm installation accident. National Business Daily. 12 October 2011. http://epaper.nbd.com.cn/shtml/mrjjxw/20111012/2415620.shtml.

Zhong, Z.W., Song, B., Loh, P.E. 2011. LCAs of a polycrystalline photovoltaic module and a wind turbine. Renewable Energy 36, 2227-2237.

© J.D.S/Shutterstock

Chapter 6

Concentrating solar power

Lead authors: Pallav Purohit, Mathieu Saurat, Garvin Heath
Contributing authors: Peter Viebahn, Anders Arvesen, Thomas Gibon

6.1 INTRODUCTION

Concentrating solar power (CSP) technologies are designed to produce high temperature heat for electricity generation or for cogeneration of electricity and heat. CSP systems utilize direct normal irradiation (DNI), which is the energy received directly from the sun and that which is not scattered by the atmosphere on a surface tracked perpendicular to the sun's rays. Areas suitable for CSP development are those with strong sunshine and clear skies, usually arid or semi-arid areas. Four main technologies have been identified during the past decades for generating electricity in the 10 kW to several hundred MW range: a) parabolic trough and b) linear Fresnel technology, which produces high pressure superheated steam at temperatures less than 500°C to power a Rankine steam cycle, c) solar tower technology, which produces steam at temperatures near 560°C for a steam cycle, or air at temperatures approaching 1,000°C or synthesis gas for gas turbine operation, and d) dish/engine technology, which can directly generate electricity via Stirling engine or other small heat-engine technologies.

Commercial CSP plants were first developed in the 1980s, with construction and operation of the 354 MW CSP installations located in the Mojave Desert in California. After years of inactivity, the CSP market has revived with an increase of more than 60 per cent per year on average in the installed capacity during 2009. Total CSP installed capacity was an estimated 4,400 MW by the end of 2014, based on 2,300 MW in Spain, 1,634 MW in the United States, 225 in India, 100 MW in United Arab Emirates, 25 MW in Algeria, 20 MW in Egypt, 20 MW in Morocco, 10 MW in China and 5 MW in Thailand (REN21, 2015). Most of this capacity was added in Spain, home to more than half of the world's CSP capacity. According to the recently published Solar Thermal Electricity Global Outlook 2016, global CSP installed capacity by the end of 2015 was 4,940 MW (Teske et al., 2016) with several new projects added in the pipeline in the Middle East and North Africa region and South Africa. CSP featured heavily in the leaderboard rankings for 2015 solar deals - investment rising by 139% on 2014 levels to reach some $6.8 billion, the highest level since 2011 (FS-UNEP Centre/BNEF, 2016). Globally, the five largest solar deals were all for CSP and were located in South Africa, Morocco and China at a total cost exceeding $5.4 billion.

Integration with low-cost thermal storage adds significant value to the energy delivered from CSP plants. Many utilities are including CSP in their power-generation portfolio to help meet government mandates for renewable power generation. In the southern European countries and the rapidly growing economies of China and India, which are highly dependent on fossil fuel imports, CSP generation is an important potential source for diversifying energy sources and increasing domestic energy supply. The establishment of preferential market conditions for renewable energies has been an important driver for CSP plants. In Spain and Algeria, CSP technologies were explicitly included in the government support scheme. The bulk of the worldwide CSP operating capacity is installed in Spain and south-western United States. Interest in CSP is on the rise, particularly in developing countries, with investment spreading across Africa, the Middle East, Asia, and Latin America. One of the most active markets is South Africa, where 600 MW of CSP projects were approved within a period of less than five years. South Africa's first 100 MW CSP project, KaXu Solar One, came online in March 2015. Two other projects, Khi Solar One (50 MW) and Bokpoort CSP (50 MW), recently became operational (Teske et al., 2016). In Morocco, Noor I, a 160 MW plant, became operational in February 2016. The next two phases of the Noor project total 350

MW and are scheduled to come online in 2018. India had planned to complete 500 MW by the end of 2013, but only eight projects amounting to 235 MW were installed by May 2016 (CSP Today, 2016). In Australia, a 44 MW plant is under construction to feed steam to an existing coal facility. Many other countries, including Argentina, Chile, and Mexico in Latin America, several countries in Europe, Israel, and China have projects under construction or have indicated intentions to install CSP plants. At present, CSP projects of approximately 6,000 MW capacities have been announced worldwide by various project developers and promoters; whereas greater than 10800 MW capacity CSP projects are at different stages of planning, development and construction.). Integrated Solar Combined Cycle (ISCC) projects have also been announced in many parts of the world.

Parabolic trough plants continue to dominate CSP deployment. Dramatic reductions in photovoltaic (PV) costs and low natural gas prices are challenging the growing CSP market, at least in the United States where several planned projects were redesigned to use utility-scale PV technologies. In the current U.S. market, the balancing provided by a large grid and relatively low solar penetration do not reward the clear advantages of CSP for delivering base-load and balancing power. Various global energy studies released in recent years provide an optimistic picture for the future application of solar power to electricity generation in the year 2050 (EREC, 2010; LBF, 2008; IEA, 2010a; Shell International BV, 2008; Teske et al., 2011). In several scenarios, solar-generated electricity represents at least 25 per cent of the total electricity generated by 2050 (LBF, 2008; Teske et al., 2011). In these high-penetration scenarios, CSP and PV are of comparable importance.

6.1.1 OBJECTIVES

Large-scale solar power plants are rapidly being developed. These new facilities will require thousands or millions of acres of land globally. An understanding of the environmental issues related to the installation and operation phases of such facilities, including the choice of specific CSP technologies, is hence desirable. This study therefore reviews environmental impacts of CSP technologies in a life cycle perspective and provides data sets for life cycle inventory (LCI) analysis for a further comparison of energy technologies. This study collects and evaluates the existing assessments of CSP technologies using life cycle assessments (LCAs), risk analyses, material flow accounting and integrated assessment methodologies.

6.1.2 SCOPE OF THE REVIEW

CSP systems provide obvious environmental advantages in comparison to the conventional energy sources, thus contributing to sustainable human development (Tsoutsos et al., 2005). Beyond its avoidance of further exhausting current stocks of non-renewable natural resources, the main advantage of CSP is related to reduced greenhouse gas (GHG) emissions, and, normally, absence of any air emissions or waste products during operation. Concerning the environment, the use of CSP has additional positive effects such as a) reduction of the emissions of the GHG, b) prevention of other air pollutants such as nitrogen oxides (NO_x), sulfur dioxide (SO_2), and particulates in addition to toxic gas emissions such as mercury, c) reclamation of degraded land, etc. From a socio-economic viewpoint, the benefits of using CSP technologies include a) an increase in regional and national energy independence, b) creation of significant employment opportunities, c) diversification and security of energy supply, d) support of the deregulation of energy markets, and e) acceleration of rural electrification in developing countries. This article reviews environmental aspects of CSP deployment and illustrates means by which CSP can be implemented to successfully address potential environmental burdens.

The aim of this work is to evaluate the performance of CSP systems from an environmental point of view by use of the LCA methodology, which is based on calculations and analysis of environmental effects. Chapter 6.2 provides a detailed overview of resource assessment (solar, land, water, etc.) and deployment potential. Chapter 6.3 provides a detailed description of CSP technologies. An in-depth review on the LCA assessment of CSP systems IS presented in Chapter 6.4. Chapter 6.5 presents a methodology for the collection of LCI data and LCA of the different CSP technologies. The social and ecological impacts and CSP systems is presented in Chapter 6.6. Finally, Chapter 6.7 presents conclusions and recommendations based on the detailed LCA and highlights the knowledge gaps.

6.2 RESOURCE ASSESSMENT

The special characteristics, conditions and design of CSP plants require a more cautious and elaborate approach of conducting deployment potential studies as well as an assessment of economic, ecological and social issues than applied for other renewable energies (Kocheril and Viebahn, 2011; Viebahn et al., 2011). As mentioned above, a strong solar resource is only one criterion for the effective deployment of large CSP systems. The land must also be relatively flat, unoccupied, and suitable for development. Furthermore, the land may be constrained to activities such as agriculture, or be ecologically protected, which is also a relevant issue for selecting CSP sites. In view of the fact that the economics of utility-scale CSP systems favour large facilities, land areas less than 1 km^2 may not be relevant (Mehos et al., 2009). Regional water scarcity is a limiting factor in the choice of wet or dry cooling technology options (EPRI, 2009). The ambient temperature and humidity parameters may have an impact on the cooling cycle and hence on overall plant efficiency.

6.2.1 SOLAR RESOURCES

The technical and economic potential of CSP is tremendous: less than 0.1 per cent of the areas suitable for the installation of solar thermal power stations worldwide would theoretically suffice to meet the total global energy demand (Steinhagen and Trieb, 2004). The potential for CSP implementation in any given geographic location is largely determined by the solar radiation characteristics at the site. Solar radiation consists primarily of direct beam and diffuse, or scattered, components. The term 'global' solar radiation simply refers to the sum of these two components. The daily variation of the different components depends upon meteorological and environmental factors such as cloud cover, air pollution and humidity and the relative earth-sun geometry. DNI is synonymous with direct beam radiation and defines the available solar resource. DNI is measured by tracking the sun throughout the sky, explaining the rationale for the design of collectors which track the sun throughout the day. The global distribution of DNI predominantly overlaps with the deserts of the world. The areas with the greatest CSP potential are located in North Africa, South Africa, the Middle East, India, Australia, North America and South America. Unlike most natural resources, solar energy is evenly distributed around the world and nearly all populated areas may be connected to the areas with excellent solar conditions.

FIGURE 6.1

Global CSP resource map indicating direct normal irradiation

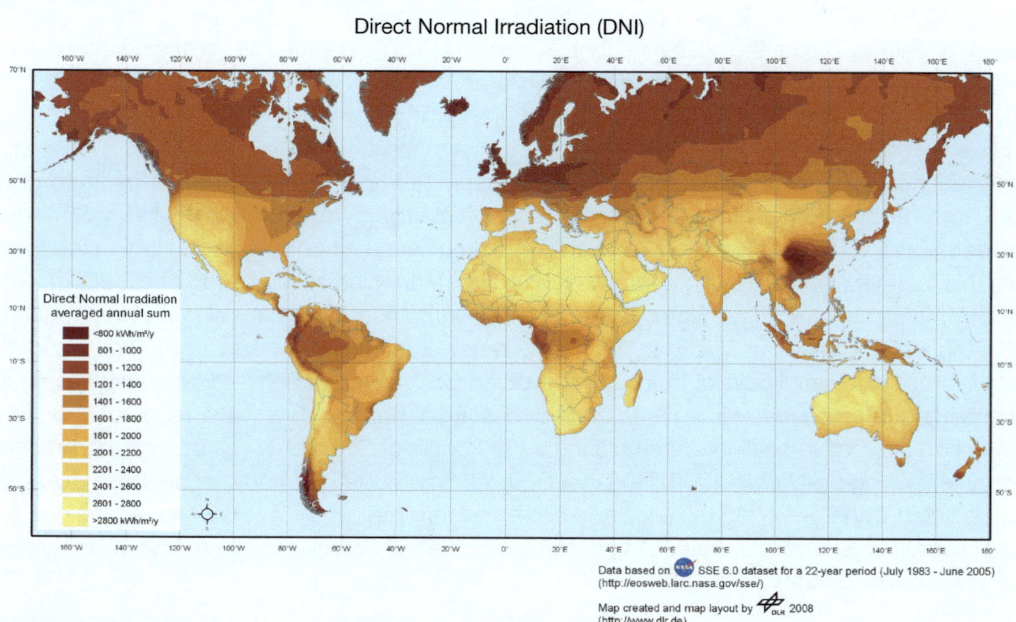

Source: http://www.dlr.de/
DNI is indicated here as an averaged annual sum in kWh/m^2/year.

6.2.2 LAND

CSP plants require a significant amount of land that typically cannot be used simultaneously for other applications. It also requires the land to be relatively flat. The potential for CSP and other renewable energy technologies have recently been assessed for different regions by considering the land use profile and its slope along with DNI. Domínguez Bravo et al. (2007) defined the capacity and generation ceiling for renewable energy technologies for Spain. For CSP, they include all areas that have a suitable slope, corresponding to a gradient below 7 per cent for slopes facing south-east to south-west or below 2 per cent for all other orientations, a suitable land use profile and average DNI values above 4.1 kWh/m²/day (1,500 kWh/m²/year). Assuming the use of parabolic trough plants with six equivalent full load hours (EFLH) of thermal storage, Domínguez Bravo et al. (2007) find a electricity generation ceiling that is more than 35 times greater than the electricity demand projected for 2050 (Domínguez Bravo et al., 2007). Plekta et al. (2007) presented a renewable energy assessment for Arizona. For CSP, they include all areas with a suitable slope, defined as a gradient less than 1 per cent, a suitable land use profile and average DNI values higher than 6.75 kWh/m²/day. Plekta et al. (2007) take Phoenix, Arizona, as the reference location, which has an annual daily average DNI of 6.9 kWh/m² (Pletka et al., 2007). They further assume a land use of 28 km²/GW. For a parabolic trough plant with six-hour storage, they predict a capacity factor of 38.8 per cent. Dahle et al. (2003), Karsteadt et al. (2005) and Dahle et al. (2008) have performed assessments similar to that of Plekta et al. (2007) for other regions in the United States. Broesamle et al. (2001) assessed the potential of CSP for North Africa and Trieb et al. (2002) for Morocco (Broesamle et al., 2001; Trieb et al., 2002). Broesamle et al. (2001) assumed a land use of 20 km²/GW of installed capacity.

The cost of land generally represents a very minor portion of total plant costs. A 100 MW CSP plant with a solar multiple[1] of one would require approximately 2 km² of land. However, the land does need to be relatively flat, particularly for linear trough and Fresnel systems. The land is also ideally near transmission lines and roads for construction traffic, and not on environmentally sensitive land. Although the mirror area itself is typically only about 25-35 per cent of the land area occupied, the site of a solar plant will usually be arid. Thus, it is generally not suitable for agriculture, but may still have protected or sensitive plant or animal species. For this kind of system, sunny deserts close to existing electricity infrastructure are ideal. As CSP plant capacity is increased, however, the economics of longer electricity transmission distances improve. If more remote siting is expected, transmission infrastructure needs will increase accordingly. Attractive CSP sites exist in many regions of the world, including southern Europe, northern and southern African countries, the Middle East, Central Asian countries, China (Tibet, Xinjan), India (in the states of Rajasthan and Gujarat), Australia, Chile, Peru, Mexico and south-western United States.

6.2.3 WATER

While CSP has great potential, a critical issue is its water consumption in the desert environments to which it is most suited. In contrast to other renewable energy technologies such as solar PV or wind, CSP requires a considerable amount of water for steam cycle cooling in recirculating wet cooling and cleaning the mirrors, a characteristic this technology shares with other thermal power technologies (Damerau et al., 2011). Coal and nuclear power plants show a similar water demand, while combined-cycle natural gas plants require only up to a fourth of the water demand (USDOE, 2009). Some renewable energy experts argue that this water demand constrains the large scale development of wet-cooled CSP in arid or semi-arid desert regions; these would either consume too much water in an area that, by definition, has very low water resources, or, when using more expensive alternative cooling systems such as dry cooling, CSP could not be cost-competitive with other energy technologies (Carter, 2009; Hogan, 2009; Woody, 2009). Damerau et al. (2011) examined water usage associated with CSP in North Africa and observed that the use of wet cooling technologies would likely be unsustainable whereas dry cooling systems, as well as sourcing of alternative water supplies, would allow for sustainable operation.

1 Solar multiple is defined as the ratio of the power capacity of the collection field to the capacity of the power block.

Today, the 94 billion m^3 of freshwater withdrawn annually in North Africa is already twice the regional renewable freshwater resources (FAO, 2010). With growing populations, economic development, as well as increasing temperatures and coastal inundation due to climate change, the future availability of water will further decrease (Abou-Hadid, 2006; Bakir and Social, 2002; Elsharkawy, 2009). de Wit and Stankiewicz (2006) project in their scenario a decrease in rainfall of 10–20 per cent by the end of the century, while a drying of 20 per cent along the African Mediterranean coast in one scenario can be found in the regional climate projections of the fourth IPCC report (de Wit and Stankiewicz, 2006; Christensen, 2007). Arnell (1999) assumes a decrease in surface run-off up to 25 mm/yr for major parts of the region until the 2050s. Changing hydrological patterns are to be especially expected in the drainage area of the Atlas Mountains (Boulet et al., 2008). It seems probable that in many locations, CSP plants will need to be dry cooled to be viable. Further details on wet and dry cooling technologies are discussed in Chapter 6.3.6.

6.2.4 GLOBAL POTENTIAL OF CSP

Hoogwijk and Graus (2008) assessed the total global technical potential of CSP in the long term at about 992 EJ. Efforts have also been made to estimate the potential of CSP systems at regional and national levels (Hang et al., 2008; Li, 2009; Hou, 2009; Kaygusuz, 2011; Viana et al., 2011). After implementing filters that account for insolation, slope, and land-use restrictions, the technical potential of the U.S. CSP market is about 11,000 GW of potential generating capacity, which is several times higher than the entire U.S. electric grid capacity, in just seven south-western states (Mehos et al., 2009). Fluri identified a total potential nominal capacity of 548 GW for parabolic trough systems in South Africa using geographic information systems (Fluri, 2009). The potential areas are assumed suitable if they receive sufficient solar radiation, are close enough to transmission lines, are flat enough, their local vegetation is not under threat and they have a suitable land use profile. Janjai et al. quantitatively investigated the potential application of CSP plants in the tropical environment of Thailand (Janjai et al., 2011). They concluded that the tropical environment of this country has sufficient potential for a utilization of the parabolic trough systems. On the basis of a detailed solar radiation and land resource assessment, Purohit et al. (2013) estimated that the maximum theoretical potential of CSP in north-western India is estimated over 2,000 GW taking into accounts the viability of different CSP technologies and land suitability criteria. The technical potential is estimated over 1,700 GW at an annual direct normal incidence (DNI) over 1,800 kW h/m^2 and finally, the economic potential is estimated over 700 GW at an annual DNI over 2,000 kW h/m^2 in north-western India.

Using the global distribution of DNI, Breyer and Knies (2009) identified potential CSP sites. The minimum solar radiation of potential sites was 2,000 kWh/m^2/y due to economic constraints by Trieb et al. (2005), whereas other studies suggested DNI of at least 1,800 kWh/m^2/year (Figure 6.1) (Trieb, 2005; EPRI, 2009; Purohit and Purohit, 2010; Ummel, 2010). The identified coherent potential CSP areas range from a minimum of about 9,000 km^2 to more than 31 million km^2 (before exclusion of unsuitable sites for CSP plants). Table 6.1 presents a technical global CSP potential of 2,946 PWh/year (Trieb et al., 2009).

TABLE 6.1

Technical CSP potential in the RECCESS world regions, classified by DNI

Region (↓)	CSP potential (PWh/year) by DNI class (kWh/m²/y)								
DNI Class (→)	2000-2099	2100-2199	2200-2299	2300-2399	2400-2499	2500-2599	2600-2699	2700-2800+	Total
Africa	102.3	138.2	139.8	141.1	209.6	204.0	178.5	346.0	1459.4
Australia	6.6	18.6	36.8	87.8	148.0	207.8	142.5	49.6	697.6
Central Asia, Caucasus	14.3	0.3	0.4	0.2	0.1	0.0	0.0	0.0	15.2
Canada	0.0	0.0	0.0	0.0	0.0	0.0	0.0	0.0	0.0
China	8.3	18.3	43.0	28.4	11.2	11.3	2.2	3.1	125.8
Central South America	31.6	20.6	24.1	20.7	6.4	3.7	5.1	11.8	124.0
India	7.9	1.1	0.6	0.8	0.4	0.0	0.1	0.0	10.9
Japan	0.0	0.0	0.0	0.0	0.0	0.0	0.0	0.0	0.0
Middle East	3.4	12.4	39.2	60.2	71.3	35.0	32.3	36.8	290.6
Mexico	1.6	3.4	3.7	5.8	15.7	7.1	1.5	1.9	40.7
Other Developing Asia	4.5	5.2	10.9	30.8	19.4	4.4	0.3	0.1	75.6
Other Eastern Europe	0.0	0.0	0.0	0.0	0.0	0.0	0.0	0.0	0.0
Russia	0.0	0.0	0.0	0.0	0.0	0.0	0.0	0.0	0.0
South Korea	0.0	0.0	0.0	0.0	0.0	0.0	0.0	0.0	0.0
EU27+	0.9	0.5	0.7	0.2	0.1	0.1	0.0	0.0	2.4
USA	14.1	17.1	21.7	16.4	23.9	8.1	2.3	0.0	103.7

Source: Trieb et al., 2009

In addition to region-specific parameters, the plant efficiency and the capacity factor have to be considered in an adequate CSP technology potential assessment (Kocheril and Viebahn, 2011). The values of these basic parameters are essential in order to classify the assessment study result according to cost, power potential, or ecological criteria. Although locations well suited to CSP facilities would of course also be suitable for utility-scale PV plants, the value provided by CSP's efficient thermal energy storage (TES) becomes a critically important attribute as total solar deployment increases. Grid analyses suggest a mix of PV and CSP provides the greatest solar deployment (Brinkman et al., 2011).

6.3 TECHNOLOGY OVERVIEW

CSP is a technology by which sunlight is focused, or concentrated, by mirrors or reflectors called "solar collectors" to heat a fluid in a receiver to a high temperature. The heat transfer fluid, usually pressurized steam, synthetic oil, or molten salt, flows from the solar receiver to a heat engine where up to 40 per cent of the energy is converted to electricity (Jacobson, 2009). A CSP plant consists of a field of solar collectors, receivers, and a power block where the collected heat is transformed to run a thermal power cycle and produce electricity. CSP technology consists of four alternative approaches: parabolic trough, power tower, linear Fresnel reflectors and dish/engine (Table 6.2 and Figure 6.2a-d). The technologies can be categorized in different ways based on the attributes of the collector and receiver systems. In general, trough and linear Fresnel systems are simpler, but less efficient than power tower and parabolic dish designs. Because overall economics depend on both capital costs and efficiency, all four designs have proponents in the marketplace, with the exception of dish engines that are still too expensive to reach competitiveness.

Parabolic trough, power tower and linear Fresnel systems usually run steam Rankine power systems that require cooling. Evaporative or "wet" cooling consumes a significant amount of water while dry or hybrid dry/wet cooling can be used in areas with limited water resources. Parabolic dish systems can be deployed with a central power cycle, but more commonly use a single, smaller Stirling-cycle engine with each dish. These are always air-cooled and the systems use water only for mirror washing. Zhang et al. (2010) explore the capacity of CSP as base load or intermediate and peak (I&P) plants, both with and without thermal storage (Zhang et al., 2010). They conclude that I&P plants, particularly those with thermal storage, could be economically competitive enough for large scale development in upcoming years. Base load plants would need more policy incentives and significant capital cost reductions to achieve the same deployment potential as I&P plants. An overview of CSP technologies and their basic properties is provided in Table 6.2.

TABLE 6.2

Different types of CSP technologies and their features

CSP Design	Receiver attributes	Collector attributes	Typical concentration ratio	Typical capacity (MW)	Typical operating temperature (°C)	Land use factor[a]	Land use efficiency[a]
Parabolic trough	Linear, moves with collector	Curved, 1-axis tracking	30-100	30-250	300-400	25-40%	3.5-5.6%
Linear Fresnel	Linear, fixed	Flat, 1-axis tracking	100-200	1-200	250-500	60-80%	4.8-9.6%
Power tower	Single point, fixed	Flat, 2-axis tracking	500-1,000	10-150	500-800	20-25%	2.5-4.0%
Parabolic dish	Single point, moves with collector	Curved, 2-axis tracking	1000-10,000	0.01-0.04	750	-	-

Source: IEA, 2008; Purohit and Purohit, 2010; Trieb et al., 2009. ©IEA, 2008, *Combined Heat and Power*. Evaluating the benefits of greater global investment, IEA Publishing. Licence: http://www.iea.org/t&c/termsandconditions/

[a]Land use factor = collector area divided by total land area. A parabolic trough system with 12% annual solar-electric efficiency, 37% land use factor and 4.5% land use efficiency was taken as reference.

6.3.1 PARABOLIC TROUGH SYSTEMS

Parabolic trough systems are line-focusing solar thermal power plants. Trough systems use the mirrored surface of a linear parabolic concentrator to focus direct solar radiation on an absorber pipe, or "receiver", running along the focal line of the parabola (Almanza and Lentz, 1998). The heat transfer fluid (HTF) inside the receiver is heated and pumped to the steam generator, which in turn is connected to a steam turbine. The absorbed solar radiation heats the HTF flowing through the receiver (EC, 2007). The HTF is then conducted through a heat exchanger, producing steam that then generates power in the turbines. Plants that incorporate TES move heat from the HTF to a storage medium, typically a blend of nitrate salts, which is stored in insulated tanks. To date, CSP plants have used synthetic oils or molten salt as heat transfer fluids (Skumanich, 2010; Kramer, 2012).The technology potential is huge but the lack of policy incentives hinders the proper development of CSP in the market and limits the opportunities of improving the technology research. Because many solar receivers use metal containment structures and nitrate-salt HTF, the operating temperature of receivers is restricted to below 650°C. Higher operating temperatures are favourable for increased thermal conversion efficiency. Many of the current studies on HTF and storage media target salt compounds. However, there are still challenges to overcome for salt to satisfy performance requirements, including high temperature stability (>650°C), low freezing point (<0°C), and material compatibility with high-temperature metal (>650°C).

FIGURE 6.2

Principles of a) parabolic trough, b) power tower, c) Linear Fresnel, and d) parabolic dish systems

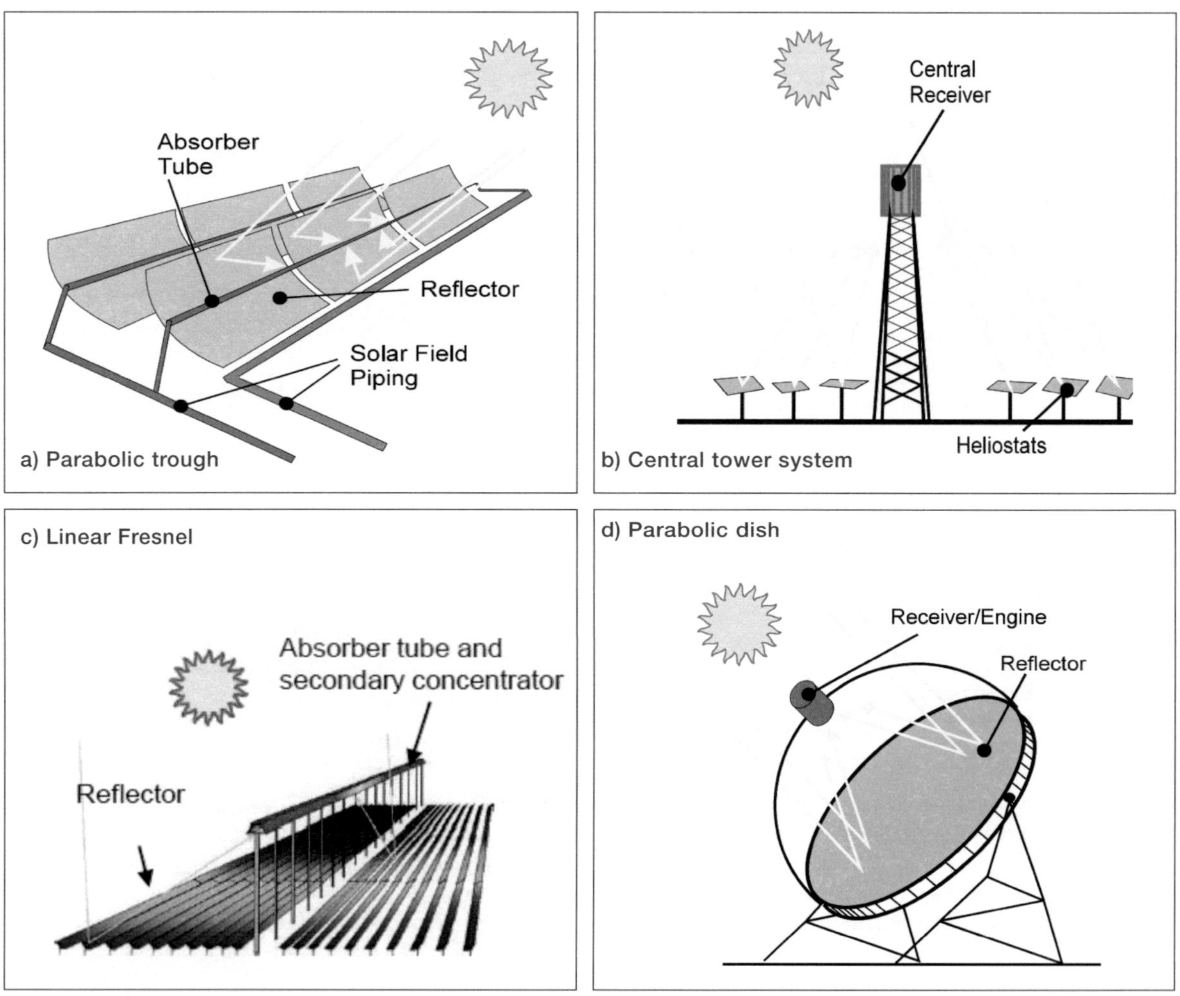

Source: http://www.solarpaces.org/

Table 6.3 presents the technical characteristics of parabolic trough systems and installed generating capacity in operation by May 2016 (Lovegrove et al., 2009; Viebahn et al., 2008a ; CSP Today, 2016). The basic principle of parabolic trough collector (PTC) based systems is presented in Figure 6.2a. A fossil fuel back-up system such as one using natural gas can assist the operation of the plant by allowing heating of the HTF as required, for instance to accelerate morning start-up or as an HTF freeze-protection measure. Some recently built plants have several hours of storage capacity, and most existing plants use some combustible fuel as a backup (IEA, 2010b). Parabolic trough is the most mature technology, and it continues to dominate the market, representing about 95 per cent of facilities in operation at the end of 2011, and 75 per cent of plants under construction by mid-2012 (REN21, 2013).

TABLE 6.3

Technical characteristics of CSP plant technologies

Technology	Annual solar-to electricity efficiency[a] (%)	Typical water consumption (m^3/MWh)[b]	Installed capacity[c] (MWe)	Currently in operation (MWe)
Parabolic trough	15-16	3.8 (wet cooled) 0.3 (dry cooled)	4180	(Spain 2,223 MW, United States 1,344 MW, Morocco 183 MW, South Africa 150 MW, India 104 MW, UAE 100 MW, Others 76 MW)
Linear Fresnel	8-10	3.8 (wet cooled) 0.3 (dry cooled)	17	(India 128 MW, Spain 31 MW, United States 11 MW, others 3.5 MW)
Central receiver	15-17	2.8 (wet cooled) 0.2 (dry cooled)	659	(United States 536 MW, Spain 51 MW, South Africa 50 MW, Others 22 MW)
Parabolic dish	20-25	0.1	1.22	(China 1.11 MW, UAE 0.11 MW)

[a]EASAC, 2011; [b]USDOE, 2012; [c]CSP Today Global Tracker
See: http://social.csptoday.com/tracker/projects/table. Accessed on 21 May 2016.

6.3.2 POWER TOWER SYSTEMS

Power tower systems, also called central receivers, use heliostats[2] to track the sun. Each heliostat tracks the azimuth and elevation of the sun in order to reflect the sunlight onto a central receiver installed at the top of the tower. In the receiver, an HTF is heated by the solar energy and is then passed either to the storage or to power-conversion systems, which convert the thermal energy into electricity (Figure 6.2b). The key features of solar power towers are that they (*a*) reflect solar energy to a punctual receiver and therefore minimize the HTF transport requirements, (*b*) achieve sunlight concentration ratios[3]of 300-1500 and working temperatures of 500-1000°C, which allow high efficiencies both in energy collection and in its conversion to electricity, (*c*) incorporate TES in the form of the molten-salt tanks, and (*d*) allow economies of scale in the energy storage and power conversion systems (DeLaquil III, 1996). Table 6.3 presents the technical characteristics of power tower systems. Towers can generate saturated or superheated steam directly by using water/steam as the HTF, or use molten salts, air or other fluids for the HTF. Direct steam generation towers can achieve higher thermal-to-electric conversion efficiency, while molten salt towers are better suited for efficient, multi-hour TES.

The earliest large-tower projects were the 10 MW Solar One and Solar Two tower demonstration projects in the Mojave Desert, which are now decommissioned. Originally called the Solar Tres Power Tower, the 19 MW Gemasolar solar thermal power plant commissioned in 2011 builds on these projects. In addition, Spain operates the 11 MW PS-10 and the 20 MW PS-20 direct-steam solar power towers. PS-20 features

2 A heliostat is a device that tracks the movement of the sun and is used to orient a flat mirror throughout the day in order to reflect sunlight onto the receiver.

3 It is the ratio between the optically active surface of the collector and the irradiated absorber's surface.

a number of significant technological improvements with respect to PS-10, the first commercial power tower. These enhancements, developed by Abengoa Solar, include a higher-efficiency receiver, various improvements in the control and operational systems, and an improved TES system. The 50 MW Khi Solar One (KSO) solar power tower system located in the Northern Cape Region of South Africa has been operating since February 2016. KSO is a super-heated steam solar tower with two hours of thermal storage. In the United States, the 392 MW Crescent Dunes solar tower project is the first utility-scale solar power plant ever built in the world with fully integrated storage Technology (SolarReserve, 2016).

6.3.3 COMPACT LINEAR FRESNEL REFLECTOR

Linear Fresnel reflectors (LFRs) approximate the parabolic shape of trough systems but by using long rows of flat or slightly curved mirrors to reflect the sun's rays onto a downward-facing, linear fixed receiver. A more recent design, known as compact linear Fresnel reflectors (CLFRs), uses two parallel receivers for each row of mirrors and thus needs less land than parabolic troughs to produce a given output (Figure 6.2c). LFR is a single-axis tracking technology but differs from a parabolic trough in that the absorber is fixed in space above the mirror field and the reflector is composed of long, near-flat mirror segments that focus collectively on the elevated receiver. Typically, water is used as the HTF and flows through the receivers to be converted into saturated steam. This system is line-concentrating, similar to a parabolic trough collector. Added advantages of CLFR include low capital costs for structural support and reflectors, fixed fluid joints, a receiver distinct from the reflector system, and long focal lengths that allow the use of flat mirrors (Purohit et al., 2013). Additionally, CLFRs are mounted close to the ground, thus minimizing structural requirements (Kalogirou, 2004). However, one challenge associated with the CLFR technology is the avoidance of shading and blocking between adjacent reflectors, which requires increased spacing between reflectors. Increasing the height of the absorber towers also reduces blocking, but increases cost. Mills and Morrison (2000) evaluated CLFR concepts suitable for large-scale solar thermal electricity generation plants (Mills and Morrison, 2000). In CLFR, it was assumed that there would be many parallel linear receivers elevated on tower structures arranged such that individual mirror rows to have the option of directing reflected solar radiation at two alternative linear receivers on separate towers (Figure 6.3). Variations of the basic CLFR concept that were evaluated include differing absorber orientation, absorber structure, the use of secondary reflectors adjacent to the absorbers, reflector field configurations, mirror packing densities, and receiver heights (Xie et al., 2011).

FIGURE 6.3

Schematic showing interleaving of mirrors in a CLFR without shading

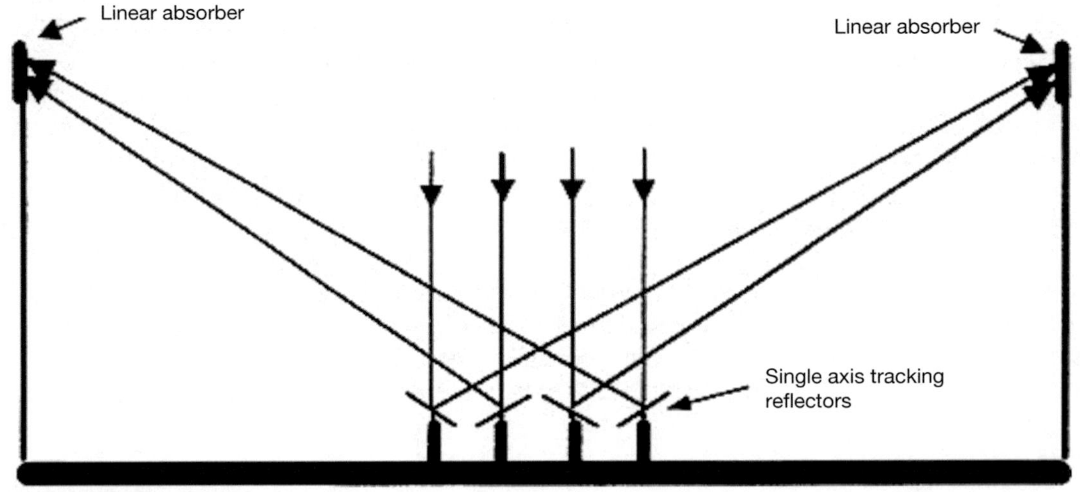

Source: Mills and Morrison, 2000

The Fresnel structure allows for a very light design and thus a decrease of the specific material consumption compared to parabolic troughs. Furthermore, land use is reduced, which is a significant advantage in highly populated or sensitive land areas. A Fresnel system requires only one-third of the area required by a parabolic trough of the same installed power. Even when accounting for the lower optical efficiency of the Fresnel design, a land use reduction of 50 per cent per produced kWh can be achieved (Viebahn et al., 2008a, 2008b). At present, CLFR technology is actively being developed by Areva Solar, which bought Fresnel developer Ausra, Inc. in 2010, and Novatec Solar. Ausra built a test plant of 1 MW in New South Wales, Australia in 2003, which pre-heats boiler feedwater for an existing coal-fired power station. Puerto Errado 1 is 1.4 MW; it was the first Fresnel-lens, solar power plant in Spain connected to the grid, (in March 2009). Puerto Errado 2 added 30 MW in February, 2012. The world's largest LFR plant, Reliance Areva CSP 1, a project of 125 MW, came on line in India in 2014, further diversifying the mix of added technologies.

6.3.4 DISH/ENGINE SYSTEMS

Dish/engine systems use mirrored parabolas to focus and concentrate sunlight onto a receiver located at the focal point of the dish (Figure 6.2d). The dish assembly tracks the sun in two axes to capture the maximum amount of direct solar energy. Several different heat engines, such as gas turbines, reciprocating steam engines, and organic Rankine engines, have been explored. Recently, most attention has been focused on Stirling cycle engines. Stirling engines are externally heated, as opposed to more familiar internal combustion engines. A typical Stirling engine has a receiver with thin tubes containing hydrogen or helium gas running along the outside of the engine's four piston cylinders and open into the cylinders. As concentrated sunlight falls on the receiver, it heats the gas in the tubes to very high temperatures, which causes hot gas to expand inside the cylinders. The expanding gas drives the pistons. The pistons turn a crank shaft, which drives an electric generator. The receiver, engine, and generator comprise a single, integrated assembly mounted at the focus of the mirrored dish. Of all the solar power generation technologies, dish engine systems are most efficient because their high optical efficiency, small area for thermal losses, and good thermal-to-electric conversion efficiency. Table 6.3 presents the technical characteristics of parabolic dishes. Dish/Stirling systems have demonstrated the highest net efficiency of any solar power generation system by converting nearly 31.25 per cent of solar radiation into electricity after accounting for parasitic power losses (Taggart, 2008). Each system is a self-contained power generator due to the modular nature of dish/Stirling systems. As a result, dish/Stirling systems can be assembled into plants ranging in size from a few kilowatts to 10 MW or more (Mancini et al., 2003). These systems can also be combined with natural gas; the resulting hybrid provides continuous power generation like conventional energy systems (Ab Kadir et al., 2010). However, the modular dish systems cannot include centralized TES as other CSP technologies do, and therefore frequently compete directly with PV solar power systems rather than other CSP technologies. Like other CSP technologies, dish/Stirling systems are applicable in regions having high DNI; the so-called Sun Belt area that includes northern Africa, Mexico, south-western United States, Australia, and the MENA region. Although an individual dish requires relatively little land area, a plant containing large numbers of dishes must occupy more area per MW than linear CSP technologies in order to avoid shading (USDOE, 2012). A significant advantage with dish/Stirling systems is that they require no water for cooling or steam cycle maintenance and only small amounts of water for mirror washing (Purohit and Purohit, 2010).

Although solar dish/Stirling technology is one of the oldest solar technologies, only a few dish/Stirling systems have been developed to commercialization: the Euro Dish from Schlaich-Bergermann und Partner (SBP), the 'SunCatcher' developed by Stirling Energy Systems (SES), and the 'PowerDish' by Infinia (Kongtragool and Wongwises, 2003). There are a number of past and current demonstration projects, mostly in Europe, Japan, Australia and the United States (Abbas, 2009; Abbas et al., 2011; Ab Kadir et al., 2010). Several realized dish/Stirling system installations are described in (Klaiß et al., 1995; Poullikkas et al., 2010). The 1-MW Hainan Nanshan Sanya Pilot project in China , is one of the largest dish/Stirling projects in operation.

6.3.5 THERMAL STORAGE SYSTEMS

Solar plants without energy storage are typically limited to a capacity factor of about 25 per cent due to the diurnal solar cycle and weather. The capacity factors for CSP plants range from 25-75 per cent, depending on design and incorporation of TES. The lower end of the capacity factor range refers to systems with no thermal storage and the upper end for those with up to 15 hours of thermal storage. Solar thermal electricity (STE) production curves fit quite well with the demand curve as it can be seen in the Figure 6.4 where the global demand corresponds to a typical summer day in Spain. The source of this data is from the Spanish Electrical System Operator, REE, and the STE generation corresponds to the whole STE park in operation in July 2012. More than 50 per cent of the plants have a storage system (ESTELA, 2012).

TES is one of the key design features that can make CSP more attractive than many other renewable technologies such as solar PV and wind, because TES can smooth the daily fluctuations associated with solar power and extend electricity production into the evening peak hours or longer (Heath et al., 2009). Depending on the receiver technology, CSP systems can store the primary energy in thermal storage media such as concrete, molten salt, phase-change materials, or ceramic materials and produce electricity by feeding the power block with the stored heat overnight. This allows CSP systems to hold energy in storage until needed by the power grid, thereby providing an on-demand power source that is not confined to an instantaneous solar or wind resource.

Molten salt TES systems are currently state-of-the-art as sensible heat thermal energy storage (SHTES) media. At the end of 2011, 62 per cent of installed CSP systems in Spain used molten salt energy storage (Lovegrove et al., 2012). Molten salt flows like liquid water with the advantage that it remains a liquid at temperatures up to several hundred degrees Celsius. Carbonate molten salt can be used at temperatures up to 850°C, although the commercial nitrate molten salt is limited to temperatures below 600°C. Current CSP plants such as Andasol 1 in Spain use a molten nitrate salt with 60 per cent sodium nitrate ($NaNO_3$) and 40 per cent potassium nitrate (KNO_3). The nitrate blend has excellent heat capacity and viscosity, but must be kept above its freezing point of approximately 220°C. Furthermore, even at a typical salt price of 1 US$/kg, the quantity required for a large solar plant makes this an expensive component. One method to reduce molten salt requirements is to use cheaper filler materials such as rocks and sand. These materials form a fill through which the molten salt flows, and are inexpensive and widely available, compared to the molten salt. Brosseau et al. (2004) recommended the use of quartzite rocks in combination with silica sand as storage filler material (Brosseau et al., 2004).

The commercial molten salt TES design used at Andasol 1 uses a two-tank system where the oil HTF heats salt pumped from a cold tank and stores the hot salt in a hot tank until needed. This is known as an indirect system because the HTF itself is not stored, but rather exchanges heat with a separate thermal storage medium. One improvement over this design is the direct two-tank system, which uses molten salt as both HTF and storage fluid (Figure 6.5). The advantage of this concept compared to the indirect two-tank TES systems is the avoidance of an expensive oil-to-molten-salt heat-exchanger, greater efficiency and flexibility in TES system dispatch, and higher operating temperatures obtainable with molten salt in comparison to oil-based HTF. Analysis indicates that trough plants operating in this fashion could produce power at 14-40 per cent lower cost than current oil-HTF designs (Turchi et al., 2010a) if they can avoid corrosion and freeze-risk issues associated with running a molten salt HTF. Torresol Energy's Gemasolar power tower in Spain uses this design. The use of molten salt as HTF is under development for parabolic trough technology, but is not yet ready for commercial development. Although there is a pilot plant in Italy, there are still some hurdles to overcome before utility-scale plants can be deployed. Some of these hurdles include identifying valves and joints that will work in a corrosive environment at high temperature and pressure, developing a robust freeze protection system, such as heat tracing in the solar field and developing a recovery system (Price and Kearney, 2003).

FIGURE 6.4

Solar thermal electricity (STE) production curves for Spain

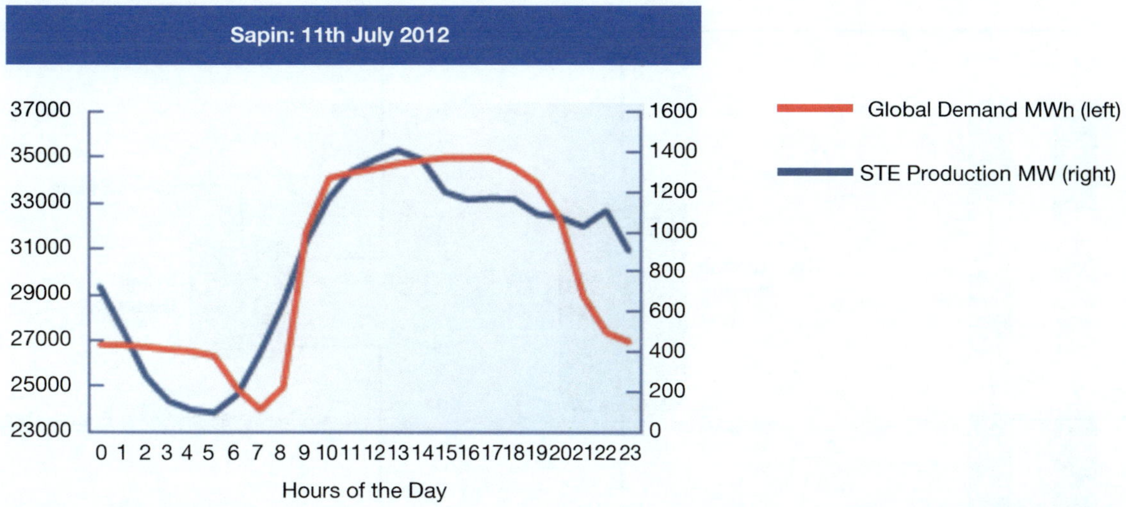

Source: ESTELA, 2012

FIGURE 6.5

Schematic of direct two-tank concept

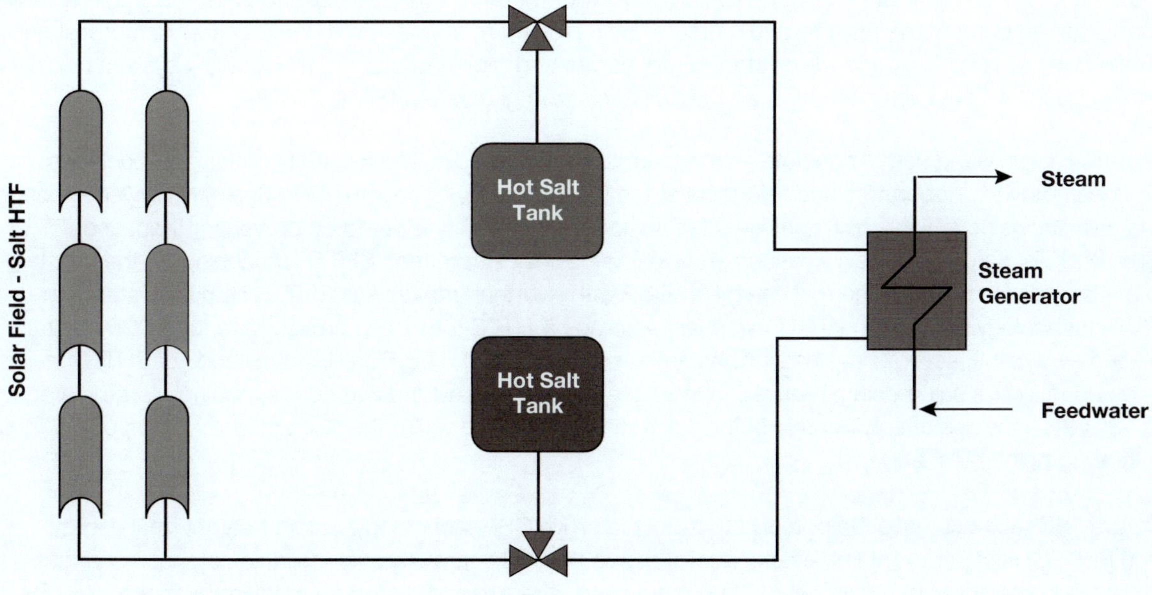

Source: Brosseau et al., 2004

FIGURE 6.6

Schematic of direct one-tank thermocline concept

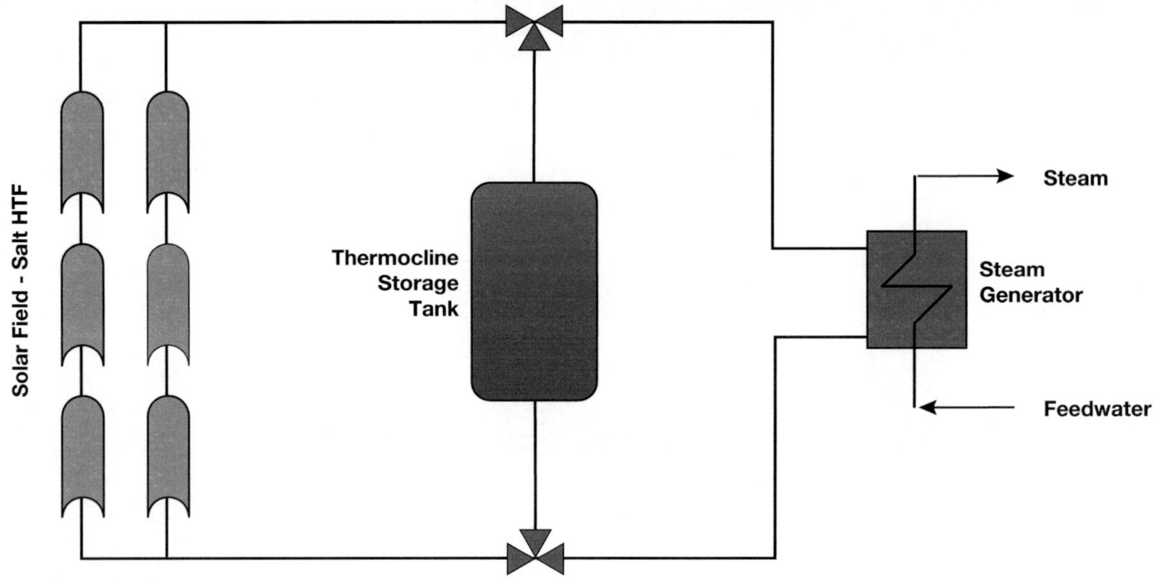

Source: (Brosseau et al., 2004)

A further cost reduction is offered by one-tank thermocline storage (Figure 6.6). The thermocline storage system uses a single tank that is only marginally larger than one of the tanks in a two-tank thermal storage system. With the hot and cold fluid in a single tank, the thermocline storage system relies on thermal buoyancy to maintain thermal stratification. A low-cost filler material, which is used to pack the single storage tank, acts as the primary thermal storage medium. By replacing molten salt with inexpensive fill material and eliminating one storage tank and the related pump, valves and piping costs, the thermocline system could potentially be 20-40 per cent cheaper than the two-tank storage system (EPRI, 2010).

Another alternative is the use of concrete as the thermal storage media. The feasibility of this approach has already been proven in laboratory and field tests at the Plataforma Solar de Almeria in Spain. However, although the material shows no critical challenges thus far, the long term stability is yet to be proven for thousands of charging and discharging cycles (Herrmann and Kearney, 2002). At present, SHTES technology is the only large-scale method for storing solar thermal energy for CSP with two-tank molten salt SHTES being the standard CSP TES technology. Sensible heat thermal energy storage (SHTES) has been proven to reduce CSP capital costs, and as a result, several new installations have implemented SHTES. The disadvantages of SHTES include the large quantities of salt medium required to store the thermal energy, the two correspondingly large storage tanks, as well as the risk of solidification of the salt if the temperature within the storage tanks drops below the salt's freezing point (Mills, 2004).

Alternative TES methods have the potential to further improve CSP technology. Latent heat thermal energy storage (LHTES) is of particular interest because it stores thermal energy through solid-liquid phase changes and, when compounds with a high heat of fusion are used, it requires less storage medium relative to SHTES, reducing capital and construction costs. Because LHTES is designed to undergo melting and solidification during sequential charging and discharging processes within a single storage unit, it eliminates the need for two separate tanks and the freezing risk encountered by two-tank SHTES. The primary disadvantage of LHTES is the low thermal conductivity that characterizes many phase-change materials (PCMs), leading to potentially slow discharging and charging rates, as well as a reduced thermodynamic efficiency as defined by low maximum cycle temperatures of the CSP power block. As a result, much research involving LHTES has been aimed at circumventing the high thermal resistance posed by the PCM.

Medrano et al. (2009) considered several LHTES designs for small-scale, low temperature applications. They tested different heat exchanger designs to reduce the thermal resistance of the LHTES (Medrano et al., 2009). It was concluded that a double-pipe PCM-HTF heat exchanger with an embedded graphite matrix worked well for low temperature, paraffin-based PCMs. Laing et al. proposed a higher temperature LHTES unit for a commercial scale CSP operation using direct steam generation (Laing et al., 2010). Their research included the design and testing of a LHTES using sodium nitrate as the PCM that was in turn enclosed in a vertically oriented tank with hexagonal finned tubes carrying water as an HTF. In this particular design, the water, pressurized to 100 bars, is vaporized in the LHTES unit to create steam for electric power production. A 1 MWh system was successfully constructed and field-tested.

Several researchers have investigated the integration of heat pipes into a PCM/HTF heat exchanger to reduce thermal resistances, increase heat transfer rates, and minimize temperature differences between the freezing or melting PCM and the HTF. This approach has been demonstrated experimentally using a paraffin-based PCM by Robak et al., resulting in a 180 per cent increase in the solidification rate and 60 per cent increase in the melting rate when several heat pipes were embedded in a PCM, thereby thermally connecting the PCM with the HTF (Robak et al., 2011). The quantitative studies of Shabgard et al. provide additional evidence that heat pipes or thermosyphons can decrease thermal resistances in high temperature LHTES (Shabgard et al., 2010). However, to the authors' knowledge, it has not been shown how this approach could be implemented at a scale commensurate with CSP.

6.3.6 WET/DRY COOLING TECHNOLOGIES

Water consumption at CSP plants is critically important because such plants are typically located in dry areas. Water is required for mirror washing, steam cycle maintenance and often cooling. Air cooling is technologically feasible but lowers the plant efficiency and increases capital costs by roughly 10 per cent (USDOE, 2009, 2012).

In arid and semi-arid regions with intense water demand, water supply is not an issue exclusive to CSP; it is relevant for siting any thermo-electric power plant. The trend is toward more freshwater-efficient cooling technologies for CSP and other thermo-electric generation. The concern about water footprint associated with CSP facilities arises from the fact that the use of wet cooling can consume more water per unit of electricity generated than traditional fossil fuel facilities with wet cooling. For example, coal-fired power plants consume 2-3 m^3/MWh, whereas CSP plants may consume up to 3.8 m^3/MWh (Table 6.3) (Burkhardt et al., 2011). Although options that reduce the freshwater consumed by CSP and other thermo-electric facilities exist, the available dry cooling alternatives, often reduce the quantity of electricity produced and increase electricity production costs. The quantity of electricity produced at these facilities, the water intensity per unit of electricity generated, and the local and regional constraints on freshwater will shape the cumulative effect of CSP deployment on water resources and the long term sustainability of CSP as a renewable energy technology. Water resource constraints may prompt the increased adoption of dry cooling technologies or cause decision makers to reject CSP facilities at certain locations (Carter and Campbell, 2009). CSP plants are already moving to dry cooling, as demonstrated by the Ivanpah and Crescent Dunes power tower projects in the United States.

Currently, most CSP plants, whether parabolic trough, linear Fresnel, or power tower, use the steam Rankine thermal power cycle. This cycle generally consists of pressurized water boiled using either direct solar heat or via an HTF, with the produced steam being used to drive a steam turbine to generate electricity. The cycle is completed and repeats with the condensing of the steam and repressurizing of the water. In the Rankine cycle, the steam is preferably condensed using external cooling water and an evaporative cooling tower. The cooling water used to condense the steam represents over 90 per cent of all water consumption at such a plant and is thus the target for reductions in water use.

FIGURE 6.7

Schematic of wet and dry cooling systems for CSP plants

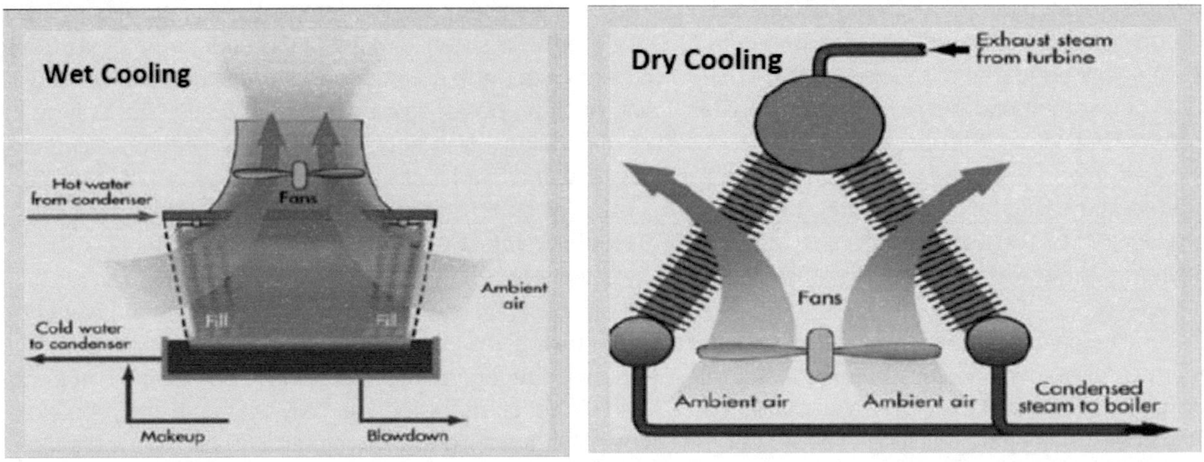

At present, dry cooling and hybrid wet/dry cooling systems are being considered for both fossil and CSP generating plants due to water limitations (Figure 6.8). The technical challenges and performance limitations facing CSP are comparable to those of new fossil and nuclear power generating plants. Dry cooling methods are becoming increasingly common for thermal power plants. The disadvantages of dry cooling include higher capital costs, higher auxiliary operating power requirements, fan noise, and an overall lower plant performance, particularly on hot days when the demand for peak power is highest (Kelly, 2006). The relative cost impact to CSP is similar to that of fossil fuel power plants (USDOE, 2009).

Dry cooling, or air cooling, reduces plant efficiency because the cooling temperature is never as low as for a wet-cooled system. In a Rankine steam cycle, the heat input is at a high temperature and rejected at a low temperature; these are called the source and sink temperatures, respectively. The efficiency of the cycle, defined as the ratio of the turbine work output to the heat input, is a function of the difference between the source and sink temperatures. Raising the source temperature or lowering the sink temperature will increase the cycle efficiency. Thus the performance of a power tower that operates at a higher steam temperature will be penalized less by air cooling than current trough plants or linear Fresnel designs. Dry cooling will reduce water consumption to near zero for the heat rejection system of a Rankine power system, leaving only a minimal amount of water for boiler blowdown, mirror washing and miscellaneous domestic plant uses. A dry-cooled trough plant requires about 80 gal/MWh for cycle make-up and mirror washing (WorleyParsons, 2008). In comparison, a wet-cooled plant requires 800 gal/MWh or more (Cohen et al., 1999).

Hybrid wet-dry systems allow the plant to maintain design or near-design performance, while having much lower water usage than a wet evaporative cooling system, albeit at a higher cooling system cost in comparison to wet cooling. A recent study conducted by NREL indicated that, for Rankine-cycle power plants, dry cooling has the potential to reduce water consumption by more than 90 per cent. In the case of parabolic trough plants, the levelised cost of electricity (LCOE) was estimated to increase approximately 2.5-8 per cent, depending on the exact site in south-western United States. Hybrid cooling was found to reduce the LCOE increase, but at a higher capital cost and operational complexity (Turchi et al., 2010b). Interestingly, the overall footprint of the dry-cooled plant did not increase because the additional solar field area required to offset the decrease in efficiency was balanced by the elimination of cooling tower evaporation ponds. Table 6.4 summarizes the advantages and disadvantages of various cooling methods.

TABLE 6.4

Advantages and disadvantages of various cooling methods

Type	Advantages	Disadvantages
Wet cooling	• Lowest cost • Low parasitic loads • Best cooling, especially in arid climates; governed by wet-bulb temperature	• High water consumption • Water treatment and blowdown disposal required
Air cooling	• Near zero water consumption • No water treatment required • Lower operation and maintenance costs	• More expensive equipment • Higher parasitic loads • Poorer cooling governed by dry-bulb temperature
Hybrid cooling	• Less water consumption • Potentially less expensive than dry cooling • Maintains good performance during hot weather	• Complicated system involving wet and dry cooling • Same disadvantages of a wet system, but to a lesser degree

Source: Turchi et al., 2010b

With technological progress, these alternatives can become more efficient than and economically as attractive as conventionally cooled plants. Even without such developments, however, the cost penalties associated with reducing water consumption appear to be relatively minor. Damerau et al. observed that the sustainability of CSP does not depend on technical limitations or major economic penalties (Damerau et al., 2011). Instead, it will likely depend on political regulation and governance to ensure an ecologically sound development that matches the appropriate technologies with different locations' precise needs.

6.3.7 LAND USE

A recent study empirically estimated land use for CSP facilities larger than 1 MW installed capacity located in the United States using a combination of project documents and analysis of satellite imagery (Ong et al., 2013). Note that this is just the land occupied by the generation facility and does not consider any land used for upstream processes. Ong et al. (2013) quantified land use requirements on both a capacity (area/MW) and generation (area/GWh) basis. Measuring on a generation basis enables consideration of differences in solar resource, variations in concentrating technology, and thermal storage options, thus providing the fairest point of comparison to other generation technologies. CSP energy production was simulated using the System Advisor Model for each solar project using inputs specific to individual facilities such as plant configuration, technology, and storage (Ong et al., 2013). Ong et al. (2013) also evaluated the impact of various CSP technologies and implementation of multi-hour thermal storage on land use requirements.

Ong et al. (2013) obtained land use data for a total of 25 built, under construction, or proposed projects representing approximately 3.7 GW capacity, alternating current. Table 6.5 summarizes the findings. 'Total area' is defined as the physical project boundary, and 'direct area' is the area covered by mirrors, turbines, thermal storage and any other systems as well as any structures that encumber the land. Direct area is always smaller than and within the total area. The capacity-based average results represent capacity-weighted averages. Figure 6.8 shows the distribution of results for CSP land use.

Table 6.5 summarizes descriptive statistics for total and direct land area for all CSP facilities in the United States, organized by CSP technology. Note that there are significantly fewer CSP installations in the United States than PV installations; this difference limits the robustness of conclusions, and in particular, there are very few projects using certain CSP technologies. On average, based on evidence from the United States, tower systems have the greatest CSP land use requirements on both a capacity and energy basis. Towers use 50 per cent more direct land than troughs on a capacity (MW) basis, and use 20 per cent more on a generation basis.

FIGURE 6.8

Distribution of estimates of total and direct land use for CSP projects in the United States

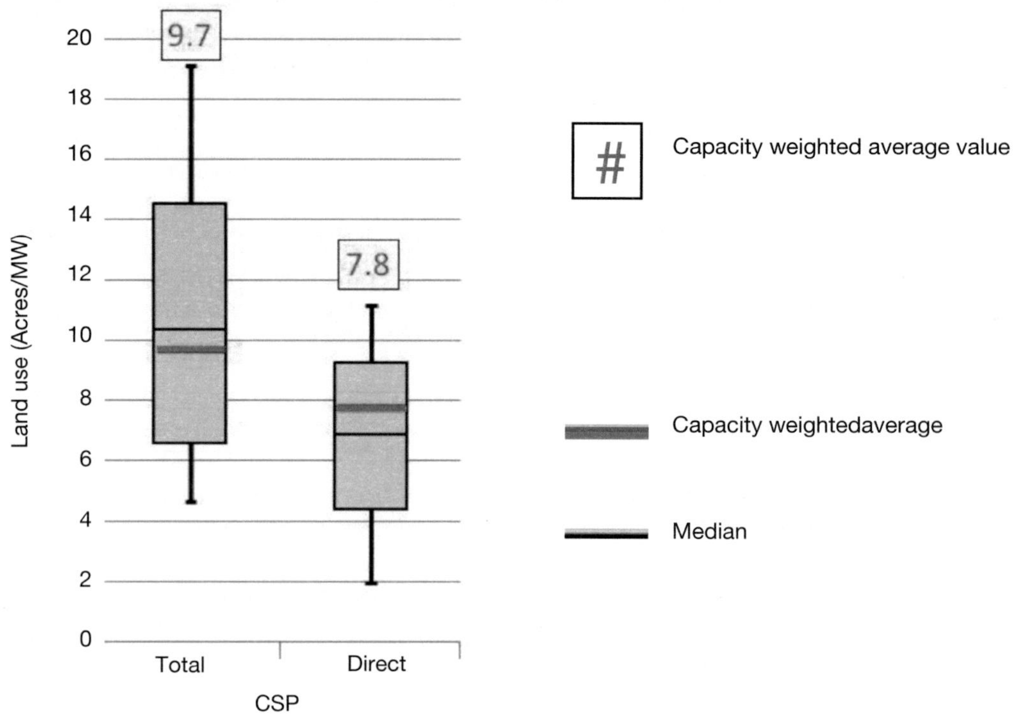

The bar ends indicate the maximum and minimum values, and the top and bottom of the box are the 75th and 25th percentile values, respectively.

Source: Ong et al., 2013

TABLE 6.5

Land used by CSP facilities in the United States

System Type	Total area				Direct area			
	Projects	Capacity (MW)	ha/MW	ha/GWh	Projects	Capacity (MW)	ha/MW	ha/GWh
All	25	374	4.0	1.4	18	2218	3.1	1.1
Trough	9	1380	3.8	1.6	7	851	2.5	1.0
Tower	14	2358	4.0	1.3	9	1358	3.6	1.1
Linear Fresnel	1	8	1.9	1.6	1	8	0.80	0.68
Dish Stirling	1	2	4.0	2.1	1	1.1	1.1	0.61

Source: Ong et al., 2013

6.4 REVIEW OF LIFE CYCLE ASSESSMENTS

For CSP plants, the environmental consequences vary depending on the technology. In general, GHG emissions and other pollutants are reduced without incurring additional environmental risks. Each square metre of CSP concentrator surface is enough to avoid the annual production of 0.25 to 0.4 tons of CO_2. The energy payback time of CSP systems can be as low as five months, which compares favourably with their lifespan of 25-30 years. Most CSP solar field materials can be recycled and reused in new plants (Geyer and Agency, 2008).

Like many other renewable energy systems, the harmful effects that result from the deployment of CSP technology are limited. Specifically, CSP produces significantly fewer GHG emissions during its lifespan when compared to traditional energy systems such as coal and natural gas (Burkhardt et al., 2011). The majority of GHG emissions originate from energy-intensive processes involved in the extraction of raw materials and the manufacture of components required in CSP plants (Burkhardt et al., 2011). Although CSP plants typically require a significant investment in specialized equipment, most solar field materials can be recycled and reused in new plants (Geyer and Agency, 2008). Land consumption and impacts on local flora and wildlife during the construction of the solar field and other facilities are the main environmental issues for CSP systems (Pregger et al., 2009). At design operating conditions, CSP technology produces approximately 100 kWh of electricity per square meter of land used each year, or approximately 400 kWh per square meter of reflective surface (Burkhardt et al., 2011).

6.4.1 SYSTEM BOUNDARIES AND METHODOLOGIES

Life cycle assessment is recognized as a holistic and standard approach for quantifying environmental impacts of renewable energy systems. LCAs account for the impacts resulting from all activities occurring over the life of a power plant, including those that are upstream and downstream from the operational phase (Burkhardt et al., 2012). In order to accurately assess the environmental impacts of a CSP plant, one must conduct a comprehensive LCA study using site-specific environmental data and a robust LCI. A comprehensive LCA should assess the impact of each life cycle phase of a CSP plant. These phases typically include manufacturing, construction, operation and maintenance, dismantling, and disposal (Burkhardt et al., 2011). It is also beneficial to disaggregate the LCA results into the primary plant subsystems in order to identify hot spots. For example, in Burkhardt et al. (2011), the life cycle impacts are disaggregated into the heat transfer fluid, solar field, TES, and power plant systems.

6.4.2 REVIEW OF LIFE CYCLE ASSESSMENTS

6.4.2.1 Overview of existing studies

As mentioned above, lifecycle assessment (LCA) is a standard method to holistically quantify the environmental impacts of renewables over their entire lifetime, including material extraction and manufacture, construction, operation and decommission. Energy use and emissions in material processing and manufacture of solar thermal systems are considerable (Chen et al., 2011). The impacts of these emissions vary according to location, and are fewer than those of conventional fossil fuel technologies. GHG emission evaluations for solar electric power plants have been carried out as early as 1990 (Kreith et al., 1990). There is large variation in the published LCAs because of the difference in technologies, plant locations, assessment scopes and performance assumptions. Table 6.6 reviews environmental assessment studies of CSP systems. Norton et al. compared the environmental impacts of different renewable energy technologies and estimated the lifecycle GHG emissions for the CSP technologies considered to be in the range of 21 to 80 g CO_2 eq./kWh_e (Norton et al., 1998). Weinrebe et al. (1998) calculate material-based GHG emissions for central receiver and parabolic trough plants of 25 g CO_2 eq./kWh_e and 17 g CO_2 eq./kWh_e, respectively. Lenzen uses process-based and input-output analyses to evaluate the life-cycle emissions of solar-thermal electricity, with results ranging from 30-150 g CO_2 eq./kWh_e (Lenzen, 1999). Variations depend mainly on plant size and capacity factor, which in turn depends on the integration with a fossil-fuelled back-up system. Combined natural gas-central tower and natural gas-parabolic troughs power plants are compared with each other, with life-cycle emissions of 202 and

196 g CO_2 eq./kWh$_e$, respectively (Lechón et al., 2008). Most of the GHGs being released during operation were found to be due to natural gas combustion.

Pehnt (2006) investigated a dynamic approach towards the LCA of renewable energy technologies and proves that for all renewable energy chains, the inputs of finite energy resources and emissions of GHGs are extremely low in comparison to the conventional system. With regard to the other environmental impacts, the findings do not reveal any clear verdict for or against renewable energies. Martin (1997) evaluated the potential GHG savings that could result from the deployment of solar generation technologies in utility systems in the United States. Total fuel cycle analyses were performed for several renewable and conventional generation technologies to estimate the total GHG emission contribution from each generation technology. The penetration of solar energy technologies can lead to substantial GHG offsets over the life of a given resource portfolio (Martin and Shaw, 2010). The TRANS-CSP study focuses on the interconnection of the electricity grid of Europe, the Middle East and North Africa (EUMENA) with the goal of supplying about 15 per cent of the European electricity demand with solar energy imports from the south by the year 2050 (Lohmann, 2006). One important result of the LCA and eco-balance is that each installation consisting of a solar thermal power plant and associated the HVDC line causes distinctly less pollution than the reference electricity mix, even when using the enhanced electricity mix of 2030 as the reference.

Lechón et al. (2008) evaluated the environmental impacts of the electricity produced in a 17 MW solar thermal plant with central tower technology and a 50 MW solar thermal plant with parabolic trough technology to identify the system improvement opportunities that would reduce their environmental impacts, and to evaluate the environmental impact resulting from compliance with the solar thermal power objectives in Spain. Environmental impacts analysed include the global warming impacts throughout the whole life cycle of the power plants, which were 202 g CO_2 eq./kWh$_e$ for central tower system and 196 g CO_2 eq./kWh$_e$ for parabolic trough technology (Lechón et al., 2008).

Jacobson (2009) reviews and ranks major proposed energy-related solutions to global warming, mortality due to air pollution, and energy security while considering other impacts of the proposed solutions, such as impacts on water supply, land use, wildlife, resource availability, thermal pollution, water chemical pollution, nuclear proliferation, and undernourishment. It is observed that use of wind, CSP, geothermal, tidal, PV, wave, and hydropower to provide electricity for transport, residential, industrial, and commercial sectors will result in the greatest benefit among the options considered. Compared to other technologies, the life-cycle GHG emissions per kWh produced determined using a mass and energy balance method, are relatively low, at 8.5 to 11.3 g CO_2 eq./kWh$_e$ (Jacobson, 2009). Cavallaro and Ciraolo (2006) provided a preliminary environmental assessment of a parabolic dish solar thermal power plant. For parabolic trough, only 13.6g CO_2 eq. is emitted for each kWh of produced electricity. Ordóñez et al. (2009) estimated life cycle GHG emissions of 89.4 g CO_2 eq./kWh for a dish/Stirling system in Spain. The GHG emissions for the first MW-class solar tower power plant in China is calculated as 0.04 kg CO_2 eq./MJ, which is similar to the results reported by (Chen et al., 2011; Lenzen, 1999).

Laing et al. (2010) carried out economic analysis and life cycle assessment of concrete TES system for parabolic trough power plants. Concrete-based storage shows an advantage over two-tank molten salt storage in terms of environmental impacts. Viebahn et al. (2011) presented a dynamic LCA of CSP in Africa and Europe. In the study, individual LCI are calculated for present, 2025 and 2050 scenarios, with the 'dynamic' LCI analysis accounting for six development steps - lifetime, up-scaling, increase of storage time, efficiencies, material learning curve and adapting background processes. Although the GHG emissions of current CSP systems operating purely on solar power have low emissions of 31 g CO_2 eq/kWh$_e$ compared with advanced fossil fuel systems with corresponding 130--900 g CO_2 eq/kWh$_e$ emissions, they could further be reduced to 18 g CO_2-eq/kWh$_e$ in 2050, including transmission from North Africa to Europe (Viebahn et al.,

2011). Burkhardt et al. (2011) analysed the expansion of parabolic trough power plant located in California in terms of GHG emissions, water consumption, cumulative energy demand, and payback time. During its lifecycle, the reference 103 MW capacity CSP plant is estimated to emit 26 g of CO_2 eq/kWh, consume 4.7 litres water/kWh, and demand 0.40 MJ/kWh of energy, resulting in an EPBT of approximately one year. The dry-cooled alternative is estimated to reduce life cycle water consumption by 77 per cent but increase life cycle GHG emissions and cumulative energy demand by 8 per cent. Synthetic nitrate salts may increase lifecycle GHG emissions by 52 per cent compared to mined salts. Switching from two-tank to thermocline TES configuration reduces lifecycle GHG emissions, particularly for plants using synthetically derived nitrate salts. Although CSP can significantly reduce GHG emissions compared to fossil-fuelled generation, dry cooling may be required in many locations to minimize water consumption.

Burkhardt et al. (2012) developed and applied a systematic approach to review LCA literature addressing CSP systems, identify primary sources of variability between these assessments, and, where possible, reduce variability in GHG emissions estimates. The median estimates were 69 and 25g of CO_2 equivalent per kWh for current parabolic trough and central receiver systems, respectively. The difference is caused by the exclusion of auxiliary electricity consumption impacts in central receiver systems due to lack of data (Burkhardt et al., 2012). Although the results of LCA of CSP are very variable and are likely to change with further technological innovations, they are already far more positive than those of fossil fuels and solar PV (Burkhardt et al., 2011). Longer operation lifetimes, upscaling power loads, increasing storage times, increasing efficiencies, reducing material use and using renewable energy in the manufacturing process may substantially reduce future environmental impacts (Viebahn et al., 2011). Life cycle GHG emissions of CSP plants with TES are strongly dependent upon the source of salts. In the event that synthetic salts must be used, it would be beneficial to utilize a thermocline design, which can mitigate the increased burdens associated with synthetic salts. The LCA methodology is used as a tool to estimate the environmental impact of TES, particularly in solar power plants (Battisti and Corrado, 2005; Oró et al., 2012; Piemonte et al., 2011). These studies suggested that incorporating of PCM substantially reduces the overall environmental impact under the experimental conditions studied.

Zhang et al. (2010) explore the capacity of CSP as base load or intermediate and peak (I&P) plants, with or without thermal storage. They conclude that I&P plants, especially those with thermal storage, could be economically competitive for large scale development in coming years. Base load plants would require more policy incentives and significant capital cost reductions to undergo the same deployment. Known critical obstacles for the large-scale rollout of CSP include the choice of back-up scheme (Lenzen, 1999) and the generating costs (Lenzen, 2010). Intensive water consumption for CSP plant operation has been reported, which is important because direct solar radiation is most intensive in dry areas. Water is required for mirror washing, the steam cycle and cooling. Air cooling is technologically feasible but lowers the plant efficiency, increases its land use and increases capital costs by 10 per cent (US DOE, 2007). Solar thermal power under direct normal insolation and with trackers, has higher land transformation than fossil-fuelled power plants (Fthenakis and Kim, 2009). Some parabolic trough systems use a heat transfer fluid called Therminol that is classified as hazardous in some U.S. states. Leakages might therefore have an important toxicity potential considering the volume of this substance used in solar thermal power plants (EPRI, 1997; Gamble and Schopf, 2010).

Desideri et al. (2013) presented a comparative analysis of CSP and photovoltaic technologies and observed that the CSP plant shows a better environmental profile than the PV plant. This study considers a CSP plant with parabolic trough collectors using water as transfer fluid, with neither thermal storage nor hybrid operation compared with a ground-mounted PV plant with a single-axis tracking system and mono-crystalline silicon modules. CO_2 emissions using GWP 100 and energy payback period related to the CSP technology were lower than those of the PV technology. Referring to the whole life cycle of the plants, Desideri et al. (2013) estimated GHG emissions at 29.9 g CO_2 eq./kWh for the CSP plant and 47.9 g CO_2 eq./kWh for the PV plant.

TABLE 6.6

Overview of environmental assessment studies of CSP systems

Source	Technology				Environmental impact categories[1]	Specific impacts
	Parabolic trough	Linear Fresnel	Central receiver	Dish-Stirling		
Becerra-Lopez and Golding (2007)	X				1,2,3,4,5,11	CO_2, CH_4, N_2O, SO_2, NO_x emissions in construction and operation
Burkhardt et al. (2011)	X				1,2,3,9	GHG emissions, water consumption, energy payback period
Cavallaro and Ciraolo (2006)				X	1,2,3,6,10,11,12	CO_2, global warming, ozone depletion, material, acidification, eutrophication, solid waste generation, etc.
Chen et al. (2011)			X		1,2,4	CO_2 emissions, energy cost
Desideri et al. (2013)	X				1,2,3	CO_2 emissions, energy payback period
EPRI (1997)	X			X	3,9,13	Land, water and critical material requirements
Fthenakis and Kim (2009)	X		X		13	Land transformation
Hernández-Moro and Martínez-Duart (2012)	X		X		2	Levelized costs of energy (LCOE)
Jacobson (2009)	X	X	X		1,2,3,13	Global warming, air pollution mortality, energy security, water supply, land use, wildlife, resource availability, thermal pollution, water chemical pollution
Kreith et al. (1990)			X		1,2,3	CO_2 emissions, energy analysis
Lechón et al. (2008)	X	X			1,2,3,4,5,6,7,9, 10,11	Global warming, abiotic depletion, ozone layer depletion, human toxicity, fresh water aquatic ecotoxicity, marine aquatic ecotoxicity, terrestrial ecotoxicity, photochemical oxidation, acidification, eutrophication
Lenzen (1999)	X		X	X	1,2,3,4	GHG costs, material inventories, monetary cost breakdown
Norton et al. (1998)	X		X	X	1,2,5,11	CO_2, SO_2, NO_x emissions
Ordóñez et al. (2009)				X	1,2,4	CO_2 emissions

Table 6.6 Overview of environmental assessment studies of CSP systems (continued)

Source	Technology				Environmental impact categories[1]	Specific impacts
	Parabolic trough	Linear Fresnel	Central receiver	Dish-Stirling		
Pehnt (2006)	X				1,2,3,5,8,11	Iron ore, bauxite consumption, CO, CO_2, N_2O, NO_x, SO_2, NMHC, HCl, NH_3, benzene and benzopyrene emissions, particles
Pihl et al. (2012)	X		X		4	Material intensity
US DOE (2009)	X	X	X	X	9	Water consumption
Viebahn et al. (2008a, 2011)	X	X			1,2,3,4,8,11,13	Primary energy demand, water, land requirements plus 71 emissions with a detailed focus on GHG, NOx, NMVOC, PM10, PM2.5, SOx (but not aggregated to impact categories)
Weinrebe et al. (1998)	X		X		1,2	GHG costs, material inventories
Zhang et al. (2012)			X		1,2,3	Energy yield ratio (EYR), Environmental loading ratio (ELR), Environmental sustainability index

Grey rows denote non-LCA studies. Key for environmental impacts: [1]CO_2 emissions; [2]Global warming, climate change; [3]Cumulative energy demand, total energy demand, primary energy, energy payback time/ratio; [4]Abiotic depletion, resource requirements, material intensity, non-renewable resource depletion; [5]Acidification; [6]Ozone depletion; [7]Human toxicity; [8]Particulate matter formation, particles/dust; [9]Ecotoxicity, freshwater and marine aquatic ecotoxicity, sediment ecotoxicity, terrestrial ecotoxicity; [10]Photochemical ozone creation, photochemical oxidation, photochemical ozone formation, photochemical oxidant formation, smog; [11]Eutrophication, nutrient enrichment; [12]Solid waste generation; [13]Land use, land occupation, land transformation.

6.4.3 RANGE OF RESULTS FOR SELECTED IMPACT CATEGORIES

6.4.3.1 Greenhouse gas emissions

The National Renewable Energy Laboratory (NREL) conducted an extensive meta-analysis on LCAs of renewable energy systems, which included CSP technology (Burkhardt et al., 2012). In the study, 42 life cycle GHG emissions estimates were identified in thirteen unique references for parabolic trough, Fresnel, power tower, and dish/Stirling technologies. Figure 6.9 displays distributions of published estimates of life cycle GHG emissions (Arvizu et al., 2011). Although the majority of published estimates for life cycle GHG emissions fall between 14 and 32 g CO_2 eq./kWh, there are certain outliers that were screened out in Figure 6.9 because they modelled plants combusting natural gas to generate additional electricity; these facilities operated as hybrid solar-natural gas facilities. For example, the maximum published estimate of life cycle GHG emissions otherwise passing quality screens are 230 and 190 g CO_2 eq./kWh for parabolic trough and power tower, respectively (Burkhardt et al., 2012).

FIGURE 6.9

Life cycle GHG emissions (g CO_2-eq/kWh) of CSP technologies

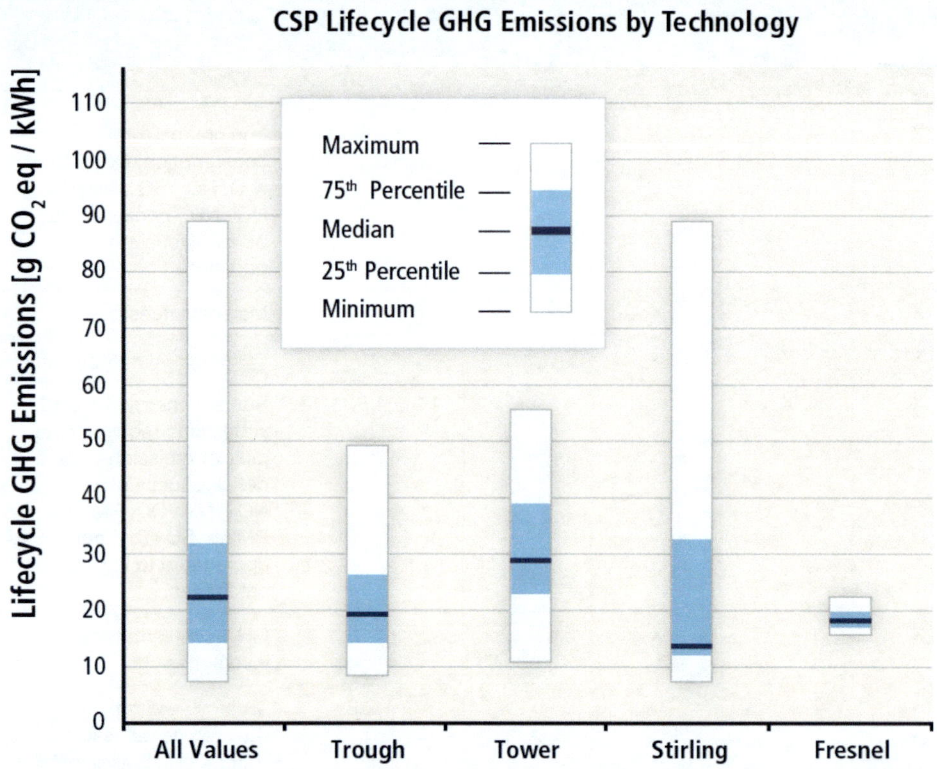

	All Values	Trough	Tower	Stirling	Fresnel
Estimates:	42	20	14	4	4
References:	13	7	5	3	1

Source: Arvizu et al., 2011

There are several key parameters pertaining to plant design and LCA system boundaries that can have a significant impact on the estimate of life cycle GHG emissions. Using a process called 'harmonization', Burkhardt et al. (2012) significantly reduced the variability in life cycle GHG emissions across quality-screened literature. The harmonization process requires that key assumptions leading to variability between life cycle GHG emissions estimates are identified and that their values are standardized. In the case of CSP, parameters that can cause such variability include the amount of on-site natural gas combustion and electricity consumption, available DNI, plant lifetime, and solar-to-electric efficiency (Burkhardt et al., 2012). After harmonization, the range of life cycle GHG emissions for trough and tower, including those that had considered hybrid facilities, were reduced by 82 per cent and their median values reduced by 17 per cent and 38 per cent, respectively. Although the work conducted by Burkhardt et al. can better inform LCA practitioners and policy makers of the general environmental impacts of CSP, including GHG emissions, the most accurate estimate of life cycle GHG emissions associated with a specific plant design is obtained by conducting a comprehensive LCA study using site-specific information (Burkhardt et al., 2012).

6.4.3.2 Air pollution

Most LCA studies actually concentrate on life cycle GHG emissions without addressing other emissions or resource use. Viebahn et al. provided 71 inventory items for different types of CSP power plants referring to the current status and to future projections in 2025 and 2050 (Viebahn et al., 2008a). A dynamic energy analysis undertaken by Becerra-Lopez and Golding (2007) to determine maximum allowable growth rates for alternative power-generation technologies and applied to the expansion of a regional system . Waste gas emissions for natural gas combined cycle (NGCC), wind turbine (WT), photovoltaic (PV), hybrid solar thermal parabolic trough (HSTPT), and solid oxide fuel cell (SOFC) power plants during the phases of construction and operation are estimated. SO_2 emissions by HSTPT plant were lower as compared to NGCC plant during the construction phase, at 6 ton/$_e$MW$_e$ and 10.5 ton/$_e$MW$_e$, respectively. However, SO_2 emissions from the HSTPT plant were higher in comparison to the WT, PV and SOFC plants, at 2.2, 4.6 and 3.3 ton/MW$_e$, respectively. NO_x emissions were very high during the construction and operation of HSTPT plant, at 8.1 and 227.5 ton/MW$_e$ in comparison to WT, PV and SOFC (Becerra-Lopez and Golding, 2007).

6.4.3.3 Water pollution and water consumption

Existing LCA studies on CSP usually do not consider emissions to water. However, neither direct nor cumulative life cycle emissions to water are expected to be significant. Water consumption, on the other hand, is an issue that has received a lot of attention. There have been large meta-analyses conducted that compare water consumed across both renewable and fossil-fuel based energy systems, in addition to independent LCA studies that consider water consumption (e.g. Burkhardt et al., 2011). Regarding meta-analyses, Macknick et al. (2011) screened and compiled estimates of consumption and withdrawal for water used to generate electricity for operation of power generation facilities only; it did not consider water used in other life cycle phases.

One LCA study evaluated the life cycle water consumption for both a wet-cooled and a dry-cooled parabolic trough plant design (Burkhardt et al., 2011). In the study, the life cycle GHG emissions, cumulative energy demand, and water consumption are estimated for a 103 MW$_e$ (net) parabolic trough plant with 6.3 EFLH of molten salt storage (indirect). The wet-cooled plant required 4.7 litres of water for each kilowatt-hour of electricity produced over its lifetime. The vast majority of the life cycle water consumption, corresponding to 72 per cent, or 3.4 l/kWh, could be attributed to the operational water requirement in the cooling tower, used for evaporative cooling. Another 10 per cent could be attributed to the energy-intensive processes associated with raw material extraction and component manufacture. In the dry-cooled design, the evaporative cooling tower is replaced by an air-cooled condenser (ACC) and wet surface air cooler; this modification reduces the power block's operational water consumption to only 23 per cent of the total, or 0.25 l/kWh. While the amount of water required for raw material extraction and manufacturing increases only slightly from 0.47 to 0.50 l/kWh, primarily due to the larger mass of materials embodied in the ACC, its contribution to the total life cycle water consumption is more significant at 45 per cent.

6.4.3.4 Soil pollution

Similarly to water, emissions to soil have thus far not been the focus of LCAs studying CSP. Most utility-scale parabolic trough systems currently in operation use a biphenyl/diphenyl oxide HTF such as *Therminol®* some states in the United States classify as hazardous. Leakages might thus have an important toxicity potential considering the volume of this substance in solar thermal power plants (EPRI, 1997). Although the amount of land occupied by a CSP plant is larger than that of a fossil fuel plant, both types of plants use about the same amount of land because fossil fuel plants use additional land for mining and exploration as well as road building to reach the mines (Fthenakis and Kim, 2009; USDOE, 2001). Trough systems require an even surface for installing the collectors, hence the need for earthwork. Central receiver systems can be built on more challenging terrain since each heliostat can be adjusted separately and therefore have a lower impact on land transformation and earth movements (Solar PEIS, 2012). Recent power tower projects in the United States have installed their heliostat fields without major land grading. Individual placement of parabolic dish systems enables greater flexibility than other CSP systems since dish systems can be placed on varied terrain with grades up to 5 per cent (TEEIC, 2010). Other solar energy systems require relatively level land with grades of 3 per cent or less.

6.5 LCA RESULTS

As of today, parabolic trough is the most commonly used CSP technology in utility-scale projects. The typical parabolic trough design uses synthetic oil HTF combined with an indirect molten salt storage system, typically with a capacity equivalent to 6 to 7.5 full load-hours of energy. The selection of dry or wet cooling technology is very location dependent. An LCA was conducted for this type of technology using LCI data based on actual commercial facilities. However, linear Fresnel technologies are also becoming commercially available. For some, central receiver technologies are the future, mainly due to the higher operation temperatures and efficiencies they can achieve. Some data from experimental plants or first commercial implementations exist. Alternative storage and HTF concepts are researched intensively but are mostly still in their infancy.

For ease of comparison to other energy-related technologies, LCI data and impact categories are normalized to a functional unit of 1 kWh generated.The life cycle impact assessment has been carried out for all nine regions of the model, however, only the results for the CSP reference region[4], Africa and the Middle East, are shown here.

6.5.1 LCA MODEL AND RESULTS

Figure 6.10 shows the environmental profile of the production of 1 kWh of electricity from a central tower plant with a 15 MW_e capacity, and from a 103 MW_e parabolic trough plant. Impact categories are compared to the environmental profile of the 2010 electricity grid mix of the Africa and Middle East region. For the tower plant: seven impact categories are under 10 per cent in 2010, while urban land occupation and metal depletion are 60 per cent and 400 per cent of the average grid mix, respectively. Metal depletion is thus the only impact that increases when using CSP-generated electricity rather than the current electricity mix. Regarding the production of 1 kWh of electricity from the 103 MW_e parabolic trough plant: out of the eighteen impact categories, the lowest sixteen are in the range 1.7–16 per cent in 2010. Land occupation and metal depletion are respectively 60 per cent and 220 per cent of the average grid mix.

FIGURE 6.10

Impacts of 1 kWh of electricity delivered at plant by Solar Tres

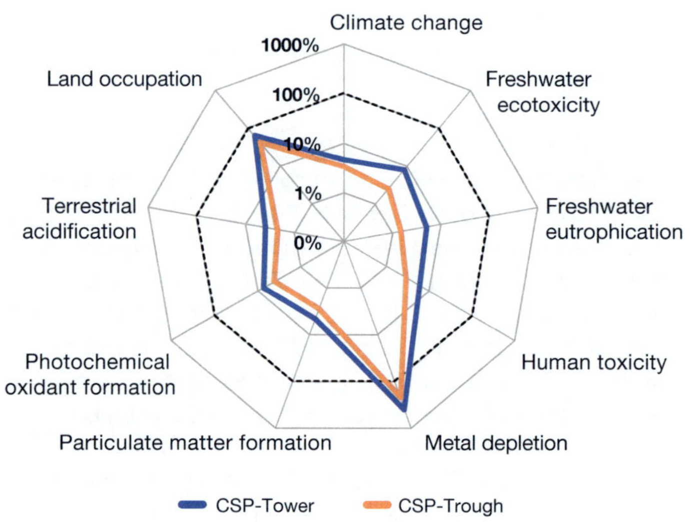

Solar Tres is a central tower 15 MW_{el} CSP plant, with 16 hour storage (two-tank indirect, mined molten salt), wet cooling, under Africa and Middle-East conditions (2400 kWh/m² direct normal insolation) and for years 2010, 2030 and 2050, compared to the grid mix of the same region in 2010.

4 Reference regions are chosen according to the highest capacity potential achievable in 2050 across regions.

6.5.2 CENTRAL TOWER

6.5.2.1 Life-cycle inventory

Central tower technologies infrastructure inventories were built from the existing plants of Solar Tres, in Spain (Viebahn et al., 2008b, 2008a) and the design for a prospective plant in Tucson, Arizona, United States (Whitaker et al., 2013). Their respective capacities are 15MW and 103MW. Non-infrastructure inputs cover the consumption of heat transfer fluid (HTF) and water for the water cooling systems. The HTF is assumed to be Therminol VP-1, a mixture of diphenyl oxide (DPO, 73.5 per cent) and biphenyl (26.5 per cent) (Burkhardt et al., 2011). The preferred cooling technology (dry or wet) is very location-dependent. The only limitation to inexpensive wet cooling option is water availability in some regions of the world. Dry cooling may save about 90 per cent of water consumption (Macknick et al., 2011), but is more energy-penalising than wet cooling, which virtually does not consume energy. Wet cooling was then chosen for both technologies, keeping in mind that water use is a major issue.

Contribution analysis

Contribution analysis improves understanding of what processes are most responsible for the environmental burden induced by the system. Figure 6.11 displays the contribution of life cycle phases to each of the environmental impact categories introduced in Figure 6.10. The next figures focus on one impact category at a time. Not all impact categories have been analysed; climate change and metal depletion are systematically selected for the contribution analysis, as well land occupation, human toxicity and freshwater eco-toxicity (which are deemed to be the most relevant categories), in comparison with the impacts of the current mix. The top ten contributing processes or sectors are shown, the rest of them being aggregated under the '"Rest' category.

Figure 6.11 shows that the collector, as well as the operation and maintenance phase, dominate all impacts but one. The collector consists mainly of steel, concrete and glass, in large amounts compared with the other elements of the plant. The operation and maintenance phase comprises electricity production, on-site transportation and fuel consumption. The 'site preparation and related' contributes to 95% of the life cycle, i.e. direct and indirect, land occupation.

Figure 6.12 shows the contribution of different processes of the central tower production system to climate change. The manufacture of fertilizers and nitrogen compounds appears as a principal contributor, due to the synthesis of nitric acid, which is a necessary precursor to the calcium nitrate and potassium nitrate components for the storage system. The large amounts of clinker used as a primary material to build the tower also contribute to the impact on climate change through the direct emission of GHG during the manufacturing process. Iron and steel are used in substantial amounts in the infrastructure, appearing as a third contributing category. Their share of impact decreases in later years due to improved industrial efficiency of iron and steel making processes. Finally, the electricity background and flat glass manufacturing categories are the two other significant contributors to impact climate change impacts.

The impact on metal depletion from the production of electricity by a central tower plant is mainly due to the extraction of the iron ore necessary to build its infrastructure (Figure 6.13). Ferronickel and copper concentrate also appear to contribute, due to their presence in the electrical installation. The impact does not decrease with time since no improvement in the use of metals during the plant construction has been modelled, and, unlike the climate change indicator, this indicator is based on mass-balance.

As seen in the environmental profile of the central tower plant, land occupation is a major issue (Figure 6.14). The direct use of land for the solar field is the main contribution to land occupation, with roughly half the impact. The "Silviculture and other forestry activities" category represents the amount of timber necessary to build the underground mines that provides hard coal and lignite for the production of steel and electricity. The third contributing category encompasses the construction processes, representing the land that is only occupied during the construction of the plant.

As seen in Figure 6.15, the impact on human toxicity has two major origins. The first is the treatment and disposal of hazardous waste that occur during the coal extraction process, which provides the raw material for background electricity production and iron and steel making. A third, minor, contribution stems from the use of copper in the power block and electrical installations at the plant. The impact peaks in 2030 because of the intensive use of fossil fuels, including the adoption of CCS, in the background energy mix forecasted at that time, followed by a higher share of renewables in 2050.

The impact on freshwater ecotoxicity, shown in Figure 6.16, is uniquely due to the discharge of toxic compounds to soil or water. Coal mining and smelter emissions are the main contributors to this impact category.

FIGURE 6.11

Contribution analysis of nine environmental impact categories from Solar Tres

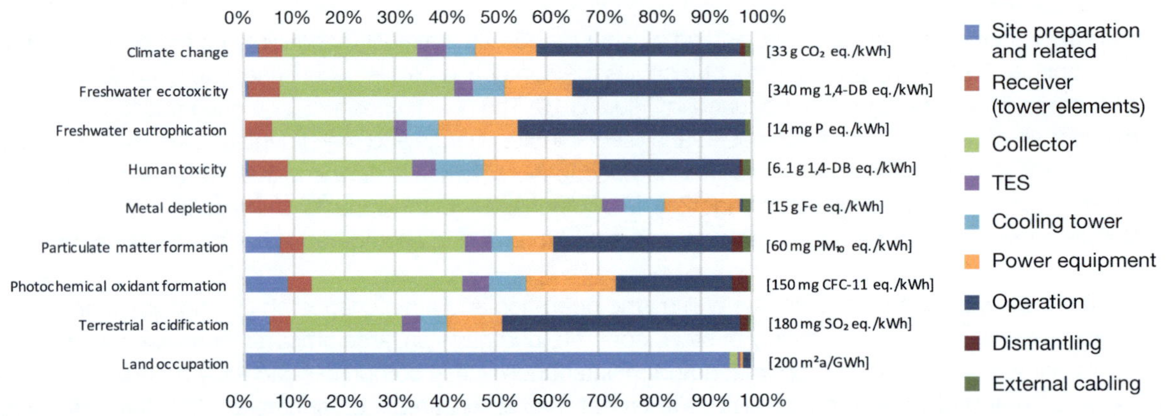

The contribution analysis of nine environmental impact categories, broken down by foreground process, for 1 kWh of electricity from Solar Tres, a central tower 15 MW_{el} CSP plant with 16 hours storage (two-tank indirect, mined molten salt), wet cooling, under Africa and Middle East conditions (2400 kWh/m² direct normal insolation).

FIGURE 6.12

Process contribution to the impact on climate change from Solar Tres

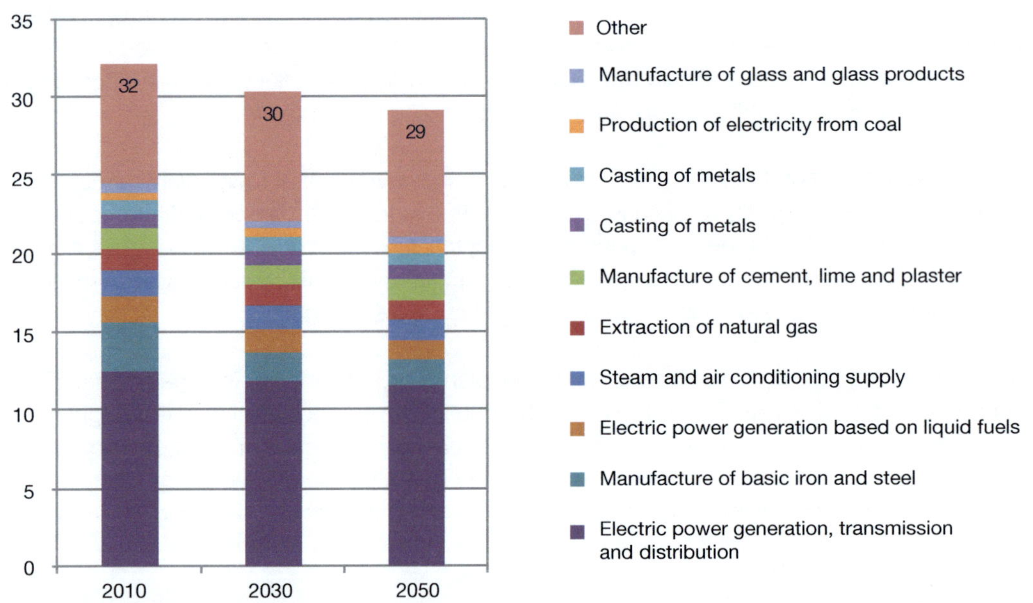

An analysis for 1 kWh of electricity from Solar Tres, a central tower 15 MW_{el} CSP plant with 16 hours storage (two-tank indirect, mined molten salt), wet cooling, under Africa and Middle East conditions (2400 kWh/m² direct normal insolation), in g CO_2-eq./kWh.

FIGURE 6.13

Highest contributing processes to the impact on metal depletion at Solar Tres

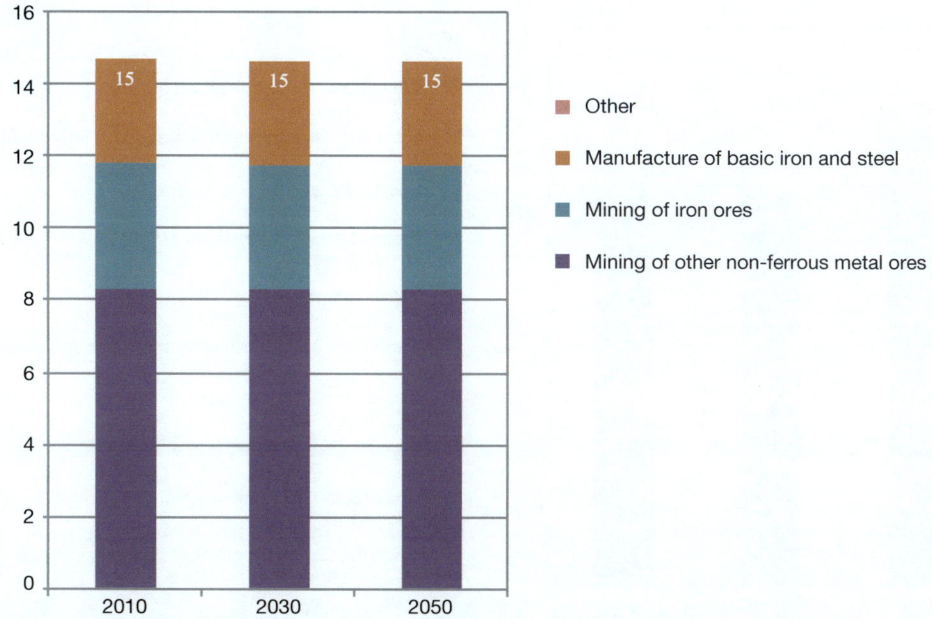

The highest contributing processes to the impact on metal depletion of 1 kWh of electricity from Solar Tres, a central tower 15 MW_{el} CSP plant with 16 hours storage (two-tank indirect, mined molten salt), wet cooling, under Africa and Middle East conditions (2400 kWh/m^2 direct normal insolation), in g iron eq./kWh.

FIGURE 6.14

Highest contributing processes to the impact on land occupation at Solar Tres

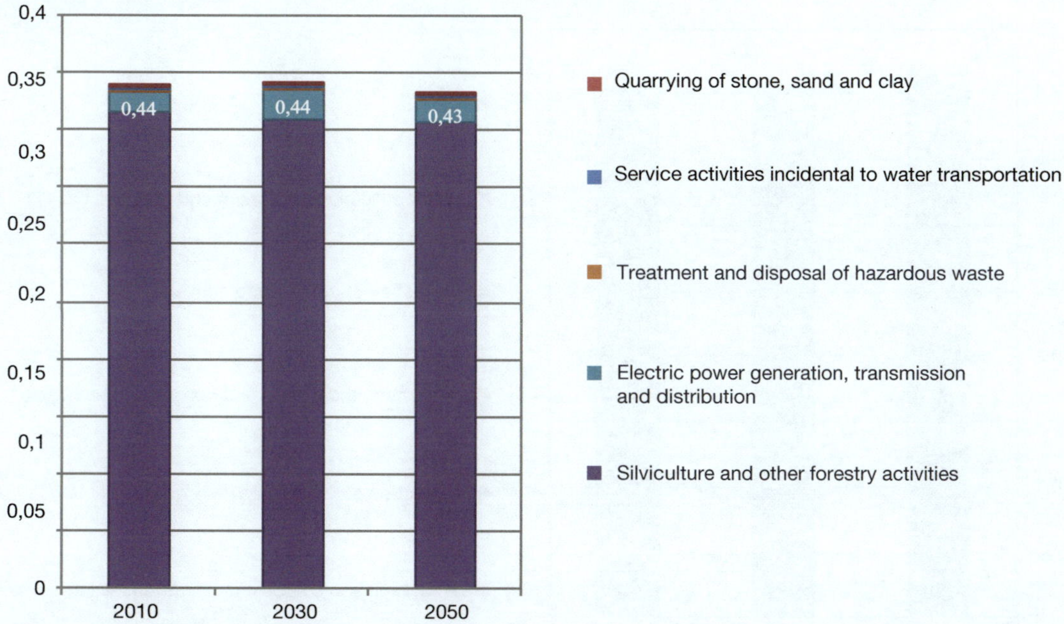

The highest contributing processes to the impact on land occupation of 1 kWh of electricity from Solar Tres, a central tower 15 MW_{el} CSP plant with 16 hours storage (two-tank indirect, mined molten salt), wet cooling, under Africa and Middle East conditions (2400 kWh/m^2 direct normal insolation), in $m^2 \cdot a$/kWh.

FIGURE 6.15

Highest contributing processes to the impact on human toxicity at Solar Tres

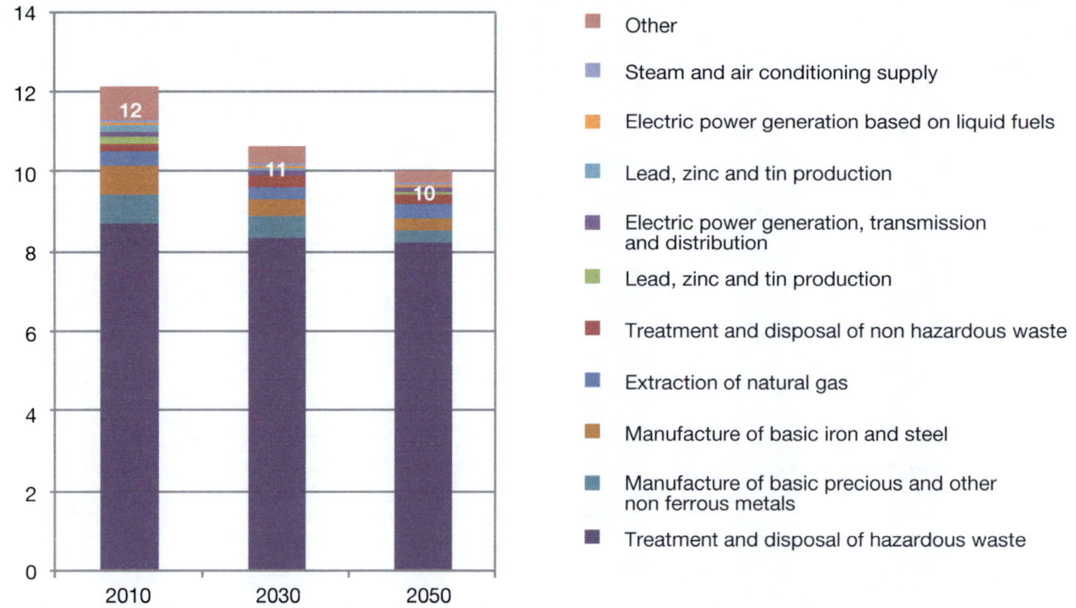

The highest contributing processes to the impact on human toxicity of 1 kWh of electricity from Solar Tres, a central tower 15 MW_{el} CSP plant with 16 hours storage (two-tank indirect, mined molten salt), wet cooling, under Africa and Middle East conditions (2400 kWh/m^2 direct normal insolation), in g 1,4-DB eq./kWh.

FIGURE 6.16

Most contributing processes to the impact on freshwater ecotoxicity at Solar Tres

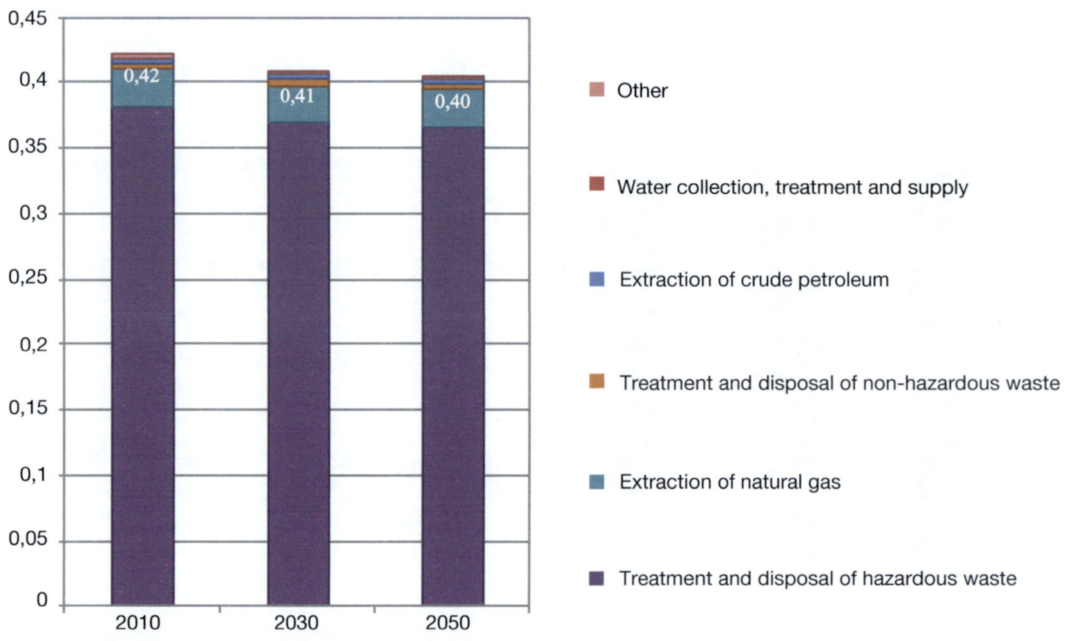

6.5.2.2 Regional comparison

This section presents a comparison of regional environmental impacts in 2010, 2030 and 2050 for a central tower solar power plant. Due to dissimilar insolation and background energy mixes, some impact categories can vary substantially across regions.

Figure 6.17 shows that comparison for the impact on climate change. Sun-rich regions show the lowest impacts as the infrastructure will be more efficient. The region 'Economies in transition' combine a relatively high carbon electricity mix and low insolation, yielding a higher contribution to climate change than the other regions.

Measurements are in g CO_2-eq./kWhFigure 6.18 shows the metal depletion impact of the same plant, built and installed in each of the nine regions. As seen in Figure 6.13, only a very slight decrease can be observed through the years. The specific technology used to build the plant has been considered identical in all three years. The only reason for variation across the different regions is changing insolation.

FIGURE 6.17

Regional comparison of impacts on climate change from the production of 1 kWh of electricity from a central tower solar power plant

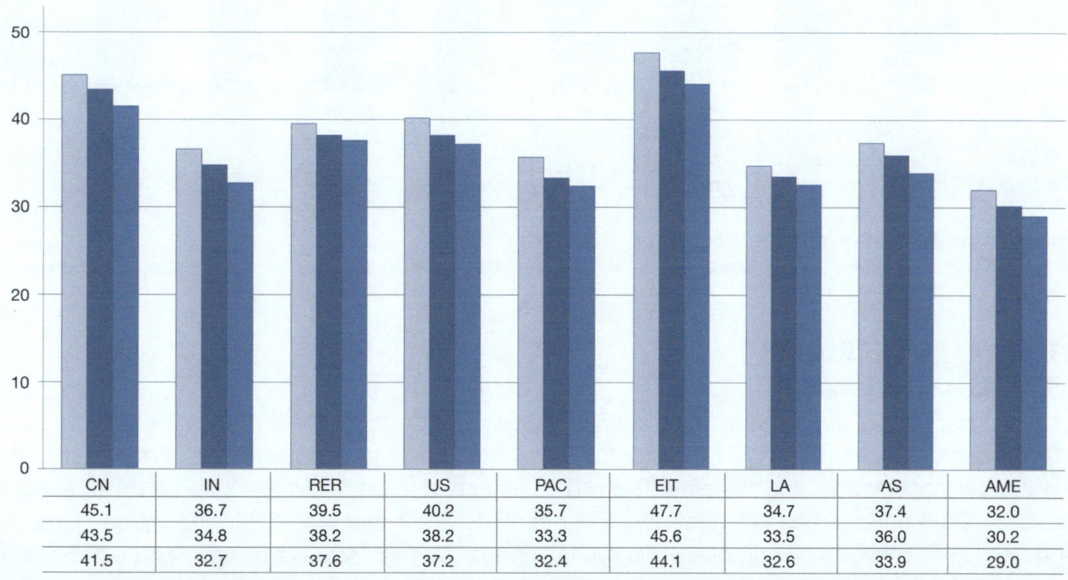

	CN	IN	RER	US	PAC	EIT	LA	AS	AME
2010	45.1	36.7	39.5	40.2	35.7	47.7	34.7	37.4	32.0
2030	43.5	34.8	38.2	38.2	33.3	45.6	33.5	36.0	30.2
2050	41.5	32.7	37.6	37.2	32.4	44.1	32.6	33.9	29.0

FIGURE 6.18

Regional comparison of impacts on metal depletion from the production of 1 kWh of electricity from a central tower solar power plant

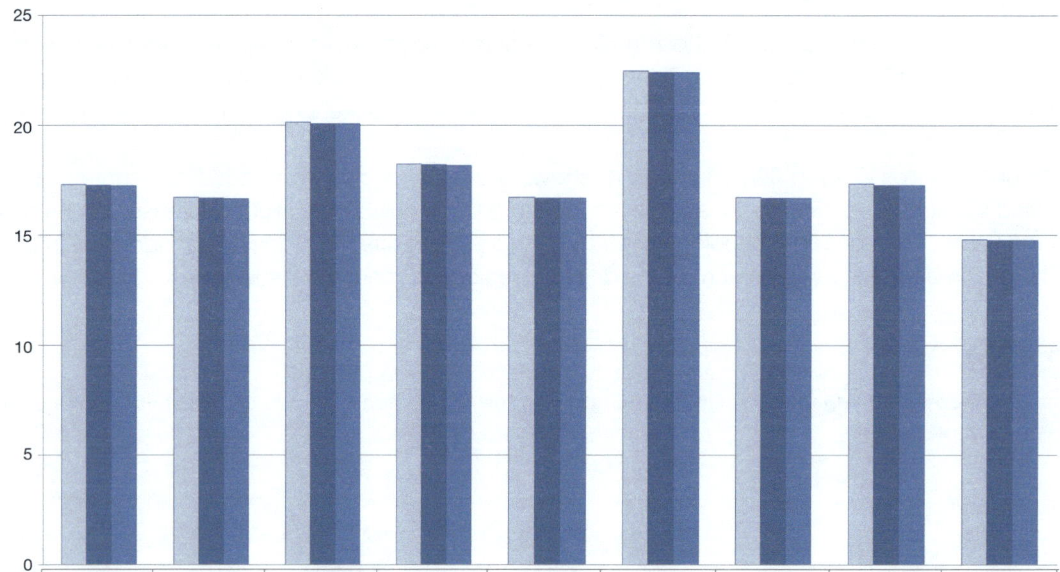

	CN	IN	RER	US	PAC	EIT	LA	AS	AME
2010	17.3	16.7	20.1	18.3	16.7	22.5	16.7	17.3	14.8
2030	17.3	16.7	20.1	18.2	16.7	22.4	16.7	17.3	14.8
2050	17.2	16.7	20.1	18.2	16.7	22.4	16.7	17.3	14.7

Measurements are in g iron eq./kWh.

6.5.3 PARABOLIC TROUGH

6.5.3.1 Life cycle inventory

Parabolic trough inventory was built from the existing plant of Daggett, California, in the U.S. (Burkhardt et al., 2011) with a capacity of 103 MW. Non-infrastructure inputs cover the consumption of heat transfer fluid (HTF) and water for the water cooling systems. The HTF is assumed to be Therminol VP-1, a mixture of diphenyl oxide (DPO, 73.5 per cent) and biphenyl (26.5 per cent) (Burkhardt et al., 2011). The typical parabolic trough design uses synthetic oil HTF combined with an indirect molten salt storage system with a capacity usually between 6-7.5 full-load-hours' equivalent of energy. Both for parabolic trough and central tower, the preferred cooling technology (dry or wet) is very location-dependent. Wet cooling was then chosen, keeping in mind that water use is a major issue.

Contribution analysis

This section presents the different results obtained for the most relevant impact categories. Figure 6.19 shows a very diverse contribution pattern for the impact on climate change. Electric power generation is the main contributor in 2010, due to the mining and combustion of fossil resources, but becomes the third contributor in 2050, as the carbon content of the background energy mix decreases. The refining of sodium nitrate for the storage system contributes significantly. Steam and air conditioning supply is a category that contains high-pressure natural gas, which is necessary for electricity production and the production of ammonia.

Figure 6.20 shows how metal depletion is impacted through the production of 1 kWh of electricity from a parabolic trough power plant. As in Figure 6.13, iron ore and copper concentrate dominate the impact on metal depletion. Infrastructure, power blocks and other electrical installations are the most important processes in that regard. However, the absolute impact of this technology is about half the impact of a central tower power plant. A less intensive bill of materials and the larger 103 MW capacity in comparison to the 15 MW central tower facility contribute to this gap.

Water depletion is a greater concern for a parabolic trough plant than for a central tower plant. Figure 6.21 shows a major contribution from the direct water inputs to the plant, including during the use phase. Water is indeed used for operation and maintenance, namely for the power block and the cleaning of the mirrors. Manufacture of heat fluid transfer and storage chemicals or mining processes are also water-consuming processes.

Figure 6.22 shows how the production of electricity from a parabolic trough plant can have human toxic effects. Treatment and disposal of hazardous waste, mainly from hard coal mining and milling, contribute to more than half of this impact. Iron and steel manufacturing emits particulate matter and metal particles into the atmosphere, which also increases the human toxicity potential impact of the electricity produced. As the share of fossil fuels in the background electricity production mix decreases, the impact on human toxicity decreases.

Terrestrial ecotoxicity from the current system has various origins. Figure 6.23 shows that inland water transport is a major one, but decreases rapidly, due to the inclusion of an assumed decrease in sulfur emissions. The casting of metals, and disposal and treatment of non-hazardous waste are the main contributing processes. Overall, the impact declines rapidly from 2010 to 2050.

FIGURE 6.19

Highest contributing processes to the impact on climate change from a parabolic trough CSP plant

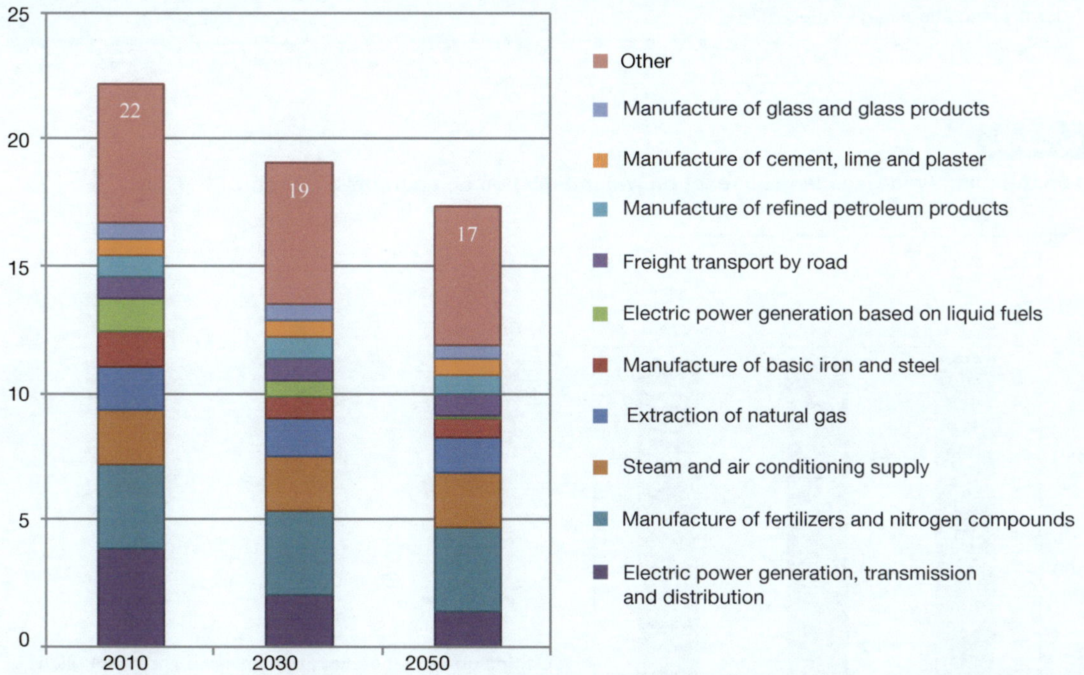

The analysis is for 1 kWh of electricity from a 103 MW parabolic trough CSP plant with 6.3 hour storage (two-tank indirect, mined molten salt), wet cooling, temporal vintage year 2010, Africa and Middle East regional background (2400 kWh/m² direct normal insolation) in g CO_2-eq./kWh.

FIGURE 6.20

Highest contributing processes to the impact on metal depletion from a parabolic trough CSP plant

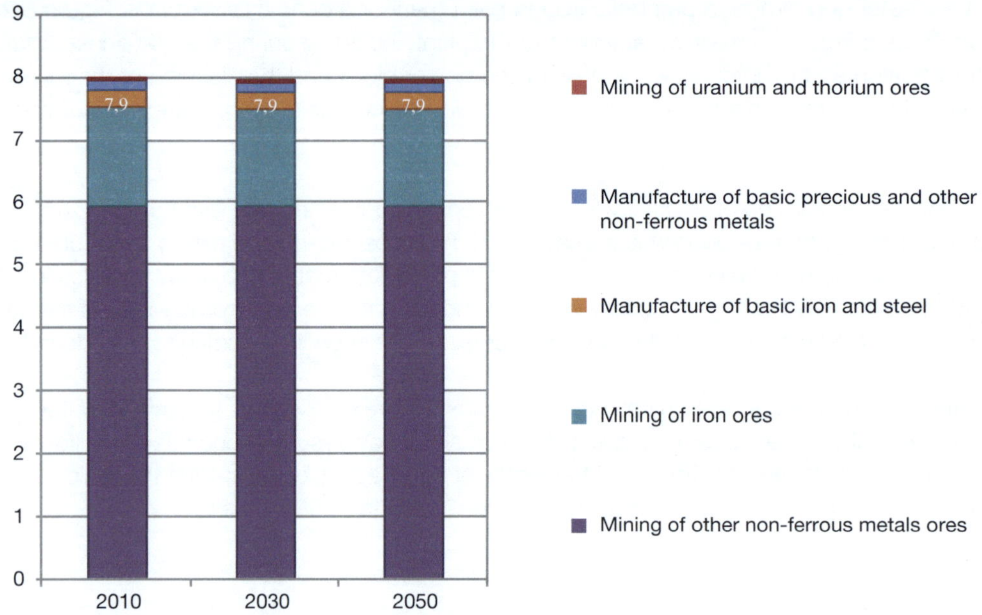

The analysis is for 1 kWh of electricity from a 103 MW parabolic trough CSP plant with 6.3 hour storage (two-tank indirect, mined molten salt), wet cooling, temporal vintage year 2010, Africa and Middle East regional background (2400 kWh/m^2 direct normal insolation) in g iron-eq./kWh.

FIGURE 6.21

Most contributing processes to the impact on water depletion at a parabolic trough CSP plant

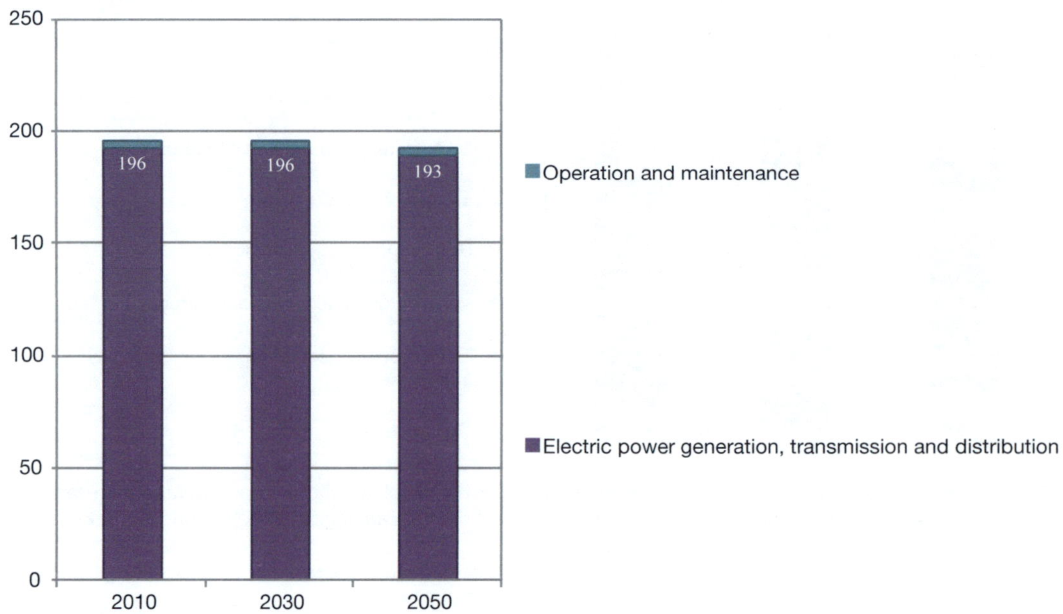

The analysis is for 1 kWh of electricity from a 103 MW parabolic trough CSP plant with 6.3 hour storage (two-tank indirect, mined molten salt), wet cooling, temporal vintage year 2010, Africa and Middle East regional background in litres of water per kWh.

FIGURE 6.22

Most contributing processes to the impact on human toxicity from a parabolic trough CSP Plant

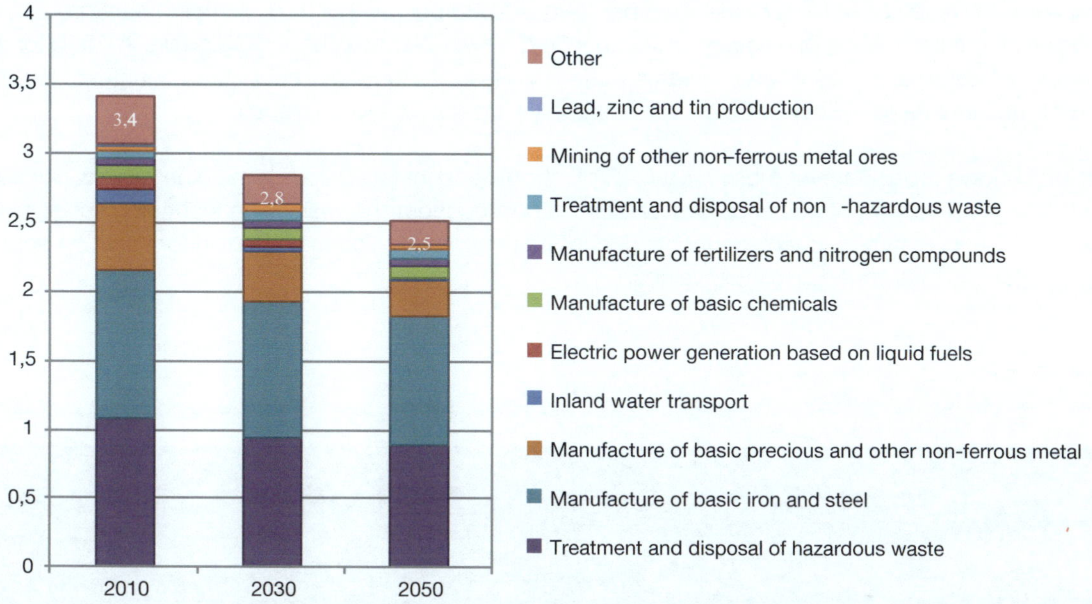

The analysis is for 1 kWh of electricity from a 103 MW parabolic trough CSP plant with 6.3 hour storage (two-tank indirect, mined molten salt), wet cooling, temporal vintage year 2010, Africa and Middle East regional background in g 1,4-DB eq./kWh.

FIGURE 6.23

Contribution of processes to the impact on terrestrial ecotoxicity from a parabolic trough CSP Plant

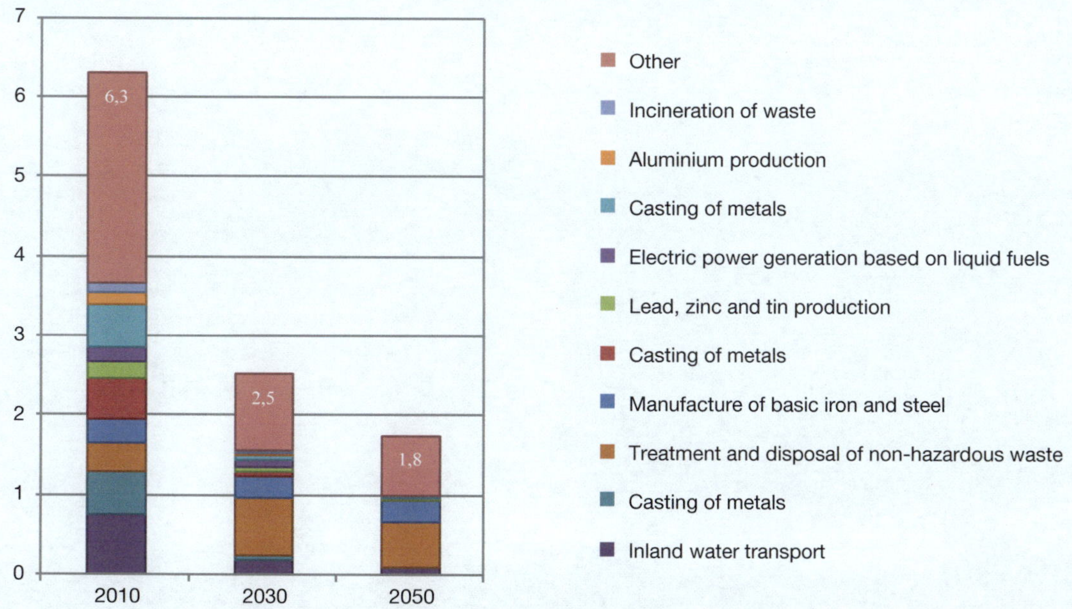

The analysis is for 1 kWh of electricity from a 103 MW parabolic trough CSP plant with 6.3 hour storage (two-tank indirect, mined molten salt), wet cooling, temporal vintage year 2010, Africa and Middle East regional background in mg 1,4-DB eq./kWh.

6.5.3.2 Regional comparison

This section presents a comparison of the electricity production from a parabolic trough plant between regions. Figure 6.24 shows a slight reduction of climate change impacts for most of the regions. India and economies in transition have higher emissions per kWh in 2030 because of their changing background electricity mix; the production of electricity from coal-fired power plants without CCS peaks in 2030 for India, and the use of fossil resources follow the same pattern in Russia and Eastern European countries. In the context of the latter region, GHG emissions are relatively high due to low insolation.

Figure 6.25 shows the impact on metal for a CSP plant built and installed across the nine regions. As seen in Figure 6.18, only a very slight decrease can be observed throughout the years. The technology used to build the plant has been assumed to remain unchanged in the three years. The main reason for variation across the different regions is thus varied insolation values.

FIGURE 6.24

Regional comparison of impacts on climate change from a parabolic trough CSP plant

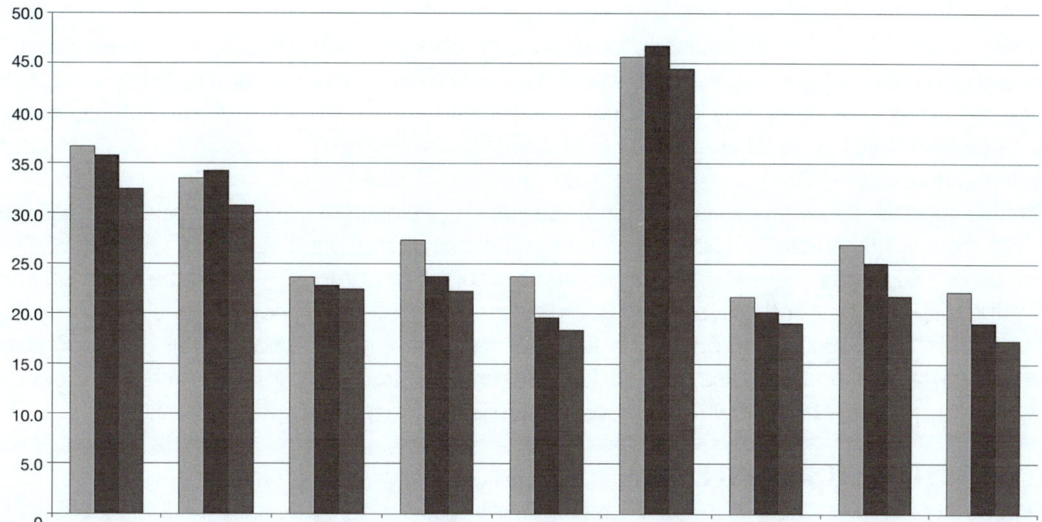

	CN	IN	RER	US	PAC	EIT	LA	AS	AME
2010	36.7	33.5	23.6	27.3	23.7	45.7	21.7	27.0	22.2
2030	35.8	34.2	22.8	23.7	19.6	46.7	20.1	25.1	19.0
2050	32.4	30.8	22.4	22.2	18.4	44.5	19.0	21.7	17.3

The comparison of impacts is from the production of 1 kWh of electricity from a parabolic trough solar power plant in g CO_2-eq./kWh. CN: China; IN: India; RER: rest OECD Europe; US: OECD North America; PAC: OECD Pacific; EIT: economies in transition and non-OECD Europe; LA: Latin America; AS: other developing Asia; AME: Africa and Middle East

FIGURE 6.25

Regional comparison of impacts on metal depletion from a parabolic trough CSP plant

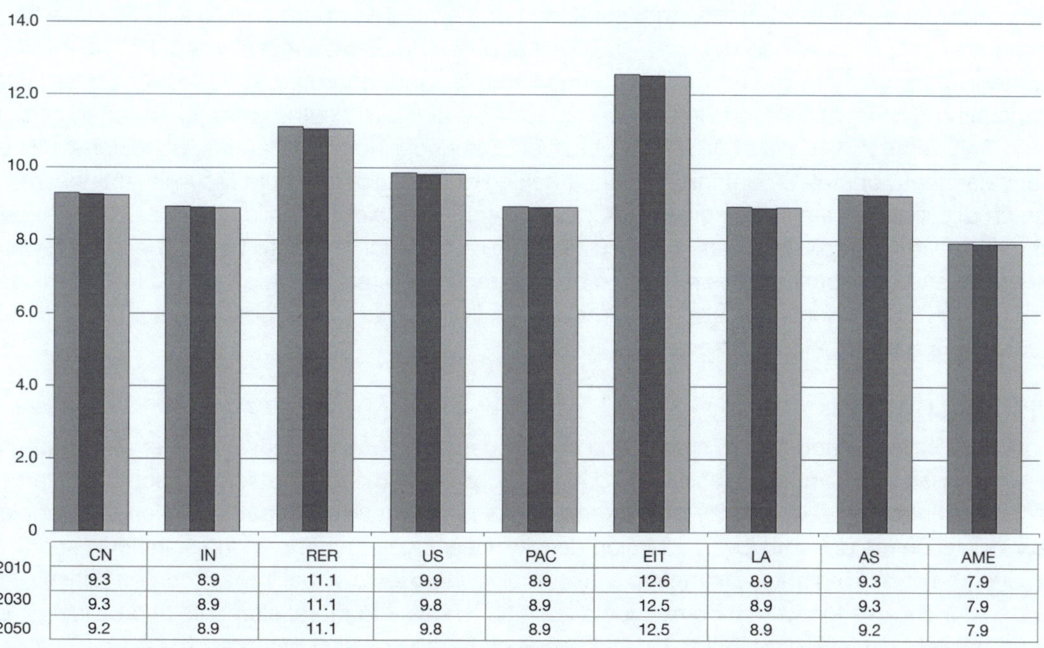

	CN	IN	RER	US	PAC	EIT	LA	AS	AME
2010	9.3	8.9	11.1	9.9	8.9	12.6	8.9	9.3	7.9
2030	9.3	8.9	11.1	9.8	8.9	12.5	8.9	9.3	7.9
2050	9.2	8.9	11.1	9.8	8.9	12.5	8.9	9.2	7.9

The comparison of impacts is from the production of 1 kWh of electricity from a parabolic trough solar power plant in g iron eq./kWh. CN: China; IN: India; RER: rest OECD Europe; US: OECD North America; PAC: OECD Pacific; EIT: economies in transition and non-OECD Europe; LA: Latin America; AS: other developing Asia; AME: Africa and Middle East

6.6 SOCIAL AND ECOLOGICAL IMPACTS FROM SOLAR THERMAL ELECTRICITY

No consensus exists for the value society should place on cleaner energy. However, in recent years, there has been progress in analysing the costs associated with environmental damage, due to several major projects evaluating the externalities of energy in the United States and Europe (Bickel and Friedrich, 2005; Gordon and Society, 2001; NRC, 2010). Solar energy has been considered desirable because it causes a much smaller environmental burden than non-renewable sources of energy. This argument has almost always been justified by qualitative appeals, although this is changing; the limited deployment of solar thermal electricity to date means that there is little empirical evidence of the environmental impacts that such a scheme may have. Similar to other SETs, solar thermal electricity systems displace or completely avoid the basic environmental benefit of the displacement or the avoidance of emissions associated with conventional electricity generation (Tsoutsos et al., 2003). During their operation, these systems have no emissions at all. Although some emissions do arise from other phases of their life cycle, primarily from construction materials processing and manufacture, these are relatively low compared to those avoided by the system operation.

6.6.1 ECOSYSTEM, FLORA AND FAUNA

A growing body of studies underscores the vast potential of solar energy development in places that minimize adverse environmental impacts and confer environmental co-benefits (McDonald et al., 2009; Cameron et al., 2012; Stoms et al., 2013; Hernandez et al., 2015). Utility-scale solar plants occupy large land areas and it is essential to assess proposed locations to minimize the impact to native flora and fauna. Taking carefully considered action during the planning, construction and operation phases can minimize the effects on vegetation, soil and habitat (IEA, 1998). Furthermore, the shade offered by the reflectors has a beneficial effect on the microclimate around the scheme and on the vegetation (Tsoutsos et al., 2005). The best situation is to avoid plant construction in ecologically sensitive areas or in areas of natural beauty. Databases of environmentally sensitive lands can be helpful, but onsite assessment is required (USDOE, 2011). The use of previously disturbed industrial or agricultural land is one way to mitigate impact on native species.

CSP systems could pose a danger to birds (Peck, 2014). In particular, central receiver systems have the potential to concentrate light to intensities that could damage eyesight or burn birds if they fly close the focal point (Hernandez et al., 2014). According to a U.S. Fish and Wildlife Service report in April 2014, 141 birds, including peregrine falcon, barn owl and yellow-rumped warbler were collected at 392 MW Ivanpah Solar Electric Generating System in California in October 2013; 47 of the bird deaths were attributed to solar flux (Belenky and Anderson, 2014). Wu et al. (2014) observed that under normal operating conditions, this should not pose any danger to operators, but failure of the tracking of the heliostat could result in stray beams that might pose an occupational safety risk on-site. Accidental collisions with the collector field is likely to be the greatest threat to birds, as collisions with glass buildings are a common threat to birds. The trend toward smaller heliostats and additional studies related to bird interactions with solar fields should further mitigate bird strike risk. Although flying insects can also be burned when flying close to the receiver area, the loss of the insect population is insignificant (Tsoutsos et al., 2005).

6.6.2 VISUAL IMPACT

CSP technologies have the potential to meet rising energy demands and decrease GHG emissions, but utility-scale solar systems have faced resistance due to public concerns among some groups. The land area requirements for centralized CSP and PV plants raise concerns about visual impacts. Although solar plants are typically located in regions with low population density, clear skies also imply long sight lines and the plants can often be seen from many kilometres away. Analysis suggests visual impact can be minimized by appropriate colour choices for system components such as fences, structural elements and buildings, but cannot be completely eliminated (Sullivan, 2011). In addition to the collector systems, the main visual impact would come from the tower of the central receiver systems. Ho et al. (2011) has studied methods to quantify glint and glare from CSP facilities. The initial data suggest the glint and glare could be an annoyance, but not a health threat. Anecdotal evidence suggests the response to the visual presence of a utility-scale solar plant is a function of the values of the individual viewer (Ho et al., 2011). As with any human endeavour, it is impossible to avoid affecting surroundings; the question largely comes down to the degree to which the local and global benefits of solar power outweigh local costs.

6.6.3 NOISE

Noise from solar power stations is insignificant in comparison to other power options, such as conventional coal, wind power generation, and gas turbines. The noise from the generating plant of large-scale trough, Fresnel and power tower plants is unlikely to cause any disturbance to the public since the power block is invariably located at the centre of the large solar field, far from the facility boundary. Noise would be generated primarily only during the day; at night, when people are more sensitive to noise, the system is typically not operating either due to a lack of energy storage or because power demand is low. Noise impacts may also be of concern in the construction phase, but impacts can be mitigated in the site selection phase and by adopting good work practices (Tsoutsos et al., 2005). The Stirling engines of standalone parabolic dish systems are a source of noise during operation, but they are unlikely to be any noisier than the stand-by diesel generating sets they generally displace. Also, new, technologically advanced Stirling engines are constructed to operate at much lower noise levels. Community engagement throughout the planning process of CSP projects can also significantly increase public acceptance of projects (Zoellner et al., 2008).

6.6.4 HEALTH AND SAFETY (OCCUPATIONAL HAZARDS)

As with any high temperature industrial process, the accidental release of heat transfer fluids such as water, oil, molten salt, or liquid metal from parabolic trough and central receiver systems could form a health hazard. Indeed a fatal accident has previously occurred in a system using liquid sodium (Tsoutsos et al., 2005). Adopting less reactive HTFs such as water/steam, air, and carbon dioxide can minimize these dangers. Central tower systems have the potential to concentrate light to intensities that could damage eyesight. Under normal operating conditions, this should not pose any danger to operators, but tracking system failures could result in stray beams that might pose an occupational safety risk on site. The results of studies such as Viebahn et al. (2008a), summarized in Table 6.7 for CSP, confirm that CSP is usually beneficial, although impacts still exist. In comparison to the figures presented for CSP, the external costs associated with fossil generation options are considerably higher, especially for coal-fired generation.

TABLE 6.7

Quantifiable external costs for CSP (US 2005 cents/kWh)

Impacts	2005	2025	2050
Health	0.65	0.1	0.06
Biodiversity	0.03	0.0	0.0
Crop yield	0.0	0.0	0.0
Material damage	0.01	0	0
Land use	N/A	N/A	N/A
Total	0.69	0.1	0.06

Source: Viebahn et al., 2008a, 2008b

6.6.5 SOCIAL IMPACTS

The construction and operation of solar thermal plants will have significant economic benefits. A large number of component inputs require specialized production, much of which is likely to be locally sourced if there is aggressive regional deployment of CSP (Stoddard et al., 2006). Construction labour is also likely to be sourced locally. Pollin et al. argue that solar power creates 5.4 direct jobs per million dollars of output, while coal generation only creates 1.9 direct jobs and oil and gas creates only 0.8 direct jobs (Pollin et al., 2009). If indirect jobs, commonly referred to as supply chain jobs, are included, then solar power creates 9.8 jobs per million dollars of output whilst coal generates 4.9 jobs and oil and gas only creating 3.7 jobs. A recent study comparing job impacts across energy technologies showed that solar PV had the greatest job-generating potential at an average of 0.87 job-years per GWh, whereas CSP yielded an average of 0.23 job-years per GWh, both of which exceeded estimated job creation for fossil technologies (Wei et al., 2010). A 100-MW CSP plant is estimated to create 455 construction jobs per year. Another estimated 3,500 jobs are created indirectly within the supply chain to support construction. According to Protermosolar, a total of 23,844 people were employed by the solar thermal electricity (STE) industry in Spain in 2010 (Protermosolar, 2011). The STE industry, according to the targets set in the plan for renewable energy in Spain, would maintain this level throughout the decade, and could sustain annual employment of nearly 20,000 jobs in 2020.

Should CSP expand to the extent that gigatons of CO_2 are avoided, approximately 460,000 permanent jobs in operations would be created by 2020. In construction, a maximum of 8.7 million construction workers would be required per year(Augustine et al., 2009). However, this is likely a high estimate, as it is a linear extrapolation of current labour requirements. Investment would be needed to provide education and training to expand the solar thermal workforce at that scale. Stoddard et al. estimated that for each 100 MW of generating capacity, CSP was estimated to generate 94 permanent jobs compared to 56 jobs and 13 jobs for combined cycle and simple cycle plants, respectively (Stoddard et al., 2006). Figure 6.26 shows jobs that could potentially be created in the CSP sector during scale-up.

FIGURE 6.26

Jobs created in the CSP industry

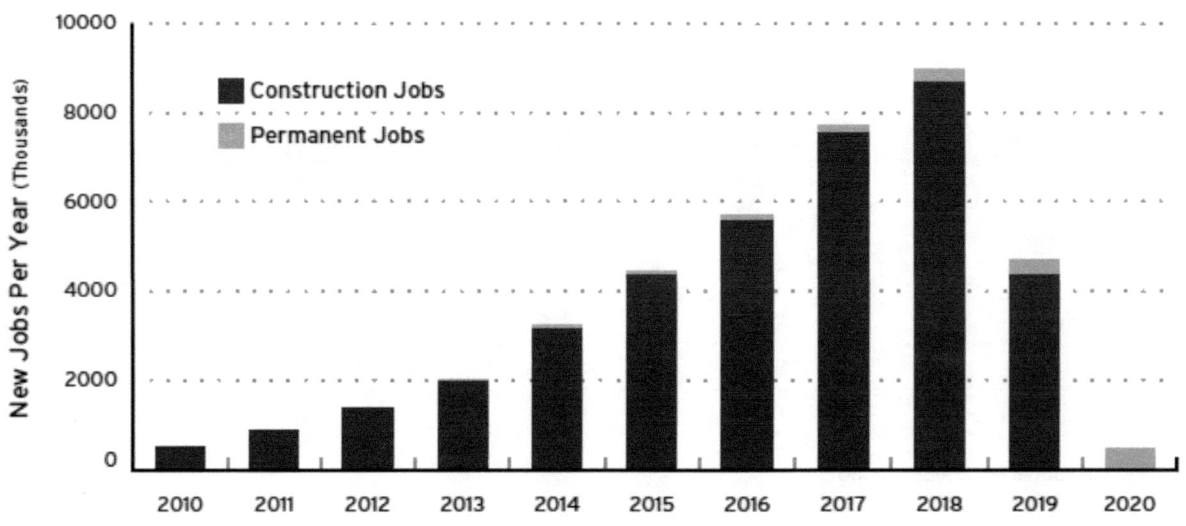

Source: Stoddard et al., 2006

6.7 DISCUSSION

6.7.1 INTERPRETATION OF RESULTS

The results shown in this chapter illustrate how several environmental impacts of CSP vary with technology, time and region. In terms of global warming, photochemical oxidant formation, and toxicity, electricity production of CSP reduces environmental impacts by an order of magnitude compared to background electricity mixes. However, metal depletion levels remain similar with an average current mix of 2010. This is due to a highly metal-intensive infrastructure: the quantity of metal required to build the balance of system is substantial in terms of iron and copper. Interestingly, in the same fashion as photovoltaic technology, a trade-off between metal depletion and fossil depletion and combustion-related impacts can be observed.

6.7.2 TECHNOLOGY-SPECIFIC ISSUES

6.7.2.1 Thermal storage

CSP technologies are still developing. Improvements can be expected in all aspects of CSP plants, that is mirrors, receivers, working fluids, power blocks, and cooling systems, as well as in automated control and maintenance systems. Special attention needs to be paid to energy storage designs. With storage available for even only a few hours, CSP plants can offer a very interesting option in countries with good direct insolation for covering evening peak loads. With larger storage capacities, CSP can become an option for consistent baseload power.

6.7.2.2 Cooling

Given the arid or semi-arid nature of environments that are well suited for CSP, a key challenge is accessing the cooling water needed for CSP plants. Dry cooling, which reduces water consumption by over 90 per cent, or hybrid dry/wet cooling systems can be used in areas with limited water resources, albeit with a slight increase in overall cost.

6.7.2.3 Power transmission

The main limitation to the expansion of CSP plants is not the availability of areas suitable for power production, but the distance between these areas and many large consumption centres. Technologies can address this challenge through efficient long-distance electricity transportation.

6.7.3 COST

The resource for CSP is massive and distributed throughout much of the world. Reducing cost is the key issue in making CSP more commercially relevant and in a position to claim a larger share of the worldwide energy market. This can only be achieved if CSP costs are reduced as CSP technology moves along the learning curves, which primarily depends on deployment and market volumes. In addition, continuous research and development efforts are required to ensure that the slopes of the learning curves do not flatten too early. Potential deployment depends on the actual resources and availability of the respective technology. However, the regulatory and legal framework in place can foster or hinder the uptake of direct solar energy applications to a large extent. Transparent, streamlined administrative procedures to incorporate CSP technologies in existing grid infrastructures can further lower the system costs.

6.7.4 SOCIAL ACCEPTANCE

The effective marketing of solar power, publicizing benefits and impacts relative to traditional power generation facilities, environmental benefits and contribution to a secure energy supply, has helped to accelerate social acceptance despite the higher costs in comparison to conventional sources. Moreover, government spending on CSPs through fiscal incentives and research and development initiatives could garner increased public support through increased quantification and dissemination of the economic impacts associated with those programs.

6.7.5 UNCERTAINTY

6.7.5.1 Representativeness

This LCA only investigates two technologies: parabolic trough and central tower systems. These two technologies do not represent the entire spectrum of CSP technologies. As described in Chapter 6.3 other CSP technologies include compact linear Fresnel reflectors and Dish/engine systems. It remains to be determined whether these technologies are relevant in a large-scale assessment, e.g. depending on their global market share. Options for dry or wet cooling were not investigated for both technologies and only some thermal storage options were looked at.

6.7.5.2 Geographic variability

Geographic variability exists at every phase of the technologies' life cycles. At production and manufacturing, background technologies and electricity mixes, transportation can make an important difference in life-cycle results. During the use phase, DNI is an influential parameter that varies with the region where the plant is built. More precisely, it varies with the latitude of that region. Similar to PV technologies, the first CSP plants will be deployed in the sunniest areas, where the energy payback ratio is expected to be the best. This pattern will probably change with policy incentives, and, while the majority of CSP plants can be found in southern United States or southern Europe, many projects are planned or already in operation phase in the Middle East (Jordan, Saudi Arabia), in Asia (China, India), in South Africa and South America (Chile).

6.7.5.3 End of life

As specific data could not be found for the end of life phase, and to ensure consistency with the scopes of other technologies, the impacts from decommissioning were not accounted for in this study. The first CSP plants will reach their end of life around 2040. It can be reasonably expected that reusing and recycling schemes will be investigated in the same way as it has recently been for PV power systems. Among all renewable energies, CSP stands out for its distinct technical features, such as dispatchability through storage and hybridization and its grid stability. In addition, CSP plants induce large positive macroeconomic effects on the economy by adding to the domestic GDP through high investments, fiscal contributions, fuel imports reduction and the creation of jobs in component manufacturing and plant construction and operation.

In conclusion, this review and assessment of CSP electricity generation technologies suggests that CSP technologies are at a critical stage in development. While current electricity generation by CSP is small relative to conventional energy sources, CSP is growing quickly and its potential environmental impacts and ability to mitigate GHG cannot be ignored. At this stage, policy makers have the opportunity to greatly influence how CSP contribute to a low-carbon energy future.

6.8 REFERENCES

Ab Kadir, M. Z. A., Y. Rafeeu, and N. M. Adam. 2010. Prospective scenarios for the full solar energy development in Malaysia. *Renewable and Sustainable Energy Reviews* 14(9): 3023-3031.

Abbas, M., B. Boumeddane, N. Said, and A. Chikouche. 2011. Dish Stirling technology: A 100 MW solar power plant using hydrogen for Algeria. *International Journal of Hydrogen Energy* 36(7): 4305-4314.

Abbas, M., Boumeddane, B., Said, N., Chikouche, A. . 2009. Techno economic evaluation of solar Dish Stirling system for stand alone electricity generation in Algeria. *Journal of Engineering and Applied Sciences* 4(4): 258–267.

Abou-Hadid, A. 2006. Assessment of Impacts, Adaptation, and Vulnerability to Climate Change in North Africa: Food Production and Water Resources. Final Report Submitted to Assessments of Impacts and Adaptations to Climate Change (AIACC). Accessed at http://www.start.org/Projects/AIACC_Project/FinalReports/FinalReports/FinalRept_AIACC_AF90.pdf

Almanza, R. and A. Lentz. 1998. Electricity production at low powers by direct steam generation with parabolic troughs. *Solar Energy* 64(1–3): 115-120.

Arnell, N. W. 1999. Climate change and global water resources. *Global Environmental Change* 9, Supplement 1(0): S31-S49.

Arvizu, D. P., L. Balaya, T. Cabeza, A. Hollands, M. Jäger-Waldau, C. Kondo, V. Konseibo, W. Meleshko, Y. Stein, H. Tamaura, and R. Z. Xu. 2011. *In IPCC Special Report on Renewable Energy Sources and Climate Change Mitigation* Cambridge, United Kingdom and New York, NY, USA: Cambridge University Press.

Augustine, C., A. Byrne, E. Gimon, T. Goerner, I. Hoffman, D. Kammen, J. Kantner, J. Levin, T. Lipman, A. Mileva, R. Muren, S. Paul, S. Sapatari, H. Thorsteinsson, and C. D. Tominks. 2009. *Gigaton Throwdown: Redefining what's possible for clean energy until 2020*. San Francisco: Gigaton Throwdown.

Battisti, R. and A. Corrado. 2005. Environmental assessment of solar thermal collectors with integrated water storage. *Journal of Cleaner Production* 13(13–14): 1295-1300.

Becerra-Lopez, H. R. and P. Golding. 2007. Dynamic exergy analysis for capacity expansion of regional power-generation systems: Case study of far West Texas. *Energy* 32(11): 2167-2186.

Belenky, L.T., Anderson, I. 2014. Center for biological diversity supplement to opposition to motion to reopen evidentiary record and scheduling order. Docket No. 09-AFC-7C, Energy Resources Conservation and Development Commission, State of California, April 2014. Available at: http://docketpublic.energy.ca.gov/PublicDocuments/09-AFC-07C/TN201977_20140407T161504_Center_Supplemental_Opposition_to_Motion.pdf. Accessed on 21st May 2016.

Bickel, P. and R. Friedrich. 2005. *ExternE-Externalities of Energy: Methodology 2005 Update*. Official Publications of the European Communities.

Boulet, G., A. Boudhar, L. Hanich, B. Duchemin, V. Simonneaux, P. Maisongrande, T. S., A. Chaponnière, and A. Chehbouni. 2008. Hydrological modelling in the High Atlas mountains with the help of remote-sensing data: milestones of the SudMed project.

Breyer, C. and G. Knies. 2009. Global energy supply potential of concentrating solar power. Paper presented at SolarPACES 2009, 15-18th September 2009, Berlin.

Brinkman, G., P. Denholm, E. Drury, R. Margolis, and M. Mowers. 2011. Toward a Solar-Powered Grid. *Power and Energy Magazine, IEEE* 9(3): 24-32.

Broesamle, H., H. Mannstein, C. Schillings, and F. Trieb. 2001. Assessment of solar electricity potentials in North Africa based on satellite data and a geographic information system. *Solar Energy* 70(1): 1-12.

Brosseau, Doug A., P. F. Hlava, and M. J. Kelly. 2004. *Testing of Thermocline Filler Materials* and Molten-Salt Heat Transfer Fluids for Thermal Energy Storage Systems in Parabolic Trough Power Plants. Pittsburgh, PA: National Energy Technology Laboratory.

Burkhardt, J. J., G. A. Heath, and C. S. Turchi. 2011. Life Cycle Assessment of a Parabolic Trough Concentrating Solar Power Plant and the Impacts of Key Design Alternatives. Environmental Science & Technology 45(6): 2457-2464.

Burkhardt, J. J., G. Heath, and E. Cohen. 2012. Life Cycle Greenhouse Gas Emissions of Trough and Tower Concentrating Solar Power Electricity Generation. *Journal of Industrial Ecology* 16: S93-S109.

Cameron, D.R., Cohen, B.S., Morrison, S.A. 2012. An approach to enhance the conservation-compatibility of solar energy development. PLoS One, 7 (6):e38437.

Carter, N. and R. Campbell. 2009. Water issues of concentrating solar power (CSP) electricity in the U.S. Southwest. R40631. Congressional Research Service.

Cavallaro, F. and L. Ciraolo. 2006. A Life Cycle Assessment (LCA) of a Paraboloidal-Dish Solar Thermal Power Generation System. Paper presented at 2006 First International Symposium on Environment Identities and Mediterranean Area, 9-12 July 2006. Corte-Ajaccio: IEEE.

Chen, G. Q., Q. Yang, Y. H. Zhao, and Z. F. Wang. 2011. Nonrenewable energy cost and greenhouse gas emissions of a 1.5 MW solar power tower plant in China. Renewable and Sustainable Energy Reviews 15(4): 1961-1967.

Christensen, J. H., B. Hewitson, A. Busuioc, A. Chen, X. Gao, I. Held, R. Jones, R.K. Kolli, W.-T. Kwon, R. Laprise, V. Magaña Rueda, L. Mearns, C.G. Menéndez, J. Räisänen, A. Rinke, A. Sarr, and P. Whetton. 2007. Regional climate projections. In Climate Change 2007: The Physical Basis. Contribution of Working Group I to the Fourth Assessment Report of the Intergovernmental Panel on Climate Change Cambridge, UK, and New York, NY: Cambridge University Press.

Cohen, G. E., D. W. Kearney, and K. G. J. 1999. Final Report on the Operation and Maintenance Improvement Program for Concentrating Solar Power Plants.

CSP Today. 2016. CSP Today Global Tracker. Available at: http://social.csptoday.com/tracker/projects. Accessed on 21 May 2016.

CSP Today. 2013. CSP Today Quarterly Update: December 2013.

Dahle, D., M. Kirby, D. Heimiller, B. Farhar, and B. Owens. 2003. Assessing the Potential for Renewable Energy on Public Lands NREL/TP-710-33530. Golden, Colorado: National Renewable Energy Laboratory.

Dahle, D., D. Elliott, D. Heimiller, M. Mehos, R. Robichaud, M. Schwartz, B. Stafford, and A. Walker. 2008. Assessing the Potential for Renewable Energy Development on DOE Legacy Management Lands. TP-670-41673. Golden, Colorado: National Renewable Energy Laboratory (NREL).

Damerau, K., K. Williges, A. G. Patt, and P. Gauché. 2011. Costs of reducing water use of concentrating solar power to sustainable levels: Scenarios for North Africa. Energy Policy 39(7): 4391-4398.

de Wit, M. and J. Stankiewicz. 2006. Changes in Surface Water Supply Across Africa with Predicted Climate Change. Science 311(5769): 1917-1921.

DeLaquil III, P. 1996. Progress commercializing solar-electric power systems. Renewable Energy 8(1–4): 489-494.

Desideri, U., F. Zepparelli, V. Morettini, and E. Garroni. 2013. Comparative analysis of concentrating solar power and photovoltaic technologies: Technical and environmental evaluations. *Applied Energy* 102(0): 765-784.

Domínguez Bravo, J., X. García Casals, and I. Pinedo Pascua. 2007. GIS approach to the definition of capacity and generation ceilings of renewable energy technologies. *Energy Policy* 35(10): 4879-4892.

EASAC. 2011. *Concentrating solar power: Its potential contribution to a sustainable energy future*. Brussels: European Academies Science Advisory Council.

EC. 2007. *Concentrating Solar Power - From research to implementation*. Luxembourg: European Commission.

Elsharkawy, H., Rashed, H., Rached, I. 2009. Climate Change: The Impacts of Sea Level Rise on Egypt. In *ISOCARP Congress 2009*.

EPRI. 1997. *Renewable Energy Technology Characterizations*. Palo Alto, California: Electric Power Research Institute.

EPRI. 2007. Running Dry at the Power Plant. *EPRI Journal*: 28-35.

EPRI. 2009. *Australian Electricity Generation Technology Costs – Reference Case 2010*. Palo Alto, California: Electric Power Research Institute.

EPRI. 2010. *Solar Thermocline Storage Systems: Preliminary Design Study*. Palo Alto, California: Electric Power Research Institute.

EREC. 2010. *RE-thinking 2050: A 100% Renewable Energy Vision for the European Union*. Brussels:

ESTELA. 2012. *Solar Thermal Electricity: Strategic Research Agenda 2020-2050*. Brussels: European Solar Thermal Electricity Association.

FAO. 2010. AQUASTAT: FAO's global information system on water and agriculture. Accessed at http://www.fao.org/nr/water/aquastat/main/index.stm.

Fluri, T. P. 2009. The potential of concentrating solar power in South Africa. *Energy Policy* 37(12): 5075-5080.

FS-UNEP/BNEF. 2016. Global Trends in Renewable Energy Investment 2016. Frankfurt School (FS)-United Nations Environment Program (UNEP) Centre and Bloomberg New Energy Finance (BNEF), Frankfurt School of Finance & Management gGmbH, Frankfurt. Available at: http://fs-unep-centre.org/sites/default/files/publications/globaltrendsinrenewableenergyinvestment2016lowres_0.pdf. Accessed on 21 May 2016.

Fthenakis, V. and H. C. Kim. 2009. Land use and electricity generation: A life-cycle analysis. *Renewable and Sustainable Energy Reviews* 13(6-7): 1465-1474.

Gamble, C. E. and M. Schopf. 2010. Heat Transfer Fluid Leaks: Break the Fire Triangle. *Chemical Engineering* 117: 26-33.

Geyer, M. and I. E. Agency. 2008. *SolarPACES Annual Report 2007*: Deutsches Zentrum fuer Luft & Raumfahrt.

Gordon, J. and I. S. E. Society. 2001. *Solar Energy: The State of the Art: ISES Position Papers*: James & James.

Hang, Q., Z. Jun, Y. Xiao, and C. Junkui. 2008. Prospect of concentrating solar power in China—The sustainable future. *Renewable and Sustainable Energy Reviews* 12(9): 2505-2514.

Heath, G., J. Burkhardt, C. Turchi, T. Decker, and C. Kutscher. 2009. LCA (Life Cycle Assessment) of Parabolic Trough CSP: Materials Inventory and Embodied GHG Emissions from Two-Tank Indirect and Thermocline Thermal Storage (Presentation). Paper presented at ASME 3rd International Conference on Energy Sustainability, San Francisco, CA.

Hernandez, R.R., Easter, S.B., Murphy-Mariscal, M.L., Maestre, F.T., Tavassoli, M., Allen, E.B., Barrows, C.W., Belnap, J., Ochoa-Hueso, R., Ravi, S., Allen, M.F. 2014. Environmental impacts of utility-scale solar energy. Renewable and Sustainable Energy Reviews, 29, pp. 766-779.

Hernandez, R.R., Hoffacker, M.K., Field, C.B. 2015. Efficient use of land to meet sustainable energy needs. Nature Climate Change, 5: 353 -335.

Hernández-Moro, J. and J. M. Martínez-Duart. 2012. CSP electricity cost evolution and grid parities based on the IEA roadmaps. *Energy Policy* 41(0): 184-192.

Herrmann, U. and D. W. Kearney. 2002. Survey of Thermal Energy Storage for Parabolic Trough Power Plants. *Journal of Solar Energy Engineering* 124(2): 145-152.

Ho, C. K., C. M. Ghanbari, and R. B. Diver. 2011. Methodology to Assess Potential Glint and Glare Hazards From Concentrating Solar Power Plants: Analytical Models and Experimental Validation. *Journal of Solar Energy Engineering* 133: 031021–031029.

Hogan, M. 2009. The secret to low-water-use, high-efficiency concentrating solar power. *Climate Progress* 29 April 2009, Accessed at http://thinkprogress.org/climate/2009/04/29/204025/csp-concentrating-solar-power-heller-water-use/.

Hoogwijk, M. and W. Graus. 2008. *Global potential of renewable energy sources: A literature assessment*. Ecofys Netherlands.

Hou, H. J., Yang, Y.P., Cui, Y.H., Gao, S., Pan, Y.X. 2009. Assessment of concentrating solar power prospect in China. Paper presented at International Conference on Sustainable Power Generation and Supply (SUPERGEN 2009), 6-7th April 2009, Nanjing, China.

IEA. 1998. *Benign energy? The environmental implications of renewables.* Paris: OECD/IEA.

IEA. 2008. *Combined Heat and Power. Evaluating the benefits of greater global investment*. Paris, France: International Energy Agency.

IEA. 2010a. *Energy Technology Perspectives 2010*: OECD Publishing.

IEA. 2010b. *Energy Technology Perspectives 2010: Scenarios and Strategies to 2050*. Paris, France: OECD/IEA.

Jacobson, M. Z. 2009. Review of solutions to global warming, air pollution, and energy security. *Energy & Environmental Science* 2(2): 148-173.

Janjai, S., J. Laksanaboonsong, and T. Seesaard. 2011. Potential application of concentrating solar power systems for the generation of electricity in Thailand. *Applied Energy* 88(12): 4960-4967.

Kalogirou, S. A. 2004. Solar thermal collectors and applications. *Progress in Energy and Combustion Science* 30(3): 231-295.

Karsteadt, R., D. Dahle, D. Heimiller, and T. Nealon. 2005. *Assessing the Potential for Renewable Energy on National Forest System Lands* Goldon, Colorado: National Renewable Energy Laboratory.

Kaygusuz, K. 2011. Prospect of concentrating solar power in Turkey: The sustainable future. *Renewable and Sustainable Energy Reviews* 15(1): 808-814.

Kelly, B. 2006. *Nexant Parabolic Trough Solar Power Plant Systems Analysis; Task 2 Comparison of Wet and Dry Rankine Cycle Heat Rejection*. Colorado: National Renewable Energy Laboratory.

Klaiß, H., R. Köhne, J. Nitsch, and U. Sprengel. 1995. Solar thermal power plants for solar countries – Technology, economics and market potential. *Applied Energy* 52(2–3): 165-183.

Kocheril, G. and P. Viebahn. 2011. Assumptions and parameters in CSP potential studies – A Meta-Analysis. Paper presented at SolarPACES 2011 – Concentrating Solar Power and Chemical Energy Systems, 20-23th September, Granada, Spain.

Kongtragool, B. and S. Wongwises. 2003. A review of solar powered Stirling engines and low temperature differential Stirling engines. *Renewable and Sustainable Energy Reviews* 7: 131–154.

Kramer, K. 2012. Interaction of Regulation and Innovation: Solar Air Heating Collectors. *Energy Procedia* 30(0): 1311-1321.

Kreith, F., P. Norton, and D. Brown. 1990. *CO_2 Emissions from Coal-Fired and Solar Electric Power Plants*. Golden, Colorado: Solar Energy Research Institute.

Laing, D., W. D. Steinmann, P. Viebahn, F. Gräter, and C. Bahl. 2010. Economic Analysis and Life Cycle Assessment of Concrete Thermal Energy Storage for Parabolic Trough Power Plants. *Journal of Solar Energy Engineering* 132(4): 10131-10136.

LBF. 2008. *How to Combat Global Warming - An ambitious but necessary approach to reduce greenhouse gas emissions*. Prepared by the Bellona Foundation for the CC8 Conference, Oslo, 5-6th June 2008.: Luxembourg Bellona Foundation.

Lechón, Y., C. de la Rúa, and R. Sáez. 2008. Life Cycle Environmental Impacts of Electricity Production by Solarthermal Power Plants in Spain. *Journal of Solar Energy Engineering* 130(2): 10121-10127.

Lenzen, M. 1999. Greenhouse gas analysis of solar-thermal electricity generation. *Solar Energy* 65(6): 353-368.

Lenzen, M. 2010. Current state of development of electricity-generating technologies: A literature review. *Energies* 3(3): 462-591.

Li, J. 2009. Scaling up concentrating solar thermal technology in China. *Renewable and Sustainable Energy Reviews* 13(8): 2051-2060.

Lohmann, S. et al. 2006. Validation of DLR-ISIS data. Accessed at http://www.pa.op.dlr.de/ISIS/.

Lovegrove, K., J. Wyder, A. Agrawal, D. Boruah, J. McDonald, and K. Urkalan. 2009. *Concentrating Solar Power in India*. IT Power.

Lovegrove, K., M. Watt, R. Passey, G. Pollock, J. Wyder, and J. Dowse. 2012. *Realising the potential of concentrating solar power in Australia: Summary for stakeholders.* Australia: Australian Solar Institute.

Macknick, J., R. Newmark, and C. Turchi. 2011. Water consumption impacts of renewable technologies: The case of CSP. In *AWRA 2011 Spring Specialy Conference*. Baltimore, MD.

Mancini, T., P. Heller, B. Butler, B. Osborn, W. Schiel, V. Goldberg, R. Buck, R. Diver, C. Andraka, and J. Moreno. 2003. Dish-Stirling Systems: An Overview of Development and Status. *Journal of Solar Energy Engineering* 125(2): 135-151.

Martin, G. R. and J. M. Shaw. 2010. Bird collisions with power lines: Failing to see the way ahead? *Biological Conservation* 143(11): 2695-2702.

Martin, J. A. 1997. A total fuel cycle approach to reducing greenhouse gas emissions: Solar generation technologies as greenhouse gas offsets in U.S. utility systems. *Solar Energy* 59(4–6): 195-203.

McDonald, R.I., Fargione, J., Kiesecker, J., Miller, W.M., Powell, J. 2009. Energy sprawl or energy efficiency: Climate policy impacts on natural habitat for the United States of America. PLoS One, 4 (8): e6802.

Medrano, M., M. O. Yilmaz, M. Nogués, I. Martorell, J. Roca, and L. F. Cabeza. 2009. Experimental evaluation of commercial heat exchangers for use as PCM thermal storage systems. *Applied Energy* 86(10): 2047-2055.

Mehos, M., D. Kabel, and P. Smithers. 2009. Planting the Seed: Greening the Grid with Concentrating Solar Power. *IEEE Power and Energy Magazine* 7(3): 55-62.

Mills, D. 2004. Advances in solar thermal electricity technology. *Solar Energy* 76(1–3): 19-31.

Mills, D. R. and G. L. Morrison. 2000. Compact Linear Fresnel Reflector solar thermal powerplants. *Solar Energy* 68(3): 263-283.

Norton, B., P. C. Eames, and S. N. G. Lo. 1998. Full-energy-chain analysis of greenhouse gas emissions for solar thermal electric power generation systems. *Renewable Energy* 15(1–4): 131-136.

NRC. 2010. *Hidden costs of energy: Unpriced consequences of energy production and use*. Washington, DC: National Research Council.

Ong, S., C. Campbell, P. Denholm, R. Margolis, and G. Heath. 2013. *Land-Use Requirements for Solar Power Plants in the United States*. Golden, Colorado: National Renewable Energy Laboratory (NREL).

Ordóñez, I., N. Jiménez, and M. A. Silva. 2009. Life Cycle Environmental Impacts of Electricity Production by Dish/Stirling Systems in Spain. Paper presented at SolarPaces 2009, Berlin, Germany.

Oró, E., A. Gil, A. de Gracia, D. Boer, and L. F. Cabeza. 2012. Comparative life cycle assessment of thermal energy storage systems for solar power plants. *Renewable Energy* 44(0): 166-173.

Peck, M. 2014. Ivanpah Solar Power Tower Is Burning Birds. IEEE Spectrum, 20th August 2014. Available at: http://spectrum.ieee.org/energywise/green-tech/solar/ivanpah-solar-plant-turns-birds-into-smoke-streamers. Accessed on 21 May 2016.

Pehnt, M. 2006. Dynamic life cycle assessment (LCA) of renewable energy technologies. *Renewable Energy* 31(1): 55-71.

Piemonte, V., M. D. Falco, P. Tarquini, and A. Giaconia. 2011. Life cycle assessment of a high temperature molten salt concentrated solar power plant. *Solar Energy* 85(5): 1101-1108.

Pihl, E., D. Kushnir, B. Sandén, and F. Johnsson. 2012. Material constraints for concentrating solar thermal power. *Energy* 44(1): 944-954.

Pletka, R., S. Block, K. Cummer, K. Gilton, R. O'Connell, B. Roush, L. Stoddard, S. Tilley, D. Woodward, and M. Hunsaker. 2007. *Arizona Renewable Energy Assessment*. Overland Park, Kansas: Black & Veatch Corporation.

Pollin, R., J. Heintz, and H. Garrett-Peltier. 2009. *The Economic Benefits of Investing in Clean Energy*. Department of Economics and Political Economy Research Institute.

Poullikkas, A., G. Kourtis, and I. Hadjipaschalis. 2010. Parametric analysis for the installation of solar dish technologies in Mediterranean regions. *Renewable and Sustainable Energy Reviews* 14(9): 2772-2783.

Pregger, T., D. Graf, W. Krewitt, C. Sattler, M. Roeb, and S. Möller. 2009. Prospects of solar thermal hydrogen production processes. *International Journal of Hydrogen Energy* 34(10): 4256-4267.

Price, H. and D. Kearney. 2003. *Reducing the Cost of Energy from Parabolic Trough Solar Power Plants: Preprint* Golden, Colorado: National Renewable Energy Laboratory.

Protermosolar. 2011. *Macroeconomic impact of the Solar Thermal Electricity Industry in Spain*. Seville, Spain: Spanish Association of Solar Thermal Industry.

Purohit, I. and P. Purohit. 2010. Techno-economic evaluation of concentrating solar power generation in India. *Energy Policy* 38(6): 3015-3029.

Purohit, I., P. Purohit, and S. Shekhar. 2013. Evaluating the potential of concentrating solar power generation in Northwestern India. *Energy Policy* 62(0): 157-175.

REN21. 2013. *Renewables 2013 Global Status Report*. Paris: REN21.

REN21. 2015. ***Renewables 2015 Global Status Report***. Paris: REN21.

Robak, C. W., T. L. Bergman, and A. Faghri. 2011. Enhancement of latent heat energy storage using embedded heat pipes. *International Journal of Heat and Mass Transfer* 54(15–16): 3476-3484.

Shabgard, H., T. L. Bergman, N. Sharifi, and A. Faghri. 2010. High temperature latent heat thermal energy storage using heat pipes. *International Journal of Heat and Mass Transfer* 53(15–16): 2979-2988.

Shell International BV. 2008. *Shell Energy Scenarios to 2050*.

Skumanich, A. 2010. CSP: Developments in heat transfer and storage materials. *Renewable Energy Focus* 11(5): 40-43.

Solar PEIS. 2012. *Chapter 3 - Update to Overview of Solar Energy Power Production Technologies, Development and Regulation*. Washington, DC: U.S.Department of Energy (USDOE) and Bureau of Land Management (BLM).

SolarReserve. 2016. A Blueprint for a low carbon planet: SolarReserve's next generation solar plant delivers power on demand with game-changing energy storage. Available at: http://www.solarreserve.com/en/newsroom/press-releases/a-blueprint-for-a-low-carbon-planet-solarreserve2019s-next-generation-solar-plant-delivers-power-on-demand-with-game-changing-energy-storage#sthash.nFu4zwiE.dpuf. Accessed on 17 May 2016.

Steinhagen, H. M. and F. Trieb. 2004. Concentrating solar power - A review of the technology. *Quarterly of the Royal Academy of Engineering* Ingenia 18.

Stoddard, L., J. Abiecunas, and R. O'Connell. 2006. *Economic, Energy, and Environmental Benefits of Concentrating Solar Power in California*. Overland Park, Kansas: Black & Veatch.

Stoms, D.M., Dashiell, S.L., Davis, F.W. 2013. Siting solar energy development to minimize biological impacts. Renewable Energy, 57: 289 -298.

Sullivan, R. 2011. Visual Impacts of Concentrating Solar Power Facilities on Desert Landscapes in the American Southwest. Paper presented at SolarPACES 2011, Granada, Spain.

Taggart, S. 2008. CSP: Dish projects inch forward. *Renewable Energy Focus* 9(4): 52-54.

TEEIC. 2010. Solar Energy System Descriptions. Accesed at http://teeic.indianaffairs.gov/er/solar/restech/desc/index.htm.

Teske, S., Leung, J., Crespo, L., Bial, M., Dufour, E., Richter, C. 2016. Solar Thermal Electricity Global Outlook 2016. Published by Greenpeace International, European Solar Thermal Electricity Association (ESTELA), and SolarPACES. Available at: http://www.greenpeace.org/international/Global/international/publications/climate/2016/Solar-Thermal-Electricity-Global-Outlook-2016.pdf. Accessed on 21 May 2016.

Teske, S., T. Pregger, S. Simon, T. Naegler, W. Graus, and C. Lins. 2011. Energy [R]evolution 2010—a sustainable world energy outlook. *Energy Efficiency* 4(3): 409-433.

Trabish, H. K. 2012. CSP 2012: Concentrated Solar Power Review. *Greentech Media*. Accessed at http://www.greentechmedia.com/articles/read/CSP-2012-Concentrated-Solar-Power-Review-2012.

Trieb, F., C. Schillings, M. O'Sullivan, T. Pregger, and C. Hoyer-Klick. 2009. Global Potential of Concentrating Solar Power. Paper presented at SolarPACES 2009, Berlin.

Trieb, F., V. Quaschning, C. Schillings, S. Kronshage, L. Brischke, and G. Czisch. 2002. *Potenziale, Standortanalysen, Stromtransport*. FVS Themenheft.

Tsoutsos, T., V. Gekas, and K. Marketaki. 2003. Technical and economical evaluation of solar thermal power generation. *Renewable Energy* 28(6): 873-886.

Tsoutsos, T., N. Frantzeskaki, and V. Gekas. 2005. Environmental impacts from the solar energy technologies. *Energy Policy* 33(3): 289-296.

Turchi, C., M. Mehos, C. K. Ho, and G. J. Kolb. 2010a. Current and Future Costs for Parabolic Trough and Power Tower Systems in the US Market In *SolarPACES 2010*. Perpignan, France.

Turchi, C. S., M. J. Wagner, and C. F. Kutscher. 2010b. *Water Use in Parabolic Trough Power Plants: Summary Results from WorleyParsons' Analyses* Golden, Colorado: National Renewable Energy Laboratory.

Ummel, K. 2010. *Concentrating Solar Power in China and India: A Spatial Analysis of Technical Potential and the Cost of Deployment*. Washington, DC: Center for Global Development.

USDOE. 2001. *Concentrating Solar Power: Energy from Mirrors* Washington, DC: United States Department of Energy.

USDOE. 2009. *Concentrating solar power commercial application study: Reducing water consumption of concentrating solar power electricity generation*. Washington, DC: United States Department of Energy.

USDOE. 2011. *Draft Solar Programmatic Environmental Impact Statement*. Washington, DC: United States Department of Energy.

USDOE. 2012. *SunShot Vision Study*. Washington, DC: United States Department of Energy.

Viana, T. S., R. Rüther, F. R. Martins, and E. B. Pereira. 2011. Assessing the potential of concentrating solar photovoltaic generation in Brazil with satellite-derived direct normal irradiation. *Solar Energy* 85(3): 486-495.

Viebahn, P., S. Kronshage, F. Trieb, and Y. Lechon. 2008a. New Energy Externalities Developments for Sustainability (NEEDS) – Final report on technical data, costs and life cycle inventories of solar thermal power plants. Stuttgart: DLR.

Viebahn, P., Y. Lechon, and F. Trieb. 2011. The potential role of concentrated solar power (CSP) in Africa and Europe—A dynamic assessment of technology development, cost development and life cycle inventories until 2050. *Energy Policy* 39(8): 4420-4430.

Wagner, W. D., R. L. Mckernan, P. A. Flanagan, and R. W. Schreiber. 1983. *Wildlife Interactions at Solar One Facility: Final Report.* Rosemead, California: Southern California Edison Company.

Wei, M., S. Patadia, and D. Kammen. 2010. Putting renewables to work: How many jobs can the clean energy industry generate in the US? *Energy Policy* 38(2): 919-931.

Weinrebe, G., M. Bohnke, and F. Trieb. 1998. Life cycle assessment of an 80 MW SEGS plant and a 30 MW PHOEBUS power tower. In *International Solar Energy Conference on "Solar Engineering" 1998*. Albuquerque, USA.

Whitaker, M. B., G. A. Heath, J. J. Burkhardt, and C. S. Turchi. 2013. Life Cycle Assessment of a Power Tower Concentrating Solar Plant and the Impacts of Key Design Alternatives. *Environmental Science & Technology* 47(11): 5896-5903.

Woody, T. 2009. Alternative Energy Projects Stumble on a Need for Water. *The New York Times*, 30 September 2009, B1.

WorleyParsons. 2008. *PLE - Beacon Solar Energy Project: Dry Cooling Evaluation*. Folsom, California: WorleyParsons Group Inc.

Wu, Z., Hou, A., Chang, C., Huang, X., Shi, D., Wang, Z. 2014. Environmental impacts of large-scale CSP plants in northwestern China. Environmental Sciences: Processes and Impacts, 16 (10), pp. 2432-2441.

Xie, W. T., Y. J. Dai, R. Z. Wang, and K. Sumathy. 2011. Concentrated solar energy applications using Fresnel lenses: A review. *Renewable and Sustainable Energy Reviews* 15(6): 2588-2606.

Zhang, M., Z. Wang, C. Xu, and H. Jiang. 2012. Embodied energy and emergy analyses of a concentrating solar power (CSP) system. *Energy Policy* 42(0): 232-238.

Zhang, Y., S. J. Smith, G. P. Kyle, and P. W. Stackhouse Jr. 2010. Modeling the potential for thermal concentrating solar power technologies. *Energy Policy* 38(12): 7884-7897.

Zoellner, J., P. Schweizer-Ries, and C. Wemheuer. 2008. Public acceptance of renewable energies: Results from case studies in Germany. *Energy Policy* 36(11): 4136-4141.

Chapter 7

Photovoltaic power

Lead authors: Joseph Bergesen, Sangwon Suh
Contributing authors: Garvin Heath, Shi Lei, Patrick O'Donoughue, Thomas Gibon

7.1 INTRODUCTION

7.1.1 BACKGROUND

The photovoltaic (PV) solar market is growing rapidly, with 37 GW of new installed capacity in 2013 for a total global capacity of 137 GW, up from only 69 GW in 2011 and only 23 GW of global capacity in 2009 (EPIA, 2012, 2014). Over the past few years, the PV market has continued to grow by over 30 per cent per year. This rapid growth was previously driven in large part by policy incentives in European countries such as Germany and Italy, and rapidly decreasing costs. In 2013, however, the demand for new PV installations was led by Asian markets, particularly China and Japan. PV electricity generation can be achieved by a diverse group of technologies. PV power plants can be deployed as utility-scale ground-mounted operations, installed on commercial and residential roofs, and integrated into building facades. The semiconductor component of a PV device can be produced from a variety of metals and materials, of which the most commonly used is silicon, the second most abundant element in the earth's crust. By-product metals such as cadmium, tellurium, indium and gallium are also used in PV device production. The generation of electricity from PV is highly dependent on local solar irradiation, which is the amount of solar energy incident on a given power plant. As a result, the life cycle environmental impacts and energy payback times of PV facilities per unit energy delivered decrease as solar irradiation increases. Additionally, the environmental impacts and costs of PV generation can vary depending on the resources required to produce different technologies, the manufacturing processes involved and the location where a PV plant is deployed.

While many studies in the literature have addressed the life cycle impacts of PV, the studies are not harmonised; they do not use comparable assumptions and data and rarely address a comprehensive list of impact categories, rather focusing on greenhouse gas (GHG) emissions and energy payback time. Furthermore, literature has yet to aggregate the impacts of large scale development of PV from 2010 to 2050. This chapter seeks to quantify the environmental effects of the long-term and large-scale development of PV electricity generation through life cycle assessment (LCA) methodology. To accomplish this task, it is first necessary to understand the state-of-the-art for PV technologies, the changing market for different PV technologies, their differing resource inputs and the processes required to manufacture PV devices.

7.1.2 GOAL AND SCOPE

7.1.2.1 Goal

The goal of this chapter is to educate and inform decision makers on the environmental implications and natural resource risks associated with the generation of electricity from PV from the present to 2050. The PV industry is diverse in technologies and is rapidly changing and growing. Thus, policy makers from the regional to the national and global levels need to be informed about the current and future potential environmental benefits, risks and trade-offs associated with using PV power as an alternative to conventional, fossil fuel-based generation.

This chapter seeks to accomplish these goals by quantifying the life cycle environmental impacts of PV manufacturing and deployment for baseline year 2010 and projecting impacts in 2030 and 2050. This will be accomplished through a hybrid life cycle assessment (HLCA) of three representative PV technologies manufactured in Asia, the United States and Europe. The technologies analysed are: polycrystalline silicon (poly-Si), the most common PV technology, cadmium telluride (CdTe), and copper indium gallium selenide (CIGS), two of the most common and rapidly improving thin-film PV technologies on the market. These technologies were selected for their maturity, representativeness, overall market share, and data availability. A comprehensive set of life cycle environmental impacts from utility-scale and distributed PV will be calculated for all nine regions in the International Energy Agency (IEA) Energy Technology Perspectives. For reference, LCA results will be compared to the impacts of average electricity generation in those regions, allowing a discussion of the possible environmental benefits, risks and trade-offs from increased generation of PV electricity. In Chapter 10, LCA results for these technologies will be compared to other low-carbon electricity sources and results will be aggregated according to the projected demand for PV electricity given by the BLUE Map scenario (IEA, 2010).

7.1.2.2 Scope

While the goal of this study is to quantify the life cycle environmental impacts of all PV development, it would be tedious to model the life cycle of every PV technology in production; therefore, the authors have chosen certain key technologies as proxies to attempt to represent varying types of PV technologies. The following technologies were chosen: (1) mono-crystalline and poly-Si PV produced in China, (2) CdTe PV produced in Asia and the USA, and (3) CIGS PV produced in Asia, Europe and the USA. In order to compare PV to other energy sources, it is also crucial to model the life cycle of the PV balance of system (BOS), which includes the frame, supports, construction and electrical infrastructure necessary to deliver PV electricity to the grid. Additionally, BOS components and other factors affecting the LCA performance of PV systems, such as performance ratio, and solar irradiation, will differ for ground-mounted utility scale PV installations and roof-mounted commercial and residential installations.

7.2 PHOTOVOLTAIC TECHNOLOGIES

PV cells generate electricity through the photoelectric effect, converting incident photons into an electric direct current (DC). The majority of PV cells are comprised of doped semiconductors that form a p-n junction. An electron-donating n-type material and a positive, electron-accepting p-type material form a forward-biased junction, creating a voltage difference. Incident photons are absorbed, exciting the electrons across the p-n junction, and thus creating an electric current. The theoretical limit of PV energy conversion efficiency is called the Shockley Queisser Limit (Shockley and Queisser, 1961) and limits single junction PV cells to an efficiency of approximately 33 per cent. However, PV cells can surpass this limit by including multiple layers of p-n junctions with different band gaps that target photons of differing energies.

The smallest components of PV systems are cells, which are interconnected with electrical contacts and sealed to form a module. Modules are the basic building blocks of PV plants, and are connected in series or in parallel to produce the desired current and voltage. This arrangement of PV modules is called a PV array. Arrays are configured to "maximize the amount of PV capacity per connection point, which minimizes electrical costs per PV Watt peak capacity (W_p)" (Mason et al., 2006). In addition, PV power plants require inverters to convert direct current to alternating current, which is used by the electrical grid. PV facilities also require transformers that increase the voltage of the electricity generated by PV to match that of the electrical grid, which differs depending on whether the PV electricity is added to the transmission or distribution levels of the grid. Thus, transformer equipment and therefore their environmental impacts may be different for utility-scale and roof-mounted applications.

7.2.1 MARKET SHARES OF PHOTOVOLTAIC TECHNOLOGIES

By the end of 2013, the total global installed capacity of PV rose to 137 GW, up from just 69 GW in 2011 (EPIA, 2012, 2014). In 2011, the European Photovoltaic Industry Association (EPIA) estimated a global module production capacity of 30 to 32 GW in 2010 for crystalline silicon modules and 3.5 GW for thin film modules (EPIA, 2011). The annual module *production* capacity is close to the annual *installed* capacity, showing the rapid growth of the PV industry. At this rate of growth, a majority of the PV capacity that will be in operation in just a few years still has not been manufactured or deployed; those PV systems will reflect future, near-term technological changes. While the industry is currently dominated by silicon modules, thin film module production is expanding, and was estimated to reach 6 to 8.5 GW in 2012. The PV production market is a global supply chain. While 50-55 per cent of the total PV system value is produced close to the end market (in Europe in 2011, this was 80 per cent), a large portion of equipment and raw inputs are produced in the United States and Europe before being shipped to Asia for module assembly. Figure 7.1 shows the changing market shares of PV technologies from 1999-2010 (Mints, 2011).

FIGURE 7.1

Market shares of PV technologies PxSi: polysilicon, Cz:: monocrystalline silicon

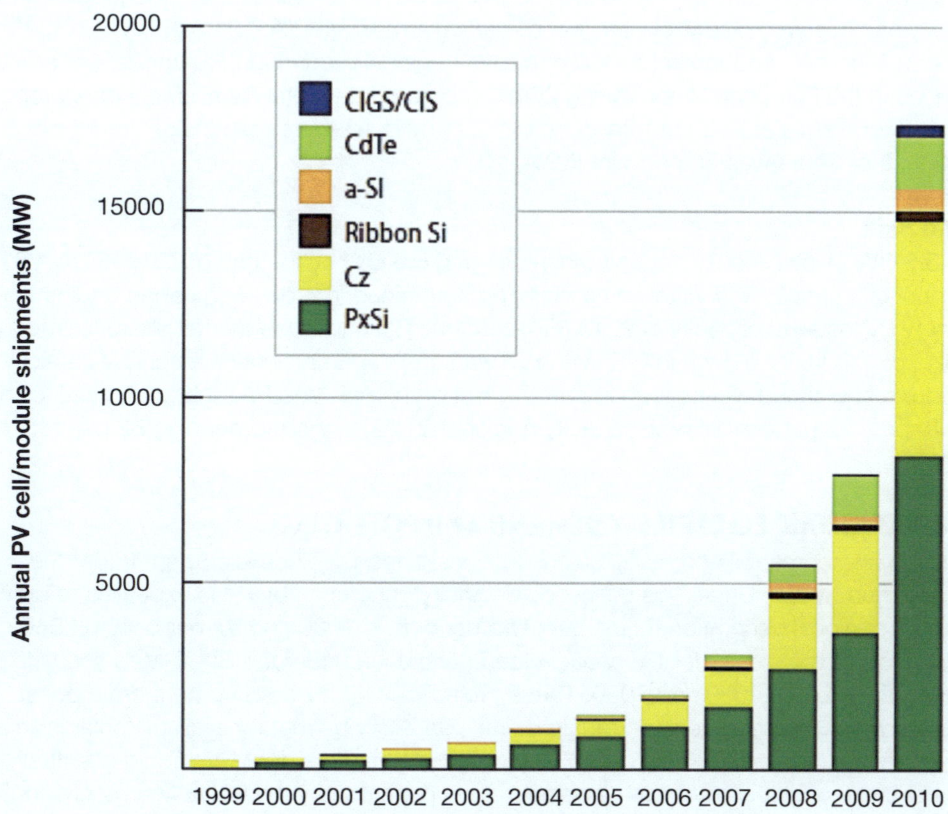

Source: Mints, 2011; Ardani and Margolis, 2011

7.2.1.1 North America

While the United States accounts for a small portion of the total PV manufacturing market, 7 per cent in 2008 (Price and Margolis, 2010), it has held significant market shares in the production of thin-film PV and polysilicon, with its two largest PV manufactures being thin-film producers of CdTe and amorphous silicon (EIA, January 2011). As of 2013, North America produced 12 per cent of the world's thin-film modules and supplied 34 per cent of the world's polysilicon feedstock (EPIA, 2013).

7.2.1.2 China

Between 2006 and 2010, China's production of PV modules grew from less than 15 per cent to more than 50 per cent of global PV production (EPIA, 2011), making China the world's largest PV module producer. China holds the majority of production capacity for poly-Si cells and modules, the most common PV technology. China's consumption of PV modules is also growing rapidly; in 2013 alone, China installed 11 GW of PV capacity, almost one third of new global installations, eclipsing Europe for the first time in annual installed capacity (EPIA, 2013). The EPIA forecasts that the total installed capacity of PV in China, currently 18 GW, could exceed 35 GW by 2016 (EPIA, 2012).

7.2.1.3 Europe

Until 2013, Europe was the largest market PV modules (80 per cent in 2010). Since then, the greater growth in PV installed capacity has occurred elsewhere the world, particularly Asia (EPIA, 2011) (EPIA, 2014). While the European share of PV module production has fallen relative to global production, 80 per cent of all PV inverters were produced in the European Union during 2010. Importantly, as total PV system costs drop, modules are accounting for a smaller share of total cost; in 2010, modules were responsible for 60 per cent of total cost while in 2005, their share was 75 per cent (EPIA, 2011).

7.2.1.4 Japan and Asia

In 2010, Japan had 3.6 GW of installed PV capacity, increasing to 6.9 GW by the end of 2012 (EPIA, 2013). Japan's module production capacity and installed capacity both continue to grow, with Japan adding an additional 7 GW of new PV installed capacity in 2013 (EPIA, 2014). The Japanese market share for module production is dropping due to faster growth in China and Taiwan, but Japan maintains a healthy capacity for both thin film modules and silicon wafers. Japan and the rest of the Asia and Pacific region (not including China) accounted for about 25 per cent of new capacity additions in 2013, while producing 62 per cent of the world's thin-film modules.

7.2.2 GLOBAL PHOTOVOLTAIC ELECTRICITY DEMAND AND POTENTIAL

This report examines the environmental and resource implications of global PV development under the BLUE Map scenario outlined by IEA. Under this scenario, electricity generation by PV is expected to grow from under one per cent of global generation to just over two per cent in 2030 and six per cent in 2050 (IEA, 2010). Table 7.1 shows the demand for PV capacity and generation under the BLUE Map scenario for each, region and the world in 2010, 2030 and 2050. These projections for PV development are modest. The following section will show how these generation requirements can easily be met by available solar irradiation, land and roof space in each region by 2030 and 2050. It is possible, therefore, that PVs could eventually provide a much greater share of global electricity generation than projected by the BLUE Map scenario. For example, seminal studies from the National Renewable Energy Laboratory (NREL) in the United States (US) have examined the feasibility of renewable energy technologies providing more than 80 per cent of electricity in the US by 2050, showing that as much 50 per cent of electricity can come from variable wind and solar generation while still meeting hourly demand (Mai et al., 2012). Furthermore, NREL's Western Wind and Solar Integration Study showed that a 33 per cent share of variable wind and solar power would be feasible in the western US, and would lead to only small increases in costs and emissions due to increased cycling of fossil fuel-based power plants (Lew et al., 2013).

TABLE 7.1

Demand for PV capacity and generation under the BLUE Map scenario (IEA, 2010)

	Installed capacity (GW)			Generation (TWh per year)		
Region	2010	2030	2050	2010	2030	2050
China	0	56	270	0	91	454
India	0	20	101	0	28	183
OECD Europe	5	85	125	4	98	168
OECD North America	1	83	148	0	148	286
OECD Pacific	2	46	123	0	52	179
Economies in transition	0	8	46	0	13	73
Latin America	0	22	154	0	43	315
Other developing Asia	0	29	135	0	49	230
Africa and Middle East	0	60	276	0	124	581
World	**8**	**410**	**1378**	**4**	**646**	**2469**
Percentage of total capacity/ generation	**0.2%**	**5.3%**	**12.6%**	**0.02%**	**2.3%**	**6.2%**

Licence: http://www.iea.org/t&c/termsandconditions/

7.2.2.1 Global solar resources

As discussed in the chapter on concentrating solar power, the technical potential of solar resources over the globe is enormous. Additionally, the locations of utility-scale PV power plants and concentrating solar power plants share similar traits, so this section will focus on the technical potential for distributed PV electricity generation on residential and commercial roofs.

Roof-mounted photovoltaics

The purpose of this study is to assess the environmental implications of the GHG mitigation scenarios. The BLUE Map scenario projects a given amount of PV energy generation in each region from 2010 to 2050. The number of PV modules and power plants needed to provide that electricity is inversely proportional to the amount of solar irradiation where the PV arrays are deployed; the higher the irradiation, the fewer the number of PV modules are required to generate the same amount of electricity.

According to the IEA BLUE Map scenario, the majority of PV development by 2050 will be distributed commercial and residential generation, which we assume to be mostly roof-mounted. To estimate the likely solar resources for roof-mounted PV throughout the world, we limited the analysis of global solar irradiation to urban areas, under the assumption that roof-mounted PV will be deployed exclusively in inhabited areas.

Solar irradiation data for the entire globe were generated from a number of sources: NASA, INPE, LABSOLAR, NREL, Natural Resources Canada, the European Commission PVGIS and the State University of New York (INPE/LABSOLAR, 2000/2001a, 2000/2001b; National Renewable Energy laboratory (NREL) Climatological solar Radiation Model, 2004; Perez Model, 2011; Kunreuther and Slovic, 1996; PVGIS (Optimal Incline) - Europe - Turkey - Portions of North Africa, 2012; Daily, 1997). Population density data were generated from Oak Ridge National Laboratory's Landscan 2009 global population data at 1 km nominal resolution (Fthenakis and Kim, 2009) and divided into 32 classes using the Geometrical Interval classification methodology. The United States Geological Survey (USGS) describes the urban threshold as 1000 inhabitants per square mile.

Consequently, several classes below this urban threshold were added to better articulate rural areas as the majority of global land is well below the urban cut-off. This population scheme was combined with solar tilt resource bins and split by simplified country boundaries provided by Natural Earth (naturalearthdata.com, 2012). The total area for each country, population and resource class was then summed.

Each IEA region was divided into classifications of irradiation and population density, and a geographic information system model was used to calculate the total available land area in each region for a given range of irradiation and population densities. Available land area was then filtered to include only urban areas and we computed a spatial average solar irradiation for urban areas in each IEA region. Table 7.2 shows irradiation values for average urban areas in each region.

TABLE 7.2

Weighted irradiation values for roof-mounted and ground-mounted PV (IEA, 2010)

	Irradiation for latitude tilted PV arrays (kWh/m²/year)								
	China	**India**	**Europe**	**North America**	**Pacific**	**Economies in Transition**	**Latin America**	**Developing Asia**	**Africa and Middle East**
Roof	1625	2061	1351	1800	1607	1486	1965	1952	2045
Ground	2308	2115	2191	2333	2507	2062	2282	2296	2471

TABLE 7.3

Area requirements for roof-mounted PV under the BLUE Map scenario (IEA, 2010)

	Area requirements for roof-mounted PV under BLUE Map scenario								
	China	**India**	**Europe**	**North America**	**Pacific**	**Economies in Transition**	**Latin America**	**Developing Asia**	**Africa and Middle East**
Roof-mounted PV generation in 2050 (TWh/year)	268	108	99	168	105	43	186	136	342
Total PV module area needed (hectares)	14	4	6	8	6	2	8	6	14
Total urban area utilized (hectares)	174	55	77	99	69	31	100	73	177
Per cent of urban area needed	3.4	1.2	3.5	4.7	12.8	2.6	9.0	1.8	5.4

Applying the assumed 2010 PV technology mix (see Chapter 7.4.5) and the corresponding energy conversion efficiency, we estimated the area necessary to generate enough PV electricity to satisfy the annual generation requirements of the BLUE Map scenario in 2050. Following the assumptions of Hofierka and Kaňuk (2009), we assumed that only approximately eight per cent of urban area is available as suitable roof space for PV installation (Hofierka and Kaňuk, 2009). Even with these strict space limitations, roof-mounted PV generation would not be limited by the availability of suitable roof space in any region. Table 7.3 shows the estimates for the roof-space and urban area required to fulfil the demand for roof-mounted generation by the BLUE Map scenario in 2050.

In this analysis, we assume that roof-mounted PV will be deployed in urban areas of average irradiation for each region. Because per-kWh PV costs are lower when deployed in areas of higher insolation, PV arrays may be installed under more productive conditions, thus reducing the total roof area required for PV. Conversely, the deployment of roof-mounted PV facilities can also be heavily affected by policies, such as feed-in tariffs, that could encourage PV installations in areas with insolation below the regional average, such as Germany in the Europe region. Such policies could potentially increase the amount of roof area needed for PV by 2050.

7.2.3 OVERVIEW OF PHOTOVOLTAIC TECHNOLOGIES AND LITERATURE REVIEW

The following section provides an overview of current and emerging PV technologies and a discussion exploring the future potential of these technologies. Comparing the overall environmental impacts of PV technologies on a life cycle basis is challenging due to the rapidly evolving nature of the PV industry, the multitude of production processes, and the difficulty in comparing results across studies. The fast growth of the PV market and technology development combined with the assumed 2030 and 2050 PV capacity in the BLUE Map scenario suggests that considering only the present technical capabilities and environmental impacts of PV technologies would be inadequate. Thus, this Chapter uses the most reliable technology roadmaps and projections available to explore the potential future environmental and resource implications of PV technologies.

7.2.3.1 Technology descriptions and life cycle impacts

A multitude of studies have analysed the life cycle impacts of PV-generated electricity, but not all studies use the same methodologies and assumptions, nor do they report a comprehensive set of environmental and natural resource impact categories. Hsu et al. (2012) and Kim et al. (2012) provide frameworks to harmonize PV LCA results across studies through a meta-analysis of crystalline silicon PV and thin film PV, respectively. In doing so, they also concisely summarize the life cycle GHG emissions as reported in literature in the last decade (Figures 7.2, 7.3). Adapting study results to use the same assumptions for performance ratio, irradiation, and BOS components, for example, is crucial in comparing LCA results across studies. Furthermore, many previous PV LCA studies only include a limited number of life cycle indicators, namely GHG emissions and energy payback time. While some studies report the life cycle emissions of specific pollutants such as cadmium, more generalized, comprehensive lifecycle indicators such as human toxicity potential are infrequently assessed. Furthermore, life cycle inventory (LCI) data for emerging technologies such as quantum dot PV (QDPV), dye sensitized PV and organic polymer PV have high uncertainty because production processes are theoretical, currently only at laboratory scale, or highly variable, that is, not standardized within the industry.

Figure 7.2 shows the distribution of LCA results in literature for GHG emissions per kWh for the most common PV technologies. Comparative studies have found that CdTe has lower emissions than poly-Si (or multi-silicon), mono-silicon and ribbon-silicon. However, it is important to note that all PV technologies analysed in these harmonization studies have overlapping distributions of results, and are close in magnitude. Among crystalline silicon technologies, ribbon has been shown to be the least GHG intensive, followed by mono-silicon and poly-Si (reported as "multi-Si" in Figure 7.2).

FIGURE 7.2

Boxplot comparing published and harmonized GHG emissions for all crystalline-silicon systems, 1700 kWh/m²/year

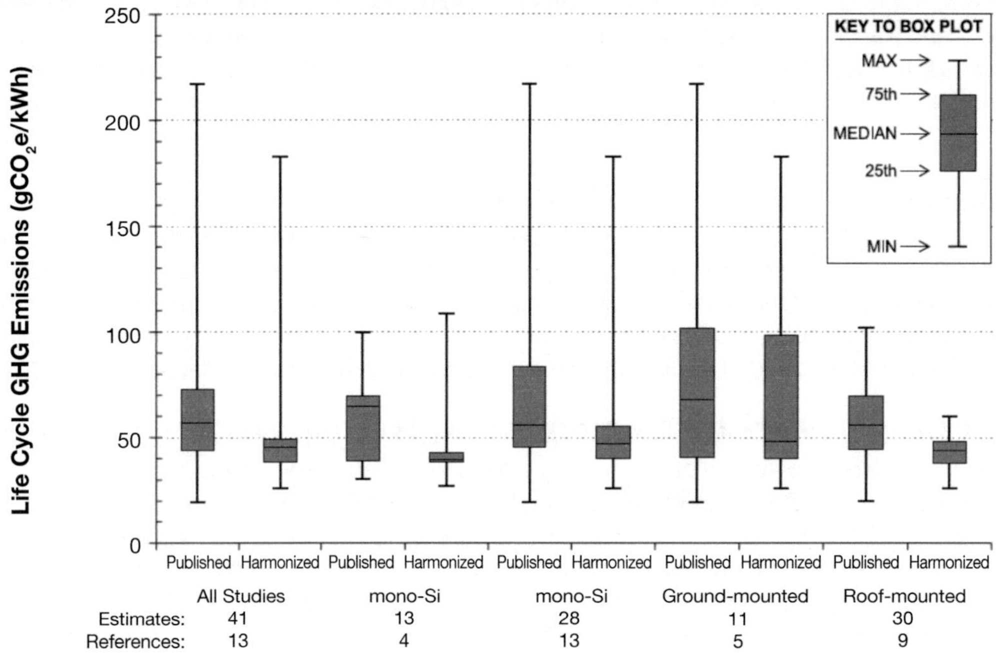

Source: Hsu et al., 2012

FIGURE 7.3

Life cycle CO_2 emissions from thin film PV life cycles as reported in literature from 1995-2010, previous to screening (left). Number in parenthesis indicates the number of scenarios. Results of harmonization presented on the right.

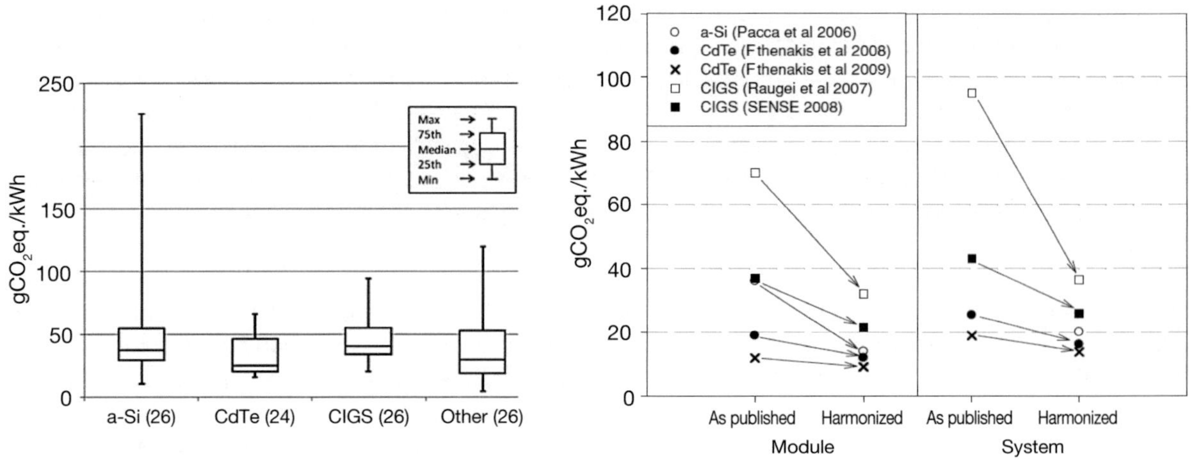

a-Si – amorphous silicon; CdTe – cadmium tellurium; CIGS – copper indium gallium selenide
Source: Kim et al., 2012

FIGURE 7.4

Median harmonized greenhouse gas emissions from PV life cycles under average irradiation in USA (1800 kWh/m²/year). Results of NREL Harmonization Studies.

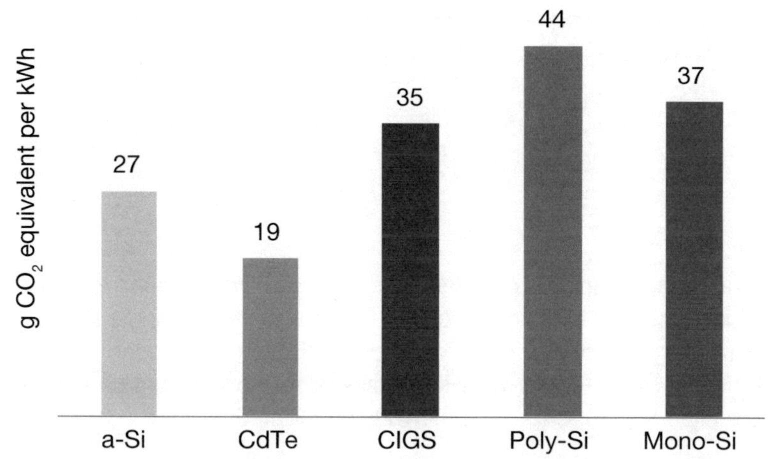

Source: Hsu et al., 2012; Kim et al. 2012

While the preceding figures show that median life cycle GHG emissions are lower for thin film technologies, particularly CdTe, the harmonization studies showed close, overlapping distributions of results for all technologies (Figure 7.2 and Figure 7.3). Figure 7.4 summarizes the median life cycle GHG emissions for major PV technologies.

7.2.3.2 Crystalline silicon photovoltaics

Crystalline silicon is the most pervasive and mature PV technology, making up 77 per cent of installed PV systems in the United States (Hsu et al., 2012). Among this category are poly- (or multi-) crystalline made from solar grade silicon semiconductor wafers. After solar grade silicon is produced, it can either be further purified to produce mono-crystalline silicon wafers or refined to produce poly-Si wafers. The further refinement of silicon to produce mono-crystalline silicon is energy intensive, but leads to more energy efficient modules. Poly-Si cells produced from the ribbon process forgo the energy intensive wafer sawing process by forming molten silicon into cells that can be simply cut rather than sawn from ingots. Crystalline silicon technologies are more mature than thin-film PV, and mono-silicon modules have generally higher efficiencies than most common thin-film modules, such as current CdTe and CIGS modules. Another advantage of silicon PV is the abundance of silicon in the earth's crust

Poly-silicon

Poly-Si PV (also referred to as multi-Si) modules typically range in efficiency from 13-16 per cent in commercial practice and have achieved module efficiencies as high as 18.2 per cent (Green et al., 2012). Poly-Si cells were originally produced from scraps of electronic grade silicon which are then re-smelted and sawed into wafers (Raugei and Frankl, 2009). However, increased PV production has induced a demand for silicon wafers that are produced especially for PV. The crystal growing and smelting steps are the most energy intensive, and some material is lost during the wafer sawing phase. In their paper, Hsu et al. give a harmonized median of LCA results for poly-Si systems of 44 g CO_2 eq/kWh, assuming average US irradiation of 1800 kWh/m²/year.

Mono-silicon

Monocrystalline, or single-crystalline, silicon modules generally have higher efficiencies but require greater energy input to manufacture than polycrystalline silicon modules. Module efficiencies are generally 1-2 per cent higher than their poly-Si counterparts, and have reached up to 22.9 per cent in the lab (Green et al., 2012). Higher efficiencies and higher manufacturing energy requirements are both a result of the additional re-crystallization step, which enhances the light-absorbing ability of the cell by producing a more pure crystal structure. Some comparative studies have shown greater GHG emissions for mono-crystalline devices compared to poly-Si (Fthenakis and Kim, 2011). The harmonization study by Hsu et al., however, shows lower median GHG emissions at 37 g CO_2 eq per kWh, compared to poly-Si, assuming 1800 kWh/m^2/year (Hsu et al., 2012).

Ribbon-silicon

Ribbon-silicon modules (ribbon-Si) are made from a proprietary process that eliminates the ingot growing and sawing processes and thereby reduces material losses and lowers the energy requirements for manufacture. Ribbon modules maintain an efficiency of around 11.5 per cent, lower than poly-Si or mono-Si (Fthenakis and Kim, 2011). In their 2011 paper, Fthenakis and Kim show GHG emissions for ribbon-Si to be somewhat lower than poly- or mono-Si at approximately 30 g CO_2 eq/kWh (Fthenakis and Kim, 2011).

7.2.3.3 Thin film photovoltaics

Thin film PV has an advantage in being able to provide a slightly lower-efficiency panel at lower cost, requiring somewhat less energy and materials for manufacture. However, thin film PV technologies can require a variety of by-product metals like tellurium and indium that may fluctuate in cost and may face supply constraints in the future (Woodhouse et al., 2012; Woodhouse et al., 2013). Cell production requires small quantities of chemicals and materials, resulting in reduced energy demand. The semiconductor materials are sputtered or co-evaporated onto a back contact, buffer layer and transparent conducting oxide layer. Thin film PV cells tend to be more flexible than their silicon counterparts, making them suited for building-integrated applications. Additionally, while current generation thin films are generally lower in efficiency than crystalline silicon PV, they may be designed for superior performance in low insolation conditions, such as overcast weather. One key issue affecting the LCA analysis of thin film devices is the allocation of emissions from the production of cadmium, tellurium and other semiconductor materials; both cadmium and tellurium are by-products of other processes such as copper and zinc smelting (Fthenakis et al., 2009). In 2008, thin films only comprised 15 per cent of the global PV market, and this number was expected to grow to as much as 34 per cent by 2012 (Price and Margolis, 2010), but has actually shrunk to 11 per cent by 2011 (Bazilian et al., 2013).

Cadmium telluride and cadmium sulfide

Cadmium telluride and cadmium sulfide (CdS) thin film modules are among the prominent thin film technologies currently produced. Typical module conversion efficiencies range from 11-13 per cent in practice (Woodhouse et al., 2012), with a record commercial module efficiency of 16.1 per cent (First Solar, 2013). Numerous studies show that CdTe PV systems have relatively low GHG emissions: 19 g CO_2-eq./kWh for ground-mount application under United States average insolation (Kim et al., 2012). Furthermore, although cadmium emissions are a public health concern, Fthenakis, Kim and Alsema have shown in their 2008 paper that life cycle cadmium emissions from CdTe PV generation are less than those from generation of conventional electricity required to manufacture crystalline silicon panels (IEA Energy Analyses, 2007). In addition, the limited tellurium supply may constrict the expansion of CdTe PV production (Zweibel, 2010; U.S. Department of Energy, 2010). However, some research suggests that thin films can nevertheless be expanded to terawatt-levels of generation through increased material efficiency in module production as well as recovery of tellurium and other absorber materials at module end-of-life (Fthenakis, 2012). Figure 7.5 shows the cell configuration for a typical CdTe module.

FIGURE 7.5

Typical cadmium telluride (CdTe) cell layer configuration. This is a generalized configuration, not representative of all CdTe panels. The panel is manufactured top down, building the various layers on top of the front glass.

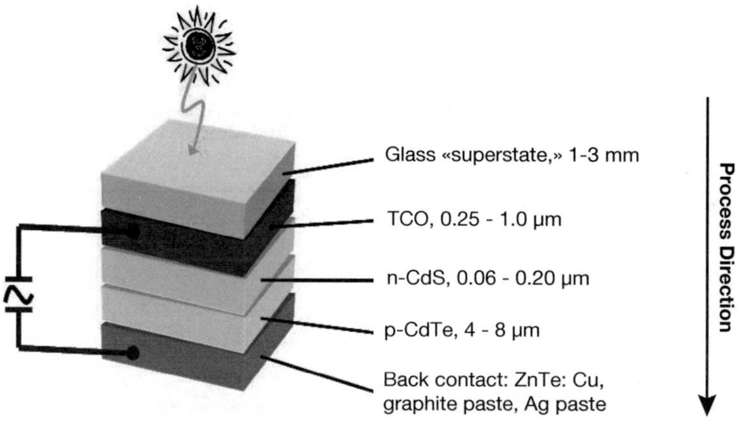

TCO: transparent conductive oxide; n-CdS: n-layer cadmium sulfide; p-CdTe: p-layer cadmium telluride; ZnTe: zinc telluride; Cu: copper, Ag: silver.
Source: Woodhouse, 2011

Copper indium gallium (di)selenide

Copper indium (di)selenide (CIS) and copper indium gallium (di)selenide (CIGS) comprise another established thin film technology. In 2010, CIGS and CIS only comprised 15 per cent of the thin film PV market, but according to the European Photovoltaic Industry Association (EPIA) CIGS was expected to equal the production capacity of CdTe and amorphous silicon by 2012 (European Photovoltaic Industry Association, 2011). Typical modules in production are around 12-13 per cent efficient (Whitney, 2014), with some commercial modules now reaching 15.7 per cent [(TSMC, 2013)]. A few past LCA studies of CIGS and CIS have been used in the ecoinvent database (Jungbluth, 2005); the most recent was based on small-scale production data in Germany circa 2006 (Sense, 2008). Currently, literature shows that CIGS PV systems have median life cycle GHG emissions of 26 g CO_2-eq/kWh, which is somewhat higher than other thin films (Kim et al., 2012). Figure 7.6 shows a typical cell configuration for CIGS modules.

FIGURE 7.6

Typical copper indium gallium selenide (CIGS) cell layer configuration

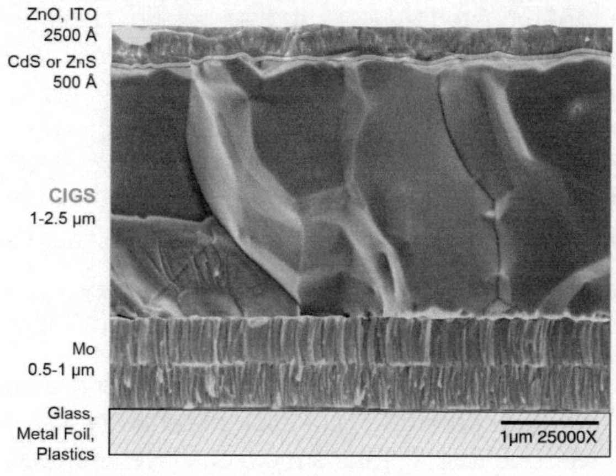

ZnO: zinc oxide; ITO: indium tin oxide; CdS: cadmium sulfide; Mo: molybdenum
Source: Goodrich, 2011

Amorphous silicon

Amorphous silicon (a-Si) PV cells are produced from un-crystallized silicon, some of which comes from waste silicon from the computer and semiconductor industries. Production capacity of a-Si is comparable to that of CdTe. Cell efficiencies range from 5-7 per cent for single junction design to 8-10 per cent for double and triple junctions (Parida et al., 2011). Although prone to some degradation in efficiency over the cell lifetime, a-Si modules are advantageous because of their low manufacturing cost and the abundance of silicon. In the harmonization studies, Kim et al. show that a-Si GHG emissions are comparable to other thin film technologies at approximately 20 g CO_2-eq/kWh.

7.2.3.4 Emerging photovoltaic technologies

Quantum dot

Quantum dot PV cells (QDPV) are based on nanotechnology and have a band gap that adjusts to different energy levels. Consequently, they have high conversion efficiency and are expected to exceed theoretical energy conversion efficiency limits due to multiple exciton generation. Azzopardi and Mutale (2010) predict GHG emissions of 2.89 g CO_2-eq/kWh at 10 per cent efficiency while Sengul and Theis (2011) predict 5 g CO_2-eq/kWh at 14 per cent efficiency.

Polymer

Organic polymer PV cells are emerging as a potential low-cost solar technology. However, efficiencies for this technology are low; current device efficiencies are in the range of 2-3 per cent, and are expected to increase to 5-7 per cent in the future. These modules also have a shorter lifetime than conventional thin film or silicon PV (Kalowekamo and Baker, 2009). Disadvantages in lifetime and cell efficiencies are eventually expected to be overcome by cost advantages, versatility, and ease of deployment. Polymer PV layers can also be used in tandem with other technologies to make hybrid PV cells with a higher overall efficiency (Azzopardi and Mutale, 2010). Roes and Patel (Roes et al., 2009) project life cycle GHG emissions in the range of 4 g CO_2-eq/kWh, excluding BOS (see Chapter 7.2.3.5) and 27 g CO_2-eq/kWh, including BOS for a five per cent efficient device. Espinosa et al. also report GHG emissions of 38 g CO_2-eq/kWh for a two per cent efficient device and 57 g CO_2e/kWh for a three per cent efficient device (2011).

7.2.3.5 Balance of system

For PV power plants, the BOS includes any components necessary for generating usable electricity that is not the PV module itself. Most importantly, this includes inverters, which convert the electricity generated from direct current (DC) to the alternating current (AC) used by the grid, frames and support structures, electric installation, and construction. For utility-scale PV, transformers are needed to feed the electricity generated by the PV facility into high-voltage transmission lines.

While the ecoinvent database includes adequate data for roof-mounted PV BOS, it does not include data on the BOS for ground-mounted nor utility-scale power plants. Mason and Fthenakis report the materials and energy required for production of the BOS and O&M of a ground-mounted, 2.5 MW utility scale PV facility in Springerville Arizona, USA (Mason et al., 2006). The plant that was studied included both thin film and crystalline silicon arrays. Results of the study showed that the BOS alone was responsible for 29 kg of CO_2-eq for each square meter of array (Mason et al., 2006). While this number is only a small portion of the total life cycle emissions of a m^2 of PV panel, the impacts of the BOS could be significant as results are aggregated to the scale of global PV development.

This study modelled the BOS and operation and maintenance (O&M) for ground-mounted, utility-scale PV facilities from the bill of materials provided in Mason and Fthenakis (Mason et al., 2006). Material and energy inputs for these processes were modelled using the ecoinvent processes. Construction materials for frames and supports were scaled based on module size while power electronics (transformers and inverters) were scaled to rated capacity of the PV system (W_{DC}).

7.2.3.6 Land use by photovoltaics

A recent study empirically estimated land use for ground-mounted PV facilities greater than 1 MW installed capacity located in the United States using a combination of project documents and analysis of satellite imagery (Ong et al., 2013). Note that this is just the land occupied by the generation facility; it does not consider any land used for upstream processes. Land use requirements were quantified by Ong et al. both on a capacity (area/MW) and generation (area/GWh) basis. Measuring on a generation basis enables consideration of differences in solar resource, tracking configurations, and various technology options, thus providing the fairest point of comparison to other generation technologies. PV energy production values were simulated using the System Advisor Model (Gilman, 2012) for each solar project using inputs specific to individual facilities such as location, array configuration, derate factor and tracking technology. Ong et al. evaluated the impact of various module types, efficiency, array configuration, and tracking type.

Ong et al. obtained land use data for a total of 192 built or under construction projects representing approximately 10.7 GW of capacity and 83 per cent of the installed and under-construction utility-scale PV capacity in the United States as of summer 2012. Table 7.4 summarizes the total and direct impact area for the solar projects evaluated, divided into systems less than 20 MW, referred to as small, and those over 20 MW, referred to as large. Total area is defined as the project boundary, while direct area is the area covered by arrays, inverters, and transformers, and any structures which encumber the land. Direct area is always smaller than and inside of the total area. The capacity-based average results represent capacity-weighted averages.

TABLE 7.4

Land use by ground-mounted, utility-scale (> 1 MW) PV systems in the United States

	Small PV	Large PV
Number of projects	126	66
Total capacity	760MW-DC	9960MW-DC
Reported total area (km^2)	21.9	284.1
Reported direct area (km^2)	9.2	50.1
Capacity total area requirements (ha/MW)	3.4 ± 0.2	3.2 ± 0.2
Generation total area requirements (ha/GWh/year)	1.6 ± 0.1	1.4 ± 0.1
Capacity direct area requirements (ha/MW)	2.4 ± 0.1	3.0 ± 0.4
Generation direct area requirements (ha/GWh/year)	1.2 ± 0.1	1.3 ± 0.2

Source: Ong et al., 2013

Tables 7.4 and 7.5 summarize the descriptive statistics for total and direct land area for all PV facilities in the United States and per PV technology. Based on the results of Ong et al., fixed-tilt systems use, on average,13 per cent less land than single axis tracking on a capacity (MW) basis, but use 15 per cent more land on a generation basis. This is due to increased generation resulting from tracking technologies. Single axis tracking systems can increase PV generation 12-25 per cent relative to fixed tilt orientations, and dual axis tracking systems can increase PV generation by 30-45 per cent (Drury et al., 2013). Ong et al. also found that dual axis flat-panel systems use more land than fixed and single axis plants both on a capacity and generation basis. Both large and small concentrating PV (CPV) systems have the lowest total and direct land use requirements on a generation basis, while having comparable land use requirements on a capacity basis.

TABLE 7.5

PV land use results

System Type	Total Area				Direct Area			
	Projects	Capacity (MW)	Hectares/ MW	Hectares/ GWh/ year	Projects	Capacity (MW)	Hectares/ MW	Hectares/ GWh/year
Small								
Fixed	51	248.7	3.1 ± 0.3	1.7 ± 0.1	42	226	2.2 ± 0.1	1.3 ± 0.1
1 Axis	55	358	3.5 ± 0.2	1.5 ± 0.1	41	198	2.6 ± 0.2	1.2 ± 0.1
2 Axis Flat Panel	4	5.4	5.1 ± 0.9	2.2 ± 0.7	4	5.4	3.8 ± 1.0	1.7 ± 0.7
2 axis CPV	4	8.9	3.7 ± 1.5	1.3 ± 0.5	4	8.9	2.8 ± 0.8	1.0 ± 0.3
Unknown	12	117	3.3 ± 0.3	1.8 ± 0.2	2	6	2.3 ± 0.2	1.3 ± 0.1
Large								
Fixed	15	2114	3.1 ± 0.3	1.5 ± 0.2	8	920	2.5 ± 0.5	1.2 ± 0.2
1 Axis	16	1935	3.8 ± 0.3	1.3 ± 0.1	6	732	3.6 ± 0.4	1.3 ± 0.1
2 Axis CPV	2	186	3.3 ± 0.3	1.1 ± 0.2	1	36	2.5	0.8
Unknown	33	5725	3.5 ± 0.4	N/A	2	140	2.8 ± 0.4	N/A

7.2.3.7 End-of-life and recycling

Concerns about the embodied energy of PV systems and the rarity of semi-conductor metals have engendered discussion about the need for recycling of PV modules and their components (Fthenakis, 2012). Due to the recent rapid growth of the PV industry, the number of PV modules currently reaching end-of-life is far less than the number of modules being produced. It is therefore likely that a large number of modules will not be recycled until 2030 and beyond. Despite these notions, some PV companies have been forward thinking regarding recycling of modules, and thin film manufacturers such as First Solar in the United States have already initiated take-back programs to recycle modules and manage end-of-life waste (First Solar, 2014).

As disposal and recycling processes are being explored and implemented by industry for both silicon and thin film modules, we are only beginning to learn about the impacts and benefits of the end-of-life phase. Müller et al. evaluate the life cycle benefits of recycling crystalline silicon solar cells at a PV recycling plant operated by Deutsche Solar AG since 2003 (2005). Figure 7.7 shows that there is a clear benefit to recycling used silicon wafers.

Furthermore, Fthenakis performed a feasibility study to show that the costs of recycling and semiconductor material recovery from thin film PV modules is not excessive (Fthenakis, 2000). In a 2010 paper, McDonald and Pearce support Fthenakis' previous claim that PV recycling is technically and commercially feasible, but qualify their results saying that policies are needed to ensure producers have the proper incentives to recycle modules, due to the low cost of landfill disposal (McDonald and Pearce, 2010). In addition, as PV production grows and semiconductor metals potentially become scarce in the future, module recycling and material recovery may become more important for ensuring that thin-film PV in particular can meet the rising demand for PV devices in the long term (Fthenakis, 2012).

FIGURE 7.7

Life cycle environmental burdens and disburdens by silicon PV recycling by Deutsche Solar AG

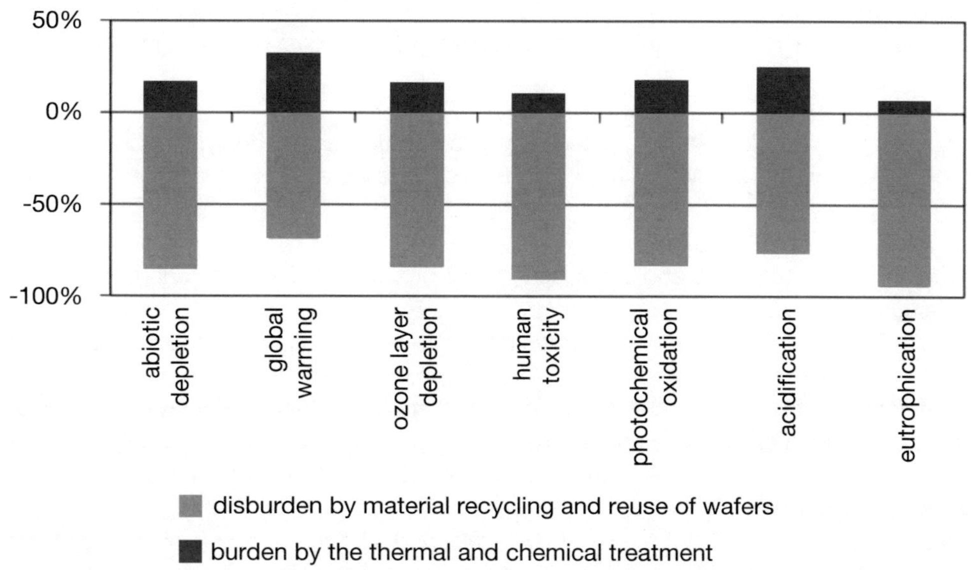

Source: Müller et al., 2005

7.3 METHODS AND DATA

7.3.1 METHODS

7.3.1.1 Hybrid analysis of photovoltaic electricity

LCA is a quantitative, scientific method for evaluating the environmental impacts of a product system. LCA analyses the potential environmental impacts of both direct emissions from manufacturing products as well as upstream supply chain emissions by a product system. Such analysis can be done using a bottom-up, traditional process-sum LCA approach, or a top-down, input-output (IO) approach. HLCA combines both approaches in order to expand the system boundary of the life cycle system and include a more complete assessment of the environmental impacts of a product (Suh et al., 2004). The hybrid method enables a complete analysis of potential impacts from PV development as it allows the inclusion of capital goods and other proprietary inputs for which only cost data is available, and are otherwise not included in a bottom-up analysis. Zhai and Williams (2010) have shown that HLCA results for embodied energy of poly-Si PV can be approximately 60 per cent higher than equivalent process-sum LCA results. This discrepancy results in a difference of 3.3 million tons of CO_2 for a 10 per cent share of PV in the total US electricity generation mix (Zhai and Williams, 2010). As mentioned above, the hybrid method allows for the easier modelling of capital goods that are generally not included in process-based LCA studies. Because PV production capacity must increase to meet the requirements of the BLUE Map scenario, new production facilities must be built, and this could potentially add to the overall environmental assessment of PV.

7.3.1.2 EXIOBASE and ecoinvent for photovoltaic electricity

As described in the methodology chapter, this hybrid analysis uses a combination of EXIOBASE and ecoinvent to model the system processes for production of crystalline silicon, CdTe and CIGS PV modules. For this analysis, ecoinvent processes were used primarily to model known physical inputs, such as regional electricity consumption, refined mineral inputs such as semi-conductor metals, and other common metals, polymers and commodities. Monetary input values were entered into EXIOBASE; these inputs were mainly for inputs that have not been previously modelled in literature or in ecoinvent and for capital goods or labour. For example, EXIOBASE sectors were used to model a few proprietary chemicals unavailable in ecoinvent, but for which cost data are available from the National Renewable Energy Laboratory (NREL) cost models and roadmaps.

7.3.1.3 System boundary

Figure 7.8 illustrates the life cycle for thin film PV technologies and delineates the system boundary for the thin film PV systems considered in this study. The following life cycle phases are included in the system boundary of the study:

1) Facility upstream: building materials and construction, site preparation, manufacturing equipment. Module production[1]
    a. Material extraction and processing: includes raw material extraction, processing, refinement, and part manufacturing; i.e., all steps during the production process that occur before input to the manufacturing facility. These steps are captured indirectly; this project inventories all of the inputs to the plant itself and relies on ecoinvent to capture the embodied burdens. The embodied emissions, energies, materials, and environmental burdens of this life cycle stage are modelled and measured by ecoinvent.
    b. Manufacturing: includes all of the energy, materials, and labour that occur within the manufacturing facility. Also includes recycling of manufacturing waste, namely the high value metals (indium, gallium, selenium, cadmium, and tellurium).[2]

2) PV facility: construction and BOS
    a. Transportation and installation: including the energy, materials, labour, and embodied burdens of those constituents.
    b. BOS includes the frame and supporting structure, inverters, wires, and associated electrical equipment for the end-use application
    c. O&M: transportation, on-site energy consumption, water used for cleaning PV arrays, and periodic replacement of power electronics throughout the 25-30 year lifetime of PV facilities.

7.3.2 DATA

7.3.2.1 Data for thin-film photovoltaics

The manufacturing cost models

NREL in the United States has developed cost models for several PV technologies. The models compile manufacturing information provided by PV manufacturers and use these data to estimate current and future PV costs per unit of energy delivered. The models are validated against financial information from industry reports and private stakeholders. Furthermore, these models are internally reviewed and verified, and have been used in seminal US government studies, namely the United States Department of Energy's Sunshot Vision Study (Margolis et al., 2012). These empirical data offer a unique basis to the life cycle analyses, which generally rely heavily on hypothetical data to generate LCI. This study leverages the wealth of information in the NREL manufacturing cost models to generate current and future estimates of the environmental burdens of thin-film PV technologies on a life cycle basis.

The NREL manufacturing cost models include all of the distinct process steps that occur in the typical manufacturing process for each specific PV technology and the costs associated with those steps. Each

1 The manufacturing cost models only report costs associated with upstream life cycle stages (capital investments). The capital cost can be utilized in lieu of direct material and energy inputs through the HLCA method.

2 Recycling energies and burdens have yet to be considered although material utilization rates are applied, i.e., the material use is scaled by the utilization rate to accommodate the amount that is sent back to the manufacturer for recycle.

PV technology has a unique model. The models generalize the manufacturing process to a certain extent; manufacturers do not implement the exact same manufacturing techniques and there is therefore some loss of resolution by aggregating data into a single model. To construct the models, data were gathered directly from manufacturers, including direct input and cost data for the manufacturing process. Manufacturers provided information pertinent to the cost of the final product, operational costs such as labour and electricity, and the capital investment costs for the building, site, land, and equipment. For some manufacturing inputs only costs were reported, such that sensitive or confidential information was protected. They also described their process stages, run times, equipment electrical demands, material utilization rates, and labour hours.

In this study, we used NREL cost model data for CdTe and CIGS, which were selected as representative thin-film technologies commonly manufactured in developed countries. The CIGS manufacturing cost model relies on data from three international manufacturers while the CdTe model uses data from a single manufacturer with facilities in the United States and Asia. NREL experts characterise these technology models as representative of industry practices for all American CIGS and CdTe manufacturers circa 2011, and have been previously used for benchmarking PV costs and analysing options for future cost reductions in seminal reports such as the Sunshot Vision Study (Margolis et al., 2012; Goodrich et al., 2012). Figures 7.9 and 7.10 depict the manufacturing process steps simulated each technology as described within the manufacturing cost models for CIGS (Whitney, 2014) and CdTe (Woodhouse et al., 2012).

Copper indium gallium selenide and cadmium tellurium life cycle model

Data extracted directly from the cost models for the 2010 base year included material inputs, electricity, and capital investments. For the 2010 base year, typical module efficiencies for CIGS and CdTe were 12 per cent and 11.6 per cent, respectively, as given by the cost models. The area of the panels was assumed to be 1.08 m^2 and 0.72 m^2 for CIGS and CdTe, respectively. Capital costs were amortized over the life of the manufacturing facility, assuming a five year lifetime for manufacturing equipment and a twenty year lifetime for the manufacturing facility. To calculate the peak capacity of a panel, we assumed 1000 $Watts_{peak}$ per m^2, yielding 154 and 84 $Watts_{peak}$ per module for CIGS and CdTe, respectively (Whitney, 2014; Woodhouse et al., 2012).

Once extracted from the manufacturing cost model and converted to consistent units, the inputs were collected into an inventory and input to a modelling framework that tabulates the environmental burdens from established LCA databases. Direct inputs to PV manufacturing were connected to appropriate processes in the ecoinvent database, allowing us to calculate the embodied burdens associated with the inputs to the manufacturing process (ecoinvent Data v2.2, 2009). The cost models were used to fill data gaps where material and energy data were not reported. Similarly, these costs were linked to the appropriate economic sectors used in the EXIOBASE multiregional IO database, which was used to calculate the embodied environmental burdens of those inputs.

System boundary and process flow diagrams

FIGURE 7.8

Process-flow diagram for the thin film PV life cycles showing the system boundary and life cycle stages included in the LCI

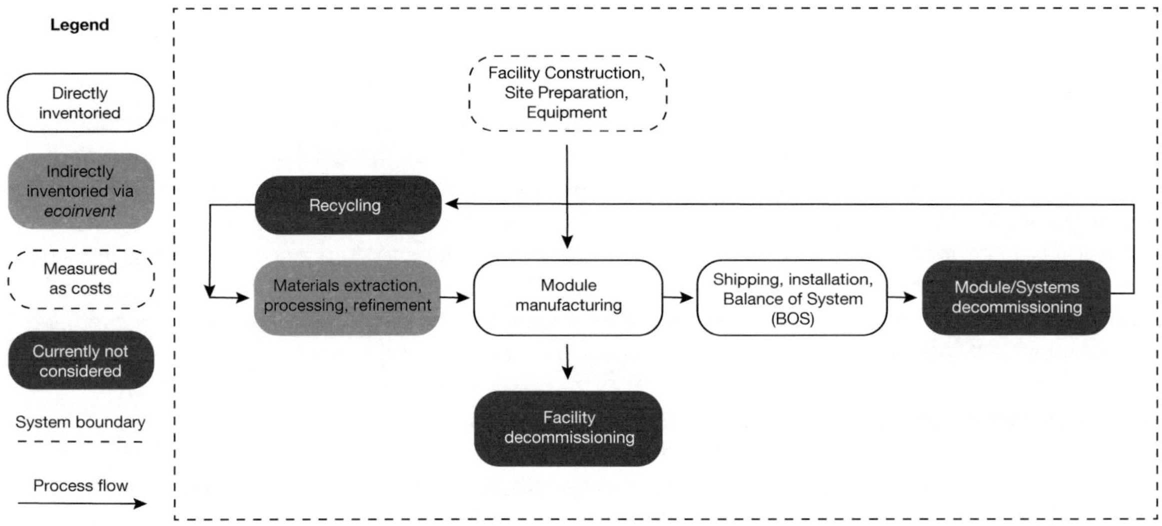

The vertical process flows represent the manufacturing facility life cycle. The horizontal process flows represent the thin film module production life cycle stages. The white boxes represent life cycle stages where material and energy inputs were directly reported in the cost model. The dashed boxes utilized costs to estimate environmental burdens. The light grey boxes represent life cycle stages where the material and energy burdens were measured indirectly through the ecoinvent database. The dark grey boxes currently are not inventoried since the cost models do not consider those stages. Future efforts will focus on these stages and other data gaps in the manufacturing cost models.

FIGURE 7.9

CIGS manufacturing flow chart showing discrete process stages as described by NREL manufacturing cost model

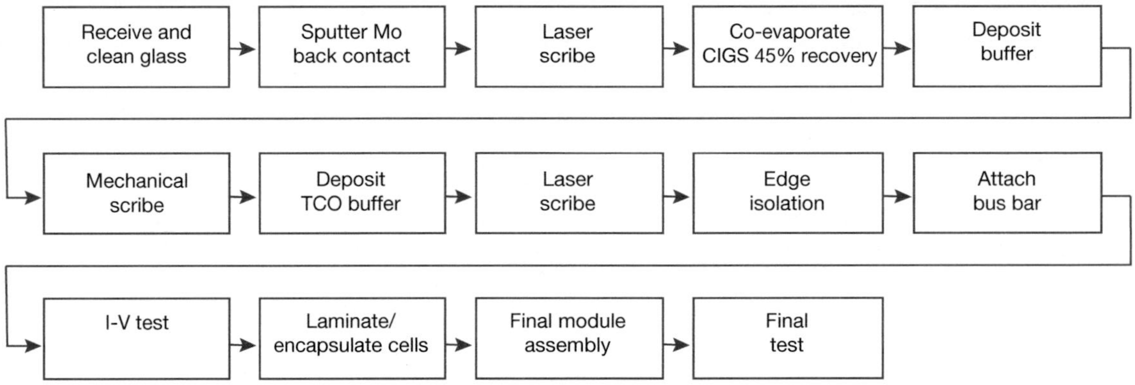

The various scribe stages make electrical connections between the layers.
Mo: molybdenum; TCO: transparent conducting oxide; I-V: current versus voltage

Source: Whitney, 2014

FIGURE 7.10

CdTe manufacturing flow chart showing discrete process stages as described by NREL manufacturing cost model

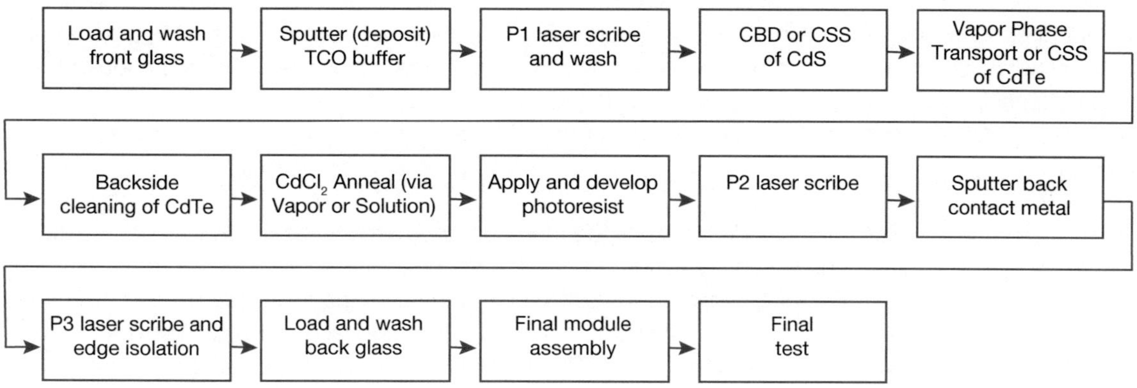

These stages are generalized to protect proprietary information. The process stages build the cell up from the glass; each layer is laid on top of the previous. TCO: transparent conducting oxide; CBD: chemical bath deposition; CSS: close space sublimation. P1, P2, P3 laser scribes make electrical connections between the layers

Source: Woodhouse et al., 2012

Deposition of thin film semi-conductor materials

The material intensity of thin film semi-conductor metals (indium, gallium, selenium, cadmium and tellurium) in PV modules was given by the manufacturing cost models, and has also been published by Woodhouse et al. (2013).

For CIGS module production, the NREL cost model gives a utilization rate of 55 per cent for coevaporated copper, indium, gallium, tellurium and molybdenum, meaning that 55 per cent of those materials are incorporated in the CIGS module and the remaining 45 per cent is available for recovery or disposal. Of that 45 per cent, approximately 25 per cent can be recovered and sold back to the supplier at 30 per cent cost. For lack of better information on the environmental impacts of processing the recovered metal, we use economic allocation to credit 30 per cent of the embodied environmental impacts of the recovered metals back to the CIGS module.

Sputtering of cadmium and tellurium for production of CdTe modules has a higher utilization rate of 70 per cent. Of the 30 per cent available for recovery or disposal, 20 per cent is recovered and sold back to suppliers.

Data gaps and uncertainty

As they were designed to accurately estimate the cost of manufacturing PVs, the manufacturing cost models do not perfectly meet the needs of LCA; the models omit materials and inputs with negligible cost. The environmental impacts of some proprietary chemicals and materials used in module manufacturing, such as the transparent conduction oxides, were estimated using environmentally extended input-output data and the costs of those materials. While the environmental impacts of chemical products can vary widely within an industry, previous assessments have shown that these material components contribute little to overall life cycle impacts (Fthenakis et al., 2008).

For CIGS, electricity demands were well documented for each process step. In some instances, the cost models provide ranges of inputs needed for manufacturing processes, e.g., 2-2.5 μm CIGS layer. The main data gap for CIGS is the absence of direct emissions data from CIGS module manufacturers. While literature shows that direct emissions from thin film manufacturing comprise a small portion of the total, the robustness of this study will rely on measuring or inferring direct emissions data from available sources or industry contacts (Fthenakis et al., 2008). Options include using IO cost data to determine the direct emissions from semi-conductor manufacturing in the United States, or applying direct emissions from the ecoinvent database for CIGS PV.

For CdTe, due to the minimal electricity demands of the manufacturing processes, there was uncertainty associated with the manufacturing electricity usage reported in the cost model. Instead, we referred to a 2011 LCA of CdTe thin film PVs by Fthenakis, which reported lifetime manufacturing electricity usage normalized per m^2 of module (Fthenakis and Kim, 2011; Fthenakis et al., 2011). A few gaps in the CdTe inventory remain. Fluorine-doped tin oxide (SnO_2), which is used in the TCO layer of the thin-film panel, is not present in the ecoinvent database. Further, to protect proprietary information, the cost of tin oxide cannot be released from the cost model. This cost was estimated using the costs of similar chemicals used in the CIGS model. Similarly, the back contact composition for CdTe is proprietary. For the back contact, potential materials can be ascertained from published CdTe studies and therefore a bounding analysis will be used to define the range of possible environmental burdens associated with the back contact layer.

7.3.2.2 Data for polycrystalline silicon photovoltaics

Although China now manufactures a majority of poly-Si PV modules, previous LCAs published in international journals have focused on the production of PV in Europe and the United States. It is well known that electricity consumed during module manufacture is a key contributor to the life cycle impacts of PV (Pacca et al., 2007). As China uses mostly coal for electricity generation, we expect that the impacts of PV produced in China would be higher than in previous assessments. In this assessment of poly-Si PV, we use data collected from Chinese producers of poly-Si and PV modules supplemented by additional data from the ecoinvent database.

The manufacturing of poly-Si PV modules, shown in Figure 7.11, includes the production of metallurgical grade silicon (MG-silicon), solar grade silicon production (SOG-silicon), ingot production, wafer sawing, cell production, and final module assembly. Each step consists of different processes and requires varying material inputs. While Chinese PV manufacturing processes are generally the same as the rest of the world, the data used here have a higher demand for process energy, and less efficient use of materials compared to poly-Si inventories used in previous studies. This could lead to higher impacts for the poly-Si modules modelled in this report.

FIGURE 7.11

Manufacturing process of silicon PV panels

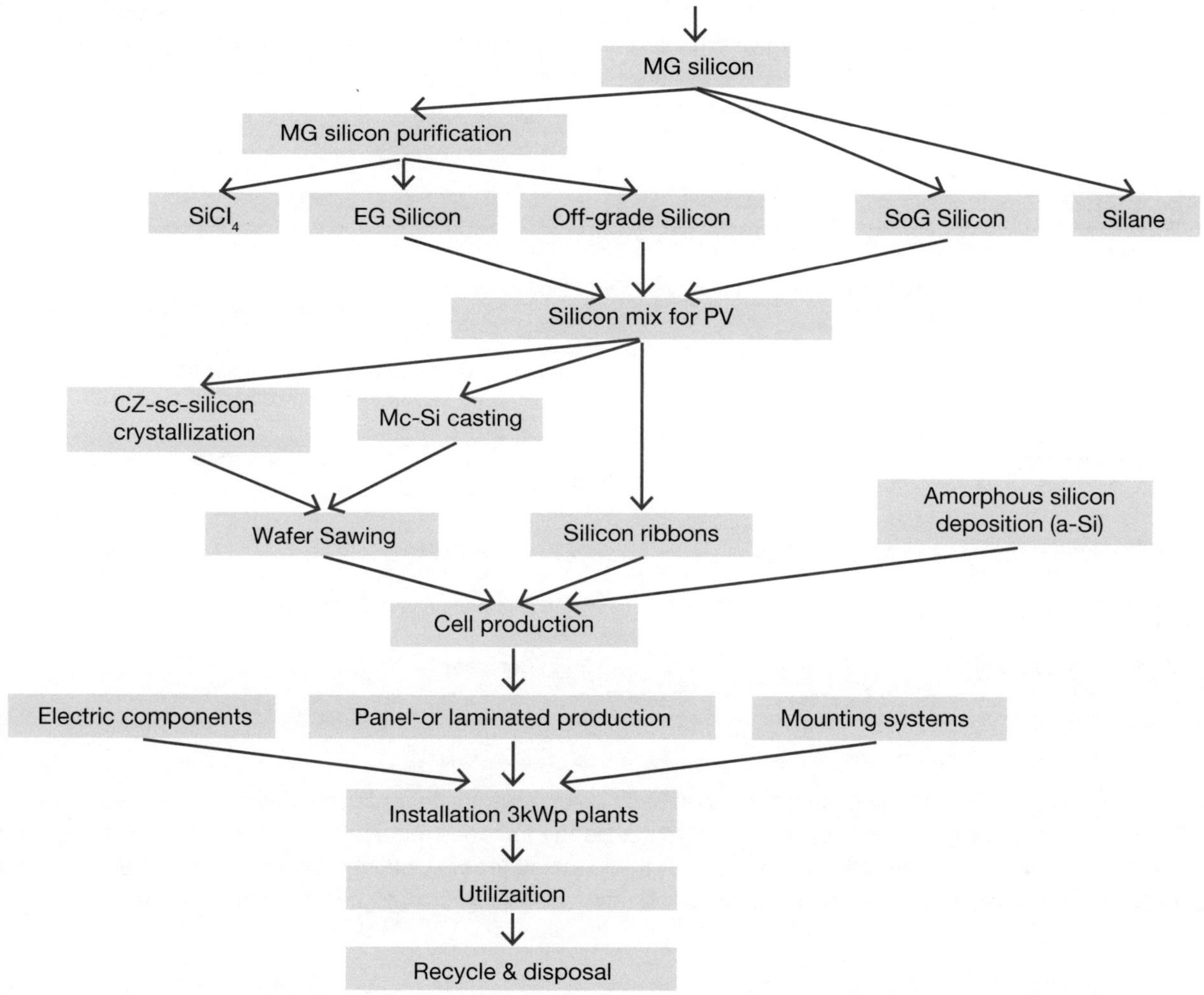

Silicon data sources

MG-silicon production and SOG-silicon production data were gathered from two factories in the Sichuan and Jiangsu provinces of China, whose names were protected for confidentiality. Data for other production steps were gathered from an international PV system integrator with several important suppliers in China, hereafter referred to as 'Company A'. Some data was gathered from field interviews in other factories and the rest was drawn from ecoinvent. The data sources are listed in Table 7.6.

TABLE 7.6

Data resources for silicon PV

Processes	Data
MG-Silicon production	Common practice from previous studies (Daishun, 2002; Diao Zhouwei, 2011); from Company A
SOG-Silicon production	EIA report for 3,000 ton poly-crystal silicon project in Sichuan; from Company A
Ingot and wafer sawing	From Company A
Cells production	From Company A and field interviews
Module assembling	From Company A

Data gathered directly from factories includes material inputs, electricity and elementary flows (emissions like CO_2, and chemical oxygen demand). For some materials the factories report contained insufficient information to ascertain the masses of those materials, and in those instances we obtained densities and other dimensions from online sources and product specification sheets to convert the given units to mass.

Silicon photovoltaics: data gaps and uncertainty

Primary data was obtained from several Chinese factories. Data not found in routine records were calculated by the authors. Table 7.7 lists important assumptions, data sources and data gaps for poly-Si module production.

For poly-Si cells, material and energy inputs are well documented. The greatest source of uncertainty stems from the data for MG-silicon production and SOG-silicon production; these data were gathered in 2009 and indicate somewhat higher material requirements than current practices or the processes found in the ecoinvent database for European PV production. For production of poly-Si modules in 2030 and 2050, we assume Chinese production of SOG-silicon ingots will achieve the same material efficiency as North American and European production that is currently reported in the ecoinvent database (ecoinvent Data v2.2, 2009).

TABLE 7.7

Assumptions for poly-silicon PV module

Processes	Item	Technique applied	Reference*
MG-silicon production	Process applied	Electricity furnace	(2)
	Reducer	Charcoal-free	
SOG-silicon production	Process applied	Modified Siemens	(2)
Ingot production	Method	Traditional	(1)
	Losses	75%	(1)
Wafer sawing	Wafer size	156mm x 156mm	(1)
	Wafer thickness	180um	(1)
	Sawing losses	89%	(1)
Etching	Solution	NaOH	Field interviews and (1)
Diffusion	Intermingle	POCl3 furnace	Field interviews and (1)
	corrosion	HF/HNO3	Field interviews and (1)
Printing			Field interviews and (1)
	Backside wirebar	Ag/Al	(1)
	Frontside wirebar	Ag	(1)
Module production	EVA thickness	0.5 mm	(1)
	TPT thickness	125 um	(1)
	PET thickness	0.2 mm	(1)
	Module size	1.63m^2	(1)
Testing	Efficiency	16%	(1)
	Service life	25 years	(1)

*(1) Diao Zhouwei, 2011, (2) ecoinvent Data v2.2, 2009.

7.4 LIFE CYCLE ASSESSMENT RESULTS

LCIs were calculated for all three PV technologies, manufactured in each of the nine IEA regions in 2010, 2030 and 2050. This chapter focuses on results for the PV reference region OECD North America. In Chapter 7.4.1 we present the 2010 baseline year results for PV technologies. We then discuss the possible technology improvements that PV technologies will undergo in the future, and how those changes may affect the environmental impacts of PV produced in 2030 and 2050. We also use contribution analysis in Chapter 7.4.2 to show the relative importance of different PV life cycle stages to key impacts. Finally, we present a regional comparison of PV generated electricity for key impact categories in Chapter 7.4.4.

7.4.1 2010 RESULTS

The following section reports the life cycle environmental and natural resource impacts of PV technologies in 2010 compared to the global average electricity mix as defined in the methodology chapter. Figures 7.12 and 7.13 show the impacts of ground- and roof-mounted PV electricity, respectively, in comparison with the global average electricity mix. CdTe and CIGS results are both consistent with existing literature, and are very similar in magnitude. As expected, thin film technologies show somewhat lower impacts than poly-Si in all categories, but both technologies have lower impacts in all categories except metal depletion. These metal depletion results suggest that increased PV generation will require more metals.

FIGURE 7.12

Impacts of ground-mounted PV technologies in 2010 compared to 2010 global average electricity mix impacts

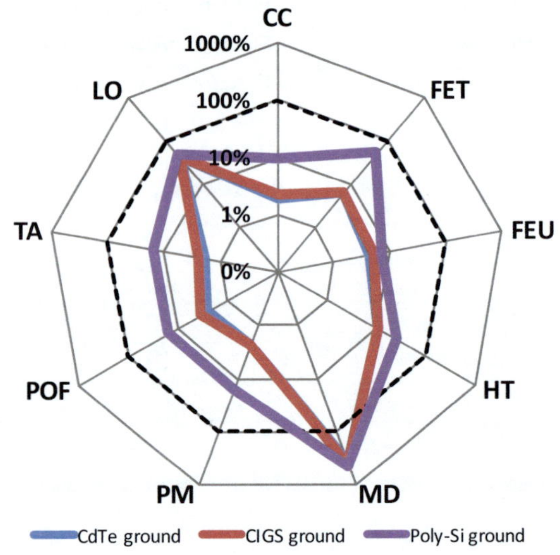

Irradiation of 2400 kWh/m²/year. CC: climate change; FET: freshwater ecotoxicity; FEU: freshwater eutrophication; HT: human toxicity; MD: metal depletion; PM: particulate matter; POF: photochemical oxidant formation; TA: terrestrial acidification; LO: land occupation. Note: small differences in the impacts of copper indium gallium selenide (CIGS) and cadmium telluride (CdTe) may be obscured by the logarithmic scale, which was necessary to present varying magnitudes of results

FIGURE 7.13

Impacts of roof-mounted PV technologies in 2010 compared to 2010 global average electricity mix impacts

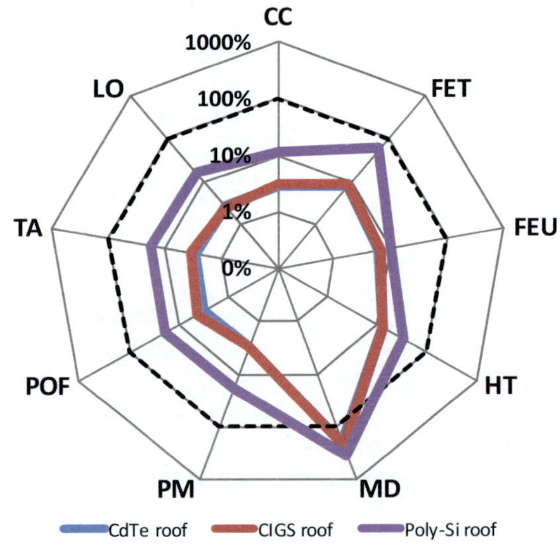

Irradiation of 2400 kWh/m²/year. CC: climate change; FET: freshwater ecotoxicity; FEU: freshwater eutrophication; HT: human toxicity; MD: metal depletion; PM: particulate matter; POF: photochemical oxidant formation; TA: terrestrial acidification; LO: land occupation. Note: small differences in the impacts of copper indium gallium selenide (CIGS) and cadmium telluride (CdTe) may be obscured by the logarithmic scale, which was necessary to present varying magnitudes of results

7.4.2 RESULTS FOR 2030 AND 2050

The following section presents the expected changes in impacts in PV technologies in 2030 and 2050.

7.4.2.1 Technology improvements in 2030 and 2050

Technology roadmaps from NREL were used to update LCI models to reflect achievable future improvements in thin film and silicon PV technologies. While it is impossible to predict all technological changes that PV technologies will undergo, the roadmaps have identified the following as feasible goals in the long term: increases in module energy conversion efficiency, reductions in material requirements for modules, also referred to as dematerialisation, and achieving economies of scale, i.e., reducing capital costs per module produced.

Table 7.8 lists the changes to PV module production and module performance assumed for 2030 and 2050, and lists the roadmaps and cost models used as a basis for these assumptions.

TABLE 7.8

Assumed technological improvements for future PV modules

		Baseline	2030	2050	Reference
CIGS	Module efficiency	12%	20.8%	25% (practical limit)	NREL CIGS cost model and roadmap (Whitney, 2014)
	CIGS layer	2 µm	1 µm (Ga:In molecular ratio = 2.3)	0.5 µm	NREL CIGS cost model and roadmap (Whitney, 2014)
	Molybdenum back contact	0.65 µm	0.5 µm	0.5 µm	NREL CIGS cost model and roadmap (Whitney, 2014)
	Optimize buffer and eliminate emitter layer	Transparent conducting oxide (TCO) ($2 per m^2)	Optimized TCO ($1.5 per m^2)	Optimized TCO ($1.5 per m^2)	NREL CIGS cost model and roadmap (Whitney, 2014)
	Glass substrate	3.2 mm glass	2.2 mm anti-reflex glass	2.2 mm anti-reflex glass	NREL CIGS cost model and roadmap (Whitney, 2014)
	Capital costs per m^2 module	$26 per m^2	$8 per m^2	$8 per m^2	NREL CIGS cost model and roadmap (Whitney, 2014)
	Direct manufacturing emissions	2.3E-8 kg Cd	0	0	ecoinvent database (Jungbluth, 2008)
CdTe	Module efficiency	11.6%	18%	24.4% (practical limit)	NREL CdTe cost model and roadmap (Woodhouse et al., 2012)
	CdTe layer	2.5 µm	1 µm	0.5 µm	NREL CdTe cost model and roadmap (Woodhouse et al., 2012)
	Optimize buffer and eliminate emitter layer	Zinc oxide (ZnO) ($2 per m^2)	Optimized ZnO ($1.5 per m^2)	Optimized ZnO ($1.5 per m^2)	NREL CdTe cost model and roadmap (Woodhouse et al., 2012)
Poly-Si	Module efficiency	16%	21%	21%	Wafer Silicon Roadmap (NREL, 2007)
	Poly-silicon wafer thickness	180 µm	120 µm	120 µm	Wafer Silicon Roadmap (NREL, 2007)
	Materials efficient ingot production	900 kg Solar grade silicon (SOG) per ingot	513 kg SOG per ingot	513 kg SOG per ingot	ecoinvent database (Jungbluth, 2008)

7.4.2.2 2030 results

Figure 7.14 and Figure 7.15 show the impacts of PV-generated electricity in 2030. Results suggest that thin film technologies will undergo 35-60 per cent reductions in environmental impacts by 2030, but only about a 20 per cent reduction in the category of metal depletion. Poly-Si shows similar reductions in environmental impacts, ranging from 35-45 per cent, and only a 13 per cent reduction in metal depletion for ground-mounted generation.

FIGURE 7.14

Impacts of ground-mounted PV technologies in 2030 compared to 2010 global average electricity mix impacts

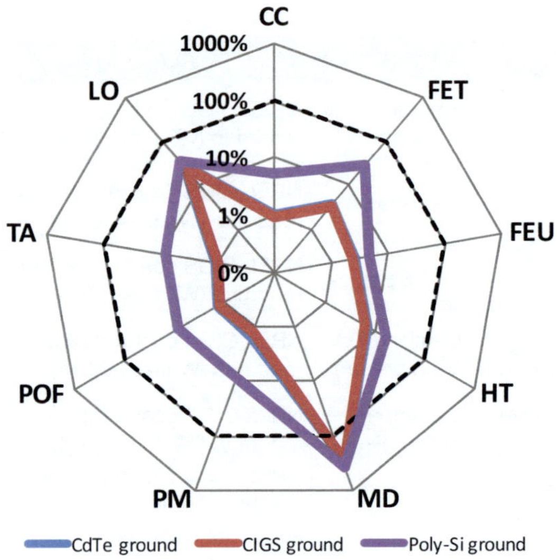

Irradiation of 2400 kWh/m^2/year. CC: climate change; FET: freshwater ecotoxicity; FEU: freshwater eutrophication; HT: human toxicity; MD: metal depletion; PM: particulate matter; POF: photochemical oxidant formation; TA: terrestrial acidification; LO: land occupation. Note: small differences in the impacts of CIGS and CdTe may be obscured by the logarithmic scale, which was necessary to present varying magnitudes of results.

FIGURE 7.15

Impacts of roof-mounted PV technologies in 2030 compared to 2010 global average electricity mix impacts

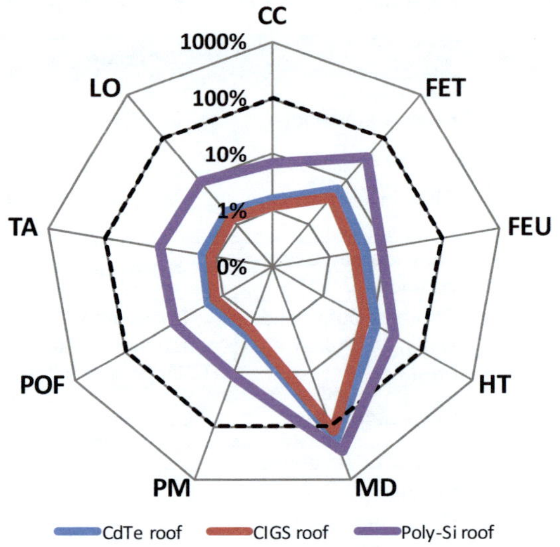

Irradiation of 2400 kWh/m^2/year. CC: climate change; FET: freshwater ecotoxicity; FEU: freshwater eutrophication; HT: human toxicity; MD: metal depletion; PM: particulate matter; POF: photochemical oxidant formation; TA: terrestrial acidification; LO: land occupation. Note: small differences in the impacts of CIGS and CdTe may be obscured by the logarithmic scale, which was necessary to present varying magnitudes of results.

7.4.2.3 2050 results

Figures 7.16 and 7.17 show the impacts of PV-generated electricity in 2050. These results suggest that thin film technologies will undergo further reductions in environmental impacts by 2050 to around 25-30 per cent from 2030 impacts, but only about a 5-10 per cent reduction in the category of metal depletion. We assume that poly-Si will reach its full efficiency potential in 2030, and only undergo minor improvements in materials efficiency by 2050, so 2050 poly-Si results only show small reductions in environmental impacts compared to 2030 impacts, ranging from 7-30 per cent, and only a 1 per cent reduction in metal depletion.

FIGURE 7.16

Impacts of ground-mounted PV technologies in 2050 compared to 2010 global average electricity mix impacts

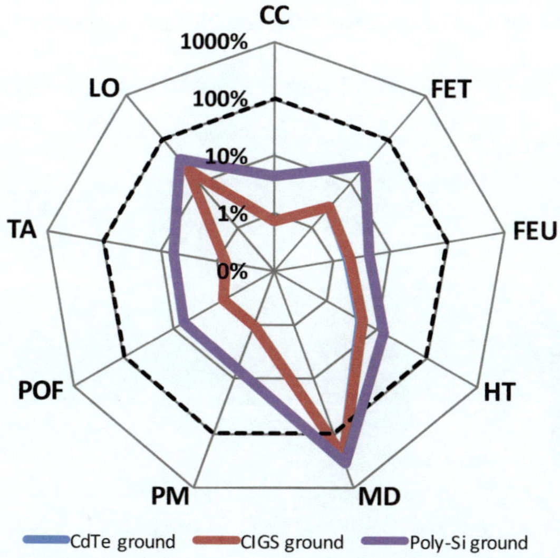

Irradiation of 2400 kWh/m²/year. CC: climate change; FET: freshwater ecotoxicity; FEU: freshwater eutrophication; HT: human toxicity; MD: metal depletion; PM: particulate matter; POF: photochemical oxidant formation; TA: terrestrial acidification; LO: land occupation. Note: small differences in the impacts of CIGS and CdTe may be obscured by the logarithmic scale, which was necessary to present varying magnitudes of results.

FIGURE 7.17

Impacts of roof-mounted PV technologies in 2050 compared to 2010 global average electricity mix impacts

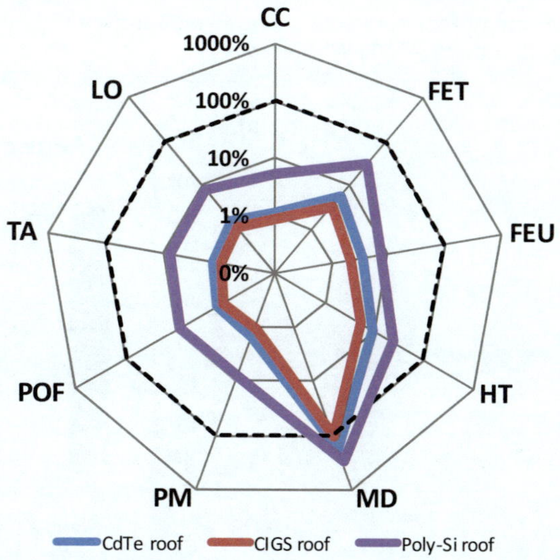

Irradiation of 2400 kWh/m²/year. CC: climate change; FET: freshwater ecotoxicity; FEU: freshwater eutrophication; HT: human toxicity; MD: metal depletion; PM: particulate matter; POF: photochemical oxidant formation; TA: terrestrial acidification; LO: land occupation. Note: small differences in the impacts of CIGS and CdTe may be obscured by the logarithmic scale, which was necessary to present varying magnitudes of results.

7.4.2.4 The effects of technological changes

The effects of different categories of technological change were decomposed for all three PV technologies. The different technological changes were assigned to three categories: background, i.e., the changing electricity mix from 2010 to 2030 and 2050, dematerialisation, and efficiency, i.e., increasing module energy conversion efficiency.

Figure 7.18 shows the effects of technological changes on decreasing the life cycle GHG emissions, human toxicity, and metal depletion of ground-mounted PV generation from 2010 to 2050. Efficiency improvements had the broadest effect on decreasing the environmental impacts of PV in most impact categories. However, metal depletion was only reduced marginally due to technological improvements.

FIGURE 7.18

Effects of technological changes on per kWh life cycle GHG emissions of ground-mounted PV facilities assuming irradiation of 2400 kWh/m²/year

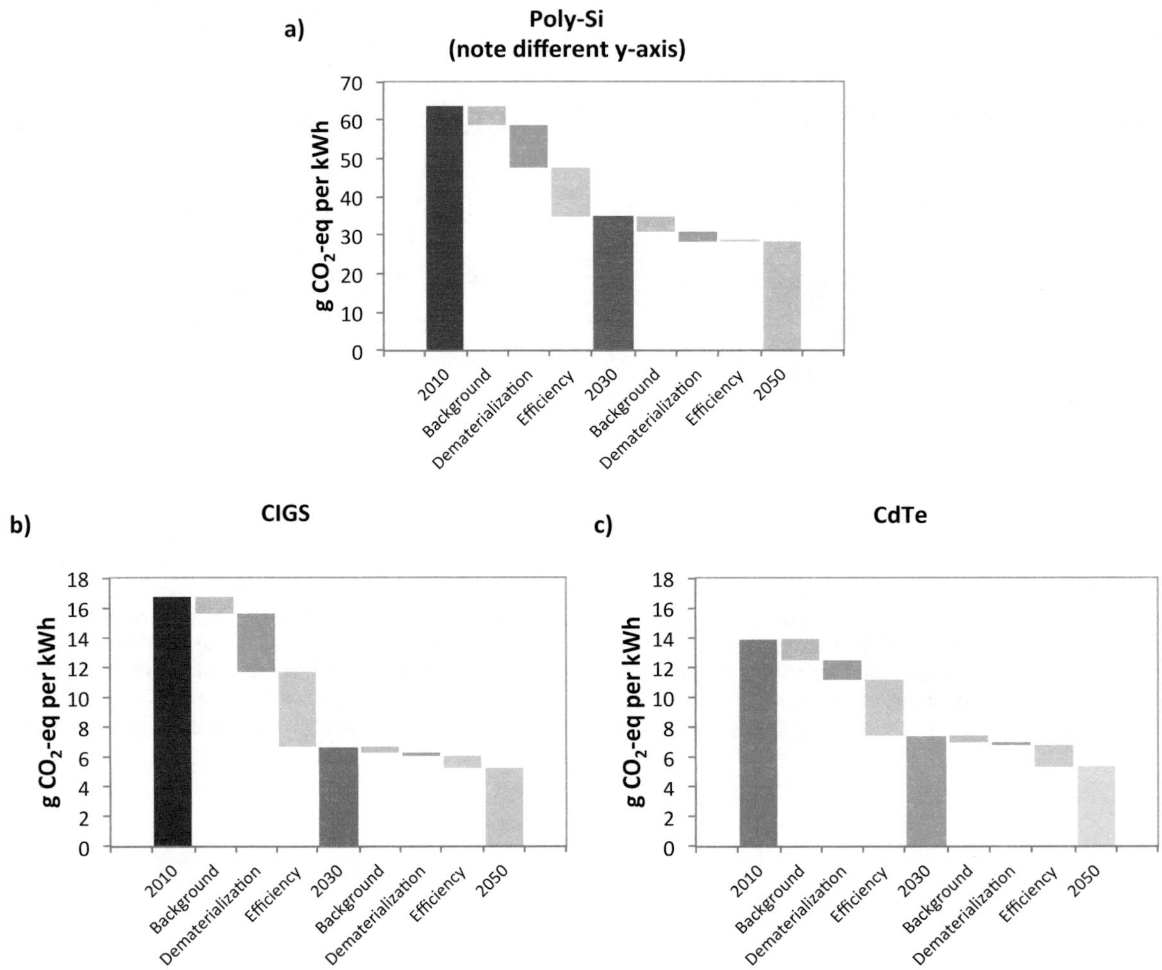

7.4.3 CONTRIBUTION ANALYSIS

This section shows the contribution of different life cycle stages of PV electricity to the overall results for selected indicators. Life cycle phases were divided into the following categories: module production including all upstream energy, materials and transportation needed for module manufacture, transformers and inverters, BOS and construction including materials production and land occupation of PV facility, and power plant operation including O&M. Figures 7.19, 7.20 and 7.22 show the contributions of those life cycle phases to overall impacts for CIGS, CdTe and Poly-Si systems in 2010.

BOS and electronic components in thin film systems show the greatest contributions to human toxicity, freshwater ecotoxicity, freshwater eutrophication, and metal depletion indicators. In general, poly-Si sees greater contributions from module production, since the Chinese manufactured silicon modules generally require more electricity and therefore primary energy and consequently cause greater environmental impact to manufacture. For all technologies, module production is responsible for the majority of life cycle GHG emissions.

The BOS and power electronics are responsible for the bulk of the metal depletion impacts for all ground- and roof-mounted technologies. Figure 7.22 shows the metals contributing to the metal depletion indicator for ground-mounted CdTe PV. Base metals like copper and manganese used in the transformers, inverters and mounting systems are of the greatest concern. It is important to note that the ReCiPe methodology does not include characterization factors for all possible metals. Of particular importance is the omission of tellurium, a by-product of copper production. Thus, the metal depletion results discussed here do not fully capture the issues of supply constraints or criticality of such metals. The availability of critical semi-conductor metals such as tellurium and indium for the production of PV over the time frame considered in the BLUE Map scenario are discussed below, in Chapter 7.5.1.1.

FIGURE 7.19

Contribution of life cycle stages to copper indium gallium selenide PV impacts in 2010

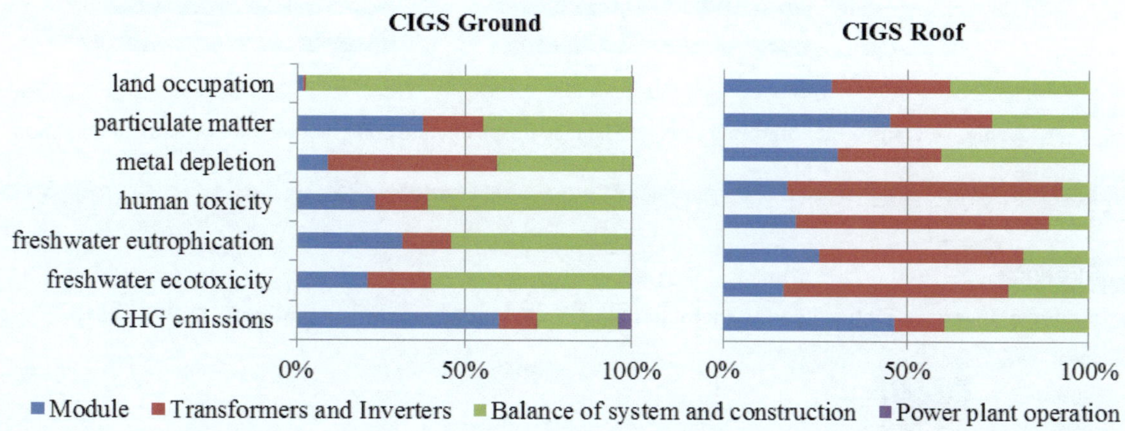

FIGURE 7.20

Contribution of life cycle stages to cadmium telluride PV impacts in 2010

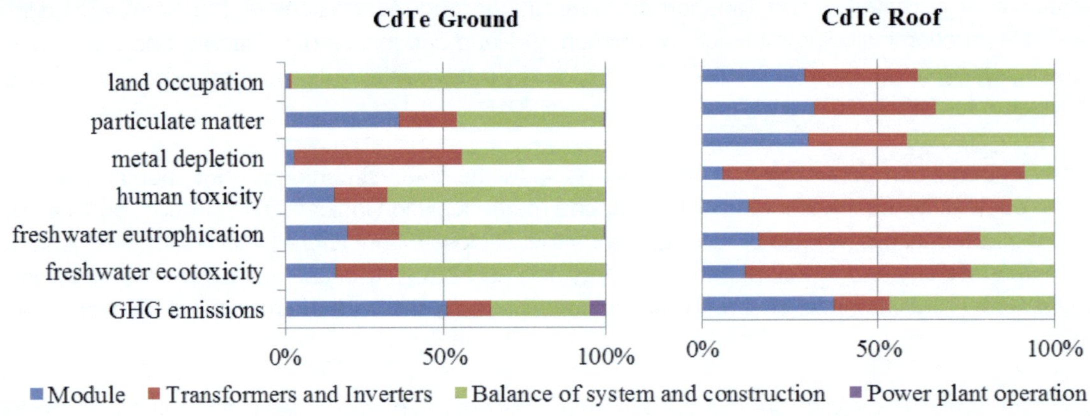

FIGURE 7.21

Contribution of life cycle stages to polycrystalline silicon PV impacts in 2010

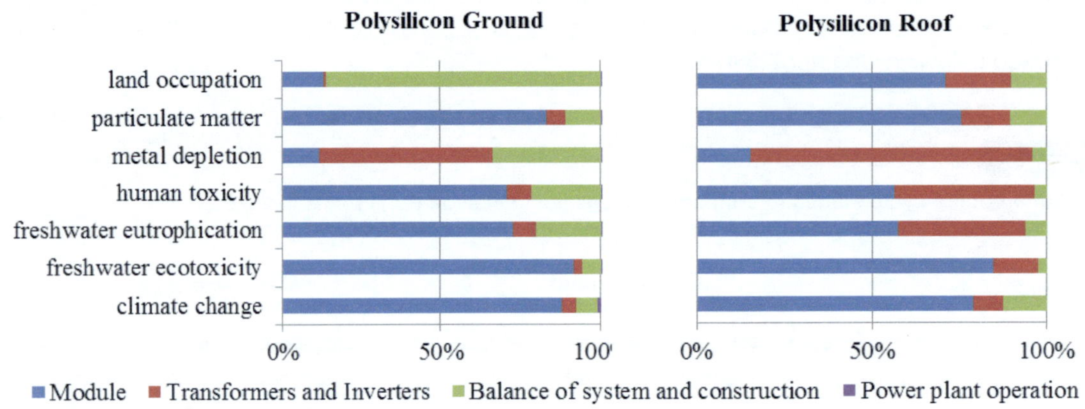

FIGURE 7.22

Contributions of various metals used to metal depletion indicator for ground-mounted CdTe PV in 2010

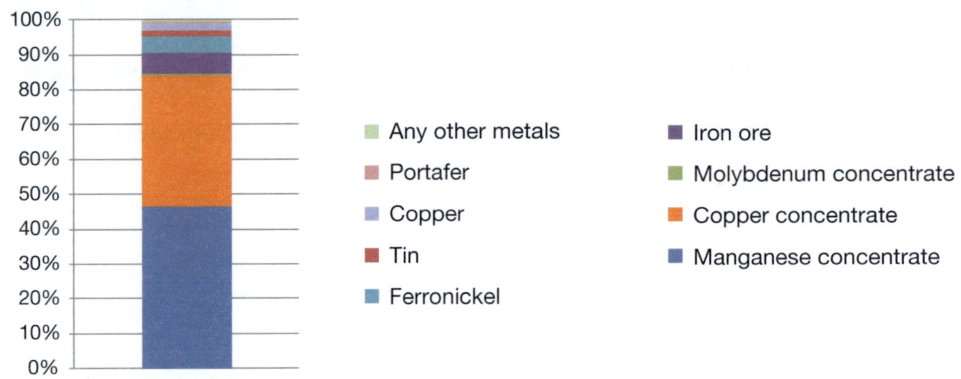

7.4.4 TECHNOLOGY COMPARISONS

In general, the life cycle impact assessment of PVs compare favourably to those of the current global electric grid mixes. In most ReCiPe category indicators, PV electricity sees a ten-fold or greater reduction in environmental impacts, with the notable exception of metal depletion; all three PV technologies modelled in this report show greater metal depletion per kWh generated than the global average grid mix.

This section compares the GHG emissions and metal depletion results of the three studied PV technologies. These results reflect the likely manufacturing location of the modules: the United States for CdTe and CIGS, and China for poly-Si. As shown in Figures 7.23 and 7.24, poly-Si PV shows higher impacts on climate change than thin film technologies. This is due to China's more carbon-intensive electricity mix, which predominantly consists of coal, as well as the higher energy requirements for the production of solar grade silicon and poly-Si modules shown by the data collected for Chinese PV production. If poly-Si PV modules were produced in the United States or Europe (with less coal-intensive electricity generation) their impacts on climate change would be closer in magnitude to those of CdTe and CIGS, as shown by the PV harmonisation studies by Kim et al. (2012) and Hsu et al. (2012). A majority of silicon PV modules, however, are now produced in China, while thin films are generally produced in Europe, the US and Japan.

Figure 7.23 shows that in 2010, both CdTe and CIGS thin films are comparable in nearly every impact category, although CdTe has a slight advantage in most impact categories. By 2030 (Figure 7.24), technological improvements cause CIGS to outpace CdTe in many impact categories. This is mainly a result of higher projected efficiency for CIGS for 2030 (20.8 per cent compared to 18 per cent for CdTe).Previous studies have shown that CdTe has an advantage over CIGS in terms of energy demand and GHG emissions (Kim et al., 2012). The results of this study suggest that CIGS has made improvements since previous LCAs of CIGS that were based on older data (circa 2006) (Sense, 2008).

The metal depletion indicator highlights an important impact for PV generation technologies—more metal is required per kWh generated in comparison with conventional, fossil fuel-based electricity. Interestingly, all three technologies have almost identical depletion impacts in 2010. This result may be surprising because different metals are used to produce each technology. However, contribution analysis shows that the majority of metal depletion impact stems from the use of metals for transformers, inverters and construction (mainly copper, iron and manganese) and rather than the rarer, more expensive by-product metals used in the semiconductor absorber layers such as cadmium, tellurium, gallium and indium. Additionally, ground-mounted systems fare worse in this category primarily due to the copper and other metals required for high-voltage transformers. Since distributed residential and commercial roof systems do not require high-voltage transformation, fewer of those metals are required.

FIGURE 7.23

Comparison of climate change impacts for ground-mounted PV systems

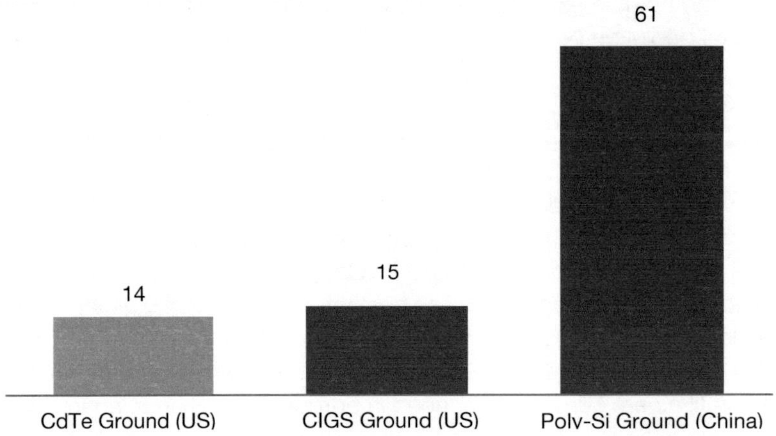

Comparison in g CO_2/kWh for US manufactured thin films (CdTe and CIGS) and Chinese poly-Si systems in 2010. Assumed irradiation is 2400 kWh/m^2/year.

FIGURE 7.24

Comparison of climate change impacts for ground-mounted PV systems

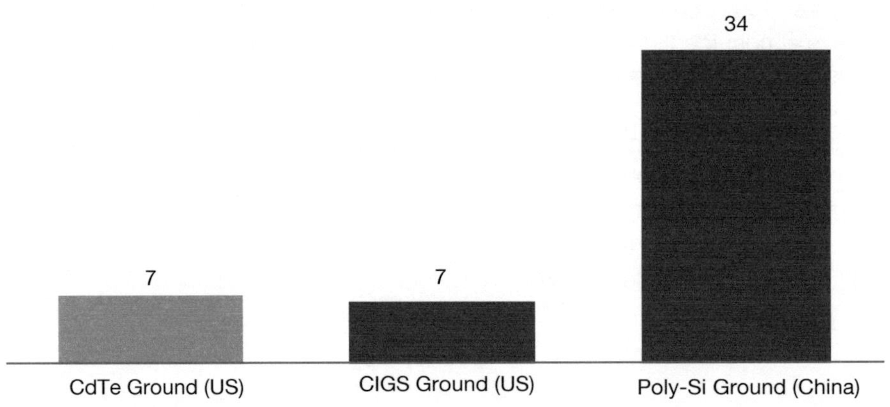

Comparison in (g CO_2/kWh for US manufactured thin films (CdTe and CIGS) and Chinese poly-Si systems in 2030. Assumed irradiation is 2400 kWh/m^2/year.

FIGURE 7.25

Comparison of metal depletion for ground-mounted PV systems in 2010

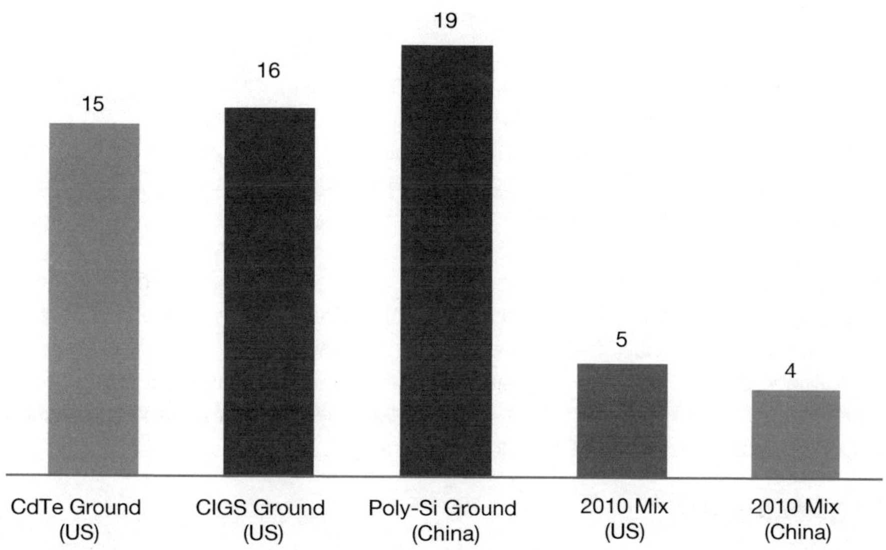

Metal depletion (g iron equivalents/kWh) for US manufactured thin films (CdTe and CIGS) and Chinese poly-Si systems compared with 2010 electricity mixes. Assumed irradiation is 2400 kWh/m²/year.

FIGURE 7.26

Figure 7.26 Comparison of metal depletion for ground-mounted PV systems in 2030

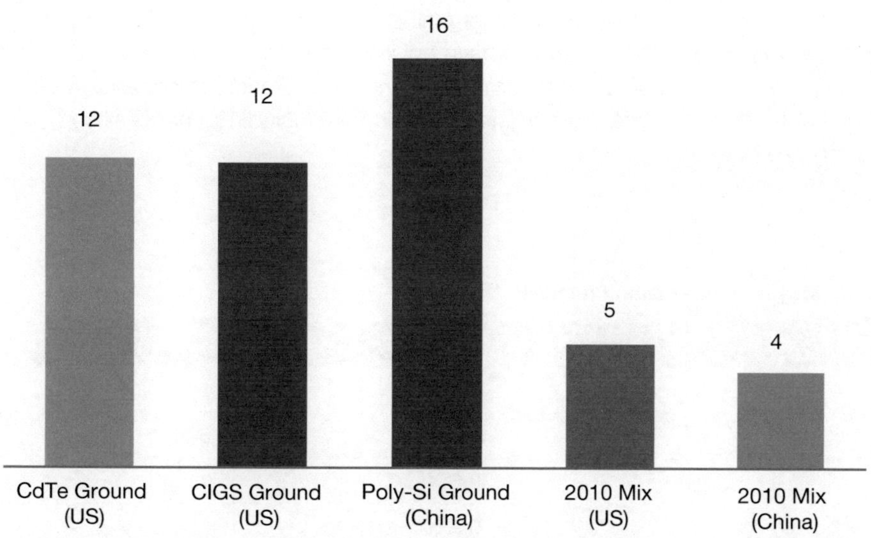

Metal depletion (g iron equivalents/kWh) for US manufactured thin films (CdTe and CIGS) and Chinese poly-Si systems compared with 2010 electricity mixes. Assumed irradiation is 2400 kWh/m²/year.

The preceding results show a relative homogeneity in metal depletion among PV technologies. While it was expected that thin films would perform better in terms of GHG emissions, it is surprising that poly-Si PV, composed of abundant silicon, would have metal depletion impacts similar in magnitude to the thin-film technologies. This result is due to large contributions to metal depletion from the PV BOS.

By 2050, the overall profile of LCIA results for PV panels remains basically the same as seen in Figure 7.25 and Figure 7.26. While showing lower life cycle impacts in almost every category compared to conventional electricity, PV electricity generation remains dependent on metal resources. While this trend is not necessarily unsustainable, it does show an impetus for material efficiency and recycling of metals used in PV panels, transformers, inverters and their other BOS components.

It is also important to note, that despite having higher life cycle GHG emissions and somewhat higher impacts in other categories (e.g. freshwater ecotoxicity), Chinese manufactured Poly-Si PV systems still offer environmental benefits relative to fossil fuel-based generation, and that all PV technologies studied have similar metal depletion impacts. Furthermore, potential improvements in process energy and materials efficiency combined with the decarbonisation of electricity generation in China can lead to drastically reduced life cycle impacts for Chinese manufactured PV, as shown in Figure 7.24.

7.4.5 REGIONAL COMPARISONS

In this section, we compare key environmental impacts of PV generation among all nine IEA regions, accounting for regional variation in solar irradiation and the regional differences in electricity mix used for manufacturing of PV modules and complete systems.

For simplicity, the graphs show the average impact of 1 kWh of PV electricity from the PV technology mix assumed for scenario analysis. As prescribed in the BLUE Map scenario, roof-mounted systems produce the majority of the PV electricity every year, a share corresponding to approximately 60 per cent. In 2010, the technology mix is weighted heavily in favour of poly-Si. For 2030 and beyond, this analysis considers one possible scenario of a transition toward thin-film technologies, which capture two-thirds of the market in 2030, and further increase their market share by 2050. While such a high penetration of thin film technologies is not assured, especially given the prevalence of low cost Chinese manufactured poly-Si modules on the market, this scenario represents a transition to higher efficiency, lower environmental impact PV technologies. This scenario best illustrates the potential reductions in environmental impacts of PV technologies that may be achievable given long term technological improvements. A further limitation of this analysis is the omission of other prominent and potentially significant emerging technologies, including: mono-silicon, amorphous silicon, quantum dot and organic polymers.

TABLE 7.9

Assumed PV technology mix for scenario analysis

Type	2010	2030	2050
Poly-Si ground	36.9 %	13.7 %	8.2 %
Poly-Si roof	52.9 %	19.6 %	11.8 %
CIGS ground	0.6 %	13.7 %	16.4 %
CIGS roof	0.9 %	19.6 %	23.6 %
CdTe ground	3.5 %	13.7 %	16.4 %
CdTe roof	5.1 %	19.6 %	23.6 %

Figures 7.27 to 7.31 compare selected environmental impacts across the nine IEA regions. Climate change and metal depletion impact categories are presented due to their general importance while the remaining categories were selected for their comparative impact to the electricity grid.

7.4.5.1 Climate change

FIGURE 7.27

Regional comparison of PV impact on climate change (g CO_2 eq/kWh)

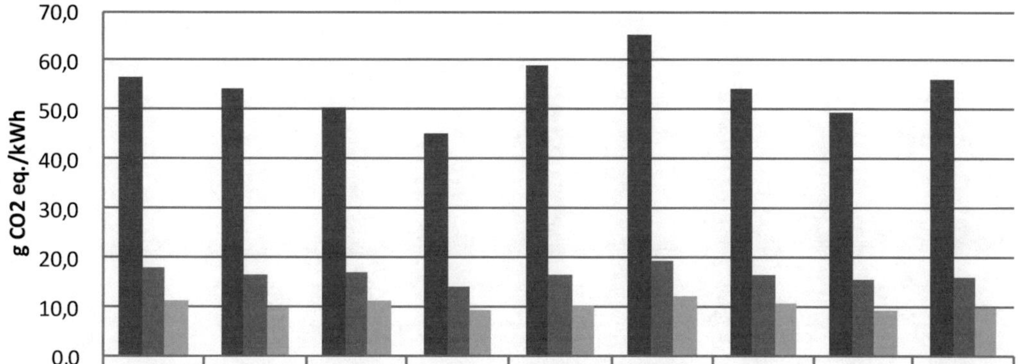

	CN	IN	RER	US	PAC	EIT	LA	AS	AME
2010	56,7	54,2	50,3	45,1	59,1	65,3	54,1	49,5	56,0
2030	18,2	16,6	16,9	14,1	16,6	19,7	16,7	15,6	16,2
2050	11,2	10,1	11,4	9,4	10,6	12,5	10,7	9,5	10,0

CN: China; IN: India; RER: OECD Europe; US: United States (OECD North America); PAC: OECD Pacific; EIT: Economies in transition and non-OECD Europe; LA: Latin America; AS: Other developing Asia; AME: Africa and Middle East

7.4.5.2 Metal depletion

FIGURE 7.28

Regional comparison of PV impact on metal depletion (g iron eq/kWh)

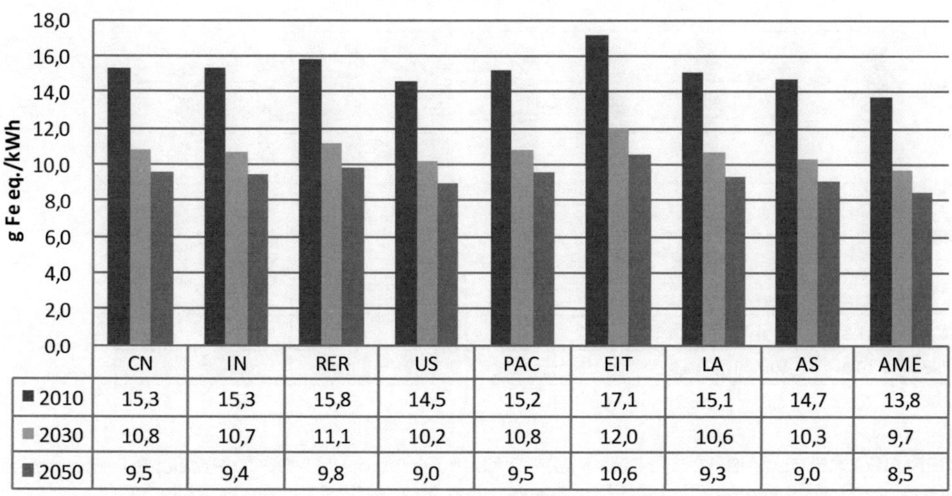

	CN	IN	RER	US	PAC	EIT	LA	AS	AME
2010	15,3	15,3	15,8	14,5	15,2	17,1	15,1	14,7	13,8
2030	10,8	10,7	11,1	10,2	10,8	12,0	10,6	10,3	9,7
2050	9,5	9,4	9,8	9,0	9,5	10,6	9,3	9,0	8,5

CN: China; IN: India; RER: OECD Europe; US: United States (OECD North America); PAC: OECD Pacific; EIT: Economies in transition and non-OECD Europe; LA: Latin America; AS:Other developing Asia; AME: Africa and Middle East

7.4.5.3 Agricultural land occupation

These results assume that ground-mounted PV systems occupy agricultural land. In reality, a ground-mounted PV facility may occupy land that would not otherwise be suitable for agriculture.

FIGURE 7.29

Regional comparison of PV impact on land occupation (10-4 m²·a/kWh)

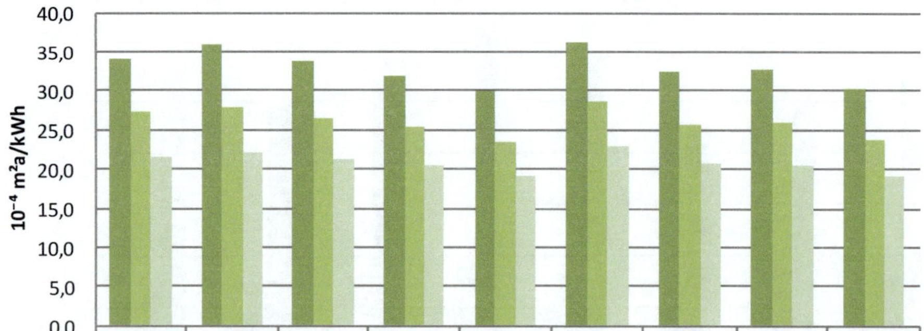

	CN	IN	RER	US	PAC	EIT	LA	AS	AME
2010	34,1	35,9	33,6	32,0	29,9	36,1	32,4	32,6	30,1
2030	27,2	27,9	26,6	25,4	23,6	28,5	25,6	25,8	23,7
2050	21,5	22,1	21,4	20,5	19,0	22,9	20,6	20,5	19,0

CN: China; IN: India; RER: OECD Europe; US: United States (OECD North America); PAC: OECD Pacific; EIT: Economies in transition and non-OECD Europe; LA: Latin America; AS:Other developing Asia; AME: Africa and Middle East

7.4.5.4 Terrestrial ecotoxicity

FIGURE 7.30

Regional comparison of PV impact on terrestrial ecotoxicity (mg 1,4-dicholorobenzene eq/kWh)

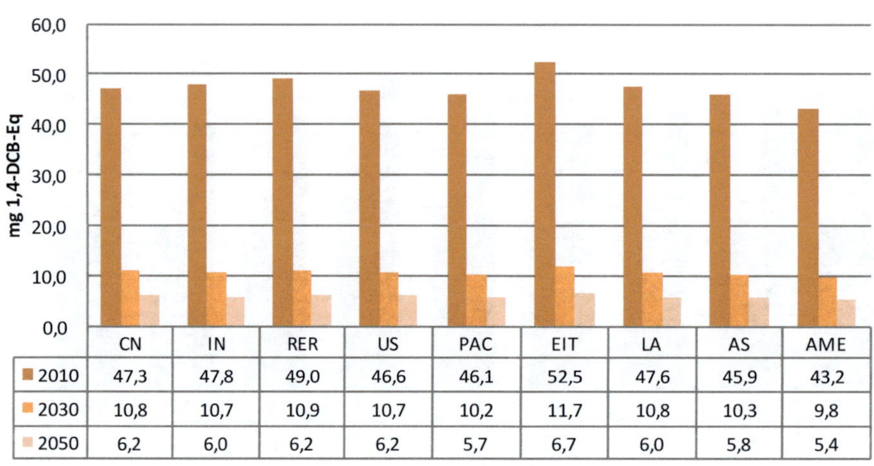

	CN	IN	RER	US	PAC	EIT	LA	AS	AME
2010	47,3	47,8	49,0	46,6	46,1	52,5	47,6	45,9	43,2
2030	10,8	10,7	10,9	10,7	10,2	11,7	10,8	10,3	9,8
2050	6,2	6,0	6,2	6,2	5,7	6,7	6,0	5,8	5,4

CN: China; IN: India; RER: OECD Europe; US: United States (OECD North America); PAC: OECD Pacific; EIT: Economies in transition and non-OECD Europe; LA: Latin America; AS:Other developing Asia; AME: Africa and Middle East

7.4.5.5 Human toxicity

FIGURE 7.31

Regional comparison of PV impact on human toxicity (g 1,4-dicholorobenzene eq/kWh)

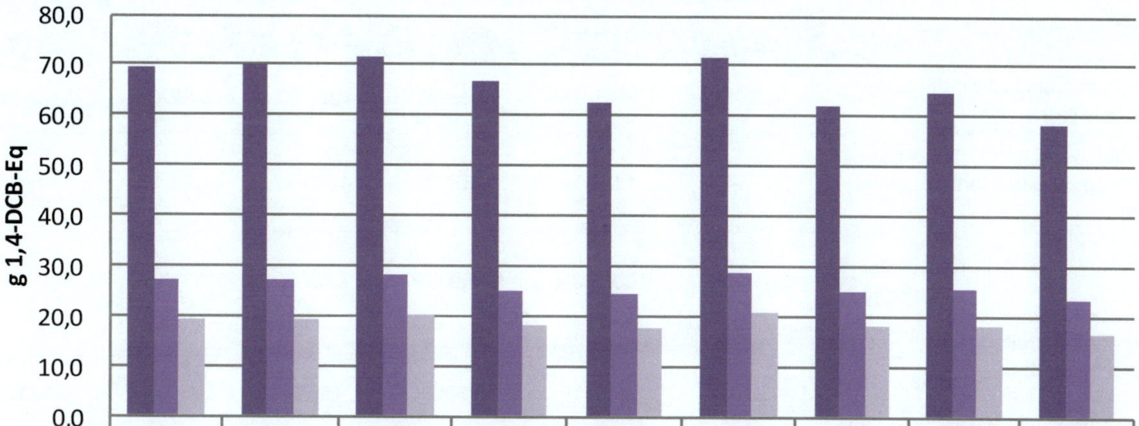

	CN	IN	RER	US	PAC	EIT	LA	AS	AME
2010	69,7	70,1	71,6	67,0	62,8	71,4	62,1	64,9	58,3
2030	27,1	27,3	28,4	25,3	24,9	28,8	25,4	25,8	23,4
2050	19,2	19,4	20,6	18,2	18,0	20,8	18,5	18,4	16,9

CN: China; IN: India; RER: OECD Europe; US: United States (OECD North America); PAC: OECD Pacific; EIT: Economies in transition and non-OECD Europe; LA: Latin America; AS:Other developing Asia; AME: Africa and Middle East

7.4.6 SENSITIVITY OF THIN FILM RESULTS TO POTENTIAL CADMIUM EMISSIONS DURING USE AND END-OF-LIFE

The LCA results presented in this chapter show that CdTe and CIGS have lower life-cycle toxic emissions than Poly-Si PV and average grid electricity in 2010. However, the presence of toxic metals, especially cadmium, in CdTe and CIGS PV suggests that thin film modules, if broken during use or not properly recycled or disposed of at end-of-life, could release cadmium or other metals to the environment under certain conditions. This section presents a sensitivity analysis in which we calculate the percentage of total cadmium contained in a thin-film module that would hypothetically need to be released in order for CdTe PV to have per kWh human toxicity, freshwater ecotoxicity and terrestrial ecotoxicity potentials equal to those of both the average electricity grid mix and poly-Si PV in the North America region. In this sensitivity case, we assume cadmium would be emitted to industrial soil, as defined by the ReCiPe method (rather than agricultural or forest soil), which was deemed the most appropriate proxy for landfill or use-phase emissions from ground-mounted PV. It is important to note that other potentially important metals contained in CdTe, CIGS and poly-Si modules are not analysed, as metals such as indium, gallium, and tellurium do not have toxicity characterization factors in the ReCiPe method. Molybdenum, selenium and copper do have characterization factors, all of which are lower than cadmium for toxicity impacts, so we use this case as a potentially worst case scenario for potential toxic releases from thin film PV. Secondly, we compute the maximum possible toxic impacts due to cadmium releases given expected breakage rates for thin film modules (0.04 per cent per year; 1.2 per cent over 30 years) (Sinha et al., 2012). Lastly, examples from literature are used to qualitatively evaluate the likelihood of such toxic releases from thin film PV at end-of-life.

TABLE 7.10

Sensitivity analysis of cadmium in CdTe modules that must be released to make toxic impacts equivalent to North American grid mix and poly-Si PV

		CdTe (ground-mounted) compared to average grid mix in North America			CdTe (ground-mounted) compared to poly-Si in North America		
		2010	2030	2050	2010	2030	2050
Cd contained in 1 m^2 of CdTe module (kg)		1.0E-02	3.9E-03	1.9E-03	1.0E-02	3.9E-03	1.9E-03
Cd contained in CdTe module per kWh generated (North America) (kg)		4.7E-05	1.2E-05	4.3E-06	4.7E-05	1.2E-05	4.3E-06
Percentage of Cd content that must be emitted to make per kWh impacts equal to the reference*	Freshwater ecotoxicity	25,000%	100,000%	280,000%	7,100%	15,000%	36,000%
	Human toxicity	1,600%	6,500%	18,000%	150%	360%	990%
	Terrestrial ecotoxicity	6.4%	30%	84%	4.1%	11%	28%

*If over 100 per cent, more cadmium than is contained in the module is needed to make CdTe PV impacts equal to grid mix impacts or poly-Si impacts.

Table 7.10 shows that for electricity generated by CdTe PV in North America, much more cadmium than is contained in a CdTe module would be needed to equal the per kWh freshwater ecotoxicity and human toxicity impacts of the average electricity grid mix or poly-Si PV. On the other hand, emission of just 6 per cent of the cadmium contained in a 2010 CdTe module would result in equivalent terrestrial ecotoxicity impacts for CdTe and the grid mix. By 2030 and 2050, reductions in the amount of materials used in the CdTe semiconductor, buffer and emitter layers mean that an increasingly greater percentage of the cadmium contained in the module would need to be released to the environment to produce the same impact as the 2010 grid mix.

The results of this analysis are sensitive to the assumed location of emissions (i.e. industrial soil, agricultural soil, freshwater or air). If cadmium is emitted to agricultural soil during use or end-of-life (an unlikely occurrence), the emission of only 9 per cent of the cadmium contained in a module would result in an equivalent human toxicity potential for CdTe and the average grid mix in 2010, and the emission of over 100 per cent of module cadmium would have the same effect in 2050.

The first sensitivity analysis does not assess the actual likelihood or risk of cadmium releases and other toxic releases from thin film PV. Sinha et al. (2012) state that CdTe modules have a 0.04 per cent annual breakage rate during use, which would translate to 1.2 per cent over 30 years of module life. Over one third of these breakages occurs during shipping and installation, and those modules were immediately returned to the manufacturer with little risk of emissions. Table 7.11 estimates the maximum possible change in toxicity results due to the release of cadmium from the 1.2 per cent broken modules over 30 years of module life. The results of this analysis suggest that even if all the cadmium contained in broken modules was released to the environment, the freshwater ecotoxicity potential of electricity from CdTe PV would increase by less than 1 per cent, and human toxicity would increase by less than 2 per cent. Terrestrial ecotoxicity would increase by 76 per cent, but results would still be 70 per cent lower than North American grid electricity and 50 per cent lower than poly-Si PV in 2010. If cadmium was instead emitted to agricultural soil or to air, human toxicity results would increase by approximately 230 per cent and 100 per cent, respectively, but would still remain lower than the human toxicity impacts of the average US grid mix in 2010. In this case, the human toxicity of CdTe would exceed that of poly-Si by 22 per cent in 2010, but would remain 26 per cent lower in 2030 and 53 per cent lower in 2050 due to the decreased amount of cadmium in modules per kWh generated.

These sensitivity results suggest that routine module breakage would have little impact on overall life cycle toxicity. Considering that one third of breakages occur during shipping and that not all cadmium contained within a module is likely to be released, this sensitivity analysis represents the worst-case scenario. These results also compliment the conclusions of Sinha et al. (2012), who found that emissions from broken CdTe modules at a commercial building were "highly unlikely" to pose a health risk to on-site workers or off-site residents. Furthermore, another study found that even in catastrophic events such as fires, cadmium emissions were very low (5 g per kg cadmium content, less than 0.01 per cent) (Fthenakis et al., 2005). Another consideration is the potential breakage of modules due to hail or other kinds of extreme weather. While peer-reviewed articles have not analysed this risk using field tests, thin film modules are required to pass certain terminal velocity hail tests and mechanical load tests performed by third party laboratories that are certified by the International Electrotechnical Commission (IEC 61646). Similar to the conditions studied in Sinha et al. (2012), cracked modules would remain mostly intact, and encapsulation by the EVA laminate would likely prevent release of elements from the semiconductor layer. Furthermore, routine power-output monitoring would detect broken modules that would then be recycled and replaced.

TABLE 7.11

Sensitivity of toxic impacts to cadmium releases from modules given expected breakage rates

2010		CdTe (ground-mounted)					
		Industrial soil			Agricultural soil		
		2030	2050	2010	2030	2050	
Cd contained in module per kWh generated (North America) (kg)		4.7E-05	1.2E-05	4.3E-06	4.7E-05	1.2E-05	4.3E-06
Maximum possible per cent increase in toxicity impacts (per kWh) that would results from typical expected module breakage*	Freshwater ecotoxicity	0.13%	0.06%	0.03%	1.03%	0.01 %	0.01%
	Human toxicity	1.33%	0.50%	0.25%	230%	86 %	43%
	Terrestrial ecotoxicity	76%	53%	34%	23%	16 %	10%

*Assumes 1.2% modules breakage over 30 years of module life, and assumes all cadmium contained in the broken module is released to industrial soil or agricultural soil (worst case scenario). In reality, only a portion of cadmium contained in broken modules would likely be released.

These sensitivity analyses do not predict the likelihood of releases of toxic metals from thin film PV, but some information can be gleaned from literature to discuss the reasonability and likelihood of the possible cadmium emissions discussed in this section. Firstly, LCA studies have focused on end-of-life recycling of CdTe and other thin film modules, a choice justified by the current prevalence of module take-back and recycling programs instituted by thin film manufacturers (Fthenakis, 2004; Raugei and Fthenakis, 2010; Raugei et al., 2012). In these studies, Fthenakis (2004) and Raugei et al. (2010, 2012) have shown that recycling is the most certain way to avoid the release of cadmium to the environment. Raugei et al. (2012) estimate Cd emissions from modules sent to municipal landfills and incineration. Citing the results of the US Environmental Protection Agency's Toxicity Characteristic Leaching Procedure (TCLP) tests, this study estimates that just 0.09 grams of Cd per kilogram of Cd content would be released to the environment (less than 0.01 per cent). Such low levels of emissions would have negligible effect on the toxicity results presented in this chapter. Similarly, Cyrs et al. (2014) reviewed previous TCLP tests of CdTe PV modules and used risk assessment methods to conclude that CdTe posed little risk at current levels of production. Another study used laboratory tests to conclude that under acidic conditions, as much as 9 per cent of Cd, and 22 per cent of module molybdenum could leach from CIGS modules disposed in landfills over a four month period, with the implication that acid rain could increase the risk of leaching toxic metals from thin films (Zimmermann et al., 2013). In such studies, however, thin film modules are broken down to much smaller pieces than would exist in an actual landfill (2 mm x 2 mm), suggesting that these studies may overestimate the amount of metals leached from disposed thin films.

7.5 DISCUSSION

7.5.1 INTERPRETATION OF RESULTS

This life cycle study of PV power generation from 2010 to 2050 shows a significant potential in environmental impact reduction relative to current grid mixes. In terms of global warming, photochemical oxidant formation, particulate formation and acidification, impact assessment results for all PV technologies considered show reductions of a factor of ten below conventional electricity.

The most notable exception is the metal depletion category, which is consistently higher for all three PV technologies in all three time periods. In a sense, PV-generated electricity can achieve substantial reductions in environmental impacts at the cost of increased metals consumption. The impacts of metal depletion, as well as human toxicity and ecotoxicity, are especially linked to the production of metals demanded by BOS and power electronics such as transformers and inverters at PV facilities. In particular, these results suggest that increased future development of PV will lead to increases in the demand for copper.

Surprisingly, PV electricity in ground-mounted utility-scale applications is comparable to conventional electricity in land occupation, due partially to the relatively large land requirements for coal production for conventional electricity. Roof-mounted PV, which makes up approximately half of the BLUE Map scenario PV development, provides a substantial reduction in land occupation in comparison with conventional electricity, since we assume that roof-mounted PV arrays do not directly occupy land as the land is already in use as a building. Based on these findings, and the assumed mixture of ground- and roof-mounted PV in the BLUE Map scenario, we are likely to see a reduction in per kWh land occupation by electricity generation due to increased generation from PV.

7.5.1.1 Scarcity of thin-film metals

While the LCA results of this chapter suggest that all three PV technologies considered could deplete metals at an increased rate in comparison with fossil fuel-based electricity generation technologies, the effects of potential supply constraints on so-called "critical" materials are not completely captured by the characterization factors for metal depletion in the ReCiPe methodology (Goedkoop et al., 2008). In particular, the thin film PV technologies considered in this report require indium (In), gallium (Ga), selenium (Se), tellurium (Te) and cadmium (Cd) that are produced mostly as by-products of other metals, namely copper and zinc. The issues surrounding the potential scarcity of these metals are well-explored and have been discussed in depth in the literature (Woodhouse et al., 2011; Woodhouse et al., 2013; Fthenakis, 2012; Anctil and Fthenakis, 2013; Du and Graedel, 2011; Elshkaki and Graedel, 2013; Graedel et al., 2013; Zweibel, 2010; US DOE, 2010).

For example, Elshkaki and Graedel (2013) indicate that tellurium availability could limit the future production of CdTe, indium availability for the production of CIGS, and silver (Ag) for the production of silicon PV technologies, where it is used as electrical contact grids. Woodhouse et al. (2011) estimate the extent to which these materials for thin films would limit production. Their study showed that current production rates of tellurium and indium would allow a maximum of 9 GW per year of CdTe module production and a maximum of 28 GW per year of CIGS production. These estimates assumed current module efficiencies and that the entire annual production is adopted for PV module production. By comparison, the BLUE Map scenario requires an average of 35 GW of new PV capacity every year from 2010-2050 to provide just six per cent of global electricity generation from PV. The expected increases in PV module efficiency and material efficiency by 2050 described in this report would enable greater annual production of thin films than reported in previous studies. Furthermore, Fthenakis discusses how recovery of tellurium from recycled CdTe PV modules, which have higher concentrations of tellurium than other devices, could result in an economically feasible source of secondary tellurium by 2050 (Fthenakis, 2012). Assuming a certain degree of technological improvements to PV, increased recovery of metals, and conservative demand for semiconductor metals by non-PV uses, Fthenakis estimates that CdTe production could reach 44-105 GW per year by 2050, and that CIGS production could reach 17-106 GW per year (Fthenakis, 2012).

While the modest demands for PV capacity under the BLUE Map scenario appear feasible, the production of particular PV technologies could be constrained if recycling and recovery of metals is not implemented. Recovery of metals would be especially important in scenarios with much higher penetrations of PV generation, such as the scenarios by NREL in the Renewable Electricity Futures study that analysed the feasibility of 80 per cent renewable scenarios in the United States (Mai et al., 2012).

7.5.1.2 Importance of balance of system to environmental impacts

Our analysis also shows the importance of including the BOS in the analysis. For PV, BOS includes all construction, supports, frames, and electronic components that connect the PV units to the grid, or make PV electricity useable by consumers. The preceding graphs showed the relative importance of the life cycle impacts of the BOS and power electronics, especially for thin film CdTe and CIGS PV technologies. Impacts of roof-mounted PV show higher contributions from power electronics in comparison with ground-mounted systems. Poly-Si systems show higher contributions to impacts by module production compared to thin film technologies, due to the somewhat higher impacts of Chinese silicon PV technologies compared to the thin film PV considered in this analysis.

7.5.1.3 Potential risk of toxic releases from thin-film PV during use and end-of-life

While the life cycle results presented in this chapter show that thin film PV technologies have lower human toxicity and ecotoxicity impacts than typical grid electricity or poly-Si PV, thin film PV modules do contain potentially hazardous materials, such as cadmium, that could affect the LCA results presented in this chapter if released into the environment. The sensitivity analyses presented in Chapter 7.4.1 indicate that, in general, more cadmium than is contained in a PV module would be needed to be released to the environment for electricity from CdTe PV to have higher freshwater ecotoxicity and human toxicity impacts than grid electricity, but higher terrestrial ecotoxicity impacts for thin films would be possible if more than 6 per cent of module cadmium were released to the soil. Furthermore, typical module breakage of 1.2 per cent would have little effect on the toxic impacts of electricity from thin film PV on average. LCA, however, is not the most suitable tool for analysing the risks of acute damages to human health and the environment from local emissions of toxic pollutants.

Since the commercialization of CdTe and CIGS, a number of studies have investigated the possible emissions of cadmium and other toxic metals and metalloids from thin film PV technologies. Some studies have researched the emissions of cadmium from a LCA perspective (Fthenakis, 2004; Raugei and Fthenakis, 2010; Raugei et al., 2012), showing that electricity from CdTe PV has lower cadmium emissions than coal electricity. Others have used environmental testing (Zimmermann et al., 2013), risk assessment (Cyrs et al., 2014), and fate and transport models (Sinha et al., 2012). Some non-regulatory environmental testing studies have shown that metals like cadmium and molybdenum can be leached out of millimetre sized pieces of thin-film modules under acid rain conditions (much smaller pieces than are likely to be found in a landfill) (Zimmerman, 2014). Cyrs et al. (2014) cite a number of regulatory tests showing that CdTe modules pass TCLP leaching tests and are thus not considered hazardous waste by the United States Environmental Protection Agency. Other studies have shown that all kinds of PV modules (thin film and silicon) have failed California's Total Threshold Limit Concentration (TTLC) and Soluble Threshold Limit Concentration (STLC) regulatory tests due to the presence of cadmium and selenium in thin film modules and silver, lead and copper in silicon modules (Fthenakis and Zweibel, 2003; Saurat and Ritthoff, 2010; Fthenakis, 2011).

No studies, however, have fully estimated the quantity of cadmium and other thin film metals likely to be leached from modules disposed in landfills, nor have they detailed all the relevant fate transport mechanisms of such metals and their eventual risk to humans. If a sizeable percentage of thin film modules are not recycled, in contrast with present day, their end-of-life emissions could be a concern as the market for thin film PV expands into the developing world (as in the scenarios analysed in this report). While current CdTe PV manufacturers in industrialized nations have well-established take-back and recycling programs for modules at their end-of-life, there is no guarantee that these programs will be successful in the future in developing nations, or that there will be an adequate demand for today's thin film PV technologies to economically incentivize take-back and recycling of used thin films from 2030-2050. Thus, if future thin film manufacturers do not continue the proactive life-cycle management practices of current companies, policy measures could be needed to ensure that the metals used in thin film PV technologies are recycled or safely disposed of so that PV technologies can fully achieve the potential environmental benefits shown in the LCA results of this chapter. Another important consideration is that cadmium is an unavoidable by-product of zinc production. Some argue that cadmium not used in PV, batteries, or other marketable products must still be disposed of, implying that sequestering cadmium in CdTe PV devices may be environmentally preferable to disposal (Raugei and Fthenakis, 2010).

7.5.2 UNCERTAINTY

While several PV technologies have been discussed, this LCA study only investigates three representative PV technologies. Thus, it is important to note the sources of variability and uncertainty that can impact the results. Data gaps were discussed earlier in the chapter, so this section will focus on other sources of variability and uncertainty.

7.5.2.1 Geographic variability of solar insolation

The performance of a PV power plant is directly proportional to the solar irradiation at the site (see framework discussed in PV harmonization studies) (Kim et al., 2012; Hsu et al., 2012). As a result, two otherwise identical PV power plants with identical production-related environmental impacts can have vastly different environmental impacts per kWh. The magnitude of these impacts depends on the amount of energy generated over the plant's lifetime of 25-30 years, which is itself dependent on the local solar irradiation.

In this analysis, ground-mounted PV plants were deployed in better locations with higher insolation within a region, and roof-mounted power plants were deployed in average urban areas. It is reasonable to assume that PV plants will be deployed first in the most optimized locations, as these sites have the shortest energy and financial payback. However, in reality, other factors such as electricity prices, policy incentives, and consumer behaviour may all play a role in determining who chooses to build a solar facility and where. For example, the analysis in this chapter may underestimate the amount of PV deployment in countries such as Germany, where solar irradiation is below average and overestimate PV deployment in sunnier countries such as Spain and Italy.

7.5.2.2 End-of-life

While end-of-life (EOL) and recycling processes are being explored in industry and in literature, there are limited data available to model LCI of PV recycling processes for all the technologies covered. Thus, neither the recycling of PV modules nor of BOS components was included in the results calculation. In reality, PV modules reaching their EOL are recycled, as made evident by the already existing implementation of recycling and take-back programs by PV manufacturers like First Solar (First Solar, 2014). Also, it is likely that the metals in BOS components and power electronics, specifically copper, will be recycled or reused.

Studies have shown that PV modules pose little or no health risks and are environmentally benign when disposed of properly (Fthenakis et al., 2005; Fthenakis, 2000). Additionally, exploratory papers have shown that there is an environmental benefit to recycling PV panels as this displaces primary production of materials used in PV (Müller et al., 2005). From this, it follows that by excluding the recycling in the results of this chapter, it is likely that environmental impacts of a PV module may be overestimated in some categories, particularly metal depletion.

PV generation also requires larger quantities of some bulk materials, particularly copper, as shown by Bergesen et al. (2014). Hertwich et al. (2014) project the materials requirements for sustaining a transition to low carbon electricity following the BLUE Map scenario, and they show that increased demand for copper, steel, aluminium and cement can be sustained by 2050 even without considering recycling. Materials used for BOS components such as steel and aluminium frames, copper wire, inverter and transformer materials could feasibly be recycled or repurposed and reused. For example, PV power plant racks and frames could theoretically be reused to support new modules after the original ones expire. Because the BOS components such as the inverter, transformers and frames contribute significantly to metal depletion, recycling and reuse at EOL for those components could significantly improve results, as could technological innovations that reduce the materials requirements of power electronics.

7.5.2.3 Omitted technologies

As discussed previously, there are a large number of commercially viable solar technologies, and this chapter only presents original research on three technologies: poly-Si, CdTe and CIGS. LCA results show similar impact profiles for these three technologies, supporting their representativeness of the world market. The most important technologies not considered were single crystalline silicon and amorphous silicon modules, which have been studied heavily in the literature, and are available in the ecoinvent database.

7.5.2.4 Technology market shares

Future market shares of technologies were modelled as a simple combination of the three technologies modelled in this chapter. Market shares will likely be different in actuality, but future market shares in this analysis assumed rapid growth and dominance of thin film technologies by 2050. Market shares of PV technologies in the future will depend heavily on technological innovation, costs per Watt, scarcity of resources and consumer preferences. Given that the BLUE Map and baseline scenarios includes only a modest percentage of PV in the electricity mixes of 2030-2050, and the similarity in impacts across technologies, this assumption is unlikely to drastically change the scenario analysis results.

7.5.2.5 The importance of technological changes

The results presented in this chapter show the importance of accounting for future technological changes when assessing the environmental implications of PV. For all technologies, impacts decreased from 2010 to 2050; these improvements are mostly attributable to improvements in module energy conversion efficiency and material efficiency. While these technological changes are plausible from an engineering perspective, it is not certain that they will occur, nor is it known the rate at which modules will achieve these efficiency improvements. For example, CIGS modules in this analysis were expected to increase in efficiency from 12 per cent in 2010 to 20.8 per cent by 2030. By 2013, CIGS and CdTe manufacturers have already produced 15.7 per cent and 16.1 per cent efficient modules, respectively, outpacing industry roadmaps (First Solar, 2013; TSMC, 2013). This example shows how PV technology can advance rapidly and unpredictably relative to the NREL roadmaps used to model technological changes in this study (Whitney, 2014).

7.5.3 POLICY IMPLICATIONS

The results discussed in this chapter present an exploratory discussion of the benefits, risks, and trade-offs of generating electricity from PV power. The global potential for PV power is enormous, and the PV industry is growing and evolving rapidly.

PVs already offer significant environmental benefits over conventional fossil fuels in most impact categories, and impacts will likely continue to decrease as the technologies mature. Thus, to understand the environmental impacts and GHG mitigation potential of PVs, policy makers must not only remain aware of the present impacts and PV technologies, but also of the future environmental trends and the potential of emerging PV technologies.

In comparison with traditional energy sources, PVs require a greater amount of metals. At larger scales, PV power will increase the demand for metals, both critical semiconductor metals as well as base metals used for power electronics (copper) and plant construction (steel, aluminium, manganese). Further research will be needed to determine the extent to which this increased metal demand will be sustainable in the long term, especially as other renewable technologies expand and more electrical grid infrastructure is built to accommodate more renewable energy sources. Metal depletion results also suggest that policies encouraging recycling or extended producer responsibility could mitigate the impact of metal depletion to some degree. However, future LCA studies and engineering research are needed to accurately assess the benefits of PV recycling.

Next, this study shows two key differences between utility-scale ground-mounted PV systems and distributed roof-mounted systems. Roof-mounted PV systems require less metal and electrical infrastructure per kWh, and occupy much less land than ground-mounted systems. However, other impacts, namely climate change, are higher for roof-mounted PV systems. These results are pertinent to policy makers who would weigh the environmental merits of feed-in-tariffs supporting distributed solar or supporting renewable portfolio standards that would incentivize utility-scale PV development. Another consideration is the potential to site utility-scale PV power plants on otherwise unusable brownfields, the benefits of which are already being considered by government agencies such as the U.S. Environmental Protection Agency (2012).

In conclusion, this review and assessment of PV electricity generation technologies suggests that PVs are at a critical stage in development. While current electricity generation by PV is small relative to conventional energy sources, PV is growing quickly and its potential environmental impacts and ability to mitigate GHGs are not negligible. At this stage of PV development, policymakers may have the opportunity to greatly influence how PV technologies contribute to a low carbon energy future.

7.6 REFERENCES

Ardani, K. and R. Margolis. 2011. *2010 Solar Technologies Market Report*. U.S. Department of Energy, Energy Efficiency and Renewable Energy.

Azzopardi, B. and J. Mutale. 2010. Life cycle analysis for future photovoltaic systems using hybrid solar cells. *Renewable and Sustainable Energy Reviews* 14(3): 1130-1134.

Bazilian, M., I. Onyeji, M. Liebreich, I. MacGill, J. Chase, J. Shah, D. Gielen, D. Arent, D. Landfear, and S. Zhengrong. 2013. Re-considering the economics of photovoltaic power. *Renewable Energy* 53: 329-338.

Bergesen, J. D., G. A. Heath, T. Gibon, and S. Suh. 2014. Thin-Film Photovoltaic Power Generation Offers Decreasing Greenhouse Gas Emissions and Increasing Environmental Co-benefits in the Long Term. *Environmental Science & Technology* 48(16): 9834-9843.

Cyrs, W. D., H. J. Avens, Z. A. Capshaw, R. A. Kingsbury, J. Sahmel, and B. E. Tvermoes. 2014. Landfill waste and recycling: Use of a screening-level risk assessment tool for end-of-life cadmium telluride (CdTe) thin-film photovoltaic (PV) panels. Energy Policy 68(0): 524-533.

Daily, G. C., ed. 1997. *Nature's services : Societal dependence on natural ecosystems*. Washington: Island Press.

Daishun, G. 2002. Production of industrial silicon from coal instead of charcocal (in Chinese). *Ferroalloy* 2: 25-27.

Diao Zhouwei, S. L. 2011. Life Cycle Assessment of Photovoltaic Panels in China (in Chinese). *Research of Environmental Sciences* 24(5): 571-579.

Drury, E., A. Lopez, P. Denholm, and R. Margolis. 2013. Relative performance of tracking versus fixed tilt photovoltaic systems in the USA. *Progress in Photovoltaics: Research and Applications*.

ecoinvent Data v2.2. 2009. edited by ecoinvent. Zurich, Switzerland.

EIA. January 2011. Solar Photovoltaic Cell/Module Manufacturing Activities 2009. Accessed at http://www.eia.gov/renewable/annual/solar_photo/archive/solarpv09.pdf.

EPIA. 2011. *Global Market Outlook for Photovoltaics Until 2015*. Brussels: European Photovoltaic Industry Association.

EPIA. 2012. *Global Market Outlook for Photovoltaics Until 2016*. European Photovoltaic Industry Association.

EPIA. 2013. *Global Market Outlook for Photovoltaics 2013-2017*. European Photovoltaic Industry Association.

EPIA. 2014. *Market Report 2013*. European Photovoltaic Industry Association.

Espinosa, N., R. Garcia-Valverde, A. Urbina, and F. C. Krebs. 2011. A life cycle analysis of polymer solar cell modules prepared using roll-to-roll methods under ambient conditions. *Solar Energy Materials and Solar Cells* 95(5): 1293-1302.

European Photovoltaic Industry Association. 2011. *Global Market Outlook for Photovoltaics Until 2015*. European Photovoltaic Industry Association.

First Solar. 2014. Prefunded Recycling Service. http://www.firstsolar.com/en/technologies-and-capabilities/recycling-services/prefunded-recycling. Accessed April 28, 2014.

First Solar. 2013. Sets CdTe Module Efficiency World Record, Launches Series 3 Black Module. http://investor.firstsolar.com/releasedetail.cfm?ReleaseID=755244. Accessed April 28, 2014.

Fthenakis, V. 2011. *Sustainability of Large Deployment of Photovoltaics: Environmental & Grid Integration Research*. Upton, New York: National Photovoltaics Environmental Research Center, Brookhaven National Laboratory.

Fthenakis, V. 2012. Sustainability metrics for extending thin-film photovoltaics to terawatt levels. *MRS bulletin* 37(04): 425-430.

Fthenakis, V. and K. Zweibel. 2003. CdTe PV: Real and perceived EHS risks. Paper presented at Prepared for the NCPV and Solar Program Review Meeting.

Fthenakis, V. and H. C. Kim. 2009. Land use and electricity generation: A life-cycle analysis. *Renewable and Sustainable Energy Reviews* 13(6-7): 1465-1474.

Fthenakis, V., W. M. Wang, and H. C. Kim. 2009. Life cycle inventory analysis of the production of metals used in photovoltaics. *Renewable & Sustainable Energy Reviews* 13(3): 493-517.

Fthenakis, V., H. C. Kim, R. Frischknecht, M. Raugei, P. Sinha, and M. Stucki. 2011. Life Cycle Inventories and Life Cycle Assessment of Photovoltaic Systems, PVPS Task 12, Report T12-02:2011., edited by IEA.

Fthenakis, V. M. 2000. End-of-life management and recycling of PV modules. *Energy Policy* 28(14): 1051-1058.

Fthenakis, V. M. 2004. Life cycle impact analysis of cadmium in CdTe PV production. *Renewable and Sustainable Energy Reviews* 8(4): 303-334.

Fthenakis, V. M. and H. C. Kim. 2011. Photovoltaics: Life-cycle analyses. *Solar Energy* 85(8): 1609-1628.

Fthenakis, V. M., H. C. Kim, and E. Alsema. 2008. Emissions from photovoltaic life cycles. *Environmental Science & Technology* 42: 2168-2174.

Fthenakis, V. M., M. Fuhrmann, J. Heiser, A. Lanzirotti, J. Fitts, and W. Wang. 2005. Emissions and encapsulation of cadmium in CdTe PV modules during fires. *Progress in Photovoltaics: Research and Applications* 13(8): 713-723.

Gilman, P. D., A. 2012. SAM 2011.12.2: General Description. NREL/TP-6A20-53437. Golden, CO: National Renewable Energy Laboratory.

Goodrich, A., T. James, and M. Woodhouse. 2012. Residential, commercial, and utility-scale photovoltaic (PV) system prices in the United States: Current drivers and cost-reduction opportunities. *Contract* 303: 275-3000.

Goodrich, A. W., M.; Noufi, R. 2011. CIGS Road Map. *NREL Technical Report (in preparation)*.

Green, M. A., K. Emery, Y. Hishikawa, W. Warta, and E. D. Dunlop. 2012. Solar cell efficiency tables (version 39). *Progress in Photovoltaics: Research and Applications* 20(1): 12-20.

Hertwich, E. G., T. Gibon, E. A. Bouman, A. Arvesen, S. Suh, G. A. Heath, J. D. Bergesen, A. Ramirez, M. I. Vega, and L. Shi. 2014. Integrated life-cycle assessment of electricity-supply scenarios confirms global environmental benefit of low-carbon technologies. *Proceedings of the National Academy of Sciences*: 201312753.

Hofierka, J. and J. Kaňuk. 2009. Assessment of photovoltaic potential in urban areas using open-source solar radiation tools. *Renewable Energy* 34(10): 2206-2214.

Hsu, D. D., P. O'Donoughue, V. Fthenakis, G. A. Heath, H. C. Kim, P. Sawyer, J.-K. Choi, and D. E. Turney. 2012. Life Cycle Greenhouse Gas Emissions of Crystalline Silicon Photovoltaic Electricity Generation. *Journal of Industrial Ecology* 16: S122-S135.

IEA Energy Analyses. 2007. Renewable Energy Costs and Benefits for Society. IEA's Implementing Agreement on Renewable Energy Technology Deployment.

IEA. 2010. *Energy Technology Perspectives 2010: Scenarios & Strategies to 2050*. Paris: International Energy Agency. Office of Energy Technology and R&D.

INPE/LABSOLAR. 2000/2001a. South America, edited by INPE (National Institute for Space Research) and LABSOLAR (Laboratory of Solar Energy/Federal University of Santa Catarina) Brazil.

INPE/LABSOLAR. 2000/2001b. Brazil, edited by INPE (National Institute for Space Research) and LABSOLAR (Laboratory of Solar Energy/Federal University of Santa Catarina) Brazil.

Jungbluth, N. 2005. Life cycle assessment of crystalline photovoltaics in the Swiss ecoinvent Database. *Progress in Photovoltaics* 13(5): 429-446.

Jungbluth, N., Tuchschmid, M., de Wild-Scholten, M. 2008. *Life Cycle Assessment of Photovoltaics: Update of ecoinvent Data v2.0*.

Kalowekamo, J. and E. Baker. 2009. Estimating the manufacturing cost of purely organic solar cells. *Solar Energy* 83(8): 1224-1231.

Kim, H. C., V. Fthenakis, J.-K. Choi, and D. E. Turney. 2012. Life Cycle Greenhouse Gas Emissions of Thin-film Photovoltaic Electricity Generation. *Journal of Industrial Ecology* 16: S110-S121.

Kunreuther, H. and P. Slovic. 1996. Science, values, and risk. *Annals of the American Academy of Political and Social Science* 545: 116-125.

Lew, D., G. Brinkman, E. Ibanez, A. Florita, M. Heaney, B. M. Hodge, M. Hummon, G. Stark, J. King, S. A. Lefton, N. Kumar, D. Agan, G. Jordan, and S. Venkataraman. 2013. *The western wind and solar integration study phase 2*. NREL/TP - 5500 - 55588. NREL.

Mai, T., R. Wiser, D. Sandor, G. Brinkman, G. Heath, P. Denholm, D. J. Hostick, N. Darghouth, A. Schlosser, and K. Strzepek. 2012. *Renewable Electricity Futures Study. Volume 1: Exploration of High-Penetration Renewable Electricity Futures*. National Renewable Energy Laboratory (NREL), Golden, CO.

Margolis, R., C. Coggeshall, and J. Zuboy. 2012. SunShot Vision Study. *US Department of Energy*.

Mason, J. E., V. M. Fthenakis, T. Hansen, and H. C. Kim. 2006. Energy payback and life-cycle CO_2 emissions of the BOS in an optimized 3·5 MW PV installation. *Progress in Photovoltaics: Research and Applications* 14(2): 179-190.

McDonald, N. and J. Pearce. 2010. Producer responsibility and recycling solar photovoltaic modules. *Energy Policy* 38(11): 7041-7047.

Mints, P. 2011. *Photovoltaic Manufacturer Shipments, Capacity, & Competitive Analysis 2010/2011*. NPS-Supply6. Palo Alto, CA: Navigant Consulting Photovoltaic Service Program.

Müller, A., K. Wambach, and E. Alsema. 2005. Life Cycle Analysis of Solar module recycling process. Paper presented at MRS Proceedings.

National Renewable Energy laboratory (NREL) Climatological solar Radiation Model. 2004. naturalearthdata.com. 2012. Natural Earth raster + vector map data: Natural earth.

NREL. 2007. *National Solar Technology Roadmap: Wafer-Silicon PV (Draft)*. Golden, Co: National Renewable Energy Laboratory.

Ong, S., C. Campbell, P. Denholm, R. Margolis, and G. Heath. 2013. Land-Use Requirements for Solar Power Plants in the United States. *National Renewable Energy Technical Reports*.

Pacca, S., D. Sivaraman, and G. A. Keoleian. 2007. Parameters affecting the life cycle performance of PV technologies and systems. *Energy Policy* 35(6): 3316-3326.

Parida, B., S. Iniyan, and R. Goic. 2011. A review of solar photovoltaic technologies. *Renewable & Sustainable Energy Reviews* 15(3): 1625-1636.

Perez Model. 2011. edited by SUNY. NREL.

Price, S. and R. Margolis. 2010. *2008 Solar Technologies Market Report*. U.S. Department of Energy, Energy Efficiency and Renewable Energy.

PVGIS (Optimal Incline) - Europe - Turkey - Portions of North Africa. 2012. edited by European Commission.

Raugei, M. and P. Frankl. 2009. Life cycle impacts and costs of photovoltaic systems: Current state of the art and future outlooks. *Energy* 34(3): 392-399.

Raugei, M. and V. Fthenakis. 2010. Cadmium flows and emissions from CdTe PV: future expectations. *Energy Policy* 38(9): 5223-5228.

Raugei, M., M. Isasa, and P. Fullana Palmer. 2012. Potential Cd emissions from end-of-life CdTe PV. *The International Journal of Life Cycle Assessment* 17(2): 192-198.

Roes, A. L., E. A. Alsema, K. Blok, and M. K. Patel. 2009. Ex-ante Environmental and Economic Evaluation of Polymer Photovoltaics. *Progress in Photovoltaics* 17(6): 372-393.

Saurat, M. and M. Ritthoff. 2010. Appraisal of Laboratory Analyses Conducted on CdTe Photovoltaic Modules. *Wuppertal Institute for Climate, Environment and Energy*.

Sengul, H. and T. L. Theis. 2011. An environmental impact assessment of quantum dot photovoltaics (QDPV) from raw material acquisition through use. *Journal of Cleaner Production* 19(1): 21-31.

Sense. 2008. *LCA analysis: Sustainability evaluation of solar energy systems, revised version*. Stuttgart, Germany: University of Stuttgart.

Shockley, W. and H. J. Queisser. 1961. Detailed Balance Limit of Efficiency of p-n Junction Solar Cells. *Journal of Applied Physics* 32(3): 510-519.

Sinha, P., R. Balas, L. Krueger, and A. Wade. 2012. Fate and transport evaluation of potential leaching risks from cadmium telluride photovoltaics. *Environmental Toxicology and Chemistry* 31(7): 1670-1675.

Suh, S., M. Lenzen, G. J. Treloar, H. Hondo, A. Horvath, G. Huppes, O. Jolliet, U. Klann, W. Krewitt, Y. Moriguchi, J. Munksgaard, and G. Norris. 2004. System boundary selection in life-cycle inventories using hybrid approaches. *Environmental Science & Technology* 38(3): 657-664.

TSMC. 2013. Solar Commercial-size Modules (1.09m^2) Set 15.7% Efficiency Record. http://www.tsmc-solar.com/Assets/downloads/en-US/TSMC_Solar_Press_Release_EN_Jun_18_2013.pdf. Accessed September 15, 2013.

U.S. Department of Energy. 2010. *Critical Materials Strategy*. U.S. Department of Energy.

U.S. Environmental Protection Agency. 2012. *Potential Advantages of Reusing Potentially Contaminated Land for Renewable Energy Fact Sheet*. U.S. Environmental Protection Agency. www.epa.gov/sites/production/files/2015-04/documents/contaminated_land_resuse_factsheet.pdf.

Whitney, E. W., M.; Fu, R.; Goodrich, A. 2014. *CIGS Roadmap*. Golden, CO: National Renewable Energy Laboratory.

Woodhouse, M., A. Goodrich, R. Margolis, T. L. James, M. Lokanc, and R. Eggert. 2013. Supply-Chain Dynamics of Tellurium, Indium, and Gallium Within the Context of PV Module Manufacturing Costs. *IEEE Journal of Photovoltaics* PP(99): 1-5.

Woodhouse, M., A. Goodrich, R. Margolis, T. James, R. Dhere, T. Gessert, T. Barnes, R. Eggert, and D. Albin. 2012. Perspectives on the pathways for cadmium telluride photovoltaic module manufacturers to address expected increases in the price for tellurium. *Solar Energy Materials and Solar Cells*.

Zhai, P. and E. D. Williams. 2010. Dynamic hybrid life cycle assessment of energy and carbon of multicrystalline silicon photovoltaic systems. *Environmental Science and Technology* 44(20): 7950-7955.

Zimmermann, Y.-S., A. Schäffer, P. F. X. Corvini, and M. Lenz. 2013. Thin-Film Photovoltaic Cells: Long-Term Metal(loid) Leaching at Their End-of-Life. *Environmental Science & Technology* 47(22): 13151-13159.

Zweibel, K. 2010. The Impact of Tellurium Supply on Cadmium Telluride Photovoltaics. *Science* 328(5979): 699-701.

Chapter 8

Geothermal power

Lead authors: Peter Bayer, Ladislaus Rybach, Jorge Isaac Martínez Corona, Thomas Gibon

8.1 INTRODUCTION

Geothermal energy is thermal energy generated and stored in the earth. Ninety nine per cent of the earth's volume has temperatures over 1,000°C, with only 0.1 per cent at temperatures less than 100°C.[1] The total heat content of the earth is estimated to be about 10^{13} EJ and is thus immense. The main sources of geothermal energy are the residual energy from planet formation and the energy continuously generated by radionuclide decay. Earth radiates heat to the atmosphere, with a thermal power of 40 million MW without experiencing any surface cooling. This amount of heat is equivalent to the thermal power of about 13,000 1-GW_e nuclear power plants. Thus, the geothermal resource base is ubiquitous and sufficiently large to be a significant energy source. Geothermal resources consist of thermal energy stored within the earth in rock, trapped steam or water. Exploitation of geothermal energy occurs through two means: electricity generation and direct use in space heating, balneotherapy, greenhouses, etc. In this study, only power generation is considered.

8.2 TECHNOLOGY DESCRIPTION

8.2.1 GEOTHERMAL POWER PLANTS

Geothermal power plants convert heat from geothermal fluid to electricity. The conversion efficiency depends mainly on the fluid's heat content as reflected by its temperature. Temperature commonly increases with depth, and is different for young, geologically active regions in comparison to older, "cooled" regions. Therefore, the most attractive sites for this technology are the few geologically young areas with very high geothermal gradients. The most productive reservoirs with large volumes of stored geothermal fluids at abnormally high temperatures thus exist at a few hundred or even thousand meters' depth. In contrast to these hydrothermal reservoirs, petrothermal production is focused on geothermal reservoirs with no or marginal amounts of water, i.e., hot and dry rock formations. These reservoirs are typically created through mechanical or chemical stimulation and exploited using engineered geothermal systems (EGS). These represent a category of rather new plant types that generate electricity from greater depth and thus can also be applied in other areas of normal geothermal gradient. However, they still have only a marginal share in the worldwide installed capacity and are therefore not further discussed in the presented study.

1 An earlier version of this chapter was published as (Bayer et al., 2013b). The main text was written by Peter Bayer and Ladislaus Rybach. The life cycle assessment of the geothermal plant was conducted by Jorge Isaac Martínez Corona and Thomas Gibon.

Geothermal power plants consist of numerous components including production and reinjection boreholes, connecting and delivery pipelines, intermediate equipment such as silencers and separators, the power house with turbines, generators and controls, and cooling towers. Each of these components has environmental effects and contributes to life cycle impacts. Some of these impacts, such as construction-related impacts, are temporary, while others occur over the entire operational lifetime of the plant, such as silencer noise. These effects and contributions are treated in Chapters 8.3.2.1 and 8.3.2.2.

Since heat loss makes it impractical to transmit high-temperature steam over long distances by pipeline, most geothermal plants are built close to the resource. Geothermal power plant capacities are typically restricted by well spacing and individual well capacity; wells must be a minimum of 200-300 m apart to avoid interference. A single geothermal well typically has a capacity of 4-10 MW, with some rare exceptions. With these constraints, geothermal power plants tend to be in the 20-60 MW range, even those associated with large reservoirs. As of 2012, the largest geothermal power unit operated with a capacity of 140 MWe at Taonga, Rotokawa geothermal field in New Zealand and is fed by only six production wells. Much smaller units in the range of 500-3,000 kW are common as binary-type plants. The following sections briefly describe the main geothermal power plant types. These characterisations are mainly based on DiPippo (1999, 2008), where more detailed descriptions can be found. The 2012 worldwide shares in produced geothermal electricity produced are shown in Figure 8.1. Average capacity for each plant type (excluding hybrid) are about 5 MW/unit for binary and back pressure plants, 30 MW/unit for flash plants and 45 MW/unit for dry steam plants (Bertani, 2012).

FIGURE 8.1

Shares of different geothermal plant technologies in global geothermal electricity production

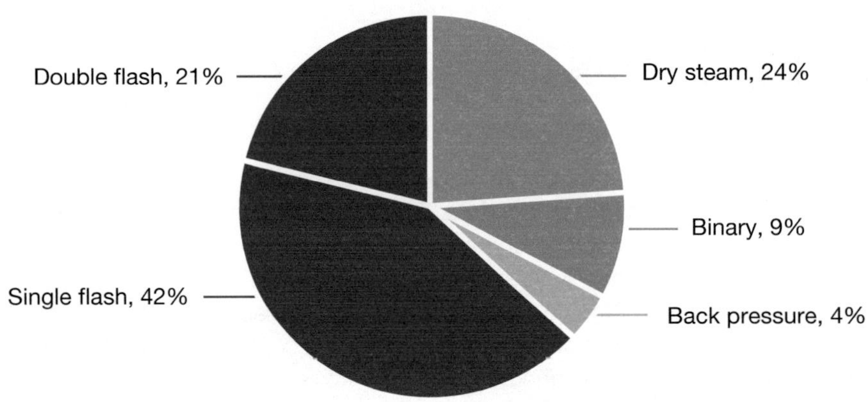

Adapted from Bertani, 2012

8.2.1.1 Direct steam plants

Direct steam plants are installed at vapour-dominated or dry steam reservoirs with temperatures of over 200°C. Dry, saturated or slightly superheated steam is produced from wells. The steam carries non-condensable gases (NCG) of variable concentration and composition, but which consists mainly of carbon dioxide (CO_2) and hydrogen sulfide. Steam from several wells is transmitted by pipeline to the powerhouse, where it is used directly in impulse or reaction turbines. Between each wellhead and the plant, in-line centrifugal cyclone separators near the wellhead remove particulates such as dust and rock bits, while drain pots (traps) along the pipelines remove condensation that forms during transmission. A final moisture remover is placed at the entrance to the powerhouse.

Figure 8.2 is a simplified flow diagram for a direct steam plant. In some countries, back pressure plants are allowed; this design exhausts directly to the atmosphere. The presence of non-condensable gases in geothermal steam makes the gas extraction system a critical plant component. These non-condensable gases typically comprise 2-10 per cent by weight of steam, but may occasionally be an even larger share. Two-stage steam ejectors with condensers are typically used, but in some cases vacuum pumps or turbo-compressors are required.

The condenser is most often direct contact-type, but can also be surface-type. The latter is preferred whenever the NCG stream must be treated or processed before release to the atmosphere, such as when emissions limits for hydrogen sulfide would be exceeded. In such cases, an elaborate chemical plant must be installed to remove the hydrogen sulfide. The steam condensate is available for cooling tower make-up as it is not recirculated to a boiler as in a conventional power plant. In fact, excess condensate, typically corresponding to 10-20 per cent of the steam by weight, is available and is usually re-injected into the reservoir. Mechanical induced draft cooling towers operating in either counter- or cross-flow configurations are mostly used for wet cooling systems, but natural-draft towers are used at some plants.

FIGURE 8.2

Schematic of a direct steam geothermal power plant for vapour-dominated reservoirs

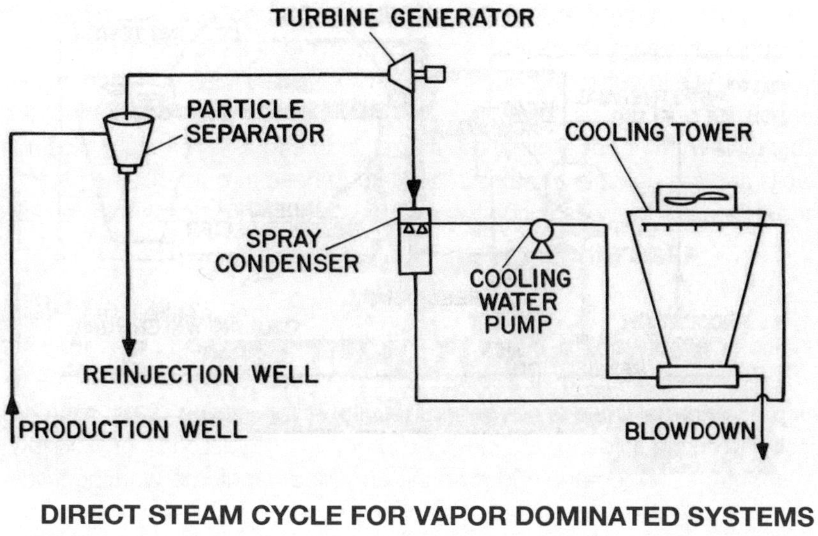

Source: Edwards et al., 1982

8.2.1.2 Flash steam plants

Dry steam reservoirs are rare, with the only known major fields being Larderello in Italy and The Geysers in the US. The most common type of geothermal reservoir is liquid-dominated. For artesian-flowing wells, the produced fluid is a two-phase mixture of liquid and vapour. The quality of the mixture, measured in the steam composition by weight, is a function of the reservoir fluid conditions, the well dimensions, and the wellhead pressure, which is controlled by a wellhead valve or orifice plate. Typical wellhead qualities may range from 10 to over 50 per cent.

The conventional approach is to separate the phases of the two-phase flow and use only the vapour to drive a steam turbine. Since the wellhead pressure is fairly low, typically 0.5-1.0 MPa, the liquid and vapour phases differ significantly in density (p_l/p_v =175-350), allowing effective separation by centrifugal action. Highly efficient cyclone separators yield steam qualities as high as 99.99 per cent.

The separated liquid phase may be re-injected, used for its thermal energy via heat exchangers for a variety of direct heat applications, or flash evaporated at a lower pressure using a control valve or orifice plate, thereby generating additional steam for use in a low pressure turbine. Plants using only primary high-pressure steam are called single flash plants while plants using both high- and low-pressure steam are called double flash plants.

8.2.1.3 Double flash plants

About 20-25 per cent more power can be generated from the same geofluid mass flow rate by using double-flash technology rather than single-flash. The two-phase flow from the well is directed horizontally and tangentially into a vertical cylindrical pressure vessel, the cyclone separator. The steam transmission lines are essentially of the same design as in dry steam plants and are usually fitted with traps.

The secondary, low-pressure steam produced by throttling the separated liquid to a lower pressure is sent either to a separate, low-pressure turbine or to an appropriate stage of the main turbine; in such a case, the main turbine would be a dual-pressure, dual-admission turbine. Figure 8.3 depicts a schematic of double-flash plants.

8.2.1.4 Binary plants

Binary plants are used for geofluid temperatures below 150°C, or for geofluids with a large share of dissolved gases. In a binary plant, the thermal energy of the geofluid is transferred via a heat exchanger to a secondary working fluid for use in a fairly conventional Rankine cycle. The secondary working fluid, having a low boiling point, evaporates and drives the turbine. The geofluid itself does not contact the moving parts of the power plant, thus minimizing, if not eliminating, the adverse effects of erosion. Most binary plants operate on pumped wells and the geofluid remains in the liquid phase throughout the plant, from production wells through the heat exchangers to the injection wells. This is a closed process that releases no emissions to the atmosphere.

A flow diagram for a typical binary plant is given in Figure 8.4. Water or air may be used for cooling depending on site conditions. If wet cooling is used, an independent source of make-up water must be found, since geosteam condensate is not available as in direct- or flash-steam plants. Due to its chemical impurities, the remaining geothermal liquid is not generally suitable for cooling tower make-up. There is a wide range of possible working fluids for the closed power cycle. Hydrocarbons with low boiling temperatures such as isobutane, isopentane and propane are good candidate working fluids, as are certain refrigerants. The optimal fluid will give high utilization efficiency together with safe and economical operation.

Binary plants are particularly well suited to modular power packages in the range of 1-3 MW per unit. Standardized, skid-mounted units can be factory-built, tested, assembled and shipped to a site for rapid field installation. Several units can then be connected at the site to match the power potential of the resource.

FIGURE 8.3

Simplified schematic of double flash plants

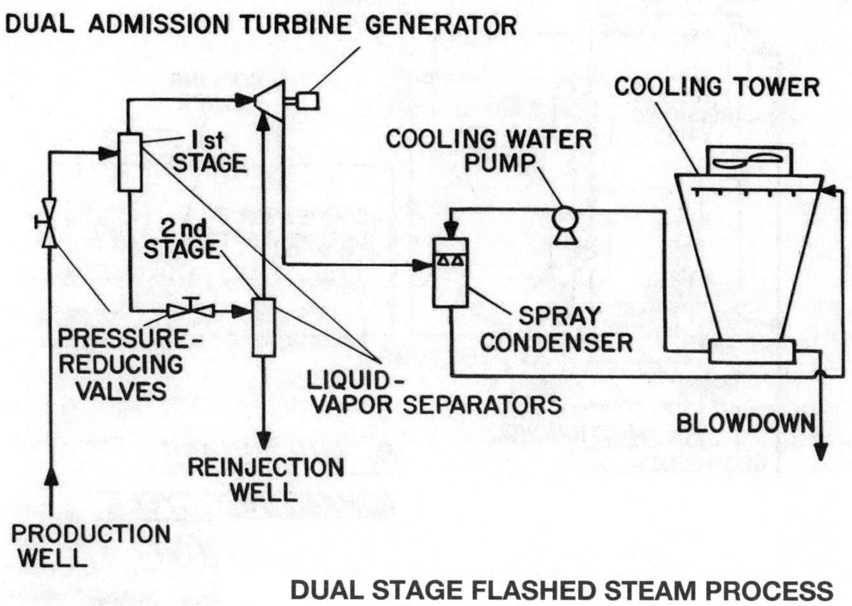

Source: Edwards et al., 1982

FIGURE 8.4

Simplified scheme of an organic Rankine-cycle binary geothermal power plant

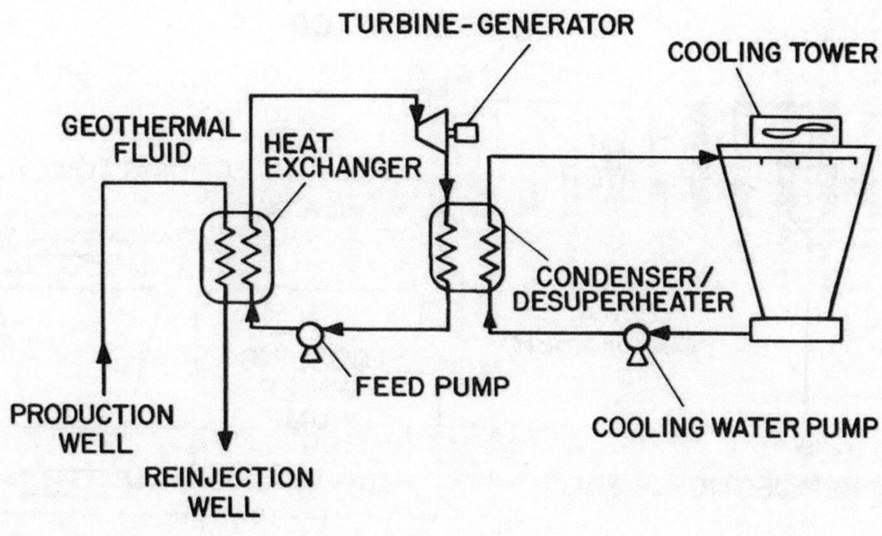

Source: Edwards et al., 1982

8.3 EMISSIONS AND ENVIRONMENTAL IMPACTS

8.3.1 SELECTION OF TECHNOLOGIES TO BE STUDIED

Current installed capacity of geothermal power worldwide is around 12 GW, and as of 2010, 24 countries have been generating geothermal power. Of these, the United States, the Philippines, Indonesia, Mexico, Italy, Iceland, New Zealand, and Japan combined produce more than 90 per cent of the global installed capacity (Figure 8.5). The deployment of existing geothermal power plants in these countries reflects the priority placement of geothermal power plants at specific regions with high geothermal gradient, typically geologically young and volcanic areas. In these areas, geothermal reservoirs can typically be accessed at shallow depths of less than 2,000 m. Even if engineered geothermal systems (EGS) are slowly on the rise, they require deep wells of several thousand meters depth. This makes them technologically more demanding and economically less attractive. Still, substantial growth rates are predicted for the next decades (Goldstein et al., 2011; Tester et al., 2006). This study is oriented at Bayer et al. (2013) and uses standard geothermal power production, especially in the United States, as a reference. Direct emissions are described independently for each plant type. Finally, emissions, resource and energy use associated with flash steam and binary plants included in the inventory are scrutinized in more detail.

FIGURE 8.5

Generated electricity (in GWh) from geothermal power plants in 2010

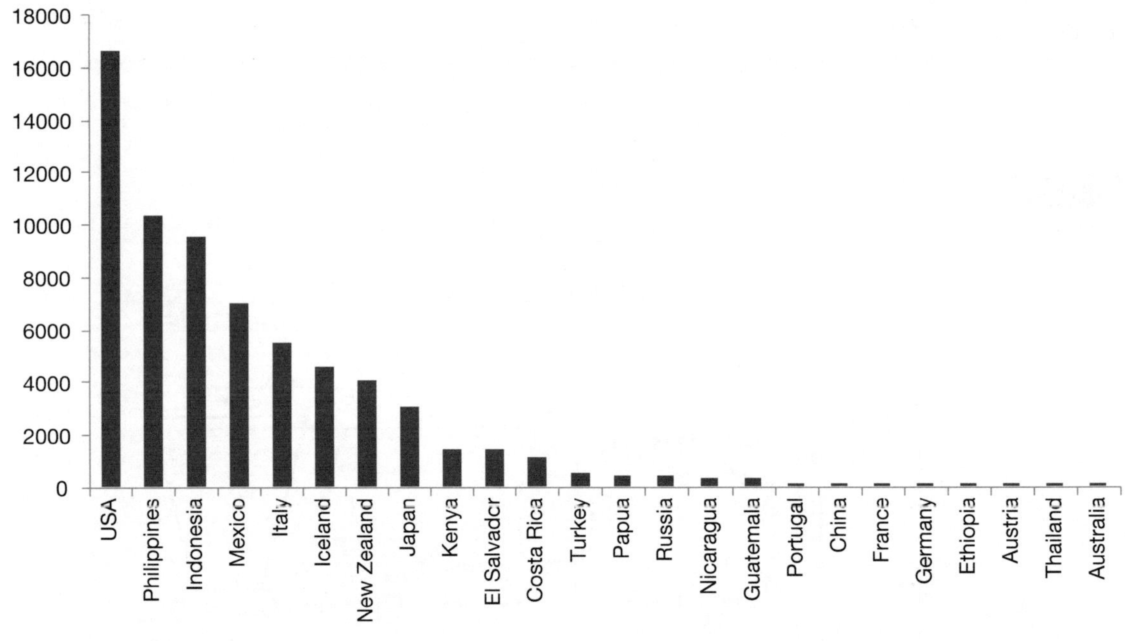

Source: Data from Bertani, 2012

8.3.2 PROCESS EMISSIONS OF GEOTHERMAL PROJECTS

8.3.2.1 Gate-to-gate impacts and benefits – direct environmental impacts

The environmental effects from geothermal electricity production are commonly categorized based on safeguard subjects, i.e., endpoint indicators such as human health, biodiversity, etc., type and pathway of stresses and emissions. A variety of publications by different authors share a similar perspective (Brophy, 1997; Heath, 2002; Kristmannsdóttir and Ármannsson, 2003). Most of these categories emphasize environmental burdens, but apart from the provision of renewable energy, geothermal activities are sometimes associated with secondary benefits as well.

We can roughly distinguish the following relevant environmental effects from geothermal electricity production:

- Land use
- Geological hazards
- Noise
- Thermal effects and emissions
- Atmospheric emissions
- Solid waste, soil and water emissions
- Water use

In the following sections, we describe the range of environmental threats as reported in the literature. This is intended to show the variety of potential impacts. We it has to be emphasize that many of these impact types are important only at some locations. If possible, specific value ranges are reviewed and compared.

8.3.2.1.1 Land use

In this general category, we include use of land as well as changes to landscape and natural features. Land surface is needed during the different life cycle stages of a geothermal power plant. This occupation may be temporary as in the construction and reclamation phases, or permanent as during operation (Rybach, 2005). Geothermal energy production is focused at the resource below the subsurface and thus the manipulation, alteration and depletion of the geothermal reservoir is associated with the use of subterranean resources. This is, however, equivalent to the fossil fuel extraction or open pit mining, and is similarly not considered a critical environmental issue.

The most detailed description of land use phases is provided by the U.S. Bureau of Land Management (BLM, 2008) (Table 8.1). The term 'Land disturbances' in the table represents land use in the western United States, and thus may be used to feed the life cycle inventory (LCI). BLM (2008) emphasizes the uncertainty in the listed values, which stems from uncertainty about timing, location, distribution of geothermal resources, as well as type of geothermal plant. Since part of the land used in the exploration and site preparation phases is only temporarily occupied, permanent land use during plant operation is estimated to be smaller, although no exact numbers are given. They estimate an average value of 0.85 km^2 for 'land disturbance' during the construction of a 50 MW power plant. The studied scenario considers six well pads with both single and multiple wells, e.g., by using advanced directional or slant drilling technology, about 0.4 km of road per well, and transmission lines 8-80 km long in a corridor about 15 m wide. However, a 50 MW power plant could also involve up to 25 production wells and 10 injection wells and would therefore have a larger land use impact.

Although holding ponds for temporary discharges during drilling can be sizeable, their contribution to the land footprint is considered minimal (Tester et al., 2006). The complete surface facility includes the power plant, cooling towers, auxiliary buildings and substation. Geothermal plants with super saline brines require huge vessels to process the brine. Tester et al. (2006) estimate 75 per cent higher land use for such plants. During the last phase of reclamation and abandonment, it is expected that power plant removal, well plugging, capping and reclamation, and site and access road re-grading will facilitate natural restoration (BLM, 2008).

TABLE 8.1

Typical land disturbances during geothermal resource development of a 50 MW power plant

Operation	Land disturbance [ha]
Exploration	8-28
Geologic mapping	negligible
Geophysical surveys	
Gravity and magnetic surveys	
Seismic surveys	
Resistivity surveys	
Shallow temperature measurements	
Road/access construction	4.0-24
Temperature gradient wells	4.0
Drilling Operations and Utilization	206-1416
Drilling and well field development	20-202
Road improvement/construction	16-129
Power plant construction	61-101[1]
Installing well field equipment including pipelines	20-81
Installing transmission lines	97-971
Well workovers, repairs and maintenance	negligible
Total	214-1490

[1] Area in thousands of square meters
Source: BLM, 2008

As revealed by the ranges in Table 8.1, maximum values for the entire land disturbances are about a factor of seven higher than the minimum. The installation of transmission lines plays a substantial role in this, whereas the power plant buildings only cover an average of about one tenth of the total land footprint. This is also reflected in the alternative land use estimates for partial or full geothermal applications estimates as listed in Table 8.2. Without wells and transmission lines, reported values range between 1,200-2,700 m^2/MW. Smaller land footprints per MW are given for high-capacity flash steam plants, particularly with the reinjection of extracted geofluids. Small-size binary power plants with cooling towers have a higher footprint. Table 8.2 summarizes land use values reported in other sources. Note that these values omit land use attributed to transmission lines. Land use of 2,300-9,700 m^2/MW is estimated for different sites, technologies and plant capacities. Tester et al. (2006), for example, suggest a value of 7,500 m^2/MW for flash-cycle Rankine-cycle power plants. On the other hand, the entire Cerro Prieto field covers 540 ha, as reported by Hunt (2001). Using this area yields 30,000 m^2/MW, which is the highest value calculated and is approximately equal to the maximum estimate from BLM (2008) (Table 8.1). Hunt (2001), however, emphasizes that this overestimates the true land use. Land may be disturbed, divided into smaller parcels, but not lost.

Still, the role of the power transmission line is never accounted for. Geothermal power plants are restricted to active geothermal areas, which may be remote, have low population density and are rarely industrialized. Transmission lines thus often stretch over long distances. Geothermal plants are on occasion strategically installed with energy intensive industries; an example is aluminium smelting in Iceland. Generalized values for transmission line length are difficult to obtain, but the 8-80 km length as used in Table 8.1 (BLM, 2008) is a reasonable range. Ultimately, whether they should be included or not will depend upon the boundary settings for the LCA.

TABLE 8.2

Land use values and ranges for geothermal power production at specific locations or as generalized values

Plant type and specifications	1000 m²/MW	Data source
181 MW Cerro Prieto field (Mexico), wells only	0.7	(Hunt, 2001)
Drill site including surface disturbances caused by excavation, construction and new roads	0.2-2.5	(Ármannsson and Kristmannsdóttir, 1992)
110-MW flash plant, excluding wells	1.3	(Goldstein et al., 2011; Tester et al., 2006)
20 MW binary, excluding wells	1.4	(Goldstein et al., 2011; Tester et al., 2006)
49 MW flash-Rankine, e.g., Salton Sea Calif., excluding wells	2.3	(Goldstein et al., 2011; Tester et al., 2006)
Single-flash plant, excluding wells	1.2	(DiPippo, 1991)
Binary plant, excluding wells	2.7	(DiPippo, 1991)
Geothermal plant, not specified	1.4	(Brophy, 1997)
56 MW flash, including wells, pipes	7.5	(Goldstein et al., 2011; Tester et al., 2006)
50 MW, no transmission line	2.3-10.3	(BLM, 2008), Table 8.1
50 MW, all installations of geothermal plant	4.3-29.7	(BLM, 2008), Table 8.1
Range for entire geothermal field, not specified	4.1-32.4	(USDOE, 2008)
180 MW Cerro Prieto field (Mexico), entire field with wells and power station	30	(Hunt, 2001)

(in thousands of square metres per megawatt installed capacity)
Source: Bayer et al., 2013

Some discussion can be found on the land quality in geothermal areas. Goldstein et al. (2011) note that new development options in countries such as Japan, Indonesia, the United States and New Zealand are constrained by land use issues in or in the vicinity of national parks and tourist areas. Often, such environments are unique (Boothroyd, 2009; Fukutina, 2012), with their own vegetation and special geothermal subsurface phenomena. In some cases, the productivity and value of such land is economically poor as in Wairakei, New Zealand (Hunt, 2001), while in other cases the land has high cultural value or is highly productive agricultural land such as in the Imperial Valley in California. Other prominent examples are geothermal power generation in Japan and Indonesia. Land use issues in these two countries often are of high relevance as many geothermal resources are located in forest conservation areas and national parks (Pasqualetti, 1980). Hunt (2001) concludes that land use impacts are not only influenced by the type and extent of the development but also the original land use prior to development.

Changes to the original landscape as a consequence of geothermal development are often seen as serious problems. The visual intrusion from surface installations often disturbs or destroys locations of great scenic quality (BLM, 2008; Kristmannsdóttir and Ármannsson, 2003). Visual disturbances are most pronounced during the drilling and site construction, for instance, when tall drill rigs are onsite. Facilities, such as buildings and pipelines used during the operation can be painted to blend in with the neighbouring environment (BLM, 2008; DiPippo, 1991). While the infrastructure built in the early days were massive, smaller and more inconspicuous installations are used today (Rybach, 2005). Still, there is little flexibility in adjusting surface constructions to the landscapes or locating at low value land, since geothermal power plants need to be built at the site of geothermal reservoirs (Hunt, 2001).

Typical surface manifestations of geothermal processes or discharges are hot or steaming ground, hot springs and pools, mud pools, fumaroles, geysers and deposits of sinter, sulfur or other minerals (Fukutina, 2012). These geothermal features are special, natural wonders that exist only in a few areas; these areas are often fragile and vulnerable to human intervention, and are thus have high environmental value. DiPippo (1991) emphasizes that care is usually taken to preserve these manifestations when they also serve as tourist attractions. However, the consequences of large-scale and long-term geofluid extraction are often difficult to predict. This inevitably causes impacts, e.g., through the permanent extinction of geysers. For example, in New Zealand, more than a hundred geysers have disappeared, which is mainly a result of geothermal developments; their recovery appears unlikely (Barrick, 2007; Fukutina, 2012). Land use effects are thus often more extensive than the occupied land itself. Barrick (2007) reports geothermal well withdrawal radiating many kilometres and causing geyser extinctions. These lateral effects from geothermal reservoir exploitation are also critical for any other competitive economic, touristic, or cultural uses of geothermal features. Spas, for example, make use of hot water and are often in conflict with geothermal energy use (Goldstein et al., 2011; Barrantes Víquez, 2006). Sometimes, however, as with the Blue Lagoon in Iceland, geothermal power production creates new features of high touristic value (Kristmannsdóttir and Ármannsson, 2003).

8.3.2.1.2 Geological hazards

Geothermal energy production is associated with extensive extraction or circulation of geofluids including steam, as well as local to large-scale manipulation of the shallow and deep ground. Consequently, a range of geological consequences represents critical hazards that are often unique to geothermal activities, such as induced seismicity. General indicators describing these environmental effects, however, are not available. The highly case-specific characteristics and diversity of the associated geological hazards hamper a general ranking. The subsequent list thus is only demonstrative and is not a comprehensive compendium of all phenomena related to geothermal developments.

The steep volcanic terrain on which geothermal facilities are commonly built is a main factor for the frequent occurrence of slumps or landslides (Maochang, 2001). The latter may also be stimulated by changes in the regional water and heat flow, and when unconsolidated sediments such as pumice are destabilized. Construction of road access in undeveloped and often rugged terrain accelerates erosion and causes loss of vegetation (Arnórsson, 2004).

Ground deformations are often observed as a consequence of reservoir pressure decline following fluid withdrawal. Subsidence is accentuated by compressible rock formations in the upper part of or above a shallow reservoir that is drained and compacted after pore pressure decline. They are more common for liquid dominated fields, which are often located in young unconsolidated volcanic rock (Hunt, 2001). For example, the subsidence rate at Wairakei geothermal field can reach 40 cm/year, at Larderello, 25 cm/year, and at Svartsengi, Iceland, 1 cm/year (Kristmannsdóttir and Ármannsson, 2003). Subsidence is often most pronounced in the early reservoir depletion stage, and reached a maximum of 0.45 m per year at Wairakei (Allis, 1990). This can yield a local loss of land through flooding and a change in contour (Fukutina, 2012), thereby altering the regional hydrological flow regime; damage to buildings and infrastructure, including the pipelines, drains, roads and well casings of the plant are additional potential consequences from subsidence (Maochang, 2001). Boothroyd (2009) also mentions a positive effect of subsidence when local wetlands are created, providing new habitats. Subsidence can be mitigated by the reinjection of spent fluids (Barbier, 1997). Ground deformations sometimes entail the creation of near-surface tensile fractures and fissures in the surrounding rock (Rissmann et al., 2012) that offer new pathways for gas and water circulation. (Pasvanoğlu et al., 2012) even report the formation of an extensive sinkhole as a consequence of intense hydrothermal water extraction in a karst regime at the Kozakli geothermal field in Turkey. As another extreme consequence, reinjection has in some cases triggered ground inflation (Kaya et al., 2011).

Under extraordinary conditions, hydrothermal explosions and well blowout may occur when steam pillows develop above the lowered groundwater surface (DiPippo, 1991; Kristmannsdóttir and Ármannsson, 2003), or due to sudden release of overburden pressure caused, for example, by an earthquake (Browne and Lawless, 2001). Nowadays, these are rare thanks to gathered experience and technological improvements. Geothermal activities are concentrated at seismically active zones, and the natural seismicity is variable. Microseismicity due to fluid injection is reported for several applications (Evans et al., 2012; Hunt, 2001; Majer et al., 2012; Rybach, 2005; Rybach and Mongillo, 2006).

In summary, the geothermal energy production is commonly focused at geologically young and active sites with fragile and sensitive environments that are often pristine and hard to access. Under such conditions, relatively large efforts are necessary in development of geothermal fields, and the risk of geological hazards is higher than elsewhere. These hazards not only threaten the geothermal installations, but also local vegetation and often specialized ecosystems, as well as the natural hydrological conditions.

8.3.2.1.3 Noise

During the life cycle of a geothermal power plant, potential noise sources include the building and drilling activities, plant operation and any deconstruction work during the land reclamation phase. While noise from construction and reclamation is considered standard noise, high noise levels of around 120 dB (muted around 85 dB) are reported for drilling (Rybach, 2005; Rybach and Mongillo, 2006). During well testing, high pressure steam is released through a silencer with noise levels of 70-110 dB (muted). The cumulative impact will depend on the total number of wells under testing, over periods that often last several months. Construction and demolition noise is mostly caused by trucks, bulldozers, graders and cranes for the time of road and power plant construction and dismantling (Kristmannsdóttir and Ármannsson, 2003; Ogola, 2005). During routine operation, the cooling towers, transformer and power house are main sources for noise. Air-cooled condensers are more sizable with higher noise emissions than water-cooled towers.

8.3.2.1.4 Thermal effects and emissions

All heat-power conversion systems produce waste heat, which can reach significant quantities. This applies to geothermal power generation, as well; the waste heat fraction depends on the conversion technology or power plant type. Geothermal power plants release considerably more waste heat per unit of power output than other power plant types due to the lower conversion efficiency. Geothermal cooling facilities such as cooling towers are, related to unit capacity (MW_e), significantly larger than for other technologies. Waste heat from geothermal power plants is released at the plant site, into the atmosphere, into ponds or natural water bodies (Rybach and Kohl, 2004).

The heat-power conversion efficiency depends mainly on the temperature of the produced geothermal fluid. Due to the considerably lower conversion efficiency of binary power plants, these plant types have comparably high amounts of waste heat, whereas direct-steam and flash-steam plants emit less heat per MW of electric power produced (Figure 8.6). When the waste heat is utilized for direct heating purposes, for example by feeding it into district heating networks, not only is the profitability of geothermal development increased, but a significant environmental benefit would also result due to the avoidance of waste heat emission to the plant surrounding.

FIGURE 8.6

Waste heat per unit electric capacity of different types of geothermal power plants

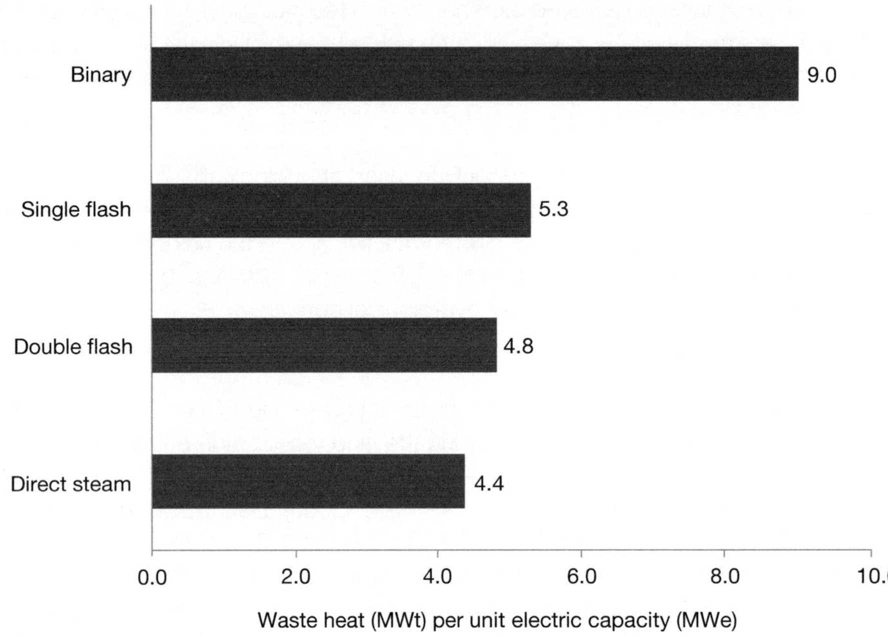

Source: DiPippo, 1991

8.3.2.1.5 Atmospheric emissions

During the life cycle of geothermal power plants, typical atmospheric emissions are from exhaust associated with transportation, use of diesel engines during road, well and plant construction (Table 8.3). In many cases, however, exhaust emissions are relatively small in comparison to fugitive emissions of steam. These emissions, common for flash or dry steam plants, are released during operation and are of paramount importance for atmospheric emissions. Geothermal gases, such as CO_2, hydrogen sulfide, and methane, with specific environmental threats are often discharged directly to the atmosphere. These non-condensable gases are released from flash steam and dry steam power plants since they do not condense at the turbine outlet in contrast to the steam, (Bloomfield et al., 2003). CO_2 is the dominant NCG constituent with around 90 per cent in geothermal fluids (Bertani and Thain, 2002). Trace amounts are reported for mercury, ammonia, radon, and boron. Downstream abatement systems are often installed to reduce concentrations of the most critical compounds, such as hydrogen sulfide. A minor source may be cooling tower drift, which was an issue at the Coso site in California, for example. If geothermal fluid is additionally used for cooling, potentially hazardous fugitive substances may be released. This was the case with boron at the Coso plant (Kagel et al., 2007); however, no negative effects were identified.

The released mixture concentrations vary substantially according to different geographical areas, as well as within a geothermal field and even during the lifetime of a power plant. For example, hydrogen sulfide increases more than CO_2 during production as demonstrated for power plants in Iceland (Kristmannsdóttir and Ármannsson, 2003). In the course of operation, a steam cap may evolve below the surface and periodically release high gaseous concentrations (Ármannsson et al., 2005). Measured, estimated, and indirectly derived numbers vary widely in the literature, with sometimes different underlying assumptions. For example, Bloomfield et al. (2003) points out that calculating CO_2 emissions from the content in the primary steam may overestimate emission factors, since fractions partition in the condensate, and the re-injection of spent fluids that also have a CO_2 component is sometimes ignored.

TABLE 8.3

Activities and selection of related pollutants during geothermal project phases

Activity	Pollutant	Project Phase	Factors
Exhaust from vehicular traffic	Carbon monoxide, carbon dioxide, oxides of nitrogen, volatile organic compounds, particulates, sulfur dioxide, air toxics	All	Vehicle-miles traveled (VMT)
Fugitive dust from vehicle traffic on paved and unpaved roads	Particulates	All	VMT, road conditions
Fugitive dust from earth-moving activities	Particulates	All	Acres disturbed, soil conditions
Exhaust from construction equipment	Carbon monoxide, carbon dioxide, oxides of nitrogen, volatile organic compounds, particulates, sulfur dioxide, air toxics	All	Volume of fuel used, engine/ abatement technology
Release of geothermal fluid vapor	Carbon dioxide, hydrogen sulfide, mercury, arsenic, boron	Exploration, drilling operations, utilization	Chemical composition of geothermal resource, duration and volume of flow testing, frequency, duration, and voume of well blow-outs, type of power plant

Source: BLM, 2008

Reported direct CO_2 emissions from geothermal power plants span a broad range. This greenhouse gas (GHG) originally stems from degassing magma, and, more rarely, from decomposition of organic sediments and metamorphic decarbonisation (Ármannsson et al., 2005). The most cited global survey on this topic is that for the International Geothermal Association (IGA) by Bertani and Thain (2002), who derived a range of 4-740 g/kWh for a large number of power plants that constitute 85 per cent of the 2001 geothermal capacity (6648 MW). The weighted average of the values is 122 g/kWh. This is very similar to the earlier estimate by (Fridleifsson, 2001), who provided a range of 3–380 g/kW while providing more details on the underlying assumptions. Often much more optimistic estimates, mostly without further background details, can be found in other literature sources. A range of 50-80 g/kWh is given by DiPippo (2008), while Kagel et al. (2007) assume 44 g/kWh, GEA (2012) lists 0-40 g/kWh, and Bloomfield et al. (2003) provides 91 g/kWh as a weighted average for American plants. For New Zealand, total geothermal electricity was about 13 PJ in 2007 (Hung and Bodger, 2009; MED, 2007), and annual CO_2 emissions from geothermal power production of 301 kt are reported (MED, 2008). This yields an emission factor of 83 g/kW for CO_2, which is comparable to the NZ average stated by Rule et al. (2009) (80 g/kWh, range of 30-570 g /kWh). For power plants in Iceland, Ármannsson et al. (2005) provided values of 152 (Krafla), 181 (Svartsengi) and 26 g/kWh (Nesjavellir) for the year 2000. According to the (USDOE, 2012), dry steam plants at The Geysers in California produce about 41 g/kWh and flash plants generate about 28 g/kWh. Binary plants ideally represent closed systems and no steam is emitted. A very large CO_2 component in the produced fluid at the Kizildere geothermal power plant (Turkey) is used for producing industrial grade CO_2 (Şimşek et al., 2005).

Emissions of CO_2 occur naturally in geothermal regions. Special interest is thus on the change of CO_2 emissions through geothermal energy use and stimulated additional or potentially accelerated CO_2 release to the atmosphere. Bertani and Thain (2002) emphasize the observed decrease of natural CO_2 emissions in the Larderello field since power plant operation began. While, for example, for Larderello, it is assumed that the anthropogenic emissions are equivalent to the original geogenic emissions, in Iceland, power plants were assumed to contribute 8-16 per cent to the national CO_2 emissions in the year 2002 (Ármannsson et al., 2005). Still, natural geothermal CO_2 emissions are high and remain high even when power plants are in operation. Dereinda and Armannsson (2010), for example, find that natural release of CO_2 at the geothermal field is about three times higher than from the vented steam from the Krafla geothermal plant in Iceland (Dereinda and Armannsson, 2008). Fridriksson et al. (2006) anticipate a six-fold increase of CO_2 emissions through operation of the Reykjanes power plant in Iceland. In New Zealand, at the Ohaaki hydrothermal field, after 20 years of geothermal power production, soil degassing was intensively measured (Rissmann et al., 2012). Main findings are that due to the low permeability of the reservoir spatial extent, the magnitude of natural CO_2 emissions has not changed. In contrast, at Wairakei, NZ, based on heat flow measurements, (Sheppard and Mroczek (2004) concluded, however, that the emissions have doubled as a consequence of geothermal energy production. Bertani (2012) expects that the emission factor per kWh will decrease with time. For binary Wairakei power plant, Rule et al. (2009) report emissions of 40 g CO_2/kWh.

Of the sulfur-bearing gaseous emissions, primary sulfur dioxide is only a minor constituent. Kagel et al. (2005) report small values of 0.159 g/kWh for flash-steam, liquid-dominated, and 9.8 x 10^{-5} g/kWh for hydrothermal dry steam at The Geysers. In contrast, hydrogen sulfide is found at higher concentrations and is often a subject for local environmental concern, because of its odour and toxicity. However, odour nuisances appear long before toxic concentrations are reached. When dissolved in water aerosols, hydrogen sulfide reacts with oxygen to form more oxidized sulfur-bearing compounds such as SO_2. Kristmannsdóttir et al. (2000) observed that hydrogen sulfide is washed out by precipitation and only a small fraction ends up as SO_2. Thus, local effects of hydrogen sulfide will be triggered by land use, rainfall, wind pattern and topography. As an example, in 2011, the 213 MW Hellisheidi power plant in Iceland emitted 13,000 tons hydrogen sulfide/ year (6.96 g/kWh) to the atmosphere; re-injection measures are underway. The emissions are suspected the reason for a 140 per cent increase of sulfur pollution in the area of Reykjavik, 30 km away (Carlsen et al., 2012; SavingIceland, 2012). Further literature values for 1991 and 1992 are in the range of 0.5-6.4 g/kWh as listed by Hunt (2001). Bloomfield et al. (2003) estimate a smaller value of 0.085 g/kWh for the weighted average for geothermal power plants in the United States, where abatement systems apparently reduce the emitted fraction. Kagel et al. (2007) emphasize that, despite increased adoption of geothermal energy since the 1970s, total hydrogen sulfide emissions have decreased by about an order of magnitude. This reflects the increased application of efficient abatement technologies (Baldacci et al., 2005; Bertani, 2012)). In the United States in particular, the observed decline in hydrogen sulfide emissions is related to the increasing use of binary plants and reinjection of geofluids.

Methane occurs at low concentrations in geothermal steam, but is of particular interest due to its high GWP. Taking again the 2007 New Zealand total geothermal electricity production of about 13 PJ (Hung and Bodger, 2009), and reported annual methane emissions from geothermal power production (MED, 2008), an emission factor of 0.85 g/kWh for methane is obtained. This is consistent with the 0.75 g/ kW given by Bloomfield et al. (2003) as the weighted average for all geothermal power plants in the United States. Alternative methane emission factors are not found, and methane emissions are seldom a focus for geothermal energy (Arnórsson, 2004).

Generalized or averaged vales for other compounds are scarce. For ammonia (NH_3), Bloomfield et al. (2003) report a weighted United States average of 0.06 g/kWh. Nitrogen oxide and particulate matter emissions from geothermal power plants are rated negligible. However, NO_x may be generated in abatement systems, which are commonly used for hydrogen sulfide oxidation. For example, Kagel et al. (2007) provide a value of 0.458 g/MWh for The Geysers dry steam field, which stems from the burning process applied in some plants with abatement (Tester et al., 2006). Another rare case is that of proximal silica deposition that has led to forest damage at Wairakei (Armstead, 1978).

Boron, ammonia, arsenic, and mercury in the atmosphere are leached by rain, thereby threatening soil and surface water surrounding power plants. Often, however, geothermal or volcanic areas are naturally burdened by deposition of pollutants, and thus the comparison to the undisturbed natural state is important to judge the net impact of steam plant operation. Also, geothermal plants are sometimes co-located with abandoned mines and the associated environmental impacts (e.g., mine drainage) have to be carefully distinguished from those arising from the geothermal plant operation. Boron, which exists as boric acid in the steam, is a critical component in the emissions at various locations, such as Larderello in Italy, and Kizildere, Turkey (Arnórsson, 2004). Case studies reveal elevated mercury concentrations in vegetation, fish, surface waters, and the atmosphere in the vicinity of geothermal plants (Arnórsson, 2004; Bacci et al., 2000). Mercury pollution in the vicinity of Larderello has been measured for decades, for example by Baldi (1988), who observed up to 1.8 µg/g in mosses at a distance as far as 0.6 km from the plant. Another example is Bacci et al. (2000), who found emission rates of 3-4 g/kWh of mercury at the Mt. Amiata geothermal power plant in Italy. Arnórsson (2004) states that mercury contents may be as a high as 0.5 ppm in geothermal steam.

While the geological regime is the main determinant of the total direct emissions of a geothermal plant, the core plant technology used, its life cycle and the application of additional procedures, such as abatement, cooling and reinjection technologies are also important factors. Steam is released from flash steam or dry steam facilities, but reinjection of the steam with spent fluid after compression of NCGs minimizes release to the atmosphere. Re-injection is considered favourable (DiPippo, 1991), but it is still not routine practice globally (Kaya et al., 2011). There are some reservoirs where full reinjection has been achieved (e.g., Coso), but more often, plants conduct partial reinjection to support and maintain productivity. Gas compression and injection will cost additional energy, which may rate the entire system uneconomical. Furthermore, gas may come out of solution in the formation and eventually increase the NCG content in the produced geofluid.

Already during initial testing, spray is often released that could damage vegetation in the surrounding area (Kristmannsdóttir and Ármannsson, 2003). In contrast to steam-based technologies, binary cycles are considered "closed systems". However, they use low-boiling fluid/gas; one commonly used fluid is isopentane (R-601a), with a GWP of about 11. Similarly to the working fluid of heat pumps, it may escape in slow fractions over time (Hunt, 2001; Saner et al., 2010). Finally, air emissions also accompany well drilling, bleeding, clean-outs and testing, and discharges can occur from line valves and waste drilling mud degassing (Rybach, 2005; Rybach and Mongillo, 2006). Exhaust is emitted from construction machinery, especially during site preparation, road construction, drilling, and plant dismantling.

Microclimatic effects can occur at sizable power plants or geothermal fields with substantial discharge of warm water vapour. Increased rainfall and fog are observed, but are rare (Maochang, 2001). However, plumes of steam may affect the visual appeal of a site (Heath, 2002).

8.3.2.1.6 Solid waste, emissions to soil and water

Especially in liquid-dominated, high temperature geothermal fields, the volumes of extracted geothermal fluids and the resulting waste can be significant. Separated and condensed fluids often accumulate in steam-based plants. Similar to gaseous emissions, geothermal fluids vary widely in composition depending on geological setting, production mode, time and technology. Concentrations rise with about the square of the salinity. For instance, the Salton Sea field in the United States is hosted by evaporate deposits. Geofluids here are highly saline with chlorine levels of approximately 155,000 ppm. In contrast, the alkaline fluids of the Krafla, Námafjall, and Nesjavellir fields in Iceland are of very low salinity, with chlorine levels of about 100 ppm (Baba, 2003; Ellis, 1978; Heath, 2002). Hydrogen sulfide, boron, ammonia and mercury are also characteristic fluid contaminants in geofluids, as are metals such as arsenic, lead, cadmium, iron, zinc, antimony, lithium, barium and aluminium (DiPippo, 1991; Heath, 2002; Kristmannsdóttir and Ármannsson, 2003). Geothermal fluids or brines can accrue in large amounts. Ideally, these fluids are fully re-injected, but for technical or economic reasons, reinjection is infrequent. As a result, decontamination is necessary before discharge, with the most common receiving body being surface water. There are numerous scenarios where insufficient control of geothermal fluids causes substantial local environmental problems, particularly when the very harmful compounds, such as mercury and arsenic are discharged. Goldstein et al. (2011) note that in the past, surface discharge of separated brines or fluids had been a problem at a few sites, but these discharges occur only rarely today, for example, in Wairakei, NZ (Fukutina, 2012) and are prohibited by environmental regulation. Robinson et al. (1995) investigated arsenic emissions in the Wairakei River and in most cases, found aqueous concentrations above the 10 μg/L WHO standard, and concentrations of around 30 μg/g in sediments. The Wairakei and Ohaaki power stations are identified as major sources for the pollution. Similarly elevated arsenic concentrations were measured in the rivers downstream of the Mt. Apo geothermal field in the Philippines (Webster, 1999). In this case, these high levels are interpreted as elevated natural background concentrations that are, for example, released from hot springs. In their global overview, Kaya et al. (2011) present a comparison of waste fluids discharged to surface water bodies at twelve plant sites (Figure 8.7). These released fluids, however, do not automatically yield hazardous waste streams. For example, Krafla is located in a remote area and the fluid run-off is released to a small stream and disappears into porous lava rock. The run-off from Svartsengi goes primarily to the Blue Lagoon spa. Cerro Prieto has a huge holding lagoon for evaporation of the brine that is not re-injected.

FIGURE 8.7

Waste water discharged to the surface from selected geothermal fields

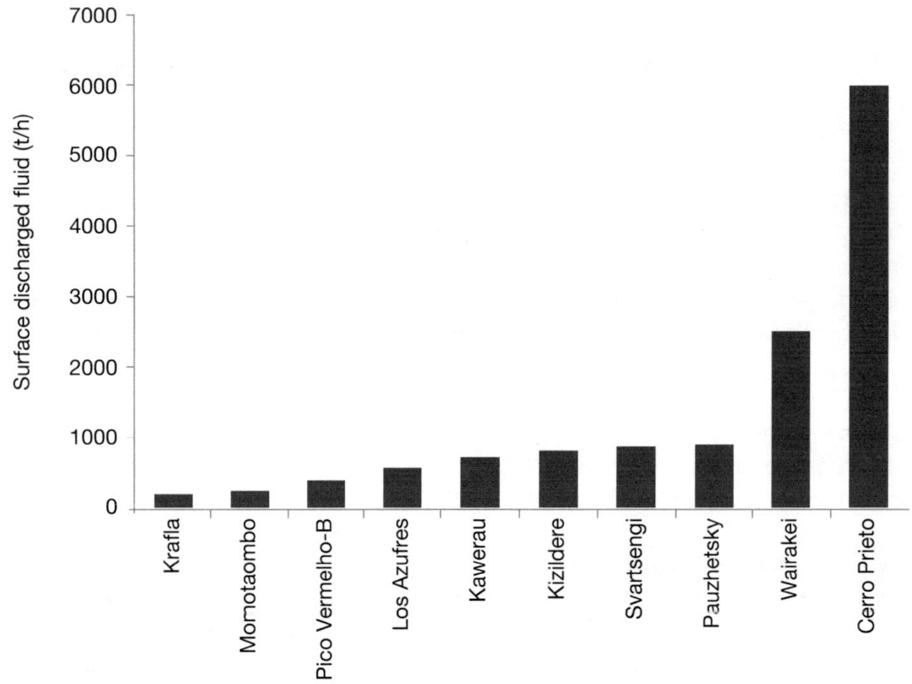

Source: Kaya et al., 2011

Hunt (2001) estimates an annual mercury emission of 50 kg from the Wairakei geothermal field into the Wairakei River. Discharge from the Kizildere power plant in Turkey leads to regional pollution of a downstream river catchment and the adjoining aquifer over an area of more than 100 km^2. As a consequence of agricultural use of the polluted water for irrigation, boron accumulates in soils and threatens natural vegetation and fruit crops. Often natural background concentrations are elevated in geothermal areas, such as boron levels in the Kizildere surrounding environment, and hence additional deposition from geothermal development makes these environments more likely to reach harmful concentrations (Koç, 2011). Koç (2011) reports harmful boron emissions of about 194 kt per year from both anthropogenic and natural sources at Buyuk in

Turkey. At the Balcova geothermal field, Turkey, both faulty reinjection, as well as discharge to surface waters threaten the environment due to high concentrations of boron, as well as of arsenic and antimony (Aksoy et al., 2009). Similarly, thermal waters from the Yangbajing geothermal field in Tibet carry high concentrations of boron, arsenic and fluorine into a downstream river. Guo et al. (2008) measured concentrations of up to 3.8 mg/l boron and 0.27 mg/l arsenic in the river, and observed health problems among inhabitants. Sometimes pollutant dispersal is retarded through storage in holding or evaporation ponds, from which they can leak in surface water bodies (Heath, 2002).

Reinjection can contaminate fresh water aquifers (Aksoy et al., 2009; Heath, 2002). Aquifers are also threatened by drilling fluids and infiltration of geothermal fluids in the case of well casing failure. Waste fluids from drilling and testing can cause gully erosion (Maochang, 2001), and depending on the composition, lead to contamination of freshwater bodies. These effects, however, are infrequently reported.

Generally, total amount of solid waste is considered small and not of environmental concern. Solid waste can be grouped as follows (Armstead, 1978; Arnórsson, 2004; Brophy, 1997; Heath, 2002; Rybach, 2005; Rybach and Mongillo, 2006):

- Drilling waste such as cuttings, cement residues, drilling muds (such as bentonite)
- Chemical deposition in pipes and vessels of the plant, scale residues with accumulated arsenic and heavy metals
- Sediments in cooling towers, possibly with mercury contamination
- Waste material, deposits, activated carbon, from treatment and abatement systems
- General waste associated with commercial operation.

Besides the limited detrimental effects of solid waste, there are also benefits: extraction of metals and minerals as by-products from geothermal energy extraction can be profitable regardless of whether the plant is for power or direct use applications. This is highly case-specific as the profitability depends on the primary mineral concentrations in the geothermal fluids. The ability to remove silica can allow for added energy extraction, reduce operation and maintenance cost and open the way for the recovery of such metals as zinc, lithium, manganese, cesium, rubidium and even precious metals such as gold, silver and platinum. In the early history of geothermal resource development, boric acid, sulfur, potassium and ammonium salts were recovered commercially until they lost economic competitiveness to other mining processes (Lehr et al., 1982). Clark et al. (2011) give an overview on by-product recovery projects, and though rate current interest in mineral extraction from geothermal fluids low. Most projects have been abandoned after feasibility assessment phase of pilot scale applications.

The mineral resources that have the greatest potential to be economically extracted are silica, lithium, and zinc. Producing a silica by-product is one possible option to address scaling issues. Silica extraction is reported for example from the power plants in Wairakei, New Zealand, as well as Mammoth Lake and Coso, both in California, and Steamboat Springs in Nevada, United States (Bourcier et al., 2003). Lithium removal from geothermal fluids is realized at Hatchobaru, Japan and at Wairakei. A site near the Salton Sea in the Imperial Valley produces one of the most metal-rich geothermal fluids in the world. The geothermal fluid has been mined for its zinc content (Bourcier et al., 2003), but this was stopped after a short period as it was unprofitable (Clark et al., 2011). Currently, new mineral extraction efforts are underway at the Featherstone plant at the Salton Sea.

8.3.2.1.7 Water use and consumption

Geothermal power plants can require considerable amount of water during both installation and operation. The use of freshwater depends on the size of the plant, the technological variant, the working temperatures and cooling mechanism, and the availability of alternative salt, sewage (i.e., "grey") or geothermal water. Geothermal fluids are not fresh water. These fluids are extracted at a rate ranging roughly between 150-1,000 m^3/hour per well (BLM, 2008). Clark et al (2011) list production rates in the range of approximately 60-80 m³/MWh for binary and 15-27 m³/MWh for flash steam power plant, which are originally from the California Division of Oil, Gas and Geothermal Resources. More conservative estimates of 96 and 74 m³/MWh are then taken for their hypothetical scenarios. It is an exception that spent geothermal fluids or condensed water is used as potable water or for agriculture (BLM, 2008; Goldstein et al., 2011); rather, the spent fluids are typically partially re-injected, treated, ponded or discharged.

Geothermal reservoirs are often overlain by shallow groundwater. The incurred pressure drop in the deep reservoir may stimulate cold downflow, especially when connecting flow paths, such as fractures, exist (Maochang, 2001). This leads to an indirect depletion of the freshwater hosted in the local shallow aquifer. Re-injection of spent geothermal fluid is a common means to avoid substantial pressure drop.

Water is consumed from the very beginning in large quantities of up to 1,000 m³/d for drilling. Clark et al. (2011) estimate the total water consumption per metre depth in well construction of around 5-30 m³, depending on geology, technology, number of liners and depth. This corresponds to water use of 8,000-55,000 m³ for a 2-km deep well. Compared with the more recent work by (Harto et al., 2013), these values may be overestimations; they present values of around 2.2 m³/m, which represents water consumption of around 4,500 m³ for a 2-km deep well. This may be associated with the fact that the latter, more realistic values account for recirculation of drilling mud, except of loss-of-circulation zones that are plugged as quickly as possible. Here, the primary use of water is in the cement used for securing casing.

During operation, water in small amounts is consumed to minimize scaling and to manage dissolved solids (Clark et al., 2011). A main determinant for water use and consumption will be the cooling technology, with the evaporation as the main water loss pathway. This means of water consumption is thus dependent on geofluid/steam inlet and outlet temperature. Binary power plants use a small amount of water, but only through air-cooling is water saved. Water cooling is often applied, is water intensive, and is a dominant factor when freshwater from rivers or aquifers is withdrawn. Due to the comparably lower steam and the higher fluid outlet temperatures, water-based cooling is rated less demanding than for alternative power plants, such as nuclear or fossil fuel-based boilers (BLM, 2008). This, however, is not supported by works that compare operational water consumption of different power plant types (Macknick et al., 2011). Fthenakis and Kim (2010) add to this discussion that geothermal power plants need more water per MWh than conventional steam plants, because of the lower heat-to-electricity conversion efficiency (8-15 per cent).

Fthenakis and Kim (2010) report a broad range of about 0-7.5 m³/MWh for dry steam plants, depending on the cooling system used, and 2.3-15 m³/MWh for withdrawal and consumption in liquid dominated systems. This upper limit is similar to the 17 m³/MWh value reported by Adee and Moore (2010) for the water-cooled Salton Sea binary plant. The most recent study by Harto et al. (2013) gives updated estimates of the values originally presented by Clark et al. (2011). These are considered most reliable, especially for characterizing the conditions in the western United States. Operational water use of air-cooled plants is very small, and associated with a variety of activities such as dust compression, maintenance and domestic needs. An average value of 0.15 m³/MWh and a range of 0.04-0.45 m³/MWh is given. Hybrid cooling represents any combination of air-cooled condensers, cooling towers, etc. Hybrid cooled plants consume water in a range of 1.1 - 6.4 m³/MWh, with an average of 3.8 m³/MWh. Water cooling means a substantial increase, requiring 2.6-14.4 m³/MWh (average of 9.1 m³/MWh) for a flash steam plant and 5.7-17.4 m³/MWh (12.9 m³/MWh) for a binary plant.

Sometimes makeup water is periodically utilized to make up for blowdown losses. In some cases, water is added to the re-injected steam condensate, such as at The Geysers, Darajat and at Laradello (Kaya et al., 2011). Make-up water does not need to be fresh water, but using low-quality water requires more frequent cycling and thus larger volumes (BLM, 2008). For example, grey water is injected at The Geysers to mitigate gradual productivity loss (Stark et al., 2005). Rybach (2005), Franco and Villani (2009) and Goldstein et al.

(2011) point out that spent geothermal fluids and/or steam condensate is preferably employed for partial cooling of steam-type power plants. Once used, cooling water is re-injected or discharged to evaporation ponds or aquifers.

The United States Environmental Protection Agency (2008) calculates a range of about 0.75-1.15 m^3/MWh of total water volume consumed for electricity generation from geothermal resources (BLM, 2008). James et al. (2012) provide a life cycle "water" withdrawal and consumption rate of 38 m^3/MWh for a flash-steam plant. However, he considers all geothermal fluid consumption due to vapour losses during flashing of the geofluid. A critical aspect here is that apparently geofluids, which are often brines, are equated with water, and the role of reinjection, discharge and evaporation is roughly considered. The use of freshwater, which is of prime interest within LCA, is not distinguished, and in some cases, low-quality water may be applied to support cooling and/or as makeup.

Two scenarios of geothermal plants are inspected by Clark et al. (2012) and Harto et al. (2013). They calculate total full life cycle water consumption of 1 m^3/MWh for a hybrid-cooled plant and 0.15 m^3/MWh for an air-cooled 10 MW binary plant. The relatively high water consumption by the binary plant in the first case stems from switching to wet cooling during hot summer daytime operation (hybrid cooling), and maintenance of reservoir pressure. In comparison, 0.15 m^3/MWh is calculated for a wet-cooled 50 MW flash steam plant, assuming that the geofluid/steam condensate is employed. A geofluid loss rate of 10 m^3/MWh is assumed, but the effect of long-term reservoir pressure decrease was not further scrutinized. Harto et al. (2013) estimate that water consumption for this flash plant may reach around 10 m^3/MWh when all of the geofluid is replaced by water.

8.3.2.1.8 Constraints

In summary, life cycle assessment (LCA) studies on geothermal power production are very rare. One reason is that the environmental impacts from certain plants are often very local and case specific, and thus generally valid conclusions from single studies can hardly be drawn. Standard geothermal power plants necessitate exceptional geological conditions. These conditions constitute a large geothermal gradient, which is found in geologically young and/or active volcanic areas, and the existence of a substantial and accessible reservoir of geothermal fluid and heat beneath. The case-specific characteristics of the potentially affected local environment, and the focus on exceptional and sometimes remote places, makes it difficult, if not impossible, to come up with a generic yet valid assessment of environmental impacts as common in LCA studies on other more industrial based renewable energy technologies.

Second, the effect geothermal power plants have on local and regional water quality and the induced change in the hydraulic regime and hazardous geological consequences are always unique. At the same time, the potentially threatened environments are unique. The value of the safeguard subject, or of natural elements with intrinsic value, is the basis for proper assessment of environmental impacts. This value, however, is not always clear. On one hand, the affected regions are often considered to have high value, as they are home to highly specialized flora and fauna, have specific and sensitive features, such as geysers that are often erased with geothermal development, and the resilience is typically low. On the other hand, undeveloped or undevelopable land, and in some cases, conditions hostile to human life, are often located far away from civilized places. As a consequence, environmental assessments sometimes conclude that these environments deserve little or no protection.

A third point is that during the last decades, geothermal energy production has evolved, and International Geothermal Association (IGA) president Horne (2011) even identifies a '"geothermal renaissance'". This has contributed to the evolution of engineered geothermal systems (EGS) that essentially stand for geothermal power plants that extract energy from much deeper, basement-type geological formations and thus are applicable under "normal" geological conditions. Aside from this, new combined cycle plants and hybrid technologies evolve. Cogeneration of heat is often coupled with electricity generation. Geothermal and solar thermal hybrids are just one example of innovative applications. The efficiency of geothermal electricity production has also increased. Instead of 1970-2000 "standard" 55 MW single-flash plants with inlet pressure of about 600 kPa and steam consumption of 8-10 kg/kWh, modern technologies work with up to 2,550 kPa and consume as little as 5 kg/kWh steam. A general LCA that evaluates current geothermal power production would need to include the different plant types and efficiencies, and even then would not be prospective enough to judge the environmental effects of future technologies.

The dynamic technological innovation and gradually increasing efficiency is hard to capture, as is the characteristic time-dependent performance of power plants. Lifetimes reported for existing plants are diverse, ranging from 30 to 80 years, and performance declines are frequently reported. Counter-measures such as reinjection, may partially overcome this, and long-term sustainable power production spanning decades is possible (see Rybach and Mongillo, 2006). Apparently, this is very case-specific, and due to the uncertainty in the geological description, is difficult to predict in the long term.

For making predictions, a critical point is that new geothermal power plants will compete for the limited number of productive and accessible hydrothermal reservoirs. While prominent and highly productive locations such as The Geysers and Wairakei are already well developed, new plants have to find new, suboptimal sites. This may balance the benefit from more efficient production and even reactivation. In several cases, such as at Wairakei and Larderello, older fields previously considered to be exhausted are now being redeveloped using more efficient energy conversion technologies.

Finally, environmental awareness is increasing. Contrary to common practice decades ago, reinjection is much more common now, air-cooling is favoured when possible, wastewater is sometimes remediated and steam emissions, especially those that carry hydrogen sulfide, are decontaminated prior to release. As positive as these improvements may be, life cycle thinking means that lingering long-term impacts from the less controlled past must nevertheless be reflected in a LCI. Further, steam emissions are especially hard to quantify using only literature studies. For instance, CO_2 is mostly unregulated as a pollutant and hence estimates rather than reliable values are found. This is also relevant for methane, and much more critical substances such as mercury, arsenic and boron.

8.3.2.1.9 Overview of existing studies, system boundaries and methodology used

There are many different categories of studies; a considerable number is dedicated to an often qualitative description or analysis of environmental burdens and benefits (DiPippo, 2008; Heath, 2002; Hunt, 2001; Kristmannsdóttir and Ármannsson, 2003; Rybach, 2005; Rybach and Kohl, 2004). Others provide guidelines for remediation, regulation or for countermeasures. Some of these are in the framework of environmental impact assessments (Maochang, 2001; Ogola, 2005), and many of them distinguish between the effects associated with different technological variants. Of the existing LCA studies, several are streamlined LCA concepts; for example, when comparing different renewable energy technologies (Pehnt, 2006), geothermal electricity production is often included for completeness. Another category of LCA-type work presents partial LCA results, such as a global warming potential (Hondo, 2005; Rule et al., 2009), water use (Clark et al., 2011; Fthenakis and Kim, 2010), or on a selected life cycle stage (Sullivan et al., 2010). In several cases, sufficient background information and complete inventory data is lacking or not easily accessible (Gerber and Maréchal, 2012; Hondo, 2005; James et al., 2012; Santoyo-Castelazo et al., 2011). In some cases, standard LCI databases such as the GEMIS database (Öko Institut, 2007) provide generic or site-specific life cycle information for direct emissions. The most comprehensive and complete LCA and qualitative studies on the environmental emissions and resource use of geothermal power production are those by Sullivan et al. (2010, 2011), Clark et al. (2011), Frick et al. (2010), Rule et al. (2009) and Karlsdottir et al. (2010) (see Table 8.4).

TABLE 8.4

Overview of LCA studies of geothermal power systems, including technologies, environmental impact categories and specific emissions considered

Source	Technology			Environmental Impact Categories[1]	Specific impacts and emissions
	Dry steam	Flash steam	Binary-cycle		
(Sullivan et al., 2010; Sullivan et al., 2011; Sullivan et al., 2012)		X	X	1 2 3 4	Consumption of aluminium, concrete, cement, bentonite, diesel, iron and steel
(Clark et al., 2011)		X	X	4	Water use
(Fthenakis and Kim, 2010)	X	X	?	4	Water use
(Hondo, 2005)		X		1	
(James et al., 2012)		X		1 2 5 6 7 8 9 10 11	Water use, Pb, Hg, NH_3, CO, NO_x, SO_2, VOC
(Rule et al., 2009)			X	1 2	
(Karlsdottir et al., 2010)		X		1 2 3 4 5 6 7 8 9 10 11	
(Pehnt, 2006)	X			1 2 3 5 8 11	Iron ore, bauxite consumption, CO, NO_x, NMHC, HC_l, NH_3, benzene and benzopyrene emissions
(Frick et al., 2010)			X (EGS)	1 2 3 5 11	Water use

[1] Impact category representations: 1: CO_2 emissions, 2: Global warming, climate change, 3: Cumulative energy demand, total energy demand, primary energy, energy payback time/ratio, 4: Abiotic depletion, resource requirements, material intensity, non-renewable resource depletion, 5: Acidification, 6: Ozone depletion, 7: Human toxicity, 8: Particulate matter formation, particles/dust, 9: Ecotoxicity, freshwater and marine aquatic ecotoxicity, sediment ecotoxicity, terrestrial ecotoxicity, 10: Photochemical ozone creation, photochemical oxidation, photochemical ozone formation, photochemical oxidant formation, smog. 11: Eutrophication, nutrient enrichment, 12: Solid waste generation, 13: Land use, land occupation, land transformation

Pehnt (2006), Frick et al. (2010) and Gerber and Maréchal (2012) examined EGS systems and thus their findings are not representative for standard geothermal electricity production. Gerber and Maréchal (2012), for example, study a hypothetical geothermal cogeneration system in Switzerland. German binary EGS plants are in the focus of Frick et al. (2010), who provide a comprehensive prospective analysis of hypothetical installations under different geological conditions, taking 1 kWh net energy at the plant as functional unit. The results are highly dependent on geological conditions such as temperature of geothermal fluid, reservoir depth and technical lifetime. Their system boundaries include drilling/construction, operation and decommissioning, and they provide a detailed insight into the inventory data and sources, for example the ecoinvent database (ecoinvent, 2012). Heat cogeneration is accounted for in different scenarios. As impact categories, demand of finite energy resources (cumulative energy demand), global warming (GWP 100), acidification, and eutrophication potential are chosen.

Hondo (2005) compares the carbon footprint of geothermal electricity production with other power generation technologies in Japan. A 55 MW, double flash steam plant at 60 per cent capacity is selected, and five exploration wells (1,500 m), 14 production and 7 re-injection wells (1,000 m), as well as additional wells each year over a total lifetime of 30 years are assumed. For the functional unit of 1 kWh electricity production, CO_2 emissions for the full life cycle of 15 g/kWh are quantified. Fugitive emissions, however, are not accounted for in this value.

The recent work by Sullivan et al. (2010) from the United States Argonne National Laboratory represents a comprehensive comparison of geothermal to other power generation alternatives in the United States. Hypothetical EGS, binary and flash-steam plants are investigated in detail, and the "plant cycle", i.e., indirect burdens from construction, material and energy provision used drilling, stimulation, construction and operation, is covered for a lifetime of 30 years with a 95 per cent capacity factor. As references, however, real cases were selected and average numbers representative for the United States were collected. The study focuses on CO_2 emissions per capacity (MW) from drilling, plant and surface construction, as well as operation. The objective is to arrive at a generally valid comparison to other power generation technologies. This study provides detailed information on materials consumed for drilling and construction. This includes aluminium, concrete, cement, bentonite, diesel, iron and steel. Sullivan et al. (2011) include, as an extension to their 2010 work, among others, hybrid geo-pressured gas and electric wells. Further, direct burdens and emissions from well-field exploration, well material and fuel requirements, on-site plant construction activities, as well as GHG emissions, are accounted. The unit provision of energy serves as the functional unit in the study, while a special focus is placed on GHG emissions and embodied energy. Embodied energy is the energy required to construct the power plant and associated infrastructure (cradle-to-gate). In Clark et al. (2011), the same authors calculate ranges of water use for the system boundaries as defined in Sullivan et al. (2010). Similarly, in another study, Fthenakis and Kim (2010) compare water use rates in United States electricity generation.

The work by Sullivan et al. (2011) is the first life cycle based study that combines the burdens from the plant cycle with direct GHG emissions, which stem from fugitive emissions of methane in the flash-steam plant. However, they also emphasize the problem in defining generally valid ranges, mainly due to the diversity and often anonymous sources of reported values.

In their comparative LCA of GHG emissions and embodied energy of renewable energy generation technologies in New Zealand, Rule et al. (2009) calculate based on the conditions at Wairakei power station. Including drilling, plant construction and 100 years of plant operation, they estimate a low value of 5.6 CO_2 g/kWh, but discuss fugitive emissions from geothermal steam release. As these operational emissions would occur naturally in the absence of the geothermal plant, they suggest excluding them in the carbon footprint calculation of their study. The presentation by James et al. (2012) compares cost and environmental impacts of different power generation technologies in the United States, including a 50 MW flash-steam geothermal plant. GHG emissions, water consumption and various air emissions are reported for construction and operation of a facility and are summarized over the entire lifetime with respect to MWh produced. CO_2 emissions (214 g/kWh) and water use (38 m^3/MWh) are dominated by fugitive gas and vapour loss from the flashing of the geofluid. Hunt (2001) mentions that IAEA reviewed three studies from 1989 and 1992 on full-energy emissions from geothermal electricity provision. He refers to a range of 20-57 g/kWh, but no further details are given.

Karlsdottir et al. (2010) give insight into their LCA study on the geothermal combined heat and power production of the double-flash Hellisheidi plant in Iceland. Focus is set on the global warming potential (CO_2, methane) and cumulative energy demand that are associated with construction and operation, ignoring energy and material flows due to construction and equipment maintenance. Electricity production is estimated to account for 35-45 g CO_2-eq/kWh in total. For the fraction of primary energy demand, a bulk value range of 0.1-0.2 is given. Further preliminary results are reported for a number of further standard life cycle impact assessment categories.

8.4 METHOD AND DATA FOR LIFE CYCLE INVENTORY COMPILATION

8.4.1 DESCRIPTION OF THE CHOSEN TECHNOLOGIES

We define two reference cases, which are oriented at the hydrothermal geothermal power plant variants distinguished by Sullivan et al. (2010, 2011) and Clark et al. (2011) in the reports by the United States Argonne National Laboratory (Table 8.5):

- Binary: A binary plant with 10 MW net power output
- Flash: A flash steam plant with 50 MW net power output

TABLE 8.5

Specification of two geothermal power plants following the settings provided by Sullivan et al. (2010, 2011) and Clark et al. (2011)

Name	Binary	Flash
Number of turbines	single	multiple
Generator type	binary	flash
Cooling	Air	evaporative
Net power output, MW	10	50
Plant lifetime, years	30	
Producer-to-injector ratio	3:1 and 2:1	
Temperature, °C	150–185	175–300
Thermal drawdown, % per year	0.4–0.5	
Number of production wells (average)	3	14.6
Number of injection wells (average)	1.2	6
Well replacement	1	1
Exploration wells	1	1
Well depth, km	<2	1.5 < 3
Flow rate per well, kg/s	60–120	40–100
Pumps for production	lineshaft or submersible	none
Distance between wells, m	800–1,600	
Location of plant in relation to wells	central	

Both cases are located under southwestern United States conditions, and operate for a lifetime of 30 years. Underlying data is collected from experts in industry and United States national laboratories. We complement these two base cases with data ranges from other studies, and try to determine roughly averaged emissions and resource use for a third "universal" case. All is expressed in the standard functional unit, which is provision of a unit (here, 1 kWh) electrical energy from a high enthalpy geothermal resource. System boundaries include exploration, drilling, well installation, surface plant construction with all buildings, operation for 30 years, and plant decommissioning and recycling. For most of these stages, Sullivan et al. (2010) offer modelled life cycle metrics, which cover energy, GHG emissions, and selected "materials used in significant quantities", such as steel, aluminium and concrete. While in the Argonne reports, the plant fence line (including background processes for material and energy provision) literally represents the system boundary, we try to discuss further implications from geothermal plant within the land use category, and water and air emissions. Pipelines that connect plants with consumers, energy demand for plant construction on site, and transport of materials are not considered; these data are highly site-specific and thus no representative data could be found.

We include cooling facilities, but exclude cogeneration of heat, which may play an important role at some sites, and will improve the overall environmental performance of a plant. EGS or hybrid technologies are not considered, since these do not represent current standard technologies.

8.4.2 MATERIAL, WATER AND ENERGY REQUIREMENTS

The materials consumed for exploration, drilling and construction are adopted from Sullivan et al (2010), Clark et al. (2011) and Rule et al. (2009). Rule et al. (2009) consider the Wairakei field in New Zealand, with different assumptions for system lifetime, well number, and well depth. They define a lifetime of the power plant of 100 years, and well lifetime of 17 years, which means 4-5 new sets of wells of average depth of 660 m drilled within the total operation time. Sullivan et al. (2010) compare their results, expressed as material mass per power output, i.e., per MW, with those by Rule et al. (2009) and Frick et al (2010). Substantial differences with the latter in the material and water consumption are attributed to the different plant types, because Frick et al. (2010) inspect the life cycle of EGS.

In summary, the values in Table 8.6 display no significant differences between the studies and technologies. Most striking is the high steel consumption for the binary plant. This stems from the large air-cooling structure that is needed for this variant. In contrast, producing high temperature steam is expected to be more demanding for the deeper wells. Further details on specific trends can be found in Sullivan et al. (2010) and are not discussed here in more detail. We obtain roughly averaged values for our universal case, which may differ significantly for specific locations. We expect that representative average values may be slightly higher. Main reasons are the defined system boundaries and excluded assets such as facilities for wastewater treatment, on-site energy and material consumption, potential failure in drilling operations, as well as rudimentary reflection of the exploration phase. Further, proper definition of capacity factors and system lifetime is crucial, yet their values can hardly be generalized. Hondo (2005), for example, assumes a capacity factor of only 60 per cent, in comparison to 93 per cent in the work by Rule et al. (2009) and 95 per cent in Sullivan et al. (2010, 2011). (ridleifsson et al. (2008) estimate an average global capacity factor of 75 per cent, and it is anticipated that this will increase to 90 per cent in the near future (Bertani, 2012). Finally, transportation is not included, which will increase the diesel consumption per kWh. Contrarily, including recycling credits for steel, for example, by including dismantling of the power plant, would reduce the calculated net resource consumption.

Sullivan et al. (2011) point out that life cycle construction information for power plants is scarce. This includes activities such as earth moving, operation of cranes, etc. Direct energy burdens that are associated with plant construction activities (on-site) and transport of material to the site are estimated to be roughly 6.8 per cent of the embodied energy, or plant cycle energy. While this quantifies the energy consumption for on-site activities and materials transportation, transport of workers is neglected. We also follow their suggestion, and assume a direct energy share of 6.8 per cent, which consists of 25 per cent electricity and 75 per cent diesel.

Material, energy and water use during operation are very site-specific, mostly incompletely reported and must often be guessed or predicted. Still, operational material consumption, with the exception of re-drilling of wells, is expected to be relatively small. Such operations include for example maintaining plant pipes, dealing with corrosion and scaling, maintenance of wastewater treatment or abatement systems (e.g., for removal of hydrogen sulfide from exhaust steam), heat carrier fluid in binary cycle, etc. Therefore, these are neglected in the overall analysis (see Table 8.7). Assets of higher priority are expected to be:

- Auxiliary energy for operation of pumps (extraction, circulation, re-injection)
- Water use for makeup, cooling, and treatment.

Operational energy consumption is not covered by the Argonne reports. For the flash plant, only injection devices are assumed, whereas for the binary system, extraction as well as injection pumping has to be considered. In Sullivan et al. (2010), the equipment is roughly included based on steel mass in the well implementation expenditures as aggregated in Table 8.6. However, operational energy consumption is ignored. Frick et al. (2010) describe in detail the relevance of feed and down-hole pump operation for binary EGS systems that circulate fluids at much greater depth than standard geothermal plants (3.8-5 km). For example, a 10 per cent capacity fraction of energy that is consumed by feed pumps (0.18 MW) and a full auxiliary energy demand of 0.5-3 kWh/(m^3/h) for running the geothermal fluid cycle are assumed. Applying the lower range to the binary plant as configured in Table 8.5, an auxiliary power need of about 1.5 per cent of the capacity is obtained (Table 8.7). This is also an average value (for range 1-2 per cent) as provided by Karlsdottir et al. (2010).

For quantifying water use by geothermal plants, some fundamental assumptions are necessary. First, here, water use only refers to the consumption of freshwater (Pfister et al., 2011). Geofluids from geothermal reservoirs or geothermal brines are not usually freshwater. Thus, for example, vapour discharge from flash steaming is not accounted for. During the life cycle, water is used directly for well and plant construction, as makeup water to balance geofluid deficits and to increase reinjection volume, and partial cooling. Indirect effects are all those associated with change of the hydro(geo)logical flow regime, and down-flow water loss from upper freshwater aquifers. The latter cannot be generalized, are rudimentarily covered in the LCA literature and are accordingly neglected within the scope of this study.

In general, the plant technology and especially the cooling type will determine total water consumption. Even air-cooling and steam condensate-based cooling will require a dominantly fresh water supply, and thus the water consumed for cooling or to balance evaporative water and geofluid loss will dominate the overall consumption. For low-water flash-steam cooling, Clark et al. (2011) shows that drilling fluid requirements are only responsible for 14 per cent of the necessary life cycle water volume. The Argonne report by Clark et al. (2011) is dedicated to the water use of United States geothermal power plants. We adopt their values for the full life cycle, which are 1 l/kWh for the air-cooled 10 MW binary plant with water-cooling in hot summers, and 0.04 l/kWh for the 50 MW flash plant using wet cooling through the steam condensate. The latter is at the lower limit in comparison to other reported values, and taking the (USEPA, 2008) range of 0.75-1.15 l/kWh as another reliable reference, 1 l/kWh appears to be a reasonable average estimate (Table 8.7). However, this value is somewhat low when compared to the values given by Fthenakis and Kim (2010), which is a minimum of 2.3 l/kWh for binary plants. In contrast, air-cooled modern facilities in a relatively cold climate may need lower volumes.

TABLE 8.6

Materials and resource consumption for exploration, drilling and plant construction for binary and flash plant scenarios (Table 8.5), Wairakei case study by Rule et al. (2009), and an approximate average (universal)

Material		Binary				Flash				Wairakei	Range (rounded)	Universal
		Plant	Well	Well-to-plant	Total	Plant	Well	Well-to-plant	Total	Total		Total
Aluminium	g/kWh	0.18			0.18					0.02	0.02-0.18	0.1
Concrete	g/kWh	1.75			1.75	0.61			0.61	2.15	0.6-1.8	1
Cement	g/kWh		0.27	0.06	0.33		0.83	0.06	0.88	0.89	0.3-0.9	0.6
Bentonite	g/kWh		0.13		0.13		0.29		0.29		0.1-0.3	0.2
Diesel	l/kWh		1.5E-04	3.9E-05	1.9E-04		1.9E-04	3.8E-05	2.2E-04	2.4E-04	1.9-2.4E-4	2.2E-04
Iron	g/kWh	0.016			0.02	0.01			0.01		0.01-0.02	0.015
Steel	g/kWh	0.88	0.41	0.06	1.35	0.10	0.97	0.05	1.12	0.65	0.6-1.4	1
Water	l/kWh				4.0E-03				4.0E-03			4.0E-03

TABLE 8.7

Water and energy use of binary and flash scenarios (Table 8.5), ranges and approximate average (universal) values during geothermal power plant operation

		Binary	Flash	Range	Universal
Transport (construction)	Fraction of total embodied energy			5-10 %	6.2 %
Water (full life cycle)	l/kWh	1	0.04	0-17	1
Auxiliary energy (operation)	Fraction of generated energy			1-2 %	1.5 %
Abatement technology					case-specific

TABLE 8.8

Atmospheric emissions from geothermal plants. Main pollutants with ranges from literature and estimated universal reference values

Substance		Range	Universal
Hydrogen sulfide (H_2S)	g/kWh	0.085-7.0	0.1
Sulfur dioxide (SO_2)	g/kWh	0.0001-0.16	0.001
Carbon dioxide (CO_2)	g/kWh	4-740	122
Methane (CH_4)	g/kWh	0.75-0.85	0.8
Ammonia (NH_3)	g/kWh	-	0.06
Nitrogen oxides (NO_x)	-	-	—
Particulate matter	-	-	—
Boron, mercury, etc.	-	-	—

8.4.3 LAND USE

According to BLM (2008), land use ranges from 4,200 to 29,300 m^2/MW capacity (see Table 8.1). Lower values denote conditions where only the power plant facilities are considered, while wells and conduits, land use from exploration activities and road construction are not included. For the boundary conditions of the scenarios given here, i.e., 30 years of operation and a capacity factor of 95 per cent, land use is 17-120 mm^2/kWh produced. An arithmetic mean of 70 mm^2/kWh is obtained (Table 8.1).

The main issues surrounding land use and geothermal power are when, if, through which effort and to what extent restoration will be possible after use. Even if "only" 30 years of operation are considered, as in the given scenario, long term consequences remain. Most precious geothermal features such as geysers and fumaroles can be permanently lost, land and aquifers are potentially contaminated, and roads remain. Land use per kWh may be less for long-term operation over, for example, 90 years as indicated by the Wairakei case study by Rule et al. (2009). However, land use may be more intense particularly when far reaching effects, such as from subsidence, extinction of geysers, etc., are included.

8.4.4 EMISSIONS

Most relevant direct emissions during plant operation originate from geothermal steam release. While such fugitive emissions may be close to zero for closed cycles such as binary plants, NCG release from vapour-based plants is common. Based on the literature as described in detail in previous chapters, ranges and "averaged" values for hydrogen sulfide, sulfur dioxide, CO_2, methane, and ammonia emission rates are listed in Table 8.9. An underlying assumption is that even if geothermal plants "accelerate" release of GHGs through geofluid and steam extraction, short term natural release rates are not significantly altered. This is in line with observations from recent measurement campaigns. Long term decreases in natural GHG emissions through the surface are not accounted for. This is equivalent to the boundary conditions applied for carbon sequestration technologies. Methane is a frequently ignored compound that is often released at low rates. However, it increases the GWP by more than 15 per cent in comparison to considering CO_2 only. In comparison, the GEMIS database (Öko Institut, 2007) estimates full life cycle direct emissions including well drilling as high as 122.4 g/kWh CO_2 and 2.7 g/kWh for sulfur dioxide. This is based on a flash steam geothermal plant in the Philippines with a reservoir at 1,000 m depth. The CO_2 emissions value corresponds well to our universal value and apparently reflects the weighted mean from the survey by Bertani and Thain (2002). SO_2 emissions, however, are much higher than estimated in Table 8.9, and this may reflect that the given value accumulates over the full life cycle, and that secondary oxidation of hydrogen sulfide is quantified in SO_2 equivalents.

Abatement technologies may generate secondary emissions such as nitrogen oxides and additional sulfur oxide. However, even though steam clean-up is considered standard practice for geothermal plants, the general impacts associated with this operation have not yet been studied. Therefore, we consider the atmospheric emissions listed in Table 8.9 to be optimistic values. In principle, critical substances such as mercury or boron, as well as high hydrogen sulfide emissions in the absence of treatment operations, and secondary release of nitrogen oxide, sulfur dioxide, etc. (treatment) should be accounted for. For mercury, which is a very toxic compound, the value of 0.5 ppm as assumed by Arnórsson (2005) could be added to Table 8.9.

Due to the scarcity of reported data, and the very case-specific variability, further emissions are not included in this study. This is especially conservative for aquatic emissions, which have led to contaminated land, surface- and groundwater bodies at some locations such as Kizildere, Wairakei, Cerro Prieto and Yangbajing. Discharge of separated brines is considered a problem of the past (Goldstein et al., 2011), but the early years of geothermal plants are also part of the life cycle we consider from today's perspective. Of particular interest are the toxic substances such as mercury, arsenic, boron, etc., which can linger as permanent contamination in natural water bodies and soil. Since current regulations aim at avoiding such emissions, however, these contaminants are expected to be commonly well controlled.

8.4.5 IMPACT ON BIODIVERSITY

A short verbal assessment of potential biodiversity impacts is favoured, because generally applicable indicators cannot be quantified with the scarce information available on this topic. Geothermal environments, their native species and populations of plants, animals and micro-organisms and the dependent ecosystems are often unique, and considered to be fragile and sensitive (Fukutina, 2012). In several cases, endemic organisms are reported. Therefore, even if geothermal power generation plays only a minor role for electricity generation worldwide, local effects on biodiversity may be substantial. Another point is the typically partial land use over great areas. Fragmentation of land and forest lowers species number and changes community composition (Barrantes Víquez, 2006). Furthermore, toxic air and aquatic emissions can pose a threat to adjacent habitats. For example, small to moderate biodiversity impacts caused by mercury and hydrogen sulfide releases have been reported by in the geothermal fields of Bagnore and Piancastagnaio in the area of Mt. Amiata. At several geothermal fields, the role of biodiversity and the need for control and compensation measures have been recognized, for example, in Berlin and El Salvador. Specific designs have been adopted to accommodate wildlife and their migration patterns as in Olkaria, Kenya, or to establich animal rescue facilities to care for endangered species.

8.4.6 SOCIAL IMPACT

Geothermal energy production is often concentrated in regions with extraordinary landscapes that are touristic attractions with mud pools, geysers, fumaroles and steaming ground, and are often remote and pristine (De Jesus, 1995). By extinction of geothermal surface features, and inducing industrial development in such regions, there is a high risk that land of high social value will be lost. This includes the prominent role of such landscapes and geothermal features for indigenous people, ethnic, religious and social groups that have traditional ties to the land, such as in New Zealand (Fukutina, 2012), San Pedro de Atacama, Chile (Correa and Vergara, 2012) or the Maasai community in Kenya (Mariita, 2002). In Bali, Indonesia, geothermal development is limited due to severe religious, cultural, as well as environmental concerns by the public. Currently, for example, the Balinese community, religious leaders and the local government do not accept a planned 165 MW plant at Bedugul in Bali (Richter, 2012). Some followers of traditional Hawaiian religious practices are convinced that geothermal power is harmful (Callis, 2012).

Whereas many natural attractions in nature parks and other protected areas can be excluded from geothermal development, geothermal fluid production wells often influence thermal springs, mainly by affecting their flow-rate. Frequently, the effects become evident only after a certain time (Rybach, 2005). Large scale hydrogeological effects from geothermal power generation may mitigate the productivity of hot springs, and thus competes with the tourism sector. However, in some cases, such as at the Blue Lagoon in Iceland, new tourist attractions are fortuitously created by geothermal development. Iceland exemplifies the presence of critical social activity groups such as SavingIceland, which criticize geothermal power production for its direct environmental and health impacts, and social consequences. The main source of resistance against geothermal power in Iceland, however, is the resulting attraction of energy-intense industries such as aluminium smelting, which build or plan their facilities close to or in combination with geothermal power plants.

Noise emissions are most critical during exploration and well drilling. Since wells may be continuously installed over the lifetime of the plant in order to increase or maintain production level, for injection, etc., sporadic noise problems are potentially present during the entire life cycle of a plant and can be mitigated by sound barriers such as adequate vegetation, or by modifying ground characteristics (López, 2001). Aside from this, the increased risk of seismic events, land subsidence or lifting, may be seen as a local social threat.

The provision of energy to remote areas, and creation of job opportunities are the positive effects of geothermal development. Local communities, however, typically have only a marginal direct employment benefit, since mostly specialized personnel are needed for exploration, drilling and plant operation (Mariita, 2002). Rather, retail trade, health care and social assistance, accommodation and food services sectors providing support for the influx of new workers often provide potential new sources of jobs for local communities (BLM, 2008).

Political and public acceptance are difficult to achieve with geothermal power projects in contrast to other energy technologies, such as photovoltaic (PV) energy, which is visually aesthetic. Geothermal facilities require dirty and noisy equipment, test drilling and steam or hot water at the surface, all of which gather apprehension rather than acceptance (Popovski, 2003).

Cataldi (1999) says, "the three main goals for social acceptance are minimisation of environmental impact, prevention of adverse effects on people's health, and creation of tangible benefits for the local populations". A Geoelec report (Reith et al., 2013) considering three case studies indicates that public acceptance of geothermal power technology can vary depending on the location and is related to the level of knowledge and the evolution of the produced disturbances. The report also indicates the importance of early-phase public involvement efforts. (Popovski, 2003) also includes many geothermal public acceptance cases. (Polyzou and Stamataki, 2010) present a very detailed social study of two cities in Greece, which reveals very interesting conclusions on geothermal energy projects, specifically how local political problems may result from the lack of information made available to the local society. According to a research document released by Geothermal Communities, a European Commission Seventh Framework project (FP7), generally the public acceptance for geothermal energy is slightly lower than for the other renewable sources of energy.

An example of a very detailed discussion on such socio-economic issues, consequences of geothermal development in less developed regions and successful mitigation of potential problems can be found in a report of the development of the Las Pailas geothermal field in Costa Rica by Barrantes Víquez (2005). Focus is set on public acceptance of the project, and social integration of new workers in the existing indigenous community. This also shows the specific nature of potential social impacts.

8.5 LCA MODELLING

8.5.1 INVENTORY

The results presented hereafter describe the environmental profile of a hypothetical geothermal plant, adapted from the LCI of the Wairakei geothermal plant (Rule et al., 2009). Direct emissions were added from several other sources, as shown in Table 8.9. Further detail on how and where these emissions occur can be found in Chapter 8.4.4.

TABLE 8.9

Assumptions for the direct emissions of the modelled plant

Emissions	Reference
Carbon dioxide (CO_2)	83 g CO_2/ kWh (Bayer et al., 2013a)
Sulfur dioxide (SO_2)	0.1587 g SO_2/kWh (Kagel et al., 2007)
Methane (CH_4)	0.75 g CH_4/kWh (Bloomfield et al., 2003)
Ammonia (NH_3)	0.06 g NH_3/kWh (Bloomfield et al., 2003)
Mercury (Hg)	4 g Hg/kWh (Bacci et al., 2000)
Heat, waste	9 MW Waste heat / MW electric capacity (Bayer et al., 2013a)

8.5.2 MODELLING RESULTS

Figure 8.8 illustrates how these direct impacts dominate the environmental profile of the plant. The assumed long lifetime of 100 years is one of the reasons for this imbalance between on-site, direct emissions and emissions embodied in infrastructure. As with hydropower projects, geothermal plants, wells and piping, are unique and their respective impacts are site-specific. Direct emissions may therefore vary a lot.

FIGURE 8.8

Process contribution to a set of environmental impacts of a 177 MW geothermal plant

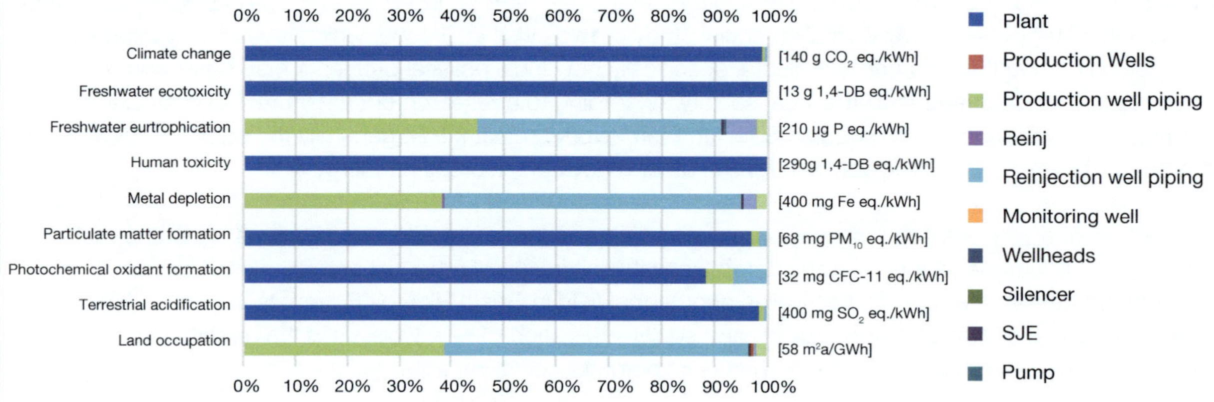

Adapted from an inventory for Wairakei, regional background OECD Pacific, 2010.

FIGURE 8.9

Environmental impacts for a 177 MW geothermal plant (Wairakei) relative to the OECD Pacific electricity mix of 2010

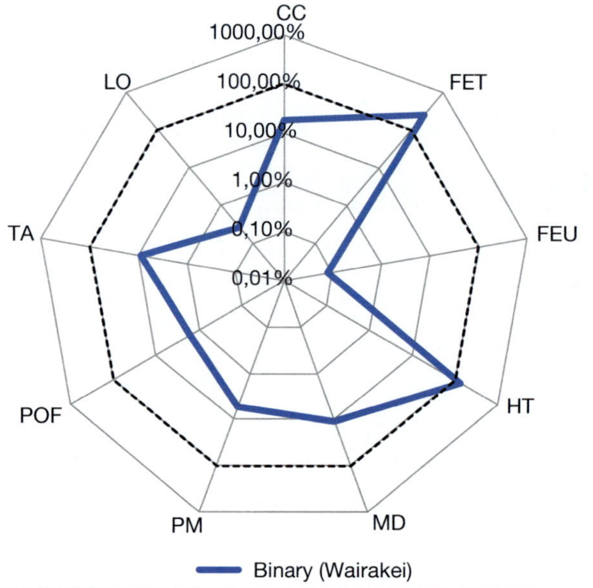

Abbreviations for the impact indicators are:
CC-climate change;
FET-freshwater ecotoxicity;
FEU-freshwater eutrophication;
HT-human toxicity;
MD-metal depletion;
PM-particulate matter formation;
POF-photochemical oxidant formation;
TA-terrestrial acidification;
LO-land occupation.

Compared with the background electricity mix of the OECD Pacific region, the background region for the assessment, 1 kWh from this geothermal plant scores lower on all indicators but two, as seen on Figure 8.9. The two toxicity indicators, freshwater ecotoxicity and human toxicity, have high values because of the direct mercury emissions of the plant. The rest of the figure shows a relatively low environmental profile, with a potential impact ranging from 0.08 per cent (freshwater eutrophication) to 19 per cent (climate change) of the impact of the average background electricity production.

8.6 OVERALL ASSESSMENT AND TECHNOLOGY-SPECIFIC ISSUES

Power generation by using near-surface, high-temperature reservoirs is an appealing, economically attractive, technologically established, but geographically limited renewable energy variant. A general overall assessment is ideally based on a representative universal case that averages the different geological, thermal, technological and local environmental conditions. Even if different technologies such as binary and flash steam plant types are distinguished, however, site-specific factors will govern their ultimate environmental performance.

There are two extremes. One is a modern binary plant, with high capacity of over 10 MW and a capacity factor of more than 90 per cent. Few wells, e.g., three wells for production and two for injection, are drilled, with a small number of new installations during the time of operation. The circulated geothermal fluid shows little temperature decline, and make-up water is grey water that does not stress local water bodies. A modern air cooling facility is applied, and due to the moderate climate, hybrid use of water cooling is not necessary. The cogenerated heat may supply a local community, fish farm or desalination facility. The plant is located at a well-developed site or on land of environmentally low value, and can be sustainably operated for far more than 30 years. This one ideal case is the environmentally benign variant, where direct emissions to the atmosphere are very small, closed system operation is well controlled and efficient, and water as well as precious land use is minimized. There are actually many real examples of this “ideal” case. Nearly all of the binary plants operating in California and Nevada in the United States meet nearly every one of these stipulations, except, of course, for a negligible geofluid temperature decline. The life cycle impacts will be dominated by the resource and energy use as well as interference with the regional hydrological regime from drilling and operating wells. Material and energy consumed for plant construction, as well as for auxiliary devices such as circulation pumps will play the most dominant role. Life cycle GHG emissions will be approximately 120 grams of CO_2 equivalents per kWh. We estimated a standard value of 1-2 per cent embodied energy, and impacts in other categories will be insignificant.

The other extreme is a flash steam plant with relatively low capacity and capacity factor, for example 20 MW capacity and factor of 60 per cent. The geothermal reservoir is deep (2000 m), and performance of the plant continuously declines. To overcome this, new wells are regularly drilled and reinjection is partially practiced. Reinjection, however, creates productivity problems such as temperature decline and thermal breakthrough. The steam released is rich in CO_2. Abatement systems to reduce hydrogen sulfide as well as toxic metal and boron concentrations are not very efficient. Sulfur dioxide and nitrogen oxides could potentially be released as secondary products from steam oxidation. Even if the discharge of used fluids and brines to surface and groundwater bodies is minimized or buffered by ponds, uncontrolled and accidental aquatic emissions have contaminated the adjacent freshwater bodies. In the hot climate, air-cooling, at least temporarily, has to be switched to water-cooling, and including the consumed make-up water, life cycle water use reaches nearly several litres per kWh. Finally, the region in which the plant operates hosts many unique geothermal surface features, is a habitat of special endemic organisms and contains land of high environmental and cultural value. Certainly, this worst case extreme is very rare. Here, environmental effects will probably be dominated by climate change impacts due to the high CO_2 content and volume of the released steam. Possible values of more than 500 g CO_2/KWh are reported. Thus, this variant has an environmental performance similar to modern fossil fuel based plants. Furthermore, contamination and destruction of precious natural freshwater bodies and pristine land are often mentioned as a consequence of geothermal energy production. In the extreme case, substantial local problems are generated, especially with respect to human and ecotoxicity, land and water use.

Our "universal" case is somewhere between these two extremes, and by comparison, a conservative variant that benefits from the lessons learned through the experience from the past. For example, it presumes that water cooling is mitigated, abatement technologies are standard and partial re-injection is applicable. Finally, the main determinant for the climate change impacts will be the direct steam release, in particular, its composition and the volumes. If this factor is not relevant, then the environmental performance of geothermal plants with respect to this category is excellent. Relevant contributions to other impact categories, such as ecotoxic effects from release of metals and boron, and land use, will be site-specific. Water use can hardly be ignored for such applications since make-up water is typically necessary; water use may be significant if reinjection is required. Even if water cooling is no longer a standard practice in the future, it has often previously been applied in today's plants, and hybrid cooling applications are often selected instead of exclusive air/steam condensate-based techniques. A critical assumption that is necessary for life cycle-based quantification of environmental effects is the operational lifetime of the facility. Ranges from a few to nearly hundred years are reported or anticipated, and of course, this parameter will determine the relative contribution of operational effects versus those from power plant implementation and dismantling. Despite this, it is assumed that wells are not only installed during the initial construction phase, but new wells are regularly drilled. Fugitive emissions per kWh do not depend on power plant life time given an averaged performance and capacity factor. As a consequence, major environmental burdens, as associated with the universal case, will not fundamentally change with assumed lifetime of the plant.

8.7 CONCLUSIONS

This overview of potential life cycle environmental effects from geothermal power plants tries to bring together scattered available information of reports and related studies. In fact, life cycle assessment (LCA) studies on current geothermal electricity production are very rare, and mainly through the recent work by the United States Argonne National Laboratory, specific data that could also be utilized in our work have been compiled. A crucial point is that general assessments are hard to make. Many technological, economic and environmental aspects of geothermal plants are fundamentally controlled by geological factors. The latter are always unique, site-specific, they can span a broad range, and often even their effects are uncertain. For example, the trend of long-term productivity of geothermal reservoirs is often unknown, concentrations of compounds in geosteam or fluid are highly variable and they may change over time. Even without the uncertainty in geo-technological performance, the case-specific variability of fugitive emissions, water and land use effects, the different plant types in use, as well as the improvements made during the last decade make it virtually impossible to define an average plant case with a corresponding representative LCI. Finally, many environmental effects cover sensitive information that is difficult to obtain or anonymously reported. This is especially problematic for atmospheric and aquatic emissions, which are sometimes discussed, but cannot be generalized per kWh based on local information only. Previous work on water consumption often does not distinguish between consumption and water throughput, and it is hard to quantify the ratio of fresh water in comparison to low-quality grey or brackish water that may be used as well.

In view of these prerequisites, data presented here is highly dependent on the reliability of a few previous studies, in which the United States perspective dominates. This also means that conditions in other countries that play a prominent role in geothermal power generation, such as the Philippines or Indonesia, are poorly reflected. It is questionable whether environmental regulations in these countries and their interpretation in practice are similar. Despite this, we tried to define a streamlined universal case, with ranges and estimated average inventory data.

A main conclusion is that, in general, environmental effects are associated mostly with emissions at the site, rather than, for instance, indirect emissions from the manufacturing process of plant components or governed by drilling activities. Atmospheric emissions, and particularly fugitive GHGs via steam release are very critical, and apparently often underestimated. For example, CO_2 emissions from geothermal plants were not regulated in the United States at the time this report was developed. This should be kept in mind when Tester et al. (2006) concluded in their prominent MIT report, "...it is highly unlikely that any geothermal power plant will be a threat to the environment anywhere in the United States, given the comprehensive spectrum of regulations that must be satisfied." Other aquatic and atmospheric emissions may carry toxic concentration levels of other critical compounds such as mercury, boron and arsenic. Despite the often only local effect in the vicinity of the plant, accidental or permanent release can represent a significant threat to the environment. More research and surveys, however, are necessary to obtain a more general picture on worldwide average emission rates and consequences.

We excluded release of these potentially critical toxic substances due to missing general data. Furthermore, abatement systems are often operated to minimize emission of polluted steam or water. They add another unknown component that is ignored here. In contrast, cogeneration of heat or minerals is neglected and this may play an interesting role at some sites. In summary, even if major environmental determinants are identified and their effects are quantified, the provided data are far from being precise or complete. We suspect that the full environmental burden of worldwide geothermal power generation is somewhat higher than we estimate here. The relevance of neglected threats, the greater availability of positive-spin literature, and the life cycle burden of environmental damages from the past are the main arguments for this conjecture. Still, this critical perspective and the claim for a more comprehensive view on life cycle burdens is not intended as an argument against this technology. It is meant as invitation for more transparent reporting and assessment of local environmental consequences, in order to demonstrate the environmental benefits of this renewable energy resource. At the few exploited geothermal fields worldwide, this method of electricity generation is technologically, economically and, in principle, environmentally, a favourable option.

8.8 REFERENCES

Adee, S. and S. K. Moore. 2010. In the American Southwest, the Energy Problem Is Water. *IEEE Spectrum*, 28 May 2010.

Aksoy, N., C. Şimşek, and O. Gunduz. 2009. Groundwater contamination mechanism in a geothermal field: A case study of Balcova, Turkey. *Journal of Contaminant Hydrology* 103(1-2): 13-28.

Allis, R. G. 1990. Subsidence at Wairakei field, New Zealand. *Geothermal Resources Council Transactions* 14, Part II: 1081-1087.

Ármannsson, H. and H. Kristmannsdóttir. 1992. Geothermal environmental impact. *Geothermics* 21(5–6): 869-880.

Ármannsson, H., T. Fridriksson, and B. R. Kristjánsson. 2005. CO_2 emissions from geothermal power plants and natural geothermal activity in Iceland. *Geothermics* 34(3): 286-296.

Armstead, H. C. H. 1978. *Geothermal Energy: Its past, present, and future contributions to the energy needs of man*: Wiley.

Arnórsson, S. 2004. Environmental impact of geothermal energy utilization. *Geological Society, London, Special Publications* 236(1): 297-336.

Baba, A. 2003. *Geothermal Environmental Impact Assessment with Special Reference to the Tuzla, Geothermal Area, Canakkale, Turkey* 5/2003. Reykjavik, Iceland: The United Nations University Geothermal Training Programme.

Bacci, E., C. Gaggi, E. Lanzillotti, S. Ferrozzi, and L. Valli. 2000. Geothermal power plants at Mt. Amiata (Tuscany–Italy): mercury and hydrogen sulfide deposition revealed by vegetation. *Chemosphere* 40(8): 907-911.

Baldacci, A., M. Mannari, and F. Sansone. 2005. Greening of Geothermal Power: An Innovative Technology for Abatement of Hydrogen Sulfide and Mercury Emission. Paper presented at World Geothermal Congress, 24-29 April 2005, Antalya, Turkey.

Baldi, F. 1988. Mercury pollution in the soil and mosses around a geothermal plant. *Water, Air, and Soil Pollution* 38(1-2): 111-119.

Barbier, E. 1997. Nature and technology of geothermal energy: A review. *Renewable and Sustainable Energy Reviews* 1(1–2): 1-69.

Barrantes Víquez, M. 2006. *Geo-Environmental Aspects for the Development of Las Pailas Geothermal Field, Guanacaste, Costa Rica*. 8/2006. Reykjavik, Iceland: The United Nations University Geothermal Training Programme.

Barrick, K. 2007. Geyser Decline and Extinction in New Zealand—Energy Development Impacts and Implications for Environmental Management. *Environmental Management* 39(6): 783-805.

Bayer, P., L. Rybach, P. Blum, and R. Brauchler. 2013. Review on life cycle environmental effects of geothermal power generation. *Renewable and Sustainable Energy Reviews* 26(0): 446-463.

Bertani, R. 2012. Geothermal power generation in the world 2005–2010 update report. *Geothermics* 41(0): 1-29.

Bertani, R. and I. Thain. 2002. Geothermal power generating plant CO_2 emission survey. *IGA News*, section 1-3.

BLM. 2008. Final Programmatic Environmental Impact Statement for Geothermal Leasing in the Western United States. http://www.blm.gov/wo/st/en/prog/energy/geothermal/geothermal_nationwide/Documents/Final_PEIS.html. Accessed.

Bloomfield, K. K., J. N. Moore, and R. M. N. Jr. 2003. *Geothermal Energy Reduces Greenhouse Gases*. Davis, CA, USA: Geothermal Research Council.

Boothroyd, I. K. G. 2009. Ecological characteristics and management of geothermal systems of the Taupo Volcanic Zone, New Zealand. *Geothermics* 38(1): 200-209.

Bourcier, W., M. Lin, and G. Nix. 2003. *Recovery of Minerals and Metals from Geothermal Fluids*. Livermore, CA, USA: Lawrence Livermore National Laboratory.

Brophy, P. 1997. Environmental advantages to the utilization of geothermal energy. *Renewable Energy* 10(2–3): 367-377.

Browne, P. R. L. and J. V. Lawless. 2001. Characteristics of hydrothermal eruptions, with examples from New Zealand and elsewhere. *Earth-Science Reviews* 52(4): 299-331.

Callis, T. 2012. Geothermal safety mulled. http://hawaiitribune-herald.com/sections/news/local-news/geothermal-safety-mulled.html. Accessed April 2012.

Carlsen, H. K., H. Zoëga, U. Valdimarsdóttir, T. Gíslason, and B. Hrafnkelsson. 2012. Hydrogen sulfide and particle matter levels associated with increased dispensing of anti-asthma drugs in Iceland's capital. *Environmental Research* 113(0): 33-39.

Cataldi, R. 1999. *Social acceptance: A sine qua non for geothermal development in the 21st century.* Pisa, Italy: Centre d'Hydrogtologie, Universiti de Neuchdtel.

Clark, C., J. Sullivan, C. Harto, J. Han, and M. Wang. 2012. Life cycle environmental impacts of geothermal systems. In *Stanford Geothermal Workshop*.

Clark, C. E., C. B. Harto, J. L. Sullivan, and M. Q. Wang. 2011. *Water Use in the Development and Operation of Geothermal Power Plants*. Argonne: Argonne National Laboratory.

Correa, L. and V. Vergara. 2012. Environmental and social impact caused by the El Tatio exploration. *Interamerican Journal of Environment and Tourism*, 126-132.

De Jesus, A. C. 1995 Socio-economic impacts of geothermal development. In *World Geothermal Congress 1995: Pre-Congress Course on Environmental Aspects of Geothermal Development*. Pisa, Italy: IGA/CNR-International School of Geothermics.

Dereinda, F. H. and H. Armannsson. 2008. *CO_2 Emissions from the Krafla Geothermal Area, Iceland*. 15/2008. Reykjavik, Iceland: The United Nations University Geothermal Training Programme.

DiPippo, R. 1991. Geothermal energy: Electricity generation and environmental impact. *Energy Policy* 19(8): 798-807.

DiPippo, R. 1999. Small Geothermal Powerplants: Design, Performance and Economics. *Geo-Heat Center* [Oregon, USA], June 1999, section 1-8.

DiPippo, R. 2008. *Geothermal Power Plants, Second Edition: Principles, Applications, Case Studies, and Environmental Impact*. Edited by R. DiPippo. Oxford: Butterworth-Heinemann.

ecoinvent. 2012. ecoinvent life cycle inventory database v3: The Ecoinvent Centre, Swiss Center for Life Cycle Inventories.

Edwards, L. M., G. V. Chilingar, H. H. Rieke III, and W. H. Fertl. 1982. *Handbook of Geothermal Energy*. Houston, London, Paris, Tokyo: Gulf Publishing Company, Book Division.

Ellis, J. A. 1978. *Environmental impact of geothermal development*. United Nations Environment Programme.

EPA, U. S. 2008. *Public Review Draft – National Water Program Strategy: Response to Climate Change*.

Evans, K. F., A. Zappone, T. Kraft, N. Deichmann, and F. Moia. 2012. A survey of the induced seismic responses to fluid injection in geothermal and CO_2 reservoirs in Europe. *Geothermics* 41(0): 30-54.

Franco, A. and M. Villani. 2009. Optimal design of binary cycle power plants for water-dominated, medium-temperature geothermal fields. *Geothermics* 38(4): 379-391.

Frick, S., M. Kaltschmitt, and G. Schröder. 2010. Life cycle assessment of geothermal binary power plants using enhanced low-temperature reservoirs. *Energy* 35(5): 2281-2294.

Fridleifsson, I. B. 2001. Geothermal energy for the benefit of the people. *Renewable and Sustainable Energy Reviews* 5(3): 299-312.

Fridleifsson, I. B., R. Bertani, E. Huenges, J. W. Lund, A. Ragnarsson, and R. L. 2008. *The possible role and contribution of geothermal energy to the mitigation of climate change*. Edited by O. H. a. T. Trittin, *IPCC Scoping Meeting on Renewable Energy Sources, Proceedings*. Luebeck, Germany: International Panel on Climate Change.

Fridriksson, T., B. R. Kristjánsson, H. Ármannsson, E. Margrétardóttir, S. Ólafsdóttir, and G. Chiodini. 2006. CO_2 emissions and heat flow through soil, fumaroles, and steam heated mud pools at the Reykjanes geothermal area, SW Iceland. *Applied Geochemistry* 21(9): 1551-1569.

Fthenakis, V. and H. C. Kim. 2010. Life-cycle uses of water in U.S. electricity generation. *Renewable and Sustainable Energy Reviews* 14(7): 2039-2048.

Fukutina, K. 2012. *The Waikato regional geothermal resource*. 2012/10. Waikato Regional Council.

Gerber, L. and F. Maréchal. 2012. Design of geothermal energy conversion systems with a life cycle assessment perspective. In *37th Stanford Geothermal Workshop*. Stanford, California, USA.

Goldstein, B., G. Hiriart, R. Bertani, C. Bromley, L. Gutiérrez-Negrín, E. Huenges, H. Muraoka, A. Ragnarsson, J. Tester, and V. Zui. 2011. *Geothermal Energy*. Edited by O. Edenhofer, et al., *IPCC Special Report on Renewable Energy Sources and Climate Change Mitigation*. Cambridge, United Kingdom and New York, NY, USA.: Cambridge University Press.

Guo, Q., Y. Wang, and W. Liu. 2008. B, As, and F contamination of river water due to wastewater discharge of the Yangbajing geothermal power plant, Tibet, China. *Environmental Geology* 56(1): 197-205.

Harto, C., J. Schroeder, L. Martino, R. Horner, and C. Clark. 2013. Geothermal Energy: The Energy-Water Nexus. Paper presented at Proceedings, 38th Workshop on Geothermal Reservoir Engineering, 11-13 February, 2013, Stanford University, Stanford, California.

Heath, M. J. 2002. Environmental aspects of geothermal energy resources utilization. In *Geothermal Energy Resources for Developing Countries*, edited by D. Chandrasekharam and J. Bundschuh: Taylor & Francis.

Hondo, H. 2005. Life cycle GHG emission analysis of power generation systems: Japanese case. *Energy* 30(11–12): 2042-2056.

Horne, R. N. 2011. What Does the Future Hold for Geothermal Energy? In *Australian Geothermal Energy Conference 2011* Melbourne, Geoscience Australia.

Hung, E. and P. Bodger. 2009. CO_2 Emission Market Share and Electricity Generation Patterns in New Zealand. In *EEA Conference & Exhibition 2009*. Christchurch, New Zealand: University of Canterbury. Electrical and Computer Engineering.

Hunt, M. T. 2001. *Five Lectures on Environmental Effects of Geothermal Energy Utilization*. 1/2000. Reykjavik, Iceland: The United Nations University Geothermal Training Programme.

James, R., T. Skone, J. Marriott, J. Littlefield, and G. Cooney. 2012. Comparing life cycle costs and environmental performance of alternative energy. In *LCA XII*. Tacoma, Washington.

Kagel, A., D. Bates, and K. Gawell. 2007. *A guide to geothermal energy and the environment*. Washington D.C., USA: Geothermal Energy Association.

Karlsdottir, M. R., O. P. Palsson, and H. Palsson. 2010. Factors for Primary Energy Efficiency and CO_2 Emission of Geothermal Power Production,. Paper presented at World Geothermal Congress 2010, 25-29 April 2010, Bali, Indonesia.

Kaya, E., S. J. Zarrouk, and M. J. O'Sullivan. 2011. Reinjection in geothermal fields: A review of worldwide experience. *Renewable and Sustainable Energy Reviews* 15(1): 47-68.

Koç, C. 2011. Effects of boron pollution in the Lower Buyuk Menderes Basin (Turkey) on agricultural areas and crops. *Environmental Progress & Sustainable Energy* 30(3): 347-357.

Kristmannsdóttir, H. and H. Ármannsson. 2003. Environmental aspects of geothermal energy utilization. *Geothermics* 32(4–6): 451-461.

Lehr, L., A. D. Allen, and R. Lease. 1982. *Potential for By-Product Recovery in Geothermal Energy Operations*. EER-TR-05-821. Vienna, VA: Energy and Economics Research Inc.

López, R. E. 2001. *Preliminary Study of Noise Propagation Behaviour at the Nesjavellir Geothermal Field, SW-Iceland*. 8. Reykjavík, Iceland: The United Nations University Geothermal Training Programme.

Macknick, J., R. Newmark, G. Heath, and K. C. Hallett. 2011. *A Review of Operational Water Consumption and Withdrawal Factors for Electricity Generating Technologies*. NREL/TP-6A20-50900. Golden, Colorado: National Renewable Energy Laboratory.

Majer, E. L., J. Nelson, A. Robertson-Tait, J. Savy, and I. Wong. 2012. *Protocol for Addressing Induced Seismicity Associated with Enhanced Geothermal Systems*. DOE/EE-0662. Washington, D.C: U.S. Department of Energy.

Maochang, H. 2001. *Possible Environmental Impacts of Drilling Exploratory Wells for Geothermal Development in the Brennisfjöl Area, SW-Iceland*. Reykjavik, Iceland: The United Nations University Geothermal Training Programme.

Mariita, N. O. 2002. The impact of large-scale renewable energy development on the poor: Environmental and socio-economic impact of a geothermal power plant on a poor rural community in Kenya. *Energy Policy* 30(11–12): 1119-1128.

MED. 2007. *2006 Energy Data: Electricity - Table 1 and Table 2*. Wellington, New Zealand: Ministry of Economic Development.

MED. 2008. *Energy Greenhouse Gas Emissions 1990-2007.* Wellington, New Zealand: Ministry of Economic Development.

Ogola, P. F. A. 2005. Environmental and Social Considerations in Geothermal Development. In *Workshop for Decision Makers on Geothermal Projects and Management, organized by UNU-GTP and KengGen*. Naivasha, Kenya.

Pasqualetti, M. J. 1980. Geothermal energy and the environment: The global experience. *Energy* 5(2): 111-165.

Pasvanoğlu, S., A. Güner, and F. Gültekin. 2012. Environmental problems at the Nevşehir (Kozakli) geothermal field, central Turkey. *Environmental Earth Sciences* 66(2): 549-560.

Pehnt, M. 2006. Dynamic life cycle assessment (LCA) of renewable energy technologies. *Renewable Energy* 31(1): 55-71.

Pfister, S., P. Bayer, A. Koehler, and S. Hellweg. 2011. Environmental Impacts of Water Use in Global Crop Production: Hotspots and Trade-Offs with Land Use. *Environmental Science & Technology* 45(13): 5761-5768.

Polyzou, O. and S. Stamataki. 2010. Geothermal Energy and Local Societies – A NIMBY Syndrome Contradiction? Paper presented at Proceedings World Geothermal Congress 2010, 25-29 April 2010, Bali, Indonesia.

Popovski, K. 2003. *Political and public acceptance of geothermal energy*. Reykjavík, Iceland The United Nations University Geothermal Training Programme.

Reith, S., T. Kölbel, P. Schlagermann, A. Pellizzone, and A. Allansdottir. 2013. *Public acceptance of geothermal electricity production*. Deliverable n° 4.4. Geoelec.

Richter, A. 2012. Indonesian government urges Bali to rethink opposition to geothermal plans. *ThinkGeoEnergy*, 4 September 2012.

Rissmann, C., B. Christenson, C. Werner, M. Leybourne, J. Cole, and D. Gravley. 2012. Surface heat flow and CO_2 emissions within the Ohaaki hydrothermal field, Taupo Volcanic Zone, New Zealand. *Applied Geochemistry* 27(1): 223-239.

Robinson, B., H. Outred, R. Brooks, and J. Kirkman. 1995. The distribution and fate of arsenic in the Waikato River system, North Island, New Zealand. *Chemical Speciation & Bioavailability* 7(3): 89-96.

Rule, B. M., Z. J. Worth, and C. A. Boyle. 2009. Comparison of Life Cycle Carbon Dioxide Emissions and Embodied Energy in Four Renewable Electricity Generation Technologies in New Zealand. *Environmental Science & Technology* 43(16): 6406-6413.

Rybach, L. 2005. Environmental aspects of geothermal development and utilization, and related legal, institutional and social implications. Paper presented at World Geothermal Congress. Short Pre- and Post-Congress Courses, 24–29 April 2005, Antalya, Turkey.

Rybach, L. and T. Kohl. 2004. Waste heat problems and solutions in geothermal energy. *Geological Society, London, Special Publications* 236(1): 369-380.

Rybach, L. and M. Mongillo. 2006. Geothermal Sustainability – A Review with Identified Research Needs. *Geothermal Resources Council Transactions* 30: 1083-1090.

Saner, D., R. Juraske, M. Kübert, P. Blum, S. Hellweg, and P. Bayer. 2010. Is it only CO_2 that matters? A life cycle perspective on shallow geothermal systems. *Renewable and Sustainable Energy Reviews* 14(7): 1798-1813.

Santoyo-Castelazo, E., H. Gujba, and A. Azapagic. 2011. Life cycle assessment of electricity generation in Mexico. *Energy* 36(3): 1488-1499.

SavingIceland. 2012. The Geothermal Ecocide of Reykjanes Peninsula. *SavingIceland*, 5 November 2012.

Sheppard, D. and E. Mroczek. 2004. Greenhouse gas emissions from the exploitation geothermal systems. *IGA News No. 55*, 11-13.

Şimşek, Ş., N. Yıldırım, and A. Gülgör. 2005. Developmental and environmental effects of the Kızıldere geothermal power project, Turkey. *Geothermics* 34(2): 234-251.

Stark, M. A., J. W. T. Box, J. J. Beall, K. P. Goyal, and A. S. Pingol. 2005. The Santa Rosa – Geysers Recharge Project, Geysers Geothermal Field, California, USA. Paper presented at In: Proceedings World Geothermal Congress.

Sullivan, J. L., C. E. Clark, J. W. Han, and M. Q. Wang. 2010. *Life Cycle Analysis Results of Geothermal Systems in Comparison to Other Power Systems*. ANL/ESD/10-5. Argonne: Argonne National Laboratory.

Sullivan, J. L., C. E. Clark, L. Yuan, J. W. Han, and M. Q. Wang. 2011. *Life Cycle Analysis Results for Geothermal Systems in Comparison to Other Power Systems – Part II*. ANL/ESD/11-12. Argonne: Argonne National Laboratory.

Sullivan, J. L., C. E. Clark, L. Yuan, J. W. Han, and M. Q. Wang. 2012. *Geothermal Life-Cycle Assessment - Part 3*. ANL/ESD/12-15. Argonne: Argonne National Laboratory.

Tester, J. W., B. J. Anderson, A. S. Batchelor, D. D. Blackwell, R. DiPippo, E. M. Drake, J. Garnish, B. Livesay, M. C. Moore, K. Nichols, S. Petty, M. N. Toksöz, and J. R. W. Veatch. 2006. *The future of geothermal energy: Impact of enhanced geothermal systems (EGS) on the United States in the 21st century*. Massachusetts Institute of Technology and U.S. Department of Energy.

USDOE. 2008. *Geothermal Technologies Program: Multi-Year Research, Development and Demonstration Plan Draft*. Washington, DC: United States Department of Energy.

USDOE. 2012. http://www1.eere.energy.gov/geothermal/geopower_landuse.html. Accessed September 2012.

USEPA. 2008. *Public Review Draft – National Water Program Strategy: Response to Climate Change*. United States Environmental Protection Agency.

Webster, J. G. 1999. The source of arsenic (and other elements) in the Marbel–Matingao river catchment, Mindanao, Philippines. *Geothermics* 28(1): 95-111.

Öko Institut. 2007. Global Emission Model for Integrated Systems (GEMIS) version 4.4. Germany.

Chapter 9

Matching supply and demand: grid and storage

Lead authors: Edgar G. Hertwich, Jan Weinzettel, Evert Bouman, Thomas Gibon, Anders Arvesen, Jaroslav Knápek

9.1 INTRODUCTION

In this report, the environmental impacts of different electricity supply technologies are analysed on a per-kWh basis and as implemented in International Energy Agency (IEA) scenarios. Energy resources, however, differ in their spatial and temporal distribution. As electricity supply needs to match electricity demand for the reason of stability, a transmission and distribution grid system is required not only to transport the electricity to customers but also to ensure an adequate quality of supply in terms of voltage, frequency, and reliability. The characteristics of resources and technologies for electricity generation, as well as the characteristics of power demand, have important implications for the design of the transmission and distribution system. A high fraction of variable renewable energy (VRE) sources such as wind and solar energy presents a special challenge to system operation; these sources may require energy storage or flexible demand. In this Chapter, we discuss implications for electricity transmission and distribution of different generation technologies and system configurations, evaluate the need for electricity storage, and review assessments of environmental impacts of such systems.

9.2 ELECTRICITY SYSTEM CHARACTERISTICS

An electricity network comprises a system in which electricity is generated, delivered to consumers and consumed. This system consists of power stations, transmission and distribution lines, transformers, converters, storage units and electrical appliances. Power stations produce electricity from other energy forms and carriers, such as fossil fuels, nuclear, wind, etc. In order to minimize transmission losses[1], electricity is typically converted to a high voltage level by transformers and transmitted at this level from power plant to distribution system, where it is converted to the distribution voltage level by other transformers. In case of long distances, usually starting at about 400-800 km, direct current (DC) is used for transmission, which lowers demand placed on conductors and removes the need for reactive power compensation, thereby offsetting the high costs of AC/DC converters.

The main function of the electricity grid is to reliably deliver electricity from the power stations to the consumers. Connecting new power plants to the electricity network requires additional infrastructure in terms of transmission lines and auxiliary equipment, such as transformers, circuit breakers, converters, etc. in order to connect the new power plant with existing transmission and distribution systems or directly with consumers. Some renewable power systems, such as rooftop PV, feed directly into the distribution grid (Figure 9.1). Such direct-feed systems, however, require some adaptation, which is not addressed in this

1 Electricity losses are given by relation $P_{loss} = R \cdot I^2$, where R is conductor resistance and I is the current flowing through the conductor, given by the relation $I = P/U$, where P is the transmitted power and U is the transmission voltage. Therefore, the higher the voltage, the smaller the current. The losses hence decrease with square of the voltage ($P_{loss} = R \cdot P^2/U^2$).

report. Some large projects are located far from electricity consumption centres. Therefore, electricity from a distributed system may have to be concentrated and transmitted over a long distance to become available for consumers. The length of the new transmission lines is determined by the distance between power plant location and the location of the consumers and may be up to several thousand kilometers; for hydropower plants in China, this distance may be over 2000 km[2]. High voltage direct current transmission lines are typically used for distances longer than 800 km (Meah and Ula, 2007) due to lower per-kilometer transmission line costs and no need for reactive power compensation, both of which compensate the additional costs of converters. Overhead transmission lines are often used in onshore rural areas due to lower price, while cables are used in built-up areas and for offshore transmission, for example, to connect offshore wind farms with onshore electricity grid.

Electricity networks can be also expanded in order to use regional variations to average the variability of renewable power sources. Should the VRE in one region have low output, other regions experiencing over-generation can transmit excess electricity.

In order to provide a benchmark of the power which it is necessary to transmit, it is possible to compare for example average German power consumption with the largest transmission projects. In 2009, Germany produced 592 TWh of electricity[3], which gives an average power consumption of about 826 W per capita[4], implying that a city of one million inhabitants needs on average nearly 1 GW of electricity supply, with considerable variations. The highest capacity transmission project being constructed is for the capacity of about 7.2 GW of high voltage direct current (HVDC) overhead line[5], which could serve a region of about ten million inhabitants. Transmitting hydropower over long distances is common in Brazil, China and India. due to inexpensive electricity and remote locations of hydropower.

The distribution system connects transmission lines with individual consumers; the connection requires conversion to residential voltage levels ranging between 110V and 240V. Large electricity consumers such as electricity-intensive industries can be connected directly to transmission systems, or to distribution systems at a higher voltage level. Current electricity storage capacity is usually nearly negligible in comparison to instantaneous power generation and due to nearly instantaneous power transmission. As a result of the minimal storage capacity, most of the power produced must be immediately consumed as it is generated; otherwise, the system will become unstable. Still, the electricity market is driven by power consumption. Therefore, this balance is typically achieved through the control of power generation in order to follow electricity consumption[6]. While consumption can differ significantly on an hourly basis, individual power plants, depending on the technology used, require considerable time to ramp up and down generation, and thereby cannot match the swings in demand. On one hand, a mix of power plants with different characteristics such as price and ramp up and ramp down times allows grid operators to optimize electricity generation costs and increases system stability. On the other hand, an electricity network consisting solely of sources with fast response would allow electricity generation to follow electricity consumption easily. These technologies, however, are often more expensive. The aim of the transmission system operators is thus to match the demand while minimizing cost. Therefore, the least expensive electricity is produced when it is available (more or less constantly in case of power plants with full control) and it is denoted as base power, while the most expensive electricity is used to cover peak consumption.

2 http://en.wikipedia.org/wiki/HVDC

3 http://www.iea.org/stats/electricitydata.asp?COUNTRY_CODE=DE

4 Total population in 2011 was 81,859,000 (http://en.wikipedia.org/wiki/List_of_countries_by_population)

5 http://en.wikipedia.org/wiki/List_of_HVDC_projects; http://new.abb.com/systems/hvdc/references/jinping---sunan

6 Although even today some electricity markets use distant control of some household appliances and communicate with large consumers in order to shape consumption according to generation.

FIGURE 9.1

Schematic electricity network

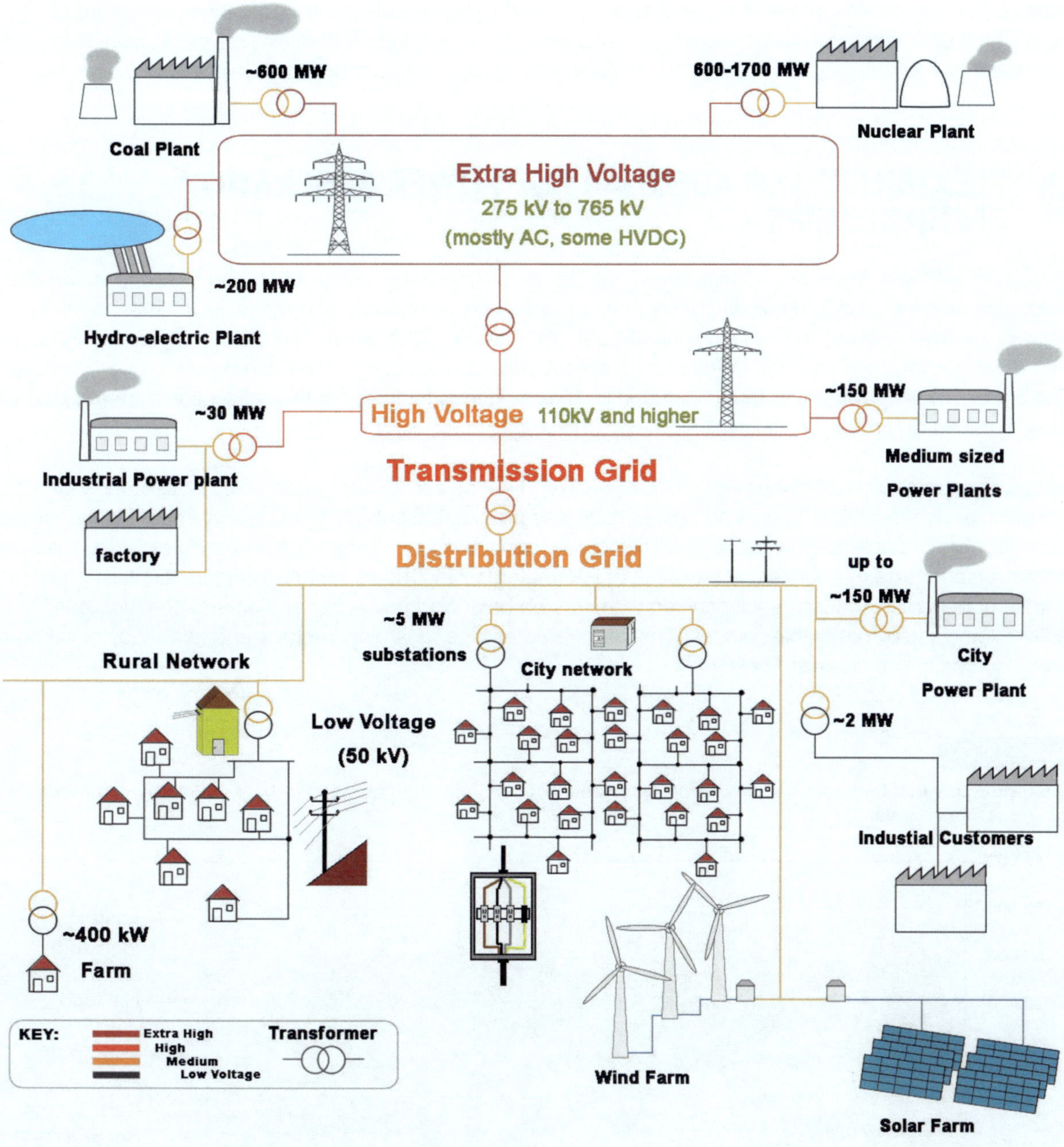

Source: http://upload.wikimedia.org/wikipedia/commons/5/56/Electricity_grid_schema-_lang-en.jpg

The typical example of the base load is a nuclear power plant, which has very high investment costs and low operation/variable costs. If running at full capacity, the total costs per electricity unit go down. It does not require too much control of output power, since it provides the least expensive electricity under these conditions. However, the cost increases as soon as the power plant stops running at full capacity. A typical example of peak power plant is a natural gas power plant, due to easily manageable output power and relatively higher costs.

9.3 FLEXIBILITY AND ADEQUACY OF POWER GENERATION TECHNOLOGIES

Power plants differ in flexibility and variability. They can be distinguished according to the extent to which their output may be controlled. Fully dispatchable power plants differ in ramp up and ramp down times, ranging from minutes (hydropower) to days (nuclear) (IEA, 2011). The output of intermittent power plants depends on external factors, such as instantaneous wind speed, solar radiation, etc. They can be considered partially dispatchable. Their maximum output is dictated by these external factors, but the production from the plants can be reduced or turned off.

The output of some intermittent power sources, such as tidal power, is very predictable, while other sources, such as solar photovoltaic (PV) or wind power, have less predictable output. Predictability improves planning and scheduling of back-up power plants and electricity consumption. Therefore, forecasting of variable power sources to some extent reduces the relevance of intermittency (Jacobson, 2009). As Figure 9.2 illustrates, intermittent generation causes additional requirements on balancing power generation and consumption, i.e. the flexibility of the dispatchable sources. The key characteristics regarding flexibility and adequacy of various power sources are summarized below.

FIGURE 9.2

Illustration of variable demand and variable renewable supply, indicating net load to be supplied by dispatchable generation resources

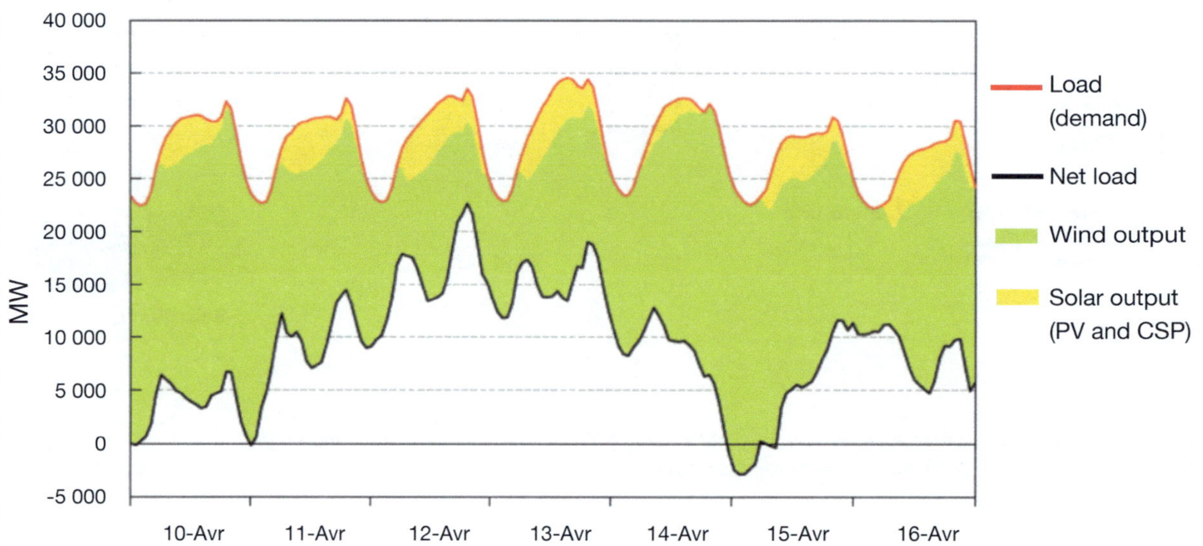

Note that power consumption differs on an hourly basis, daily basis, over the week, but also seasonally.
PV = photovoltaic; CSP = concentrating solar power.

Source: IEA, 2011. Copyright OECD/IEA 2011, Harnessing Variable Renewables - A Guide to the Balancing Challenge, IEA Publishing. License: http://www.iea.org/t&c/termsandconditions/.

Nuclear power: Due to their high capital costs and low fuel costs, nuclear power plants are usually designed to cover base load. They have a slow ramp rate. Power production can be curtailed, however, and technology is available to make nuclear power load-following (Pouret et al., 2009).

Coal-fired power: Slow response and high capital costs make coal a typical base-load power plant. However, some degree of flexibility is possible especially when following predictable swings in demand (Bugge et al., 2006); the synthesis gas storage in integrated gasification combined cycle plants offers additional flexibility. Until recently, plants with CCS have been seen as pure base-load, but recent research explores how these plants can also be built to offer more flexible operation (Chalmers et al., 2009; Cohen et al., 2012; Kang et al., 2012; Kang et al., 2011; Lin et al., 2012; Rezvani et al., 2012).

Natural gas power: Single-cycle gas turbines are often used for peak production and balancing. However, they have low fuel efficiency. Combined cycle plants are more capital intensive, but can also be operated with some degree of flexibility. Flexible operation strategies for post-combustion CCS also apply for gas fired plants.

Wind power: Wind speeds fluctuate at all relevant time scales for power system planning, scheduling and operations (Holttinen et al., 2011). Integration over larger geographical areas removes some variability due to reduced correlation in wind speeds over larger distances. Prediction accuracy for future wind power production decreases with the distance in time and increases with area considered (Wiser et al., 2011). There has been substantial research exploring the integration of wind power into existing power grids at various levels of demand (IEA, 2011). Reserves are needed to respond to unpredictable loss of power output from wind. Costs and emissions are associated with the operation of these reserves (Holttinen et al., 2011; Fripp, 2011). The capacity credit[7] of wind power depends both on the area covered by the grid and the penetration level of the technology (Hasche et al., 2011), but is often around 5-30 per cent of the conventional capacity (Figure 9.3). Differences in the correlation of wind speed and peaks in electricity demand lead to regional differences in capacity credit. Another important factor is the targeted reliability of the power system (IEA, 2011).

Solar power: PV power plants also have a high variability at various scales; however, the variability decreases and hence predictability increases with connected area (Sims et al., 2011). Capacity credit also depends on the joint occurrence of high demand (cooling loads) with high insolation (Pelland and Abboud, 2008). The issues are the same for concentrating solar power (CSP) except that this technology offers the option to store heat for later power generation, which substantially improves the adequacy and capacity credit for this technology (Sioshansi and Denholm, 2010; Madaeni et al., 2012). Figure 9.3 indicates that at low market penetration, the capacity credit of solar power is higher than that of wind power.

Geothermal energy: Geothermal electricity usually supplies base-load power, but it can also be used to meet variable loads (Sims et al., 2011), at increased costs. As for other power plants with high capital costs and low variable costs, the cost of electricity increases if the power plant is not running at full capacity.

Hydropower: Reservoir-based hydropower offers short-term to long-term storage and presents in many ways an ideal power balancing opportunity. These plants are constrained by the required water regime below the dam (Acker et al., 2012; Heide et al., 2011; Pete et al., 2010; Robitaille et al., 2012). Run-of-the-river hydropower plants can have significant seasonal, monthly and daily variations, but run comparatively steady on shorter time scales. Some plants may have diurnal storage for meeting daily peak demand during periods of low water availability (Sims et al., 2011; Hug-Glanzmann, 2011).

Bioelectricity: Bioenergy plants can be regulated in a similar manner as coal-fired power plants. In practice, bioelectricity generation often occurs in combined heat and power plants, in which case the heat load often determines electricity production (Sims et al., 2011). Such production is more easily predicted but can still vary

7 The capacity credit is the percentage of conventional capacity that a given turbine can replace. (http://www.termwiki.com/EN:Capacity_credit)

considerably. In some grids, the capacity credit can be high if heat and electricity demand are covariant. In the future, cogeneration involving electricity, heat, fuels, and cooling may offer additional flexibility (Li et al., 2012).

FIGURE 9.3

Capacity credit for wind (top panel) and solar (lower panel) energy in selected regions as a function of penetration (Western United states for solar power)

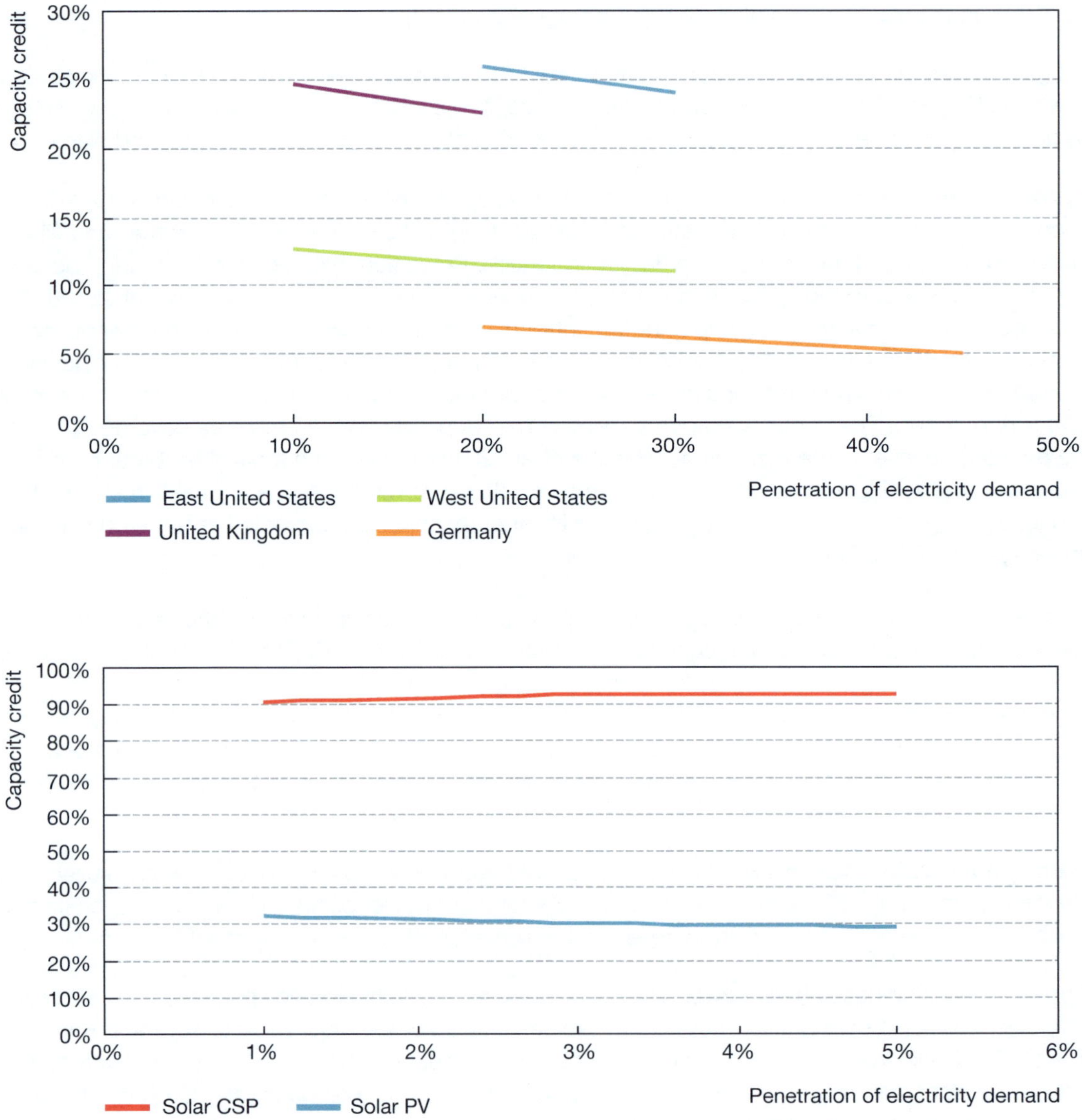

Source: Copyright OECD/IEA 2011, Harnessing Variable Renewables - A Guide to the Balancing Challenge, IEA Publishing. License: http://www.iea.org/t&c/termsandconditions/

9.4 SOLUTIONS FOR A HIGH SHARE OF VARIABLE RENEWABLES

The grid integration and balancing requirements of VRE supply have been discussed for a long time (Sørensen, 1978) and recently have come into a broader focus. The International Energy Agency (IEA, 2011) and the Intergovernmental Panel on Climate Change (IPCC) (Sims et al., 2011) have described these challenges and reviewed our current state of knowledge with respect to electricity system characteristics that allow for a high share of variable renewable power sources.

The findings can briefly be summarized as follows:

- Larger grid area and higher capacity transmission enhance the ability of electricity systems to absorb a larger fraction of VRE because of the flexibility offered by different generation sources in such a grid and the averaging out of timing of demand peaks and VRE supply (Kempton et al., 2010; Trötscher and Korpås, 2011; Holttinen et al., 2011).

- Different VRE sources can, to some degree, compensate each other's variation as their variability is usually not strongly correlated (Fusco et al., 2010; Heide et al., 2011).

- The ability to store energy at time scales ranging from minutes to weeks enhances the potential integration and reduces the (over)capacity required to provide a stable electricity supply (Göransson and Johnsson, 2011; Williams et al., 2012; Nyamdash et al., 2010; Connolly et al., 2012; Huang et al., 2011; Esteban et al., 2012).

- Flexible timing of demand through load shedding (often in heavy industry) or turning on and off demand as foreseen in smart grids (hot water heaters, charging of electric vehicles, washing machines and freezers) can contribute significantly to system flexibility and allow for a higher penetration of renewables (Kempton and Tomic, 2005; IEA, 2011). Additional flexibility can be introduced when the demand for other energy carriers, especially heat and cooling, is optimized together with that of electricity (Lund et al., 2012; Lund and Mathiesen, 2009).

The IEA and IPCC assessments find that a penetration of VRE much higher than today is feasible but that it requires a number of adjustments and additional elements such as more robust grids, more interconnections, backup power, overcapacity and curtailment, more dynamic regulation of fossil sources, demand response, and energy storage. These adjustments and additional elements all cause additional emissions connected to the construction of additional infrastructure or the operation of the system (Mathiesen et al., 2009; Pehnt et al., 2008).

Williams et al. (2012) assessed how California can reach the greenhouse gas emission (GHG) reduction of 80 per cent below the 1990 level by 2050. They highlight the role of electricity, which needs to be de-carbonized and it has to replace carbon based energy carriers in other sectors, e.g. for transport activities. The de-carbonization is reached by a high share of renewables, nuclear and carbon capture and storage. The authors created four scenarios: blended, high nuclear, high renewable and high carbon capture and storage. New energy storage capacity and transmission lines are identified based on the geographical conditions of new power plants and load centers. It is shown that high renewable scenario requires about 20 per cent more miles of new transmission lines than other scenarios and about three times more power capacity of energy storage for the system balancing. However, the results display only power not energy for the storage capacity, and only kilometers of new transmission lines without power rating. These missing data prevent a detailed assessment of the related environmental burdens.

Williams et al. (2011) noted that high adoption of renewable energy, corresponding to 74 per cent, "created challenges to operability and reliability". The solutions consist of energy storage (utility-scale energy storage and thermal storage for solar thermal generation), smart charging of electric vehicles, international trade, or overcapacity and "spilled" electricity. Twelve GW of electricity storage capacity would be required, while only about 1.2 GW is currently available in the form of pumped storage hydro (PSH). Additional opportunities for pumped storage hydro are limited due to environmental reasons and a lack of suitable sites. Williams et al. (2011) concluded that, "Breakthroughs in energy storage will be necessary to make the high renewables scenario viable". Furthermore, this scenario required more land use due to solar thermal and PV power and almost 50,000 km of new transmission lines due to the geographical distribution of appropriate sites for renewable power generation and power consumption.

Heide et al. (2010) analysed the relation between seasonal variation in solar and wind power across Europe and in the electricity demand in Europe and found an optimal seasonal mix of wind and solar power for Europe to be 55 per cent wind energy and 45 per cent solar in a 100 per cent wind-solar electricity mix. In a follow-up article Steinke et al. (2013) examined the need for grid extension and storage with respect to the reduction of the required backup energy. They found that even with an ideal European grid 20 per cent of back-up energy generation is required, which is exceeds the available energy from biomass. The back-up energy is doubled if no appropriate grid and storage is available. Both technologies have to be considerably extended in order to limit the back-up energy at 10 per cent, which the authors assumed as energy available from biomass.

Demand side management: The purpose of demand side management is to adjust the load curve by controlling electricity consumption, which helps the operators to match electricity supply and demand. It is already part of current electricity networks. Household electric heaters with storage systems, such as boilers with reservoirs and storage heaters, are typical appliances that can be controlled by system operators. It is sufficient for the users to have an agreement that these appliances are turned on during a defined period each day. Another demand side management options are the electricity intensive industries, which do not depend on instantaneous power supply. There are also incentives that aim to shift power consumption out from peak hours to night times through price differentiation. Demand side management has a big potential, which is addressed by studies focusing on smart grids. Utilizing demand side management together with a high contribution of intermittent power sources, the control of the electricity network could be potentially reversed, such as the demand would follow the intermittent supply and not *vice versa*. There are obvious problems related to this shift. For many applications, the users need electricity supply immediately. On the other hand, this demand can be supplied from storage or by decreasing electricity consumption elsewhere. In critical situations, e.g. in Japan after the tsunami in Japan in 2011, the power consumption was strictly controlled in order to keep the electricity network stable with the available power sources. There is a strong need to shift our paradigm regarding the operation of the electricity network. It can be expected that environmental burdens related to this shift will result from electronics manufacturing, giving the increased use of controls. Closely related research areas are smart grids and intelligent buildings, which are systems that can control loads.

In the end, it is important that the electricity delivered to customers has low upstream emissions. It is hence important to consider the development not only of individual generation sources but also of entire electricity systems. To ensure low emissions electricity systems, a systematic evaluation of grid infrastructure, system operation, and balancing or back-up power and storage is required. The integration of life cycle assessments (LCAs) in power system analysis allows for finding low-impact solutions for specific grids (Göransson and Johnsson, 2011). Such analysis is today in its infancy. As we are analyzing the sustainability of future energy systems on a larger macro level, we are not attempting such integration. Rather, our aim is to get an understanding of the overall importance of grid and balancing issues and the magnitude of overall impacts in future electricity systems with a large share of variable renewable energy sources. For this purpose, we review studies of the impact of elements of the electricity system apart from generation, with a focus on grid infrastructure and energy storage. We discuss ways to integrate assessments of impacts of these components in a macro-level analysis.

9.5 ELECTRICITY MARKETS

The period until the mid-1990s was characterized by vertically integrated utilities with regional monopolies of power generation, distribution, and sale. One or a few power utilities dominated in each country. These utilities were in charge of achieving a balance between power supply and demand. Utilities were regulated as natural monopolies. Since then, the electricity industry in various regions has been split into generation, transmission and distribution. While the regulated monopoly model has been retained for transmission and distribution, electricity markets have been introduced for generation, with the idea that generators should compete for providing the market with electricity.

In the European Union, the liberalization of the power market started in the late 1990s with EU Directive 96/92/EC[8], which led only to a minimum opening of the power markets of individual EU member states. The "second energy package" (EU Directive 2003/54/EC[9]) required the functional and legal splitting of power generation and distribution, as one recognized that vertically integrated firms that were in part monopoly, in part competitive power generator had an unfair advantage over independent generators as they could cross-subsidize generation. The legislation defined the time line of power market full liberalization for big, middle and small size power consumers. Final electricity consumers thus got the right to choose among supplier of the electricity. The directive also defined the rules for internal market functioning including rules for transmission and distribution fees determination by independent national regulatory bodies.

The failure of some EU countries to open up their national power markets to outside competitors lead to the "third liberalization package" which was passed in 2009[10]. This package has been primarily aimed at reinforcement of power market liberalization and at increasing of the effectiveness of EU internal power market functioning. A third package introduced, based on the fact that there was no general agreement, three different options for national power market opening: 1) the full unbundling with the ownership separation of individual activities (power generation and power distribution and trade), 2) establishment (by national governments) of independent system operator (ISO) which ISO should bear the control on investment and business activities of power transmission system and 3) creation of independent transmission operators (ITO) through legal separation of transmission system from vertically integrated companies.

As mentioned previously, electricity is a very special commodity which requires meeting demand with the supply at any instant. A shortage of supply first impact the quality of power (voltage, frequency) and can ultimately lead to a black-out. A market for electricity hence requires therefore an overarching institution which ensures the balance between supply and demand and that deals with unexpected events. The creation of electricity markets hence lead to two different, but closely related, markets – a market with electricity as a commodity and a market with ancillary services which are needed to keep power system operation and to ensure required level of quality and security of electricity supply. Ancillary services include back-up power and voltage control. These are offered by power producers who comply with technical and commercial terms and conditions. Offering ancillary services usually requires installation of special technical and communication devices. The market of ancillary services is separated from the market with the electricity as a commodity and can be organized by a transmission system operator.

The present electricity market in the EU is based on the idea of an energy-only market, which means that only MWh are subject of trade on regional energy exchanges. Since electricity generation requires planning ahead while demand can be predicted or scheduled only imprecisely, electricity can be purchased 1) as futures

8 Directive 96/92/EC of the European Parliament and of the Council of 19 December 1996 concerning common rules for the internal market in electricity

9 Directive 2003/54/EC of the European Parliament and of the Council of 26 June 2003 concerning common rules for the internal market in electricity and repealing Directive 96/92/EC

10 http://ec.europa.eu/energy/gas_electricity/legislation/legislation_en.htm. Third energy package included also Directive 2009/72/EC 13 July 2009 concerning common rules for the internal market in electricity and repealing Directive 2003/54/EC

(month, quarter, year, all in base load and peak load), 2) on a day-ahead market, or 3) on an intra-day-market. The price of electricity is defined at all markets, in general, as the balance between the supply and demand. Power producers are placing their bids (e.g., for Day-Ahead-Market they place amount of offered electricity for each hour of the day). Bids of all participating producers are sorted from the lowest offer to highest. The resulting diagram of cumulated power offers is called "merit order". It starts with the lowest bids (i.e. lowest cost power producers in terms of variable cost) and ends with the highest ones. The power is purchased either directly by large consumers or by companies selling to small customers, including commerce and households.

If bids for electricity delivery are not accompanied in a given time period with the physical delivery of electricity, it results in occurrence of power deviation, which is basically the difference between the expected (agreed) electricity delivery and the real delivery. Bidders are, in general, responsible for the economic consequences of power deviations (somebody else has to react to this situation, for example, the subject offering the ancillary services, and should be compensated for this) and should bear their cost.[11]

Power generation from VRE sources, especially wind and PV power, can be estimated with relatively high accuracy only in the range of hours ahead. Introduction of the Day-A-Head Market and especially Intra-Day Market enabled effective participation of VRE sources on the power market. VRE participation on these two short term markets can significantly reduce power deviations resulting from VRE penetration into power grid and can also significantly reduce needs for balancing power, compared to a situation where VRE is treated as negative demand. This market participation potentially results in the decrease of prices of power deviations.

In cases where VRE sources have the right-of-way, e.g., through Feed-In-Tariffs or other mandatory purchasing requirements, the so called "merit order effect" (Bode and Groscurth, 2011) occurs, i.e., the merit order curve is shifted right[12] and fossil sources with high variable generation costs are pushed out of the market. VRE power plants, especially wind and PV, have a marginal cost of power generation close to zero, so it is natural that VRE power plants are used first and, in a fact, move the merit order curve right. The problem is that both VRE and fossil sources also need to earn back their fixed costs, which they can only do in periods when they are not marginal suppliers, so that electricity prices are higher than their marginal generation costs. Guaranteed feed-in-tariffs for renewables ensure that VREs get a good electricity price even in periods when the spot price on the exchanges is zero, while fossil power plants do not earn any money in such situations. In countries with such rules, in particular in Germany, fossil generators experience economic difficulties[13]. Economic problems in Europe are also connected to an overcapacity of generation sources and a loss of competitiveness of gas vis-a-vis coal. Nonetheless, there is a concern that energy-only markets do not create sufficient economic inventiveness for new investments into conventional power plants needed among other as a back-up capacity and the source of ancillary services. That is why the idea of "capacity" markets is now widely discussed within the EU[14]. This idea is basically aimed at a remuneration of investors into new power generating capacities based on offered (guaranteed) power capacity. Depending on how such a market is designed, electricity storage companies and demand-response aggregators could also participate in this market. It is, however, not clear that such a protected market, which can be seen as a subsidy for fossil power plants, is the appropriate response to the effect of a protected market or subsidy scheme for VRE.

11 Power generation using the RES could be excluded from this rule – e.g. legislation in the Czech Republic explicitly assumes in Feed –In-tariff scheme transfer of responsibility of power deviation to (obligatory) purchaser.

12 The influence of merit order effect can be seen in recent years on the market with electricity. Prices of electricity have fallen down from more than 80 EUR/MWh in the second half of 2008 to app. 34 EUR/MWh in 2014 (values taken from Prague Energy Exchange).

13 This can be documented e.g. on the example on the gas fired and combined cycle gas fired power stations. They realized, in many case, operational losses and they operation cannot contribute to the recovery of initial investment. Good example of this situation is the combined gas power plant Pocerady in the Czech Republic (currently being in testing phase of operation) which is realizing app. 5-10 EUAR of operational loss per each MWH generated.

14 http://theenergycollective.com/adamjames/237496/energy-nerd-lunch-break-how-capacity-market-works-and-why-it-matters

9.6 ENVIRONMENTAL IMPACTS OF TRANSMISSION AND DISTRIBUTION SYSTEMS

In this section, we investigate the environmental impacts of different elements of electricity transmission and distribution systems, i.e., power lines and cables, transformers and switchgear, as well as energy storage. Our review focuses on LCA of these system elements. For transmission lines, however, we also investigate site-specific aspects. Considerations of how to integrate these system elements and how to assess the impacts of the electricity transmission, distribution and storage infrastructure required to achieve mitigation scenarios are discussed in a subsequent section.

9.6.1 IMPACTS OF POWER LINES ON WILDLIFE

The impact of overhead power lines on birds has long been a concern (Bevanger, 1998). There are indications that power lines have potential impacts on the populations of large birds such as raptors and storks (Jenkins et al., 2010; Janss, 2000; Martin and Shaw, 2010; Rubolini et al., 2005; Barrientos et al., 2011; Sundar and Choudhury, 2005). These impacts are potentially larger than those of wind power plants (Sovacool, 2009). When measuring total bird kills, domestic cats and collisions with cars, building windows and telecommunications towers are of comparable importance to power lines (Sovacool, 2013). Affecting smaller populations of birds higher up on the food chain, the impact of power lines in terms of species endangerment is potentially larger. It is possible to reduce fatalities by marking power lines, but the size of this effect varies substantially across case studies (Barrientos et al., 2011; Barrientos et al., 2012; Jenkins et al., 2010). Kaluga et al. (2011) examine possibilities for reduction of white stork deaths due to collision with transmission lines. They claim that technical modification may reduce mortality due to electrocution to zero.

Land use associated with overhead transmission lines is of two types: (a) land for piles construction and (b) land under and near the lines. The first type of land is fully occupied as the built up land, while the second type of land can be further utilized under several constraints, such as maximum height of trees and often a ban of construction activity. While the second type of land is a continuous belt about 30 meters wide, the first one is considerably smaller.

Power lines generate audible noise through the corona effect. Audible noise emissions may have a negative impact on big carnivores (Doukas et al., 2011), leading to habitat fragmentation. The total impact of noise, electromagnetic field, land use and land fragmentation on ecosystems is not yet fully understood. The creation of tropospheric ozone was studied already thirty years ago by Droppo (1981), who concluded that ozone concentration due to HVDC transmission lines are relatively small up to ±550 kV in comparison to the background concentrations.

9.6.2 LIFE CYCLE ASSESSMENT OF TRANSMISSION AND DISTRIBUTION SYSTEM ELEMENTS

LCA addresses mostly emissions and the use of resources. LCAs show that the largest share of environmental impact from current electricity grids is associated with losses, i.e. the energy that is required to overcome the resistance and impedance in the power lines and transformers (Jorge et al., 2012b; Jorge et al., 2012a). In 2007, these losses accounted for between 5 per cent (Japan) and 26 per cent (India) of the electricity generated (Table 4.2, IEA, 2010). Building a strong and efficient grid can hence substantially reduce environmental impacts of the electricity system. Electricity grid operations are also required to match supply and demand and react to unforeseen events, such as the loss of lines, generation units, or unforeseen changes in demand. Back-up power and spinning reserves providing reliability and compensating for unforeseen events can cause substantial emissions. Larger, interconnected grids, as well as the ability of system operators to shed loads, to schedule flexible loads, and to store energy, increase reliability and reduce balancing requirements and thus environmental impacts (IEA, 2011). Significant investments in grid capacity, interconnectedness and flexibility (smart grid) will reduce losses and ease the integration of variable renewable electricity sources into the grid.

This section focuses on the environmental impacts of the manufacturing of grid components, as well as the construction and operation of the grid. We include energy storage in batteries, as we have seen that this is considered as a solution for situations with a high share of variable renewable sources and distributed generation. We do not explicitly discuss or account for grid benefits.

There has been relatively little attention to electricity transmission and distribution in the LCA of energy systems. The widely used ecoinvent database includes a life cycle inventory (LCI) of the electricity grid in Switzerland, broken down by high-voltage, medium-voltage and low-voltage grids (Dones et al., 2007; Frischknecht et al., 2007). For other countries, the Swiss results are extrapolated but country-specific electricity losses are taken into account. Other analyses focus often only on CO_2 and energy impacts (Harrison et al., 2010) or address individual components or options, such as the choice between overhead lines and underground cables (Bumby et al., 2010; Jones and McManus, 2010). Analysis by Cigre (2004) indicates that for Denmark, (high-voltage) transmission accounts for 2 per cent of the total GHG impacts of electricity, while (lower voltage) distribution accounts for 8 per cent, mostly associated with grid losses.

Data for the present analysis is based on LCI compiled by ecoinvent (Frischknecht et al., 2007) and Jorge et al. (Jorge et al., 2012b; Jorge et al., 2012a) from a range of sources, often from industry. For the transmission grid, we use an initial analysis of the Norwegian grid (Jorge and Hertwich, 2012) to provide rough estimates of the inventory of transmission systems. As shown by Jorge et al. (2012b) losses are the biggest contributor to all environmental impacts examined except for metal depletion under the assumption of European average electricity mix. Even though renewable power sources reduce environmental impacts per unit of electricity lost, Jorge and Hertwich (2012) show that losses are important also in Norway, where the electricity mix is dominated by hydropower.

9.6.2.1 Power lines and underground cables

An overview of upstream environmental impacts of material production and transport for transmission lines was provided by Jorge et al. (2012b). An assessment of environmental burdens from the construction phase was not found in the scientific literature. It is possible to quantify environmental impacts from the manufacturing of the transmission lines. Each connection point will further require a transformer, a circuit breaker and, in the case of DC lines, a power converter. Due to a rapid development in power electronics, the auxiliary equipment might differ for future electricity networks. However, it can be assumed that electricity will be transmitted through conductors also in future. It can be further assumed that losses will be kept as low as possible due to physical constraints on heat dissipation, and economic and environmental importance of lost energy. The losses during power transmission are related to the resistance of the conductor and the current flow. The most common conductors are copper and aluminum, which have a high conductivity. Resistance can be reduced through increasing the cross sectional area of the line, while the current can be reduced through going to higher voltage. An example of the trade-off between environmental impacts during manufacturing and operation given different conductor sizes is presented in Figure 9.4.

9.6.2.2 Subsea cables

Subsea cables have so far mostly been considered in LCA as part of offshore wind farms (Weinzettel et al., 2009; Wagner et al., 2011), even though they can have a substantial importance for electricity trade. Alternating current can only be used for short distances in subsea cables, because the charging current of the self-capacitance of the cables causes substantial losses at distances over 70 km. Hence, longer distance subsea cables require a conversion of electricity to high voltage direct current (HVDC).

Birkeland (2011) provides a LCI based on the 450 kV Norway-Netherlands (Norned) cable. Nes (2012; Arvesen et al., 2014) provides an assessment of a meshed North Sea grid designed to transmit electricity from large, far-offshore wind farms to different countries in the North Sea basin, depending on demand and domestic generation.

FIGURE 9.4

Trade-off between environmental impacts from production of conductors and electricity losses in a 1000-km, 800-kV HVDC line with a capacity of 10GW, assuming a 40 year lifetime

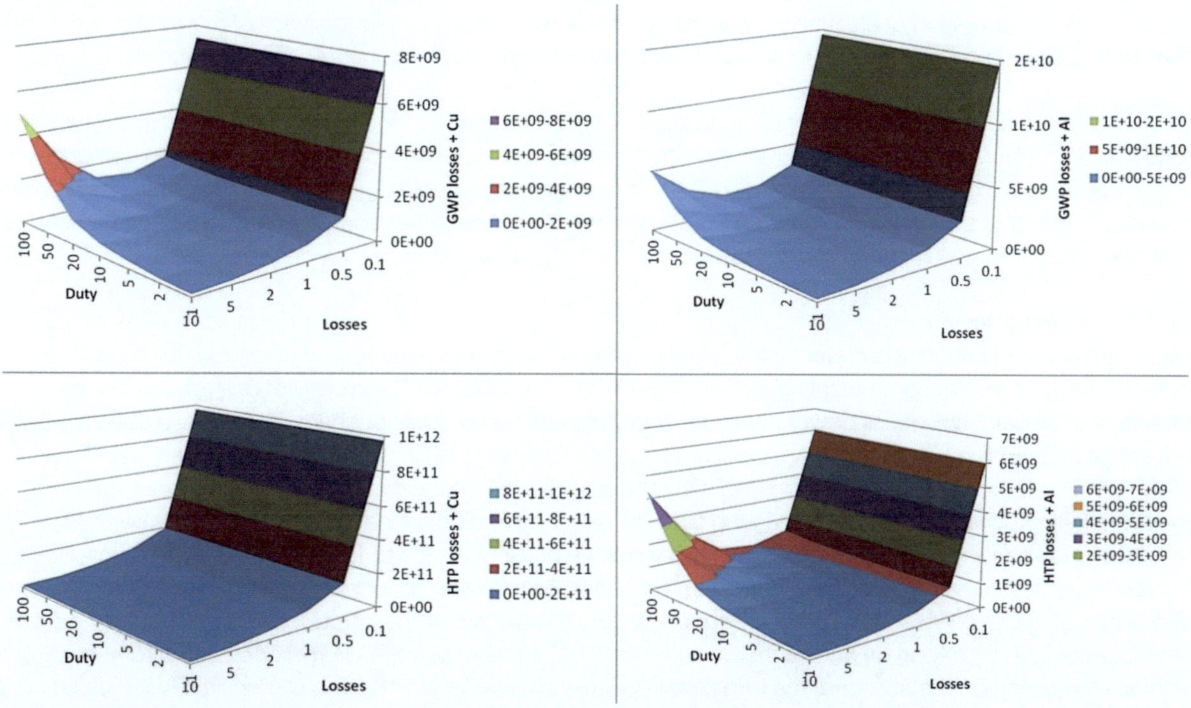

Global warming potential (GWP) and human toxicity potential (HTP) are shown for copper and aluminium conductors.

FIGURE 9.5

Impacts per kWh of electricity transmitted in a meshed North Sea grid from far offshore wind power plants, following a scenario of the WINDSPEED project (Nes, 2012).

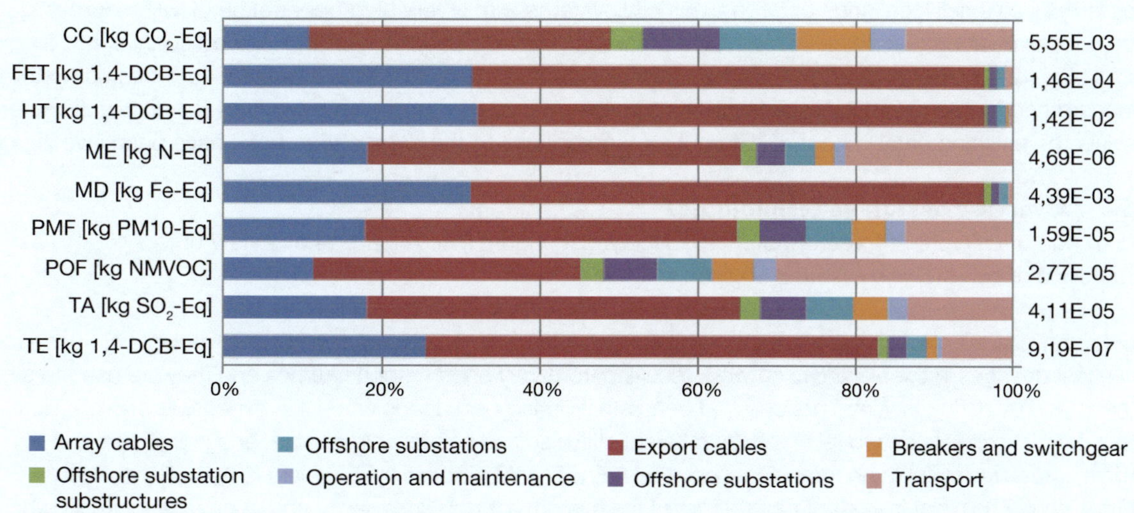

The analysis of the cables indicates that the cable manufacturing itself causes most of the environmental impact, but that vessels used to lay the cables and inspect them also contribute appreciably (Arvesen et al., 2013).

Based on the consideration of electricity produced by offshore wind power plants, Nes estimates that the entire North Sea grid contributes an equivalent of 5.5 g CO_2e/kWh of electricity delivered, accounting for one eighth of the total climate impacts of far-offshore wind power. The grid has even larger contributions to metal depletion and metal-related human- and eco-toxicity. A significant share of the impacts is associated with metal production and inputs to or waste from metal production. Wind power has higher impacts in these categories than power from fossil power plants (Nes, 2012). In other categories, cable laying and equipment transport are important. Given that the grid will also be used to trade electricity from other sources between countries when it is not fully loaded with wind electricity, the total impact is likely to be distributed among more kWh.

9.6.2.3 Backup power

Dispatchable power capacity is currently the most practiced means of ensuring that electricity supply follows electricity demand and of compensating for variations in VRE production. The modifications required in the operations of dispatchable fossil power plants from the introduction of a larger share of VRE have been modelled in electrical grid studies (Fripp, 2011; Meibom et al., 2011; Troy et al., 2010), but have rarely been included in LCA. Fripp (2011) estimates the spinning and standing reserves of natural gas power required to address the variability of wind power generation, assuming a set of wind power plants located in the United States. He shows that reserve requirements reduce dramatically with a better grid due to the averaging of wind conditions across a larger surface area. Averaging across an area of 500 km^2, the impact of operating the reserves are on the order of 25 g CO_2/kWh of wind power. Pehnt et al. (2008) investigate the introduction of offshore wind power to the German grid by relying on an electricity market model to investigate the altered operation of other power stations. Depending on the scenario, the additional systems emissions are 18-70 g CO_2/kWh of wind electricity introduced to the system. On the other hand, the reduction of emissions is substantial because the introduction of wind power displaced mostly coal-fired power. Overall, the need to keep more spinning reserves and other back-up systems introduces additional environmental impacts that can be of a similar magnitude as the life cycle impacts of the power generation technology itself.

9.6.3 UTILITY-SCALE ELECTRICITY STORAGE

Electricity storage can be useful and economical as a means to mediate between differences in the periods of generation and demand. Most forms of electricity storage, however, are characterized by high investment costs and substantial losses. Therefore, the only form of electricity storage that is widely applied by electric utilities is PSH. In small local grids or large electricity systems with a very high penetration of VRE sources, it can be necessary and economical to introduce other storage technologies (Göransson and Johnsson, 2011; Williams et al., 2012). Subsequently, there has been a recent surge in the interest of stationary electricity storage (Hedegaard and Meibom, 2012; Makansi, 2010; Evans et al., 2012; Sundararagavan and Baker, 2012). Here, we review the environmental impact of different storage options to the degree that these have been investigated.

9.6.3.1 Overview of storage technologies

A wide range of different storage options are displayed in Figure 9.6. A general overview of technical parameters of these storage systems is given in Table 9.1.

The only high-capacity electricity storage technology currently in large-scale use is PSH. PSH plants are used to balance fast changes in the electricity network resulting from very short response times and they are used to store the cheap energy during low load periods and generate it during peak loads when it is expensive. The classical concept of PSH consists of two reservoirs at different altitudes and a reversible turbine, or a pair of turbines. Therefore, it requires suitable geographical conditions. Energy efficiency of a PSH plant is typically between 70-80 per cent. The most important parameters of the system are power capacity and energy capacity. Power capacity determines how much electricity a PSH plant can deliver to the network instantaneously, while energy capacity determines for how long it can deliver the electricity without being refilled. Existing plants have a power

TABLE 9.1 List of technical parameters for various storage technologies

Technology system	Power rating	Discharge time	Self discharge/ parasitic loss	Suitable storage duration	Energy res. power density				Lifetime (years)	Cycle life	Cycle efficiency	Space requirements m²/kWh	$/kW	$/kWh
					Wh/kg	W/kg	Wh/L	W/L						
PHS	100-500 MW	1-24 h+	Very small	Hours-months	0.5-1.5		0.5-1.5		30-60	10000-30000	70-87	0.02	600-2000	5-100
CAES	5-300 MW	1-24 h+	Small	Hours-months	30-60		3-6	0,5-2	20-40	8000-12000	70-89	0.01	400-800	2-50
Lead acid	0-20 MW	seconds-hours	0.1-0.3 %	Minutes-days	30-50	75-300	50-80	10-400	5-20	200-1800	70-90	0.058	300-600	200-400
NiCd	0-40 MW	seconds-hours	0.2-0.6 %	Minutes-days	50-75	150-300	60-150		10-20	1500-3000	60-83	0.03	500-1500	800-1500
NaS	0.05-8 MW	seconds-hours	~20 %	Seconds-hours	150-240	150-230	150-250		5-15	2500-4500	75-90	0.019	100-300	300-500
Li-ion	0-100 kW	minutes-hours	0.1-0.3 %	Minutes-days	200-500	150-315	200-500		5-15	1000-10000	90-100	0.03	1200-4000	600-2500
Fuel cells	0-50 MW	second-24 h+	Almost zero	Hours-months	500-3000	500+	500-3000	500+	5-15	1000+	20-66	0.003-0.006	10000+	
Vanadium redox battery	0.03-3 MW	seconds-10h	Small	Hours-months	10-30		16-33		5-30	12000+	75-85	0.040	600-1500	150-100
ZnBr	0.05-2 MW	seconds-10h	Small	Hours-months	30-50		30-60		5-30	2000+	66-80	6-26	700-2500	150-100
SMES	0.1-10 MW	Milliseconds-8 s	10-15 %	Minutes-hours	0.5-5	500-2000	0,2-2,5	1000-4000	20-30	20000-100000+	95-98	6-26	200-300	1000-10000
Flywheel	0-250 kW	Milliseconds-15 min	100%	Seconds-minutes	10-30	400-1500	20-80	1000-2000	15-20	20000-100000	89-95	0.03-0.06	250-350	1000-5000
Capacitor	0-50 kW	Milliseconds - 60 min	40 %	Seconds-minutes	0.05-5	~100000	2-10	100000+	5	50000-100000	60-70	0.04	200-400	500-1000
Supercapacitor	0-300 kW	Milliseconds-60 min	20-40 %	Seconds-minutes	2.5-15	100000+	2.5-15	1000+	8-20	100000+	84-98	0.04	100-300	300-2000

PHS; pumped hydro storage;
CAES: compressed air storage;
NiCd: nickel cadmium;
NaS: sodium sulfur;
Li-ion: lithium ion;
ZnBr: zinc bromide;
SMES: superconducting magnetic energy storage.

Source: Data are based Chen et al., 2009 and Beaudin et al., 2010.

capacity of up to 3 GW[15], while energy capacity varies considerably and depends mostly on natural conditions. It can be very high for systems with large natural reservoirs such as lakes, and it is usually lower for smaller, purely human-made reservoirs. Environmental consequences of PSH are not explicitly addressed in the scientific literature to date, but studies of regular hydropower plants give an indication of the types of impact to be expected from PSH. However, the use phase relates to land use, while construction and equipment manufacturing causes indirect impacts. It can be expected that environmental impacts from electricity losses during charge-discharge cycle will be considerable, but depend on the electricity generation technology to power the pumps.

FIGURE 9.6

An overview and classification of important energy storage technologies

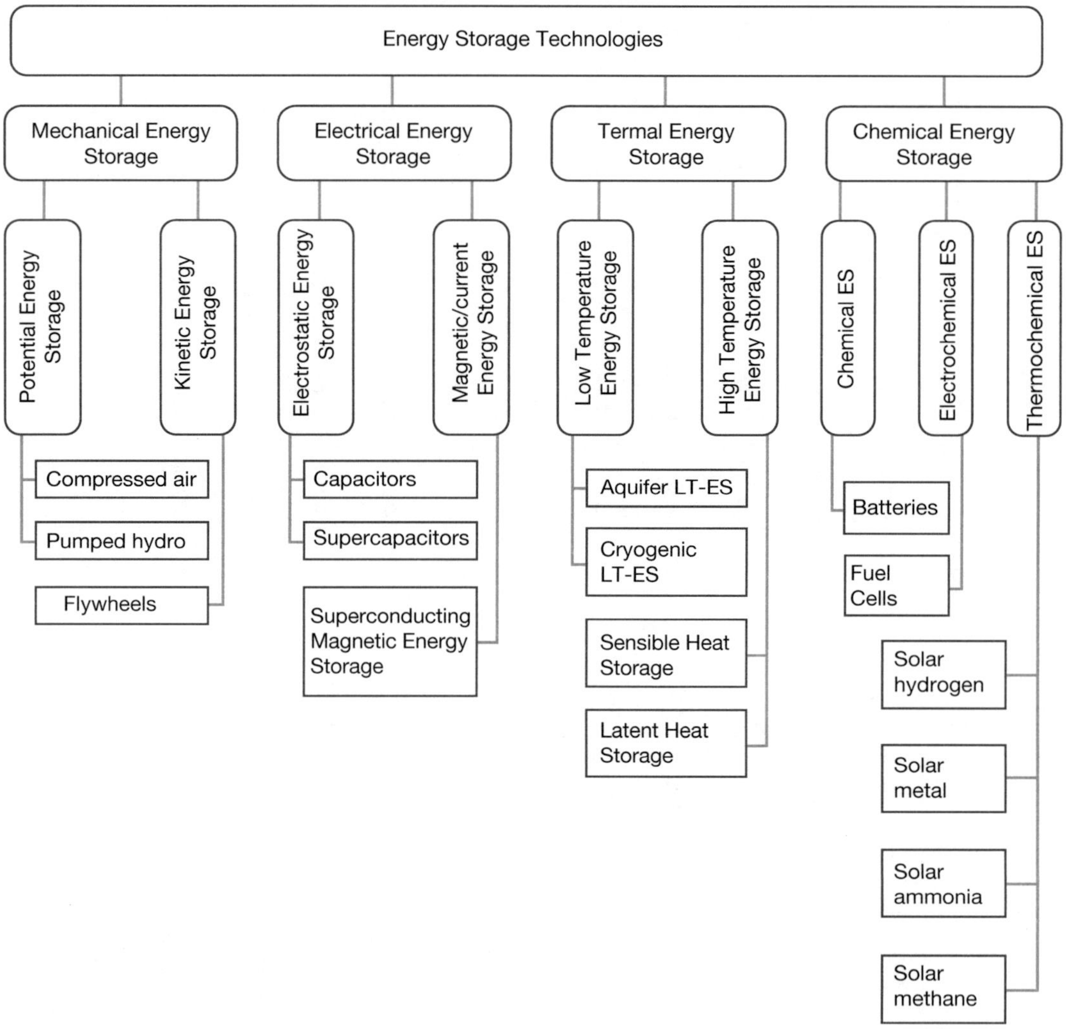

Source: Evans et al., 2012

15 http://en.wikipedia.org/wiki/List_of_pumped-storage_hydroelectric_power_stations, accessed 20.Dec.2013

Batteries are the most common electricity storage elements for portable electrical appliances and electronics. They transform electricity into chemical energy using different concepts. While current research and development activities, driven by mobile applications, focus on reducing weight and volume per unit of stored energy, these features would not necessarily be required for electricity network batteries. The aim for stationary batteries would rather be low environmental impact and cost. Environmental consequences of large-scale adoption of Nickel and Li-ion metal batteries have been discussed in connection to electric vehicles (Majeau-Bettez et al., 2011). It has been also proposed to use batteries in electric vehicles to balance the electricity network using smart grids, both on the supply and demand side. Furthermore, these batteries could be used in grids when their operational parameters drop below an acceptable level for use in electric vehicles. The main disadvantages of these batteries are high cost and critical metal content, which may limit their current use and the use of electric vehicles as well. However, new technologies are being developed for batteries intended only for use in electrical grids.

Compressed air is a promising energy storage concept due to the use of air as a low-cost storage medium. However, similarly to pumped storage hydro, it requires suitable geological conditions, since the air is stored underground in large-scale projects. While PSH energy storage is common in current grids, compressed air energy storage is very rare. A compressed air energy storage (CAES) plant is located in Huntorf, Germany, and has been running since 1978 with a capacity of 290 MW and storage capacity of 3 hours. A considerable longer storage capacity of 26 hours is achieved in a 110 MW system which is located in McIntosh, Alabama (Succar, 2008). Though CAES requires a suitable geological location, it is possible to construct the air storage and both CAES plants mentioned here use a solution-mined salt cavern as air storage location. In addition to the utility scale CAES systems, a small scale adiabatic 2 MW, 500 MWh storage plant was commissioned in Gaines, Texas in late 2012 (General Compression, 2014). A demonstration plant operating on the adiabatic CAES principle is scheduled for completion in 2016 (RWE Power AG, 2010).

Several recent papers suggest using hydrogen as a storage medium for remote autonomous systems powered by wind or solar energy. During an electricity overproduction phase, hydrogen is produced by electrolysis, compressed and stored for later use in a fuel cell for electricity generation. The disadvantage of hydrogen is the low energy efficiency of the storage system, at slightly above 50 per cent, when electrolysis is applied for hydrogen production, in comparison to 75-85 per cent for PSH, batteries or compressed air (Hammerschlag and Mazza, 2005). The biggest advantage is the longer time horizon over which energy can be stored. In addition, it is independence from geological conditions. While the environmental consequences of this system have not been assessed, it can be assumed that most of the related impacts will result from electricity losses. The economic performance of this system will most likely be determined by electricity variations and fuel cell costs.

Ulleberg et al. (2010) describe a demonstration project using hydrogen, battery and flywheel for energy storage in an autonomous system powered by a small-scale wind power plant supplying approximately ten households. The most suitable energy storage technology will depend on the distance from possible sites for PSH, compressed air, the costs of batteries, fuel cells and electricity. A storage facility operates with electricity losses and it requires additional equipment, such as transformers, power converters and circuit breakers, depending on storage type and grid requirements.

9.6.3.2 Pumped storage hydropower

The environmental impacts of pumped hydro storage are characterized (Bauer and Bolliger, 2007) as part of the development of the ecoinvent LCI database. The Bauer and Bolliger model pumped storage electricity production as conventional reservoir hydropower[16] with an input of electricity to account for the pumping energy demand. The energy efficiency used for PSH is 70 per cent. Other required inputs for electricity generation at a pumped storage power plant include production and disposal of lubricating oil, and various land transformation and occupation inputs. The lost energy is released as waste heat. Furthermore, small

16 Note: Another hydropower technology described in ecoinvent is a run-of-the-river power plant, for which the construction of a dam and reservoir lake are not required.

amounts of N_2O and biogenic methane are emitted during production. The additional equipment necessary for PSH, such as pumps or reversible turbines, is not included in the ecoinvent inventories; the material demand is assumed to be equal to a reservoir plant. There are small differences in the inventories corresponding to Swiss, alpine (France, Italy, Austria) and rest-of-Europe regions.

9.6.3.3 Batteries

Sullivan and Gaines (2012) published a review of LCI data for batteries. Five different technologies are investigated: nickel metal hydride (NiMH), lead acid (PbA), nickel cadmium (NiCd), sodium sulfur (Na/S) and lithium ion (Li-ion). The article focuses on material composition, cradle-to-gate energy demand, and cradle-to-gate emissions. The average cradle-to-gate embodied energy and average cradle-to-gate emissions are shown in Figure 9.7 and Figure 9.8 for all battery types.

FIGURE 9.7

Average cradle-to-gate energy demand of different battery technologies reviewed in Sullivan and Gaines (2012)

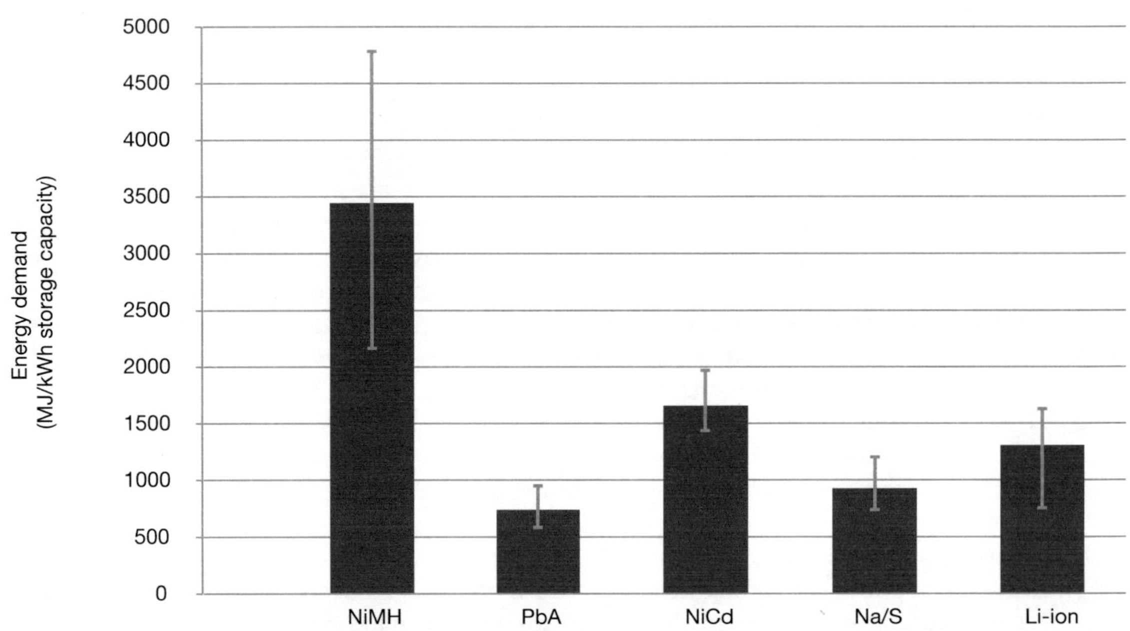

NiMH: nickel metal hydride; PbA: lead acid; NiCd: nickel cadmium; Na/S: sodium sulfur; Li-ion: lithium ion.
The full range of reported energy demand is shown by the error bars.

FIGURE 9.8

Cradle-to-gate emissions for different types of batteries

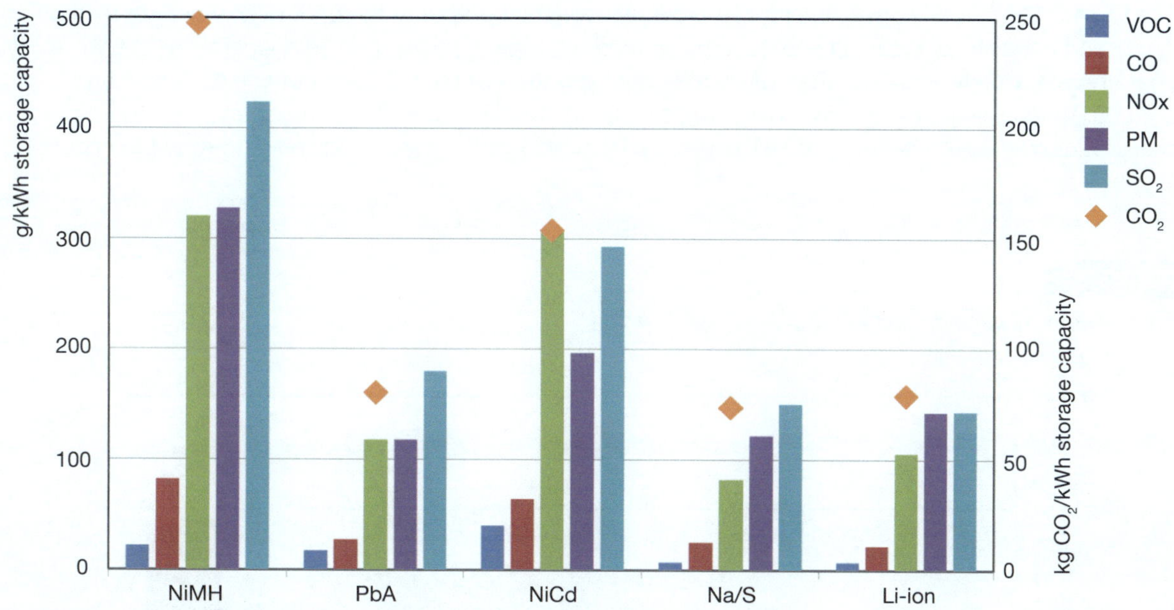

Emissions are given in g/ kWh storage capacity battery for VOC: volatile organic compounds; CO: carbon monoxide; NO_x: nitrogen oxides; PM: particulate matter; SO_2: sulfur dioxide. Emissions for carbon dioxide are given in kg CO_2/ kWh storage capacity on the secondary axis (Sullivan and Gaines 2012).

The results in Figure 9.8 are LCI numbers; no impact assessment was performed. For a specific inventory of NiMH and Li-ion batteries for electric vehicles, Majeau-Bettez et al. (2011) show GHG emissions in the range of 200-350 kg CO_2-eq/kWh storage capacity (2011). Using results from batteries for transportation purposes as an estimate of impacts from batteries for static grid-balancing purposes might introduce an overestimation. Batteries intended for electric vehicles may have larger impacts from the use of specialty materials in, for example, battery casing due to the need for light-weighting. These battery types may also have a shorter lifetime, which also increases impacts. It should be noted that one of the most significant emissions sources is electricity for producing the batteries. Battery production therefore has a potential to become cleaner as the electricity production mix becomes cleaner.

Main parameters influencing the environmental impacts of batteries used to reduce intermittency of renewable energy sources are the required storage capacity and power rating. Leadbetter and Swan (2012) define three types of grid service duration categories: short (up to 1 min), medium (minutes to hours) and long (hours to days). These categories are then coupled to the battery energy to power ratio, *E/P* (kWh/kW). The order of magnitude of the *E/P* ratios for the short, medium and long categories is respectively 0.01, 1 and 10 kWh/kW. Leadbetter and Swan report *E/P* ratios for lead acid (0.13-0.5 kWh/kW), Li-ion (0.025-0.6 kWh/kW), sodium sulfur (6 kWh/kW) and vanadium redox (1.5-6 kWh/kW) batteries. Based on these *E/P* ratios, Li-ion power cells are recommended for short duration service, PbA and Li-ion are recommended for medium duration, and Na/S and vanadium redox batteries are recommended for long duration services (Leadbetter and Swan, 2012).

Rydh and Sandén (2005) model the energy and material requirements for a stand-alone PV-battery system used to power an air conditioning unit. The system has a power rating of 50 kW and an *E/P* ratio between 9-10 kWh/kW. Depicted in Figure 9.9 is the battery weight requirement for several battery technologies (Rydh and Sanden, 2005). It can be seen that the range of battery weight needed to provide continuous power differs by an order of magnitude. Based on actual wind speed data and solar PV output, Rugolo and Aziz (2012) calculate that the storage capacity needed for a 1 MW wind turbine and a 1 MW PV solar array is 49 MWh and 14.5 MWh, respectively, in order to provide a constant output to the grid. Even when corrected for efficiency and capacity factor, wind energy requires more storage capacity than solar energy, as the variability is larger; there are longer periods when wind does not blow than when sun does not shine (Rugolo and Aziz, 2012). When divided by specific energy densities, e.g. 125 Wh/kg for a Li-ion battery, the wind system requires 392 tons of Li-ion battery per MW power rating while the PV system requires 116 tons Li-ion battery/MW.

FIGURE 9.9

Battery weight requirement per kW of power output rating

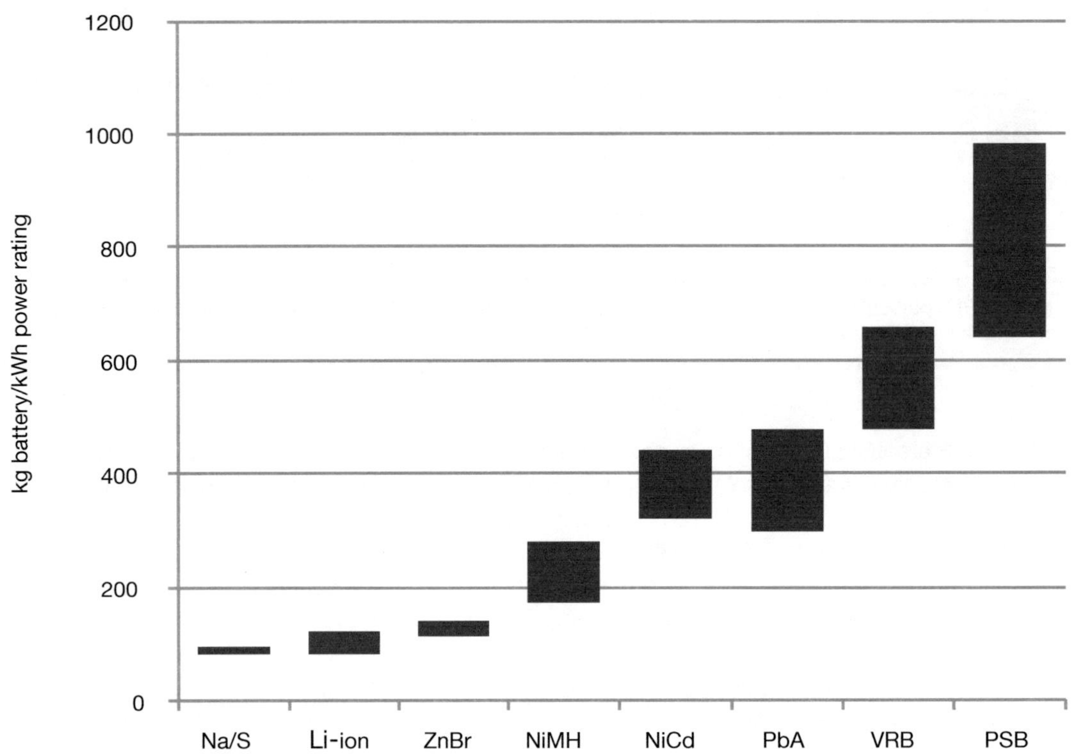

Na/S: sodium sulfur; Li-ion: lithium ion; ZnBr: zinc bromide; NiMH: nickel metal hydride; NiCd: nickel cadmium; PbA: lead acid; VRD: vanadium redox battery; PSB: polysulfide bromide
Source: Rydh and Sanden, 2005

9.6.3.4 Compressed air energy storage

Denholm and Kulcinski (2004) describe the energy requirement and GHG emissions association with air storage. In the CAES concept, excess energy is used to compress the air and store it underground under high pressures. When needed, the air is heated and released through high and low pressure expanders to generate electricity. The heating requires combustion of natural gas in a gas turbine. Therefore, on a per kWh basis, the total GHG emission related to operation (excluding the primary electricity generation) is relatively high when compared to GHG emissions from renewable power generation and reported to be 292 g CO_2-eq/kWh (Denholm and Kulcinski, 2004). However, when the GHG emissions related to CAES systems are evaluated in conjunction with the (renewable) power generation on a constant production basis, i.e. as a balancing system for intermittent wind power production, the life cycle GHG emissions are in the range of 66-104 g CO_2-eq/kWh, depending on the total system operating capacity factor (Denholm, 2005). From a life cycle perspective, the combustion of natural gas in the gas turbine is shown to be the major contributor to the environmental impact of wind power plus compressed air systems for particulate matter emissions, photochemical oxidation potential, terrestrial acidification, and GHG emissions. An alternative concept in which no gas is burned, called adiabatic CAES, makes use of a separate heat storage for heat that is released during compression of the air. The impacts are considerably lower for adiabatic CAES than for conventional CAES. Investigating the storage of offshore wind energy, Bouman et al. (2015) find that the adiabatic CAES system adds 50-100 per cent to the impacts of wind energy. These impacts are sensitive to the size and corresponding material requirements of the heat storage (Bouman et al., 2014).

9.7 CONCLUSIONS

Substantial investments in transmission and distribution are required to upgrade grids, meet unmet demand, and supply a growing population. According to the IEA Energy Technology Perspectives, annual investments in transmission and distribution are on the order of $210 billion for the Baseline and $300 billion for the BLUE Map scenarios, compared to $360 billion to $500 billion in power generation. However, in the Energy Technology Perspectives 2012, transmission and distribution investments for the two- and four-degree scenarios are comparable and investments in smart grids are profitable. In any case, questions about costs do not yet answer the question asked in this report: "What are the resource requirements and environmental impact caused by grid expansion and the grid integration of variable renewable energy?"

Our review indicates that each of the investigated response strategies to accommodate the generation of VRE sources, grid expansion, flexible operation of fossil power plants, and battery-based energy storage, causes GHG emissions that are in the same range or even higher than the life cycle emissions of the renewable generation sources investigated, primarily wind and PV power. However, this conclusion is highly dependent on the expected design of the electricity network, its required stability and the adopted strategy for matching supply and demand. No life cycle assessments have been performed on an optimal combination of these technologies resulting in a very-low GHG emissions system. The emissions from these strategies are large enough to substantially affect which combination of renewable energy sources has the lowest emissions. In fact, since the factors affecting production from different VRE sources are independent, this provides an argument for combining different sources rather than relying on a single source (Heide et al., 2011). It is also an argument for adopting sources with a higher capacity factor, e.g., offshore instead of onshore wind power. Even with these strategies, however, the GHG emissions associated with systems including VRE appear to be lower than those of alternative generation systems, such as fossil fuel power with carbon capture and storage. Further research is required to investigate other environmental impacts and identify desirable strategies to deal with the issue of intermittency.

9.8 REFERENCES

Acker, T. L., J. T. Buechler, K. Knitter, K. J. Conway, and R. Noteboom. 2012. Wind integration impacts on hydropower and system balancing operations in the Grant County PUD. *Wind Engineering* 36(1): 81-96.

Arvesen, A., C. Birkeland, and E. G. Hertwich. 2013. The Importance of Ships and Spare Parts in LCAs of Offshore Wind Power. *Environmental Science & Technology* 47(6): 2948-2956.

Arvesen, A., R. Nes, D. Huertas-Hernando, and E. Hertwich. 2014. Life cycle assessment of an offshore grid interconnecting wind farms and customers across the North Sea. *The International Journal of Life Cycle Assessment* 19(4): 826-837.

Barrientos, R., J. C. Alonso, C. Ponce, and C. Palacín. 2011. Meta-Analysis of the Effectiveness of Marked Wire in Reducing Avian Collisions with Power Lines. *Conservation Biology* 25(5): 893-903.

Barrientos, R., C. Ponce, C. Palacín, C. A. Martín, B. Martín, and J. C. Alonso. 2012. Wire marking results in a small but significant reduction in avian mortality at power lines: A BACI designed study. *Plos One* 7(3).

Bauer, C. and R. Bolliger. 2007. *Wasserkraft (Hydropower)*. Uster: ecoinvent Centre.

Beaudin, M., H. Zareipour, A. Schellenberglabe, and W. Rosehart. 2010. Energy storage for mitigating the variability of renewable electricity sources: An updated review. *Energy for Sustainable Development* 14(4): 302-314.

Bevanger, K. 1998. Biological and conservation aspects of bird mortality caused by electricity power lines: A review. *Biological Conservation* 86(1): 67-76.

Birkeland, C. 2011. Assessing the Life Cycle Environmental Impacts of Offshore Wind Power Generation and Power Transmission in the North Sea (Thesis) Energy and Process Engineering, Norwegian University of Science and Technology, Trondheim.

Bode, S. and H.-M. Groscurth. 2011. The impact of PV on the German Power Market. *Zeitschrift für Energiewirtschaft* 35: 105-115.

Bouman, E. A., M. M. Øberg, and E. G. Hertwich. 2014. Environmental impacts of balancing offshore wind power with Compressed Air Energy Storage (CAES). [submitted].

Bugge, J., S. Kjær, and R. Blum. 2006. High-efficiency coal-fired power plants development and perspectives. *Energy* 31(10-11): 1437-1445.

Bumby, S., E. Druzhinina, R. Feraldi, D. Werthmann, R. Geyer, and J. Sahl. 2010. Life cycle assessment of overhead and underground primary power distribution. *Environmental Science and Technology* 44(14): 5587-5593.

Chalmers, H., M. Lucquiaud, J. Gibbins, and M. Leach. 2009. Flexible operation of coal fired power plants with postcombustion capture of carbon dioxide. *Journal of Environmental Engineering* 135(6): 449-458.

Chen, H., T. N. Cong, W. Yang, C. Tan, Y. Li, and Y. Ding. 2009. Progress in electrical energy storage system: A critical review. *Progress in Natural Science* 19(3): 291-312.

Cigré. 2004. *Life Cycle Assessment (LCA) for Overhead Lines*. Paris: International Council on Large Electric Systems (Cigré).

Cohen, S. M., G. T. Rochelle, and M. E. Webber. 2012. Optimizing post-combustion CO_2 capture in response to volatile electricity prices. *International Journal of Greenhouse Gas Control* 8: 180-195.

Connolly, D., H. Lund, B. V. Mathiesen, E. Pican, and M. Leahy. 2012. The technical and economic implications of integrating fluctuating renewable energy using energy storage. *Renewable energy* 43: 47-60.

Denholm, P. and G. L. Kulcinski. 2004. Life cycle energy requirements and greenhouse gas emissions from large scale energy storage systems. *Energy Conversion and Management* 45(13-14): 2153-2172.

Denholm, P. K., Gerald L, Holloway, Tracey. 2005. Emissions and Energy Efficiency Assessment of Baseload WInd Energy Systems. *Environmental Science & Technology* 39: 1903-1911.

Dones, R., C. Bauer, R. Bolliger, B. Burger, M. Faist, Emmenegger, R. Frischknecht, T. Heck, N. Jungbluth, and A. Röder. 2007. *Life Cycle Inventories of Energy Systems: Results for Current Systems in Switzerland and other UCTE Countries*. Dübendorf, CH: Swiss Centre for Life Cycle Inventories.

Doukas, H., C. Karakosta, A. Flamos, and J. Psarras. 2011. Electric power transmission: An overview of associated burdens. *International Journal of Energy Research* 35(11): 979-988.

Droppo, J. G. 1981. Field Determinations of HVDC Ozone Production Rates. *Ieee Transactions on Power Apparatus and Systems* 100(2): 655-661.

Esteban, M., Q. Zhang, and A. Utama. 2012. Estimation of the energy storage requirement of a future 100% renewable energy system in Japan. *Energy Policy* 47: 22-31.

Evans, A., V. Strezov, and T. J. Evans. 2012. Assessment of utility energy storage options for increased renewable energy penetration. *Renewable and Sustainable Energy Reviews* 16(6): 4141-4147.

Fripp, M. 2011. Greenhouse Gas Emissions from Operating Reserves Used to Backup Large-Scale Wind Power. *Environmental Science & Technology* 45(21): 9405-9412.

Frischknecht, R., M. Tuchschmid, and M. Faist Emmenegger. 2007. *Strommix und Stromnetz [Electricity mix and electricity grid]*. Uster.

Fusco, F., G. Nolan, and J. V. Ringwood. 2010. Variability reduction through optimal combination of wind/wave resources – An Irish case study. *Energy* 35(1): 314-325.

General Compression. 2014. Texas Dispatchable Wind Project. http://www.generalcompression.com/index.php/tdw1. Accessed 21 March 2014.

Göransson, L. and F. Johnsson. 2011. Large scale integration of wind power: Moderating thermal power plant cycling. *Wind Energy* 14(1): 91-105.

Hammerschlag, R. and P. Mazza. 2005. Questioning hydrogen. *Energy Policy* 33(16): 2039-2043.

Harrison, G. P., E. J. Maclean, S. Karamanlis, and L. F. Ochoa. 2010. Life cycle assessment of the transmission network in Great Britain. *Energy Policy* 38(7): 3622-3631.

Hasche, B., A. Keane, and M. O'Malley. 2011. Capacity value of wind power, calculation, and data requirements: The Irish power system case. *IEEE Transactions on Power Systems* 26(1): 420-430.

Hedegaard, K. and P. Meibom. 2012. Wind power impacts and electricity storage: A time scale perspective. *Renewable energy* 37(1): 318-324.

Heide, D., M. Greiner, L. von Bremen, and C. Hoffmann. 2011. Reduced storage and balancing needs in a fully renewable European power system with excess wind and solar power generation. *Renewable energy* 36(9): 2515-2523.

Heide, D., L. von Bremen, M. Greiner, C. Hoffmann, M. Speckmann, and S. Bofinger. 2010. Seasonal optimal mix of wind and solar power in a future, highly renewable Europe. *Renewable energy* 35(11): 2483-2489.

Holttinen, H., P. Meibom, A. Orths, B. Lange, M. O'Malley, J. O. Tande, A. Estanqueiro, E. Gomez, L. Söder, G. Strbac, J. C. Smith, and F. Van Hulle. 2011. Impacts of large amounts of wind power on design and operation of power systems, results of IEA collaboration. *Wind Energy* 14(2): 179-192.

Huang, S., B. M. S. Hodge, J. Xiao, G. V. Reklaitis, and J. F. Pekny. 2011. The Effects of Electricity Storage on Large Scale Wind Integration. *Computer Aided Chemical Engineering* 29: 1879-1883.

Hug-Glanzmann, G. 2011. Predictive control for balancing wind generation variability using run-of-river power plants.

IEA. 2010. *Energy Technology Perspectives 2010: Scenarios and Strategies to 2050*. Paris, France: OECD/IEA.

IEA. 2011. *Harnessing Variable Renewables: A Guide to the Balancing Challenge*. Paris: International Energy Agency.

Jacobson, M. Z. 2009. Review of solutions to global warming, air pollution, and energy security. *Energy and Environmental Science* 2: 148 - 173.

Janss, G. F. E. 2000. Avian mortality from power lines: A morphologic approach of a species-specific mortality. *Biological Conservation* 95(3): 353-359.

Jenkins, A. R., J. J. Smallie, and M. Diamond. 2010. Avian collisions with power lines: A global review of causes and mitigation with a South African perspective. *Bird Conservation International* 20(3): 263-278.

Jones, C. I. and M. C. McManus. 2010. Life-cycle assessment of 11kV electrical overhead lines and underground cables. *Journal of Cleaner Production* 18(14): 1464-1477.

Jorge, R. S. and E. G. Hertwich. 2012. Environmental evaluation of power transmission in Norway. *Applied Energy* (APEN-D-12-00589R1).

Jorge, R. S., T. R. Hawkins, and E. G. Hertwich. 2012a. Life cycle assessment of electricity transmission and distribution - part 2: transformers and substation equipment. *International Journal of Life Cycle Assessment* 17(2): 184-191.

Jorge, R. S., T. R. Hawkins, and E. G. Hertwich. 2012b. Life cycle assessment of electricity transmission and distribution-part 1: Power lines and cables. *International Journal of Life Cycle Assessment* 17(1): 9-15.

Kaluga, I., T. H. Sparks, and P. Tryjanowski. 2011. Reducing death by electrocution of the white stork, *Ciconia ciconia*. *Conservation Letters* 4(6): 483-487.

Kang, C., Z. Ji, and Q. Chen. 2012. Review and prospects of flexible operation of carbon capture power plants. *Dianli Xitong Zidonghua/Automation of Electric Power Systems* 36(6): 1-10.

Kang, C. A., A. R. Brandt, and L. J. Durlofsky. 2011. Optimal operation of an integrated energy system including fossil fuel power generation, CO_2 capture and wind. *Energy* 36(12): 6806-6820.

Kempton, W. and J. Tomic. 2005. Vehicle-to-grid power implementation: From stabilizing the grid to supporting large-scale renewable energy. *Journal of Power Sources* 144(1): 280-294.

Kempton, W., F. M. Pimenta, D. E. Veron, and B. A. Colle. 2010. Electric power from offshore wind via synoptic-scale interconnection. *Proceedings of the National Academy of Sciences* 107(16): 7240-7245.

Leadbetter, J. and L. G. Swan. 2012. Selection of battery technology to support grid-integrated renewable electricity. *Journal of Power Sources* 216: 376-386.

Li, Z., E. Larson, and R. H. Williams. 2012. Fossil Energy Systems. In *Global Energy Assessment*, edited by IIASA. Cambridge: Cambridge University Press.

Lin, Y. J., C. C. Chang, D. S. H. Wong, S. S. Jang, and J. J. Ou. 2012. Control Strategies for Flexible Operation of Power Plant Integrated with CO_2 Capture Plant. *Computer Aided Chemical Engineering* 30: 237-241.

Lund, H. and B. V. Mathiesen. 2009. Energy system analysis of 100% renewable energy systems—The case of Denmark in years 2030 and 2050. *Energy* 34(5): 524-531.

Lund, H., A. N. Andersen, P. A. Østergaard, B. V. Mathiesen, and D. Connolly. 2012. From electricity smart grids to smart energy systems: A market operation based approach and understanding. *Energy* 42(1): 96-102.

Madaeni, S. H., R. Sioshansi, and P. Denholm. 2012. Estimating the capacity value of concentrating solar power plants: A case study of the southwestern United States. *IEEE Transactions on Power Systems* 27(2): 1116-1124.

Majeau-Bettez, G., T. R. Hawkins, and A. H. StrØmman. 2011. Life cycle environmental assessment of lithium-ion and nickel metal hydride batteries for plug-in hybrid and battery electric vehicles. *Environmental Science and Technology* 45(10): 4548-4554.

Makansi, J. 2010. Bulk storage could optimize renewable energy. *Power* 154(9).

Martin, G. R. and J. M. Shaw. 2010. Bird collisions with power lines: Failing to see the way ahead? *Biological Conservation* 143(11): 2695-2702.

Mathiesen, B. V., M. Münster, and T. Fruergaard. 2009. Uncertainties related to the identification of the marginal energy technology in consequential life cycle assessments. *Journal of Cleaner Production* 17(15): 1331-1338.

Meah, K. and S. Ula. 2007. Comparative evaluation of HVDC and HVAC transmission systems. Paper presented at 2007 IEEE Power Engineering Society General Meeting, PES.

Meibom, P., R. Barth, B. Hasche, H. Brand, C. Weber, and M. O'Malley. 2011. Stochastic optimization model to study the operational impacts of high wind penetrations in Ireland. *IEEE Transactions on Power Systems* 26(3): 1367-1379.

Nes, R. N. 2012. Life cycle assessment of an offshore electricity grid interconnecting Northern Europethesis, Energy and Process Engineering, Norwegian University of Science and Technology, Trondheim.

Nyamdash, B., E. Denny, and M. O'Malley. 2010. The viability of balancing wind generation with large scale energy storage. *Energy Policy* 38(11): 7200-7208.

Pehnt, M., M. Oeser, and D. J. Swider. 2008. Consequential environmental system analysis of expected offshore wind electricity production in Germany. *Energy* 33(5): 747-759.

Pelland, S. and I. Abboud. 2008. Comparing photovoltaic capacity value metrics: A case study for the city of Toronto. *Progress in Photovoltaics: Research and Applications* 16(8): 715-724.

Pete, C. M., T. L. Acker, G. Jordan, and D. A. Harpman. 2010. Western wind and solar integration study hydropower anaysis: Benefits of hydropower in large-scale integration of renewables in the Western United States.

Pouret, L., N. Buttery, and W. Nuttall. 2009. Is nuclear power inflexible? *Nuclear Future* 5(6): 333-341.

Rezvani, S., D. McIlveen-Wright, Y. Huang, A. Dave, J. D. Mondol, and N. Hewitt. 2012. Comparative analysis of energy storage options in connection with coal fired Integrated Gasification Combined Cycles for an optimised part load operation. *Fuel* 101: 154-160.

Robitaille, A., I. Kamwa, A. H. Oussedik, M. De Montigny, N. Menemenlis, M. Huneault, A. Forcione, R. Mailhot, J. Bourret, and L. Bernier. 2012. Preliminary impacts of wind power integration in the Hydro-Quebec system. *Wind Engineering* 36(1): 35-52.

Rubolini, D., M. Gustin, G. Bogliani, and R. Garavaglia. 2005. Birds and powerlines in Italy: An assessment. *Bird Conservation International* 15(2): 131-145.

Rugolo, J. and M. J. Aziz. 2012. Electricity storage for intermittent renewable sources. *Energy & Environmental Science* 5(5): 7151-7160.

RWE Power AG. 2010. *ADELE - Adiabatic compressed-air energy storage for electricity supply*. Cologne, Germany: RWE Energy.

Rydh, C. J. and B. A. Sanden. 2005. Energy analysis of batteries in photovoltaic systems. Part I: Performance and energy requirements. *Energy Conversion and Management* 46(11-12): 1957-1979.

Sims, R., P. Mercado, W. Krewitt, G. Bhuyan, D. Flynn, H. Holttinen, G. Jannuzzi, S. Khennas, Y. Liu, M. O'Malley, L. J. Nilsson, J. Ogden, K. Ogimoto, H. Outhred, Ø. Ulleberg, and F. van Hulle. 2011. Integration of Renewable Energy into Present and Future Energy Systems. In *IPCC Special Report on Renewable Energy Sources and Climate Change Mitigation*, edited by O. Edenhofer, et al. Cambridge, United Kingdom and New York, NY, USA: Cambridge University Press.

Sioshansi, R. and P. Denholm. 2010. The value of concentrating solar power and thermal energy storage. *IEEE Transactions on Sustainable Energy* 1(3): 173-183.

Sovacool, B. K. 2009. Contextualizing avian mortality: A preliminary appraisal of bird and bat fatalities from wind, fossil-fuel, and nuclear electricity. *Energy Policy* 37(6): 2241-2248.

Sovacool, B. K. 2013. The avian benefits of wind energy: A 2009 update. *Renewable energy* 49(0): 19-24.

Steinke, F., P. Wolfrum, and C. Hoffmann. 2013. Grid vs. storage in a 100% renewable Europe. *Renewable Energy* 50: 826-832.

Succar, S. W., Robert H. 2008. Compressed Air Energy Storage: Theory, Resources, And Applications For Wind Power.

Sullivan, J. L. and L. Gaines. 2012. Status of life cycle inventories for batteries. *Energy Conversion and Management* 58: 134-148.

Sundar, K. S. G. and B. C. Choudhury. 2005. Mortality of Sarus Cranes (*Grus antigone*) due to electricity wires in Uttar Pradesh, India. *Environmental Conservation* 32(3): 260-269.

Sundararagavan, S. and E. Baker. 2012. Evaluating energy storage technologies for wind power integration. *Solar Energy*.

Sørensen, B. 1978. On the fluctuating power generation of large wind energy converters, with and without storage facilities. *Solar Energy* 20(4): 321-331.

Troy, N., E. Denny, and M. O'Malley. 2010. Base-load cycling on a system with significant wind penetration. *IEEE Transactions on Power Systems* 25(2): 1088-1097.

Trötscher, T. and M. Korpås. 2011. A framework to determine optimal offshore grid structures for wind power integration and power exchange. *Wind Energy* 14(8): 977-992.

Ulleberg, Ø., T. Nakken, and A. Eté. 2010. The wind/hydrogen demonstration system at Utsira in Norway: Evaluation of system performance using operational data and updated hydrogen energy system modeling tools. *International Journal of Hydrogen Energy* 35(5): 1841-1852.

Wagner, H. J., C. Baack, T. Eickelkamp, A. Epe, J. Lohmann, and S. Troy. 2011. Life cycle assessment of the offshore wind farm Alpha Ventus. *Energy* 36(5): 2459-2464.

Weinzettel, J., M. Reenaas, C. Solli, and E. G. Hertwich. 2009. Life cycle assessment of a floating offshore wind turbine. *Renewable energy* 34(3): 742-747.

Williams, J. H., A. DeBenedictis, R. Ghanadan, A. Mahone, J. Moore, W. R. Morrow III, S. Price, and M. S. Torn. 2012. The technology path to deep greenhouse gas emissions cuts by 2050: The pivotal role of electricity. *Science* 335(6064): 53-59.

Williams, J. H., A. DeBenedictis, R. Ghanadan, A. Mahone, J. Moore, W. R. Morrow III, S. Price, M. S. Torn. 2011. Supporting Online Material for The Technology Path to Deep Greenhouse Gas Emissions Cuts by 2050: The Pivotal Role of Electricity. Science Express, 24 November 2011, page 41. On DOI: 10.1126/science.1208365. Accessed 5 May 2016: http://science.sciencemag.org/content/sci/suppl/2011/11/23/science.1208365.DC1/Williams.SOM.pdf

Wiser, R., Z. Yang, M. Hand, O. Hohmeyer, D. Infield, P. H. Jensen, V. Nikolaev, M. O'Malley, G. Sinden, and A. Zervos. 2011. Wind Energy. In *IPCC Special Report on Renewable Energy Sources and Climate Change Mitigation*, edited by O. Edenhofer, et al. Cambridge, United Kingdom and New York, NY, USA: Cambridge University Press.

Chapter 10

Comparison of technologies' life cycles

Lead authors: Thomas Gibon, Edgar G. Hertwich

10.1 INTRODUCTION

This Chapter presents comparisons between technologies based on the results of the life cycle modelling and the qualitative evaluation of site-specific impacts conducted in this assessment. Original inventory data were collected from bottom-up studies focused on specific local conditions[1]. The inventories were adapted to other regions through changes in the modelling of the background, representing primarily the energy mix in different world regions, as well as changes in the foreground, such as the amount of solar irradiation received by solar energy systems in different regions. Not all relevant factors could be adjusted. The inventory modelling conducted for this study is not intended to replace bottom-up studies from different regions. A further guide to the interpretation of these results will be given below. A description of the inventory for individual technologies are presented in Chapters 3-8, including descriptions of technologies, assumed operating conditions and other modelling assumptions. Life cycle assessment (LCA) results broken down by the most contributing process are also presented in these chapters. The overall inventory modelling methodology was outlined in Chapter 2.

The goal of this Chapter is therefore to compare electricity production technologies on the widest possible basis, and to highlight the environmental benefits, drawbacks and trade-offs of each technology. For the purpose of comparison, the charts present primarily results as obtained for Europe. There are many dimensions to the results of the inventory modelling, including different impacts, the contribution of various processes and life cycle stages to these impacts, and differences in impacts across regions and for the years 2010, 2030 and 2050. Not all of these results are displayed in this section. The results for different regions and times can be explored in an online tool.[2]

In Chapter 10.2, we qualitatively discuss the site-specific ecological impacts that are inadequately covered by LCA methods. In Chapter 10.3, we compare the individual technologies in terms of selected environmental indicators per unit electricity generated. The indicators are greenhouse gas (GHG) emissions, land use, and endpoint-level indicators based on the life cycle impact assessment method - ReCiPe. These methods aggregate potential impacts of different environmental mechanisms on a common endpoint; human health, species, and mineral resources. In Chapter 10.4, we provide more detailed midpoint results for the individual technologies. In Chapter 10.5, we scale up the deployment of the selected technologies to International Energy Agency (IEA) energy scenarios while accounting for technology improvements and the feedback effects of cleaner electricity mixes on the production of future power plants. We compare IEA Baseline and BLUE Map scenarios in terms of selected midpoint and endpoint indicators. In Chapter 10.6, we discuss uncertainty and limitations of the work undertaken here. Finally, in Chapter 10.7, we draw conclusions on the effect of deploying clean technologies on GHG emissions, as well as non-climate environmental and resource-related concerns.

1 PV: China and USA; CSP: Spain, North Africa and USA; Wind: Europe; Hydropower: Chile; Fossil fuel-based power and CCS: USA;

2 https://public.tableau.com/views/ElectricityTechnologyComparisonsPNAS2014/UnitDashboard

10.2 SITE-SPECIFIC IMPACTS

All land use and infrastructure have the potential to cause local ecological impacts by affecting the suitability of an area to serve as habitat for various species. The degree to which an electricity production technology affects species depends in part on the area required to install and operate a technology; this indicator is covered in our LCA and reported below. It also depends on the ecological value of the area affected and the sensitivity of important species to ecological interference. While life cycle impact assessment methods for the ecological impacts of land use (Koellner et al., 2013; De Baan et al., 2013) and water use (Kounina et al., 2013; Verones et al., 2013) have recently been proposed, these have not yet been applied here; these methods are novel and untested, and do not yet cover all relevant damage pathways for the technologies under investigation. Furthermore, these impacts are naturally site-specific, while we attempt generic, continent-wide assessments in this work. We summarize here important findings from the technology-specific Chapters 3-9, reflecting the literature on ecological impacts.

Fossil fuels: Land disturbance associated with fossil fuel extraction, as well as soil and water pollution associated with mining are important concerns. These impacts are concentrated in regions where the fuels are extracted (Chapter 3.4.1). Impacts typically associated with fossil fuel power, such as emissions from extraction or combustion, are well captured by the LCA.

Hydropower: Dams disrupt the free flow of water, associated flooding and nutrient deposition, and act as barriers to the migration of species. Globally, dams – not all used primarily for hydropower – are seen as an important reason for the significant biodiversity loss in fresh waters. Not all hydropower projects have adverse impacts and some impacts may be positive for some species. Measures to mitigate impacts exist (Chapter 4.2).

Wind power: Collisions of birds and bats with wind turbines are a concern. Some studies sound alarm about the impacts on bats. Regarding birds, the impact on raptors and other large birds is an issue, while the impacts on other bird populations is small compared to other causes of concern (Chapter 5.5).

Solar power: Potential impacts from land use, including collisions, habitat loss and concerns about toxic emissions from disposed photovoltaic (PV) solar cells and concentrating solar power (CSP) mirrors are mentioned in the literature (Chapter 6.6.1). With proper management and site selection, these impacts can be relatively small.

The **water consumption** of fossil fuel-fired and other thermal power plants and hydropower causes ecological concerns especially in dry regions (Chapters 3.4 and 6.2.3).

10.3 IMPACTS PER UNIT ELECTRICITY

10.3.1 ENDPOINT RESULTS AT A GLANCE

Global average endpoint results for different technologies (averaging across both location and specific technologies) are presented in Figure 10.1. Renewable power sources have lower pollution-related human health and ecological impacts per unit of power produced than coal-fired power plants or the current electricity mix, but require more materials. Note that the pollution and land use required to produce the materials is included in the other indicators. Natural gas has low land use but high pollution impacts on ecosystems.

FIGURE 10.1

An overview of the life cycle results of different technology groups compared to the global average mix of electricity production technologies

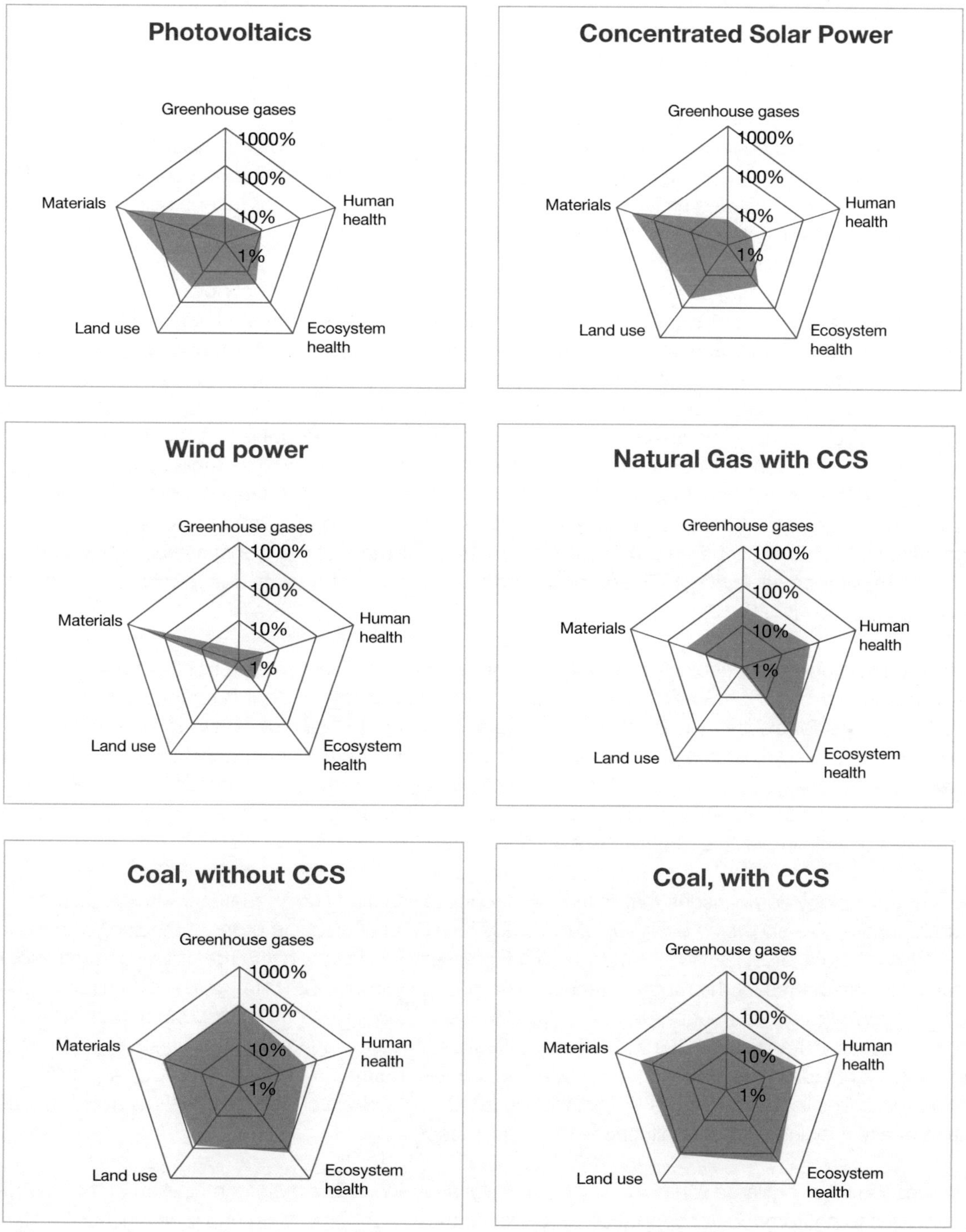

The indicators are the use of the bulk materials (cement, iron and steel, copper, and aluminium), land occupation, GHG emissions, and the endpoint indicators for human health and ecosystems (excluding the contributions of climate change and land use, but including the production of materials).

10.3.2 CLIMATE CHANGE

Figure 10.2 shows the life cycle climate change impact from the production of 1kWh of electricity delivered to the grid from different energy technologies. The modelling reflects conditions in Europe in 2010. The results clearly show that coal power has the highest global warming effect compared to the other technologies, followed by natural gas. The lowest emission technologies use renewable energy supplies, such as many, but not all, hydropower projects, wind power, PV and CSP.

Note that the emissions investigated here are largely from process-based life cycle inventories (LCIs), which suffer cut-off errors because they cannot account for all inputs required by a system. For renewable technologies, the cut-off error can be equally large as the emissions included in the inventory, but is unlikely to be larger than that of the included emissions (Arvesen and Hertwich, 2011). For technologies with large direct emissions, such as fossil fuels, cut-off errors tend to be small.

CO_2 capture and storage (CCS) substantially lowers the emissions from fossil fuel power plants, but not as much as one might expect. A significant portion of the remaining emissions are fugitive emissions from the production, transport and storage of coal and natural gas. However, the full extent of these emissions, especially for natural gas, has not been recognized until recently, and the average emissions are still contested. However, GHG emissions depend strongly on the precise fuel type and the technologies employed (Chapter 3). A peculiar case in the above comparison is hydropower. Hydropower plants are more site-sensitive; life cycle emissions associated with the same technology (dam, turbine) implemented at different sites vary significantly. In addition, biogenic methane and CO_2 emissions of the reservoirs vary widely. Some hydropower reservoirs even display a net uptake of CO_2, while other hydropower reservoirs are associated with methane emissions of over 2 kg CO_2 eq./kWh. A current estimate of the global average methane emissions from hydropower is 50 g CO_2 eq./kWh, with a substantial uncertainty due to measurement and sampling issues (Hertwich, 2013).

Variable renewable energy technologies do not have the same characteristics as fossil fuel-based power. In particular, wind and solar power do not necessarily produce electricity when it is needed. Design and operational changes for electricity grids are required to accommodate these power sources. At low penetration rates, these adjustments are not a problem, depending on the characteristics of the local grid and the other power sources. For a high share of variable renewables, substantial adjustments, grid extension, back-up power, or energy storage are required. The impact of installing and operating these technologies can be on par with or even larger than that of the electricity production itself.

We present technology comparisons that reflect both regional variations and variations within a specific technology. Figure 10.3 shows an overview of the study results, in decreasing order of median value for the life cycle emissions of GHG. No life cycle impact increases significantly over time. For each technology and year, error bars represent the range of results over the nine regions. CSP life cycle impacts range from 17 g CO_2 eq./kWh (parabolic trough, Africa and Middle East in 2050) to 95 g CO_2 eq./ kWh (parabolic trough, Economies in Transition in 2010). PV power life cycle impacts span from as low as 5.3 g CO_2 eq./ kWh (ground-mounted copper indium gallium selenide module, Europe in 2050) to 83 g CO_2 eq./kWh (roof-mounted poly-silicon, Economies in Transition in 2010). For solar technologies, the low direct normal irradiation of a few regions strongly influences life cycle impacts.

Hydropower shows a very large variation in GHG emissions per kWh. The main contribution of the maximum results appears to be from the transportation for the construction, the operation and maintenance of one of the dams modelled here. The rest of the reservoir dams show the lowest results in Latin America, with life cycle GHG emissions as low as 2.6 g CO_2 eq./kWh in 2050.

Despite a portfolio with different gearbox and foundation technologies (conventional or rare earth permanent magnet and gravity-based or steel foundation, respectively), and location (onshore or offshore), wind power

FIGURE 10.2

Comparison of the GHG emissions of different electricity supply technologies, modelled for 1 kWh produced in Europe

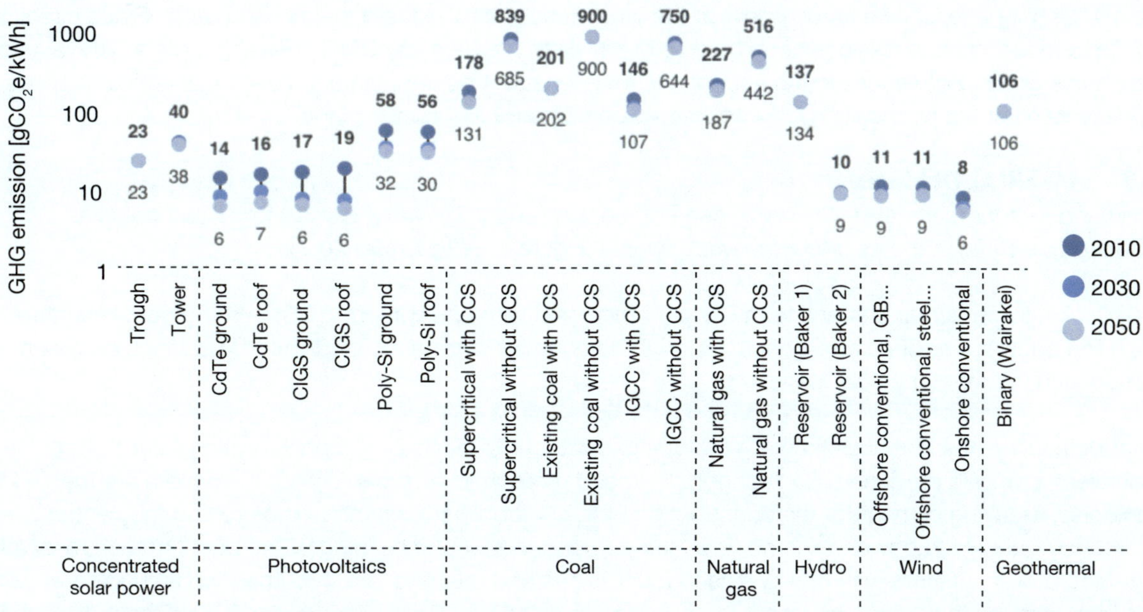

FIGURE 10.3

Distribution of life cycle GHG emissions for renewable technologies, in g CO_2 eq./kWh_{el}. Quartile values and extrema are given.

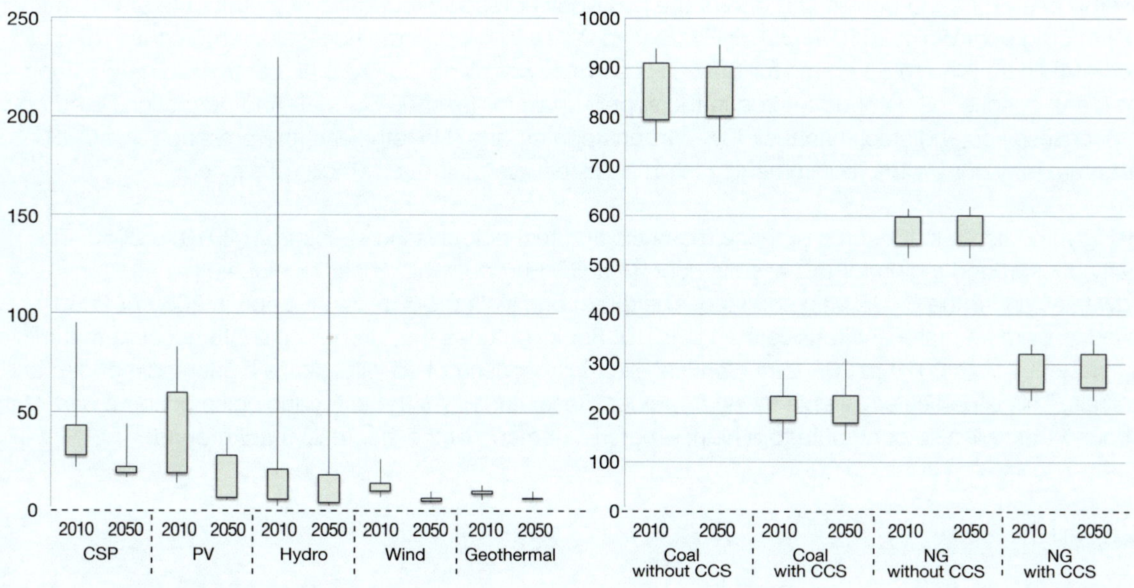

CSP: concentrating solar power; PV: photovoltaic.

results show little variation. The highest emissions are found for an onshore conventional wind farm in China in 2010, whereas the lowest impacts are found for onshore wind farms in all regions in 2050.

Figure 10.3, right pane, shows an overview of the study results for the fossil fuel power plants. Since direct emissions and fuel chain emissions contribute to nearly all of the life cycle GHG emissions, not much variation is seen in this graph. Differences in the economies that construct power plants, such as in the electricity mix used, have little impact on the life cycle GHG emissions of fossil fuel power plants.

10.3.3 MATERIAL DEMAND

The demand for four materials (aluminium, cement, copper, and iron) was analysed for all technologies. Results are presented in the next four following figures (Figure 10.4 to Figure 10.7).

For 2010, aluminium demand spans from 0.1 g (natural gas, with and without CCS) to 5.0 g (roof-mounted thin-film PV) per kWh, as seen on Figure 10.4. High demand in PV is associated with the balance of system (roof-mounting frame) of the solar panels.

Cement demand varies from less than 1 g (natural gas without CCS) to 24 g (concentrating solar power, central tower) per kWh produced. Six technologies need more than 10 g per kWh. Two of them are the CSP technologies, requiring cement for the construction of the parabolic trough infrastructure and the central tower, respectively. Hydropower also shows a high demand, due to the building of the dam for each reservoir. Finally, wind power, offshore with gravity-based foundations and onshore, are the other two technologies with such a high demand in cement, necessary to build the foundations of the wind turbine. Other technologies do not require as much; fossil-based electricity without CCS is especially low in cement demand, in comparison.

Copper demand is associated with cabling and connection to grid, hence, with decentralized electricity production. Figure 10.6 shows that PV and wind power have the highest copper demand in the presented set of technologies, with more than 1 g per kWh produced in 2010. CSP also requires copper, to a lesser extent.

Iron (and steel) is the core material of most technologies' infrastructure. Hence, the demand in iron and steel is the highest of all four materials assessed here. As seen on Figure 10.7, concentrating solar power, ground-mounted PV, wind power, and one of the hydropower reservoirs require high amounts of iron (namely, from 19 to 53 g per kWh in 2010 in OECD Europe). In 2010, all wind farms necessitate more than 40 g of iron per kWh, increasing to 53 grams for the CSP central tower plant. Due to their centralized nature, long lifetimes, and despite the demand for iron in the supply chain for fossil fuels, coal- and natural gas-fired power plants need less iron and steel than their low-carbon counterparts. Nonetheless, the presence of a CCS infrastructure may double the requirements in iron and steel, per unit of electricity produced.

Surprisingly, the demand for some of these materials and technologies increases from 2010 to 2050. The electricity background region for this assessment is OECD Europe, which has the highest penetration of renewables under the BLUE Map scenario. Therefore, one kWh produced in Europe in 2030 or 2050 becomes as or more material-intensive than in 2010. For inventories that do not take into account material or energy efficiency from 2010 to 2050, an identical electricity consumption will entail a higher indirect demand in materials. This is particularly visible for hydropower (Reservoir 1). As these technologies become part of the background they will also contribute to a higher material demand of the electricity background.

FIGURE 10.4

Comparison of the aluminium demand, in g, of different technology sources, for 1 kWh in Europe

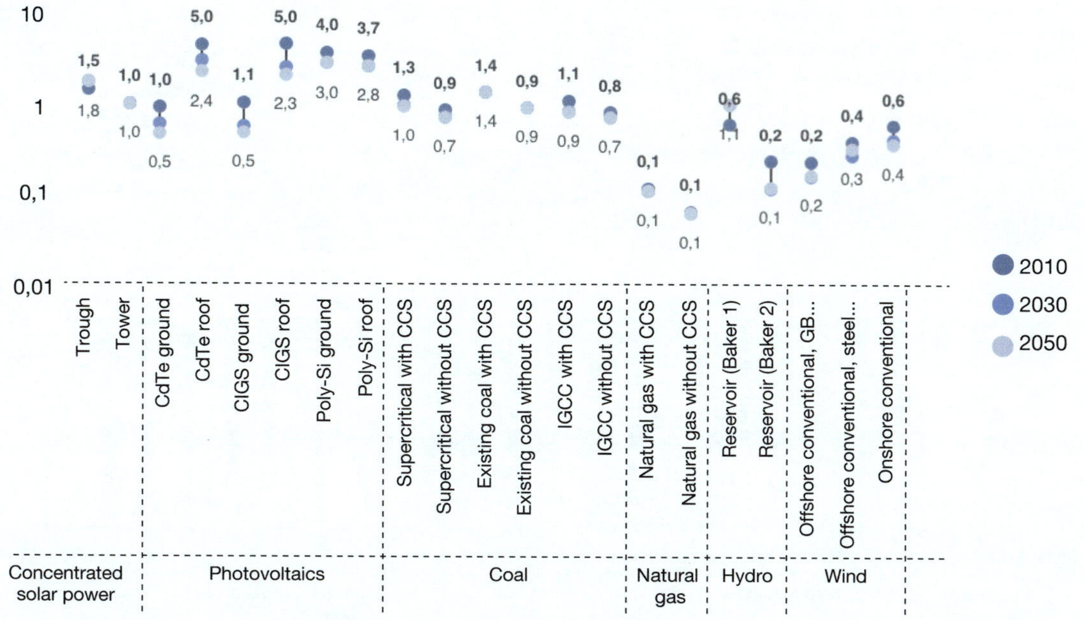

CdTe: cadmium telluride; CIGS: copper indium gallium selenide; Poly-Si: polycrystalline silicon;
IGCC: integrated gasification combined cycle; CCS: carbon dioxide capture and storage; GB: gravity base

FIGURE 10.5

Comparison of the cement demand, in g, of different technology sources, for 1 kWh in Europe

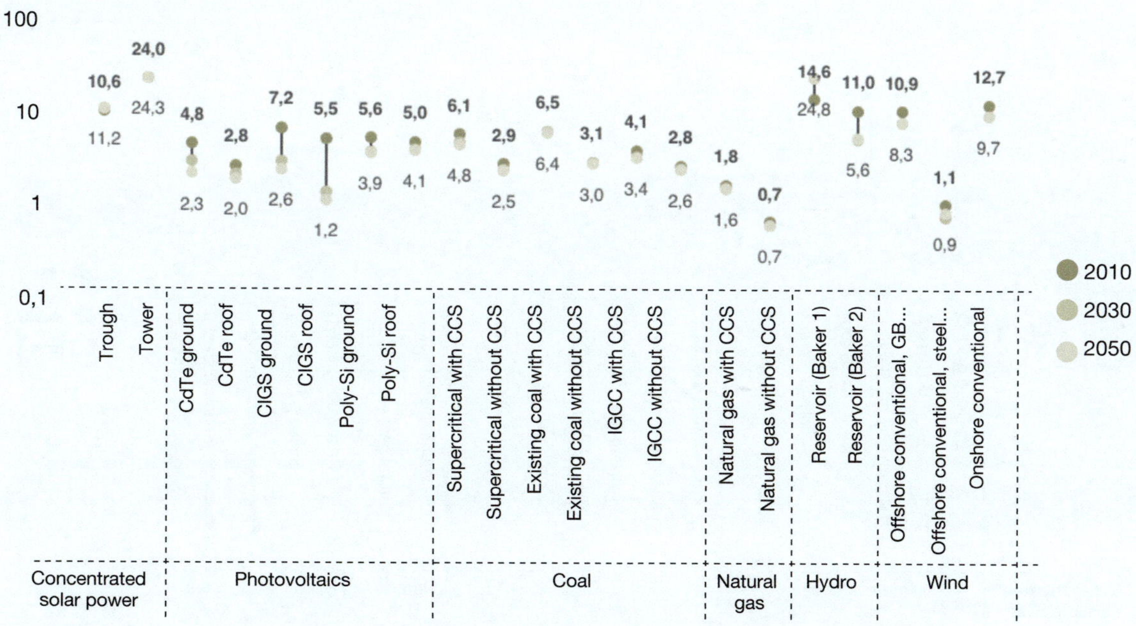

CdTe: cadmium telluride; CIGS: copper indium gallium selenide; Poly-Si: polycrystalline silicon;
IGCC: integrated gasification combined cycle; CCS: carbon dioxide capture and storage; GB: gravity base

FIGURE 10.6

Comparison of the copper demand, in g, of different technology sources, for 1 kWh in Europe

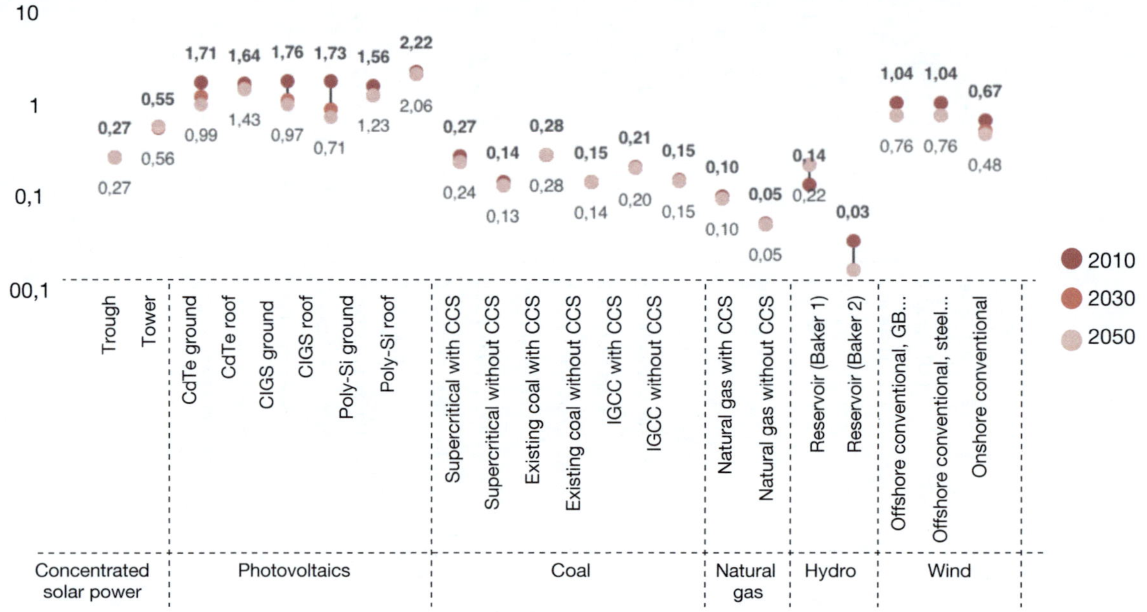

CdTe: cadmium telluride; CIGS: copper indium gallium selenide; Poly-Si: polycrystalline silicon;
IGCC: integrated gasification combined cycle; CCS: carbon dioxide capture and storage; GB: gravity base

FIGURE 10.7

Figure 10.7 – Comparison of the iron demand, in g, of different technology sources, for 1 kWh in Europe

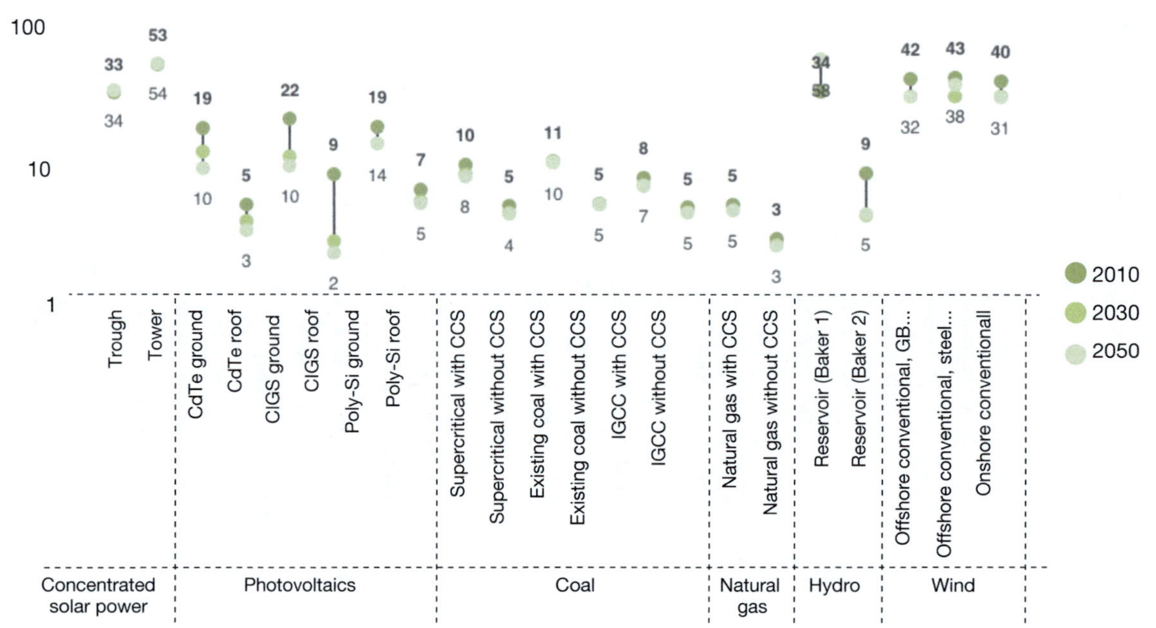

CdTe: cadmium telluride; CIGS: copper indium gallium selenide; Poly-Si: polycrystalline silicon;
IGCC: integrated gasification combined cycle; CCS: carbon dioxide capture and storage; GB: gravity base

10.3.4 METAL DEPLETION

FIGURE 10.8

Comparison of the impact on metal depletion, in g Fe eq., of different technology sources, for 1 kWh in Europe

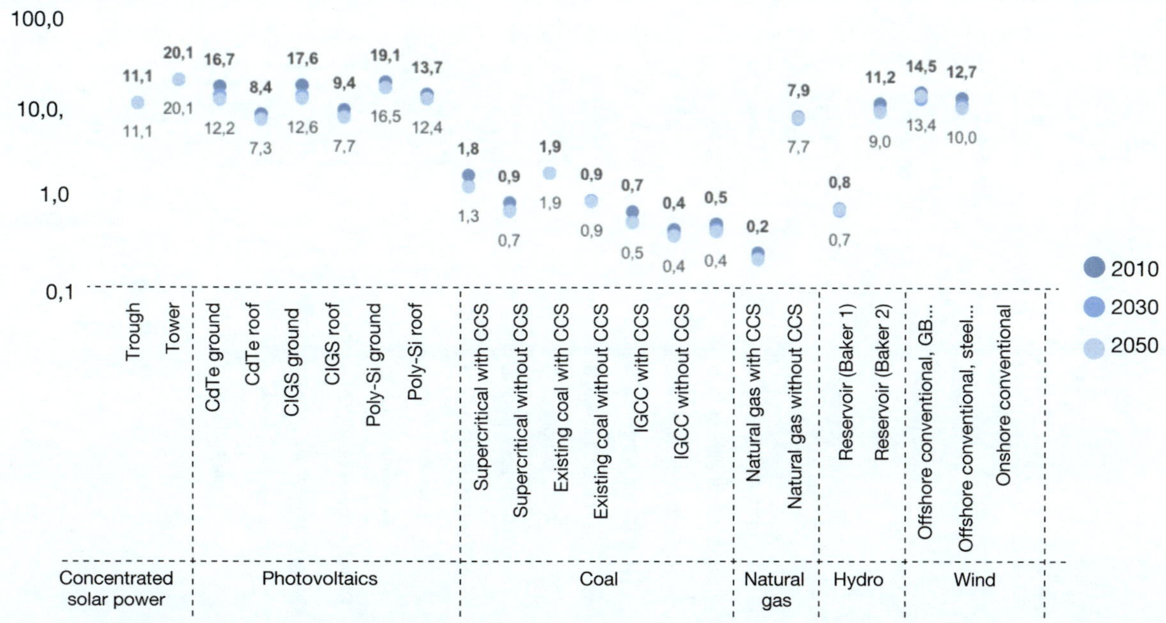

CdTe: cadmium telluride; CIGS: copper indium gallium selenide; Poly-Si: polycrystalline silicon;
IGCC: integrated gasification combined cycle; CCS: carbon dioxide capture and storage; GB: gravity base

Figure 10.8 shows a comparison of the life cycle metal depletion from the production of 1 kWh from different technologies in Europe in 2010. The technologies using the most metal resources are solar PV and wind, whereas fossil fuels and hydropower require substantially less metal per unit of electricity produced. A first reason is the different lifetimes; short lifetimes are correlated with higher life cycle metal use per unit produced. A second reason is the intensive use of copper, tin, manganese concentrate and rare earth metals in solar PV and wind technologies. In the wind category, the use of rare earth permanent magnets increases the impact on metal depletion, as well as the use of steel rather than gravity-based foundations.

10.3.5 LAND OCCUPATION

FIGURE 10.9

Comparison of the impact on land occupation in terms of m² per MWh/a of electricity production from different technology sources, in Europe, in 2010

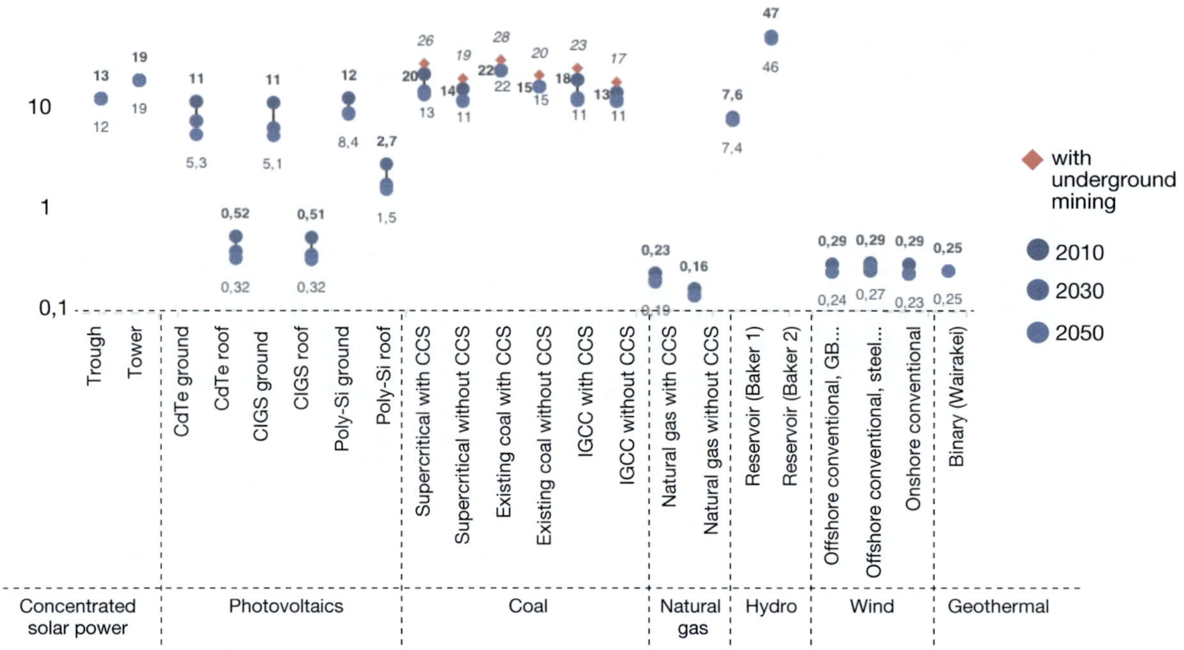

Land occupation is measured in square meter-annum (m²a) per MWh electricity produced, i.e. one square meter occupied over a year per MWh or one square meter for producing one MWh per year. For coal power, the light green bars represent the difference between the land occupation associated with power plants utilizing with underground mining (high) and with surface mining (low).

Figure 10.8 shows a comparison of life cycle results of agricultural and urban land occupation of different power plant technologies. Coal-fired power plants produce electricity with high land use impacts because of the surface occupied by surface mines for coal and the timber used to build the wooden underground infrastructure in underground mines, represented by the additional impact represented by the light green bars in Figure 10.9. Land use impacts for coal power are therefore mostly indirect. By contrast, land use associated with CSP and PV power is mostly direct, i.e. the plants themselves use most of the land occupied over the life cycle. Roof-mounted systems have a low impact on land occupation, as the land occupied by the PV installation is not included; all land occupation is allocated to the service provided by the building on which the system is mounted. The apparent land use of CSP is higher than for PV because a lower fraction of the area is effectively used to intercept sunlight. Although wind turbines and their associated infrastructure occupy only a small land area, the entire wind park occupies approximately 100 m²a/MWh. While the presence of wind parks restricts some forms of land use, this land can still be used for agriculture or left to wildlife. The investigated hydropower plants show a low level of agricultural or urban land occupation, the reservoirs occupy a large area.

10.3.6 ENDPOINT RESULTS FOR HUMAN HEALTH

The human health endpoint impact assessment method used here relies on the characterisation factors of ReCiPe 1.08. Climate change is excluded from the consideration, as it is treated as a separate topic. For fossil fuel-fired power plants, the human health impact from climate change is larger than that from other sources (Singh et al., 2012). The following impact pathways are included in human health impacts: human toxicity, ionising radiation, ozone depletion, particulate matter formation and photochemical oxidant formation. The unit for the impact on human health is disability-adjusted life year (DALY) per kWh of electricity produced by each technology.

Figure 10.10 shows the contribution of different impact categories to the impact on human health. Emissions of particulate matter and substances contributing to particulate matter formation, such as SO_2, NO_x and partially combusted hydrocarbons, contribute substantially to the overall human health impact. Further, we can also see a high health impact from toxic substances; metals such as arsenic, mercury and manganese are particularly important contributors.

FIGURE 10.10

Figure 10.10 – Human health impact of electricity production modelled for Europe in 2010

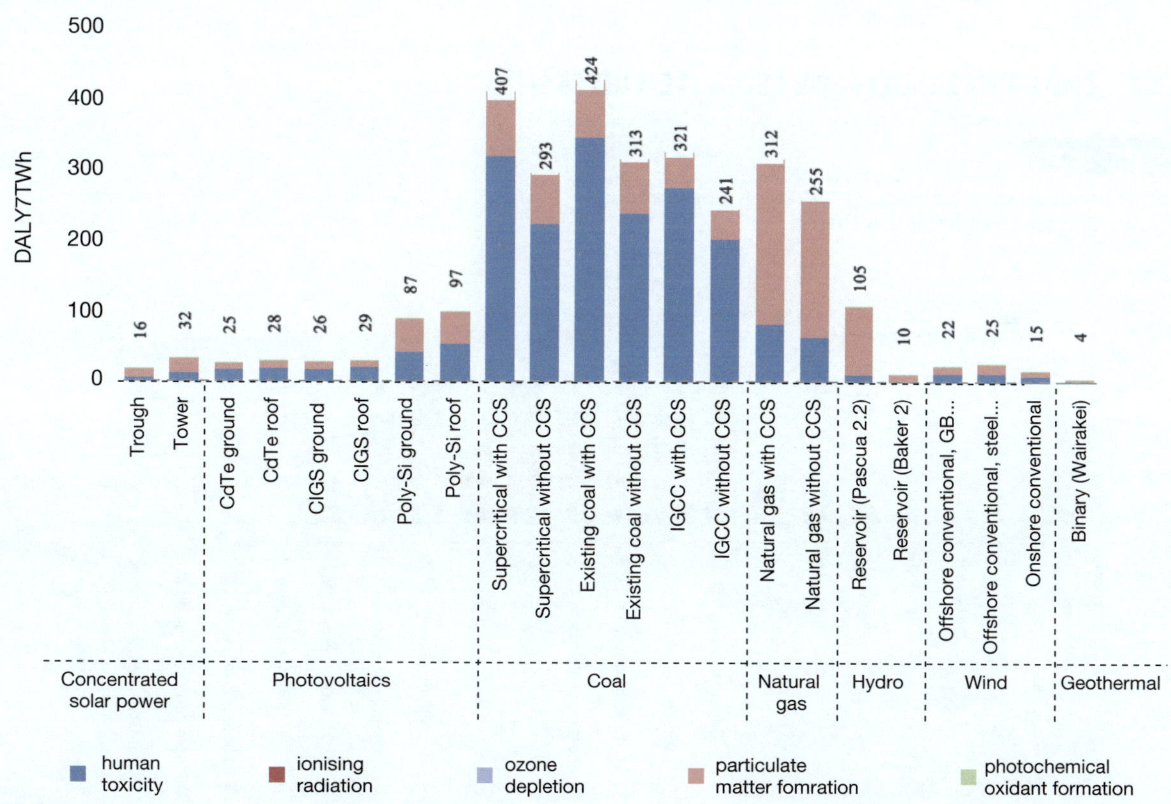

The measure is disability adjusted life years (DALY) per TWh of electricity generated following different damage pathways according to the ReCiPe (H) impact assessment method.
Abbreviations: CdTe – Cadmium telluride, CIGS – Copper indium gallium selenide, Poly-Si – Polycrystalline silicon, CCS – CO_2 capture and storage, IGCC – integrated gasification combined cycle, GB – gravity-based foundation.

One TWh of electricity produced from coal without CCS has an impact of 249 to 668 DALYs, whereas plants equipped with CCS have higher impacts, 320 to 984 DALYs. The human health impact from particulate matter is well documented and the results provided here are in broad agreement with other studies, such as the burden of disease work by the World Health Organisation (WHO) (Lim et al., 2012). By contrast, the human health impacts from toxic compounds are surprising. The impacts from this study align with those of other LCAs using the same impact assessment methodology. The public health literature, however, implies that health impacts from metal pollution are not on the same order of magnitude as the impacts from particulate matter. Recorded health impacts from arsenic, for example, are related to groundwater with naturally high levels of the metal, not to anthropogenic emissions. The life cycle impact assessment method adopted in this study differs from that used by WHO (Mathers et al., 2009). The combination of the long time horizon of emissions from abandoned mines and the assumption of a linear, no-threshold dose-response relationship in the LCA characterization method is the cause of the high human toxicity impact determined in this study. WHO is currently conducting work to better include the health effects associated with exposure to toxic chemicals in its burden of disease work. The high health impact may possibly be spurious, but it could also be that the true impacts are indeed higher than suggested by WHO.

The WHO burden of disease assessment also shows considerable impact from air pollution through photochemical oxidants (Lim et al., 2012). However, a significant portion of this air pollution comes from transportation, poorly regulated stationary combustion, and even fugitive releases and natural sources; power plants contribute comparatively little.

10.3.7 ENDPOINT RESULTS FOR ECOSYSTEM DIVERSITY

FIGURE 10.11

Ecosystem impact of electricity production from various technologies in species-year per TWh electricity produced, background Europe 2010

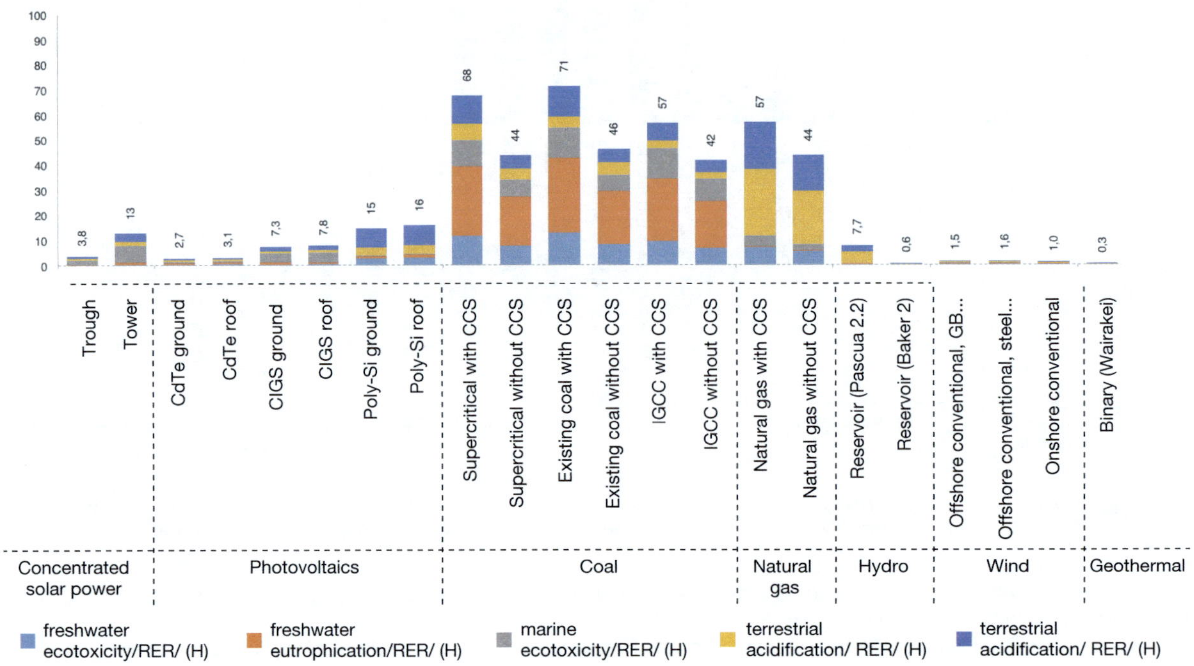

Abbreviations: CdTe: cadmium telluride, CIGS: copper indium gallium selenide, poly-Si: polycrystalline silicon, CCS: carbon dioxide capture and storage, IGCC: integrated gasification combined cycle, GB: gravity-based, RER: Rest of economic region, H: Hierarchical perspective in ReCiPe.

The ecosystems diversity endpoint impact assessment method used here relies on the characterization factors of ReCiPe 1.08. Ecosystem damages from climate change and habitat change or land occupation are not presented. For fossil fuel-fired power plants, the ecosystem diversity impacts would be larger than those from the damage pathways considered here. The following damage pathways are considered in ecosystem diversity impacts: terrestrial acidification, freshwater eutrophication, terrestrial ecotoxicity, freshwater ecotoxicity and marine ecotoxicity. The unit for the impact on ecosystem health is the "loss of species during a year" (species.yr), per kWh of electricity produced by each technology.

Figure 10.11 indicates that all impact categories contribute appreciably to the overall ecosystem impact, unlike in the situation for human health. The largest impacts are associated with the fossil fuel-based technologies. Fossil fuel-based power plants with CCS have higher impacts than those without CCS because the former require more fuel and more infrastructure per unit electricity produced (Chapter 3 and Singh et al., 2011, 2012). Other climate mitigation technologies, however, lead to a substantial reduction in all damage pathways compared to fossil fuel-based power. Among the renewable energy technologies, the impact of PV is highest, followed by CSP, wind, and hydropower.

Different technologies exist for the harnessing various energy sources such as coal, wind or PV; each of these technologies exhibit in environmental and health impacts. These differences indicate that there may be an opportunity to reduce impacts by choosing the right technology. The differences, however, may well be within the margin of error of the present assessment.

10.3.8 COMPARISON WITH LITERATURE

The Special Report on Renewable Energy (SRREN) (IPCC, 2011) and Fifth Assessment Report (AR5) (IPCC, 2014) of the Intergovernmental Panel on Climate Change (IPCC) give ranges of results for similar technology categories.

The sample of technologies is quantitatively and qualitatively different in each source. Results for CSP technologies in this study are slightly higher than the set of results gathered in previous literature. However, the results for the broad range of PV technologies in this study seem to correspond with the IPCC literature reviews. A much wider range of hydropower plants has been reviewed in the AR5, with GHG emission values surpassing 2000 g CO_2 eq./kWh. In contrast, the present study only picked two plants of 360 and 660 MW, hence the narrow range of results in Figure 10.12. The results for wind power are comparable to literature values, with a broader range for the references with most sources, but a median between 11 g and 15 g CO_2 eq./kWh. Nuclear and geothermal energy differ because of the lack of available data in the present study, results are here for the sake of comparison with other technologies, but this study did not build detailed inventories for these technologies.

The comparison of fossil fuel results shows more discrepancies (right-hand panel on Figure 10.12). The emission gap between non-CCS plants and their CCS-equipped counterparts is lower in this present study than in the reported literature of the IPCC reports. Furthermore, the emissions from non-CCS coal-fired plants are found to be 20 per cent lower in the present study. The non-CCS gas results are in line with the IPCC data, while the CCS-equipped gas plants show higher GHG emissions in the present study.

FIGURE 10.12

Comparison of life cycle GHG emissions, between this study's results (Hertwich et al., 2015) and the IPCC AR5 (Bruckner et al., 2014) and SRREN (Moomaw et al., 2011; Sathaye et al., 2011) results

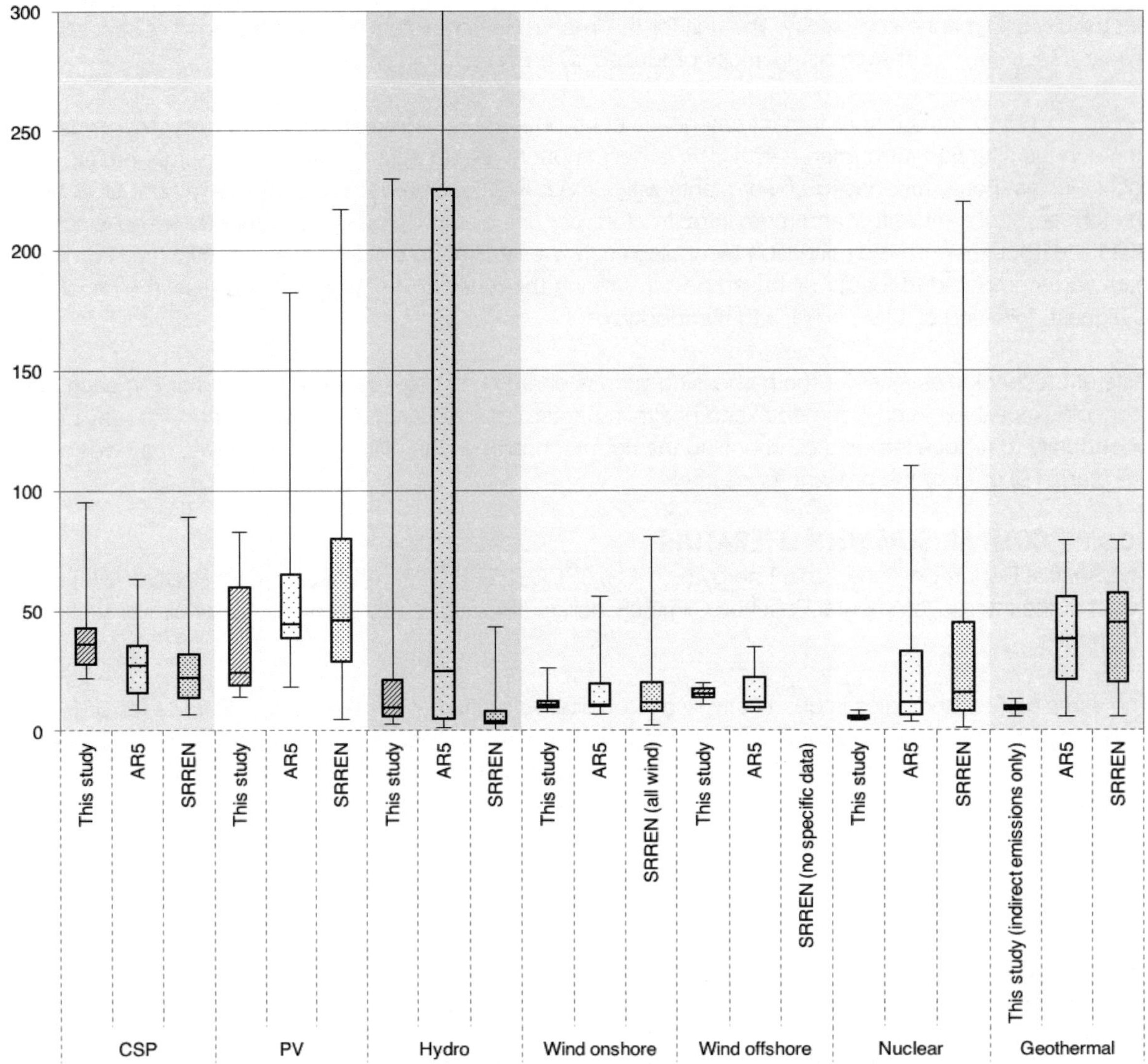

Low carbon electricity production technologies are shown in this left panel

Figure 10.12, continued

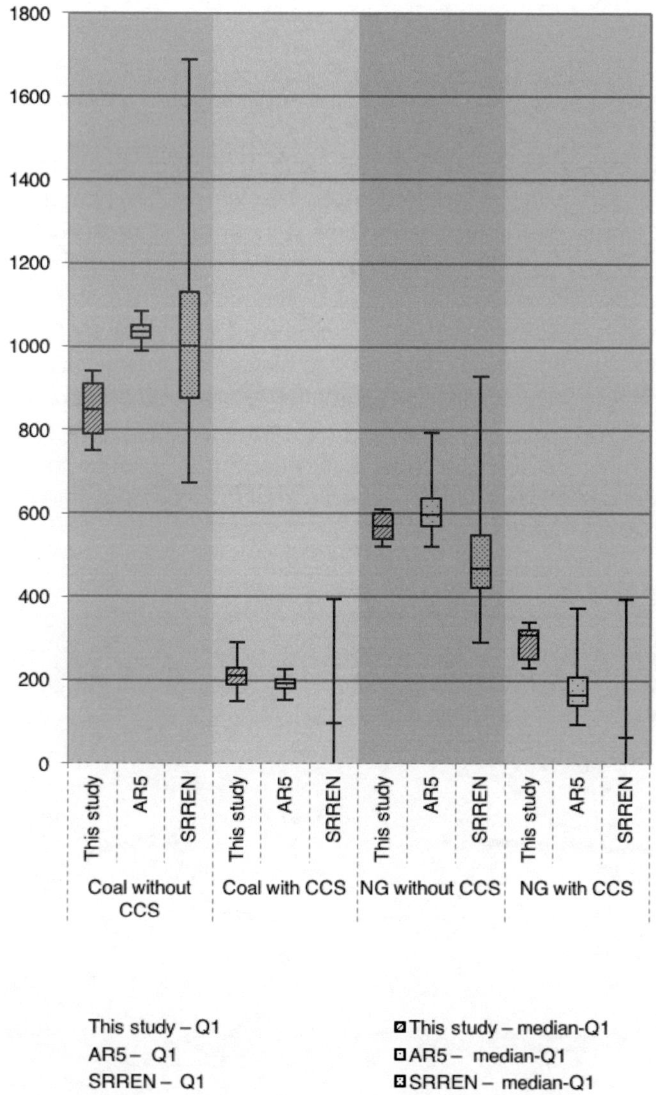

Fossil fuel electricity production shown in this panel

10.4 TECHNOLOGY RESULTS

This section presents technology results, compared to the environmental profile of the global electricity mix in 2010.

10.4.1 FOSSIL FUELS

The LCAs show a clear trade-off between climate change mitigation and other environmental impacts. Existing coal-fired power plants generally have higher impacts than supercritical and integrated gasification plants and much higher emissions than natural gas combined cycle plants (Figure 10.13). The GHG emissions of these modern power plants with CSS are on the order of 22-26 per cent of existing coal-fired power plants. Modern plants with CCS have lower particulate matter and photochemical oxidant formation impacts than current coal fired plants, but higher emissions than modern plants without CCS; these impact categories constitute the most important threats to human health. However, modern plants with CCS increase freshwater ecotoxicity and eutrophication compared to current plants without CCS. The comparison of modern plants with and without CCS indicates that CCS increases almost all impact categories by 20-60 per cent compared to the non-CCS alternatives. Natural gas combined cycle plants have higher NO_x emissions than coal-fired plants, resulting in higher terrestrial acidification potential. NO_x emissions also contribute to marine eutrophication, which is not shown here. As Figure 10.13 shows, the most important contributors to environmental impacts are the operations of the power plant itself (for climate change, human toxicity, particulate matter formation and water use) and the extraction and refining of the fossil fuel (for land occupation, eutrophication, and freshwater ecotoxicity).

FIGURE 10.13

LCA results for fossil fuel-fired systems modelled as if implemented in China in 2010 and normalized to the existing global power mi.

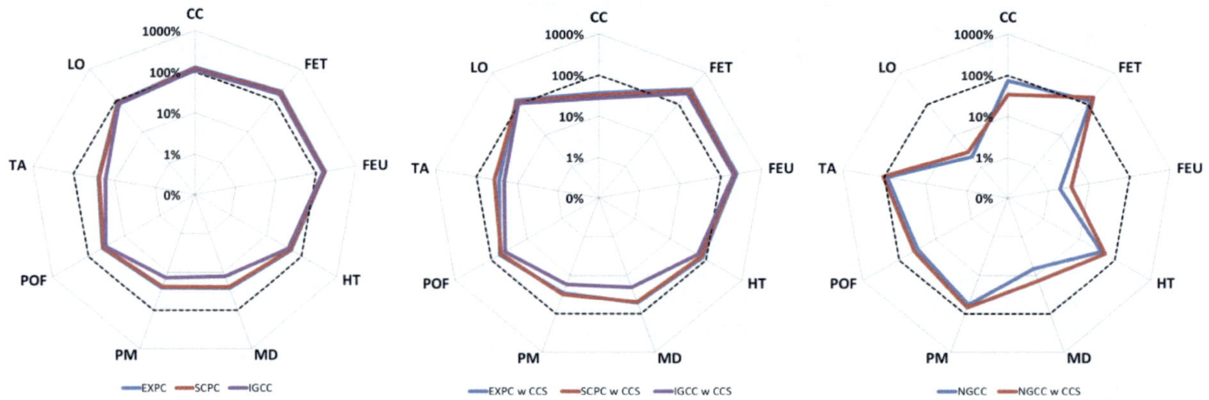

Abbreviations for the impact indicators: CC: climate change; FET: freshwater ecotoxicity; FEU: freshwater eutrophication; HT: human toxicity; MD: metal depletion; PM: particulate matter formation; POF: photochemical oxidant formation; TA: terrestrial acidification; LO: land occupation. The technologies included are EXPC: existing pulverized coal; SCPC: supercritical pulverized coal; IGCC: integrated coal gasification combined cycle; NGCC: natural gas combined cycle; CCS: CO_2 capture and storage.

10.4.2 HYDROPOWER

The material and energy required to build hydropower plants is very site-specific. Both reservoir volume and head of a hydropower plant can vary by many orders of magnitude. The LCI used in this study are based on two reservoir hydropower plants in Chile that have a lower land use, and therefore fewer GHG emissions than the global average. One of these plants, however, is located at a site so remote that the transport of construction materials to the plant site contributes substantially to the climate change impact. Figure 10.14 illustrates how different the impact of two similar plants from the same water shed can be.

10.4.3 WIND POWER

Wind power scores one to two orders of magnitude better than the reference for all the assessed impact categories except metal depletion (Figure 10.15). It should be noted that the land use indicator results include area occupied by individual turbine and power transmission elements of wind farms but does not account for inter-element spacing. If the total wind farm area was considered, land use would be about two orders of magnitude higher.

10.4.4 CONCENTRATING SOLAR POWER

With two exceptions, indicator results shown in Figure 10.16 indicate that CSP has a far superior performance compared to the reference electricity mix. The two exceptions are metal depletion burden, for which CSP has higher impact than the reference electricity mix and land use, for which CSP and reference values are comparable. The area occupied by CSP plants typically cannot be combined with larger wildlife or other human uses, but CSP plants may provide valuable habitat for smaller animals and various plants, and may be used for grazing.

FIGURE 10.14

LCA results for two different hydropower plants implemented in Latin America, normalized to the global average electricity mix

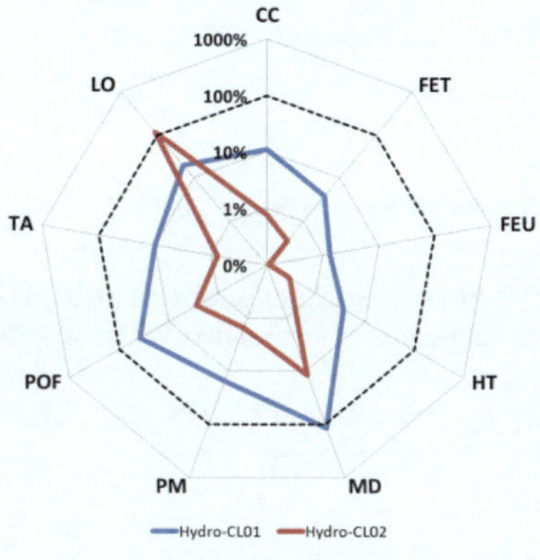

Abbreviations for the impact indicators: CC: climate change; FET: freshwater ecotoxicity; FEU: freshwater eutrophication; HT: human toxicity; MD: metal depletion; PM: particulate matter formation; POF: photochemical oxidant formation; TA: terrestrial acidification; LO: land occupation.

FIGURE 10.15

LCA results for OECD Europe onshore and offshore wind power systems normalized to global electricity mix

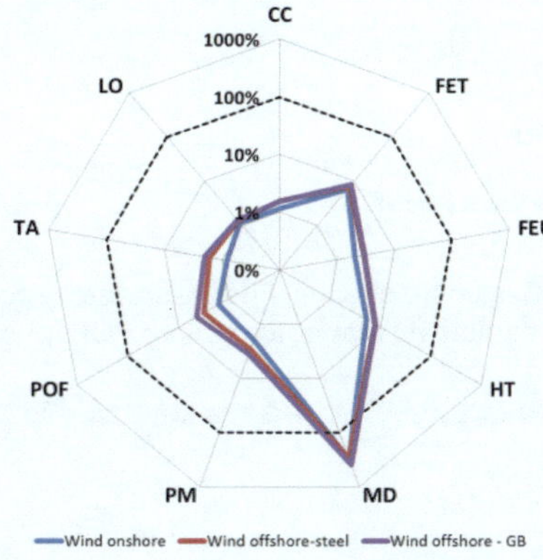

Abbreviations for the impact indicators: CC: climate change; FET: freshwater ecotoxicity; FEU: freshwater eutrophication; HT: human toxicity; MD: metal depletion; PM: particulate matter formation; POF: photochemical oxidant formation; TA: terrestrial acidification; LO: land occupation.

FIGURE 10.16

LCA results for Africa and Middle East CSP trough and tower systems normalized to global electricity mix

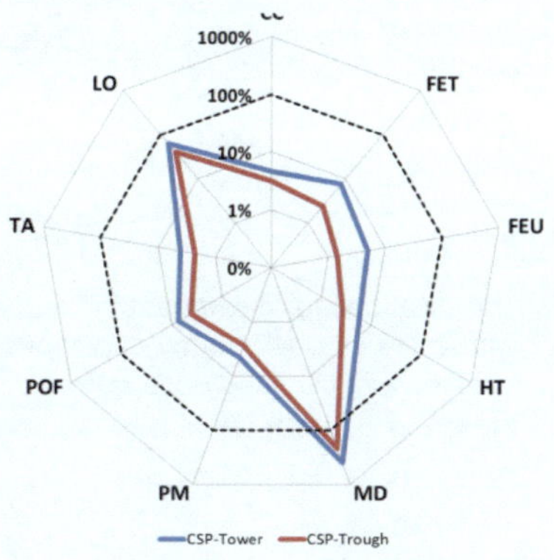

Abbreviations for the impact indicators: CC: climate change; FET: freshwater ecotoxicity; FEU: freshwater eutrophication; HT: human toxicity; MD: metal depletion; PM: particulate matter formation; POF: photochemical oxidant formation; TA: terrestrial acidification; LO: land occupation.

10.4.5 PHOTOVOLTAICS

LCA of electricity from PV technologies shows clear environmental benefits in terms of climate change, particulates, ecotoxicity, human health and eutrophication relative to fossil fuel technologies. However, PV electricity requires a greater amount of metals, particularly copper. Additionally, roof-mounted PV demands aluminium for support frame construction. Polycrystalline silicon, cadmium telluride and copper indium gallium selenide ground-mounted and roof-mounted systems demonstrate environmental and resource impacts of similar magnitude, despite their differing technological compositions. By 2030 and 2050, all three PV technologies will show drastic improvements in impacts and metal consumption due to expected increases in material efficiency in module production, increased module conversion efficiency and changes in the electricity grid. Figure 10.17 shows the environmental impacts in 2010 for PV technologies relative to the global mix.

FIGURE 10.17

Life cycle impacts for PV technologies in 2010 implemented in the OECD North America region, and normalized by the emissions associated with the present global power mix.

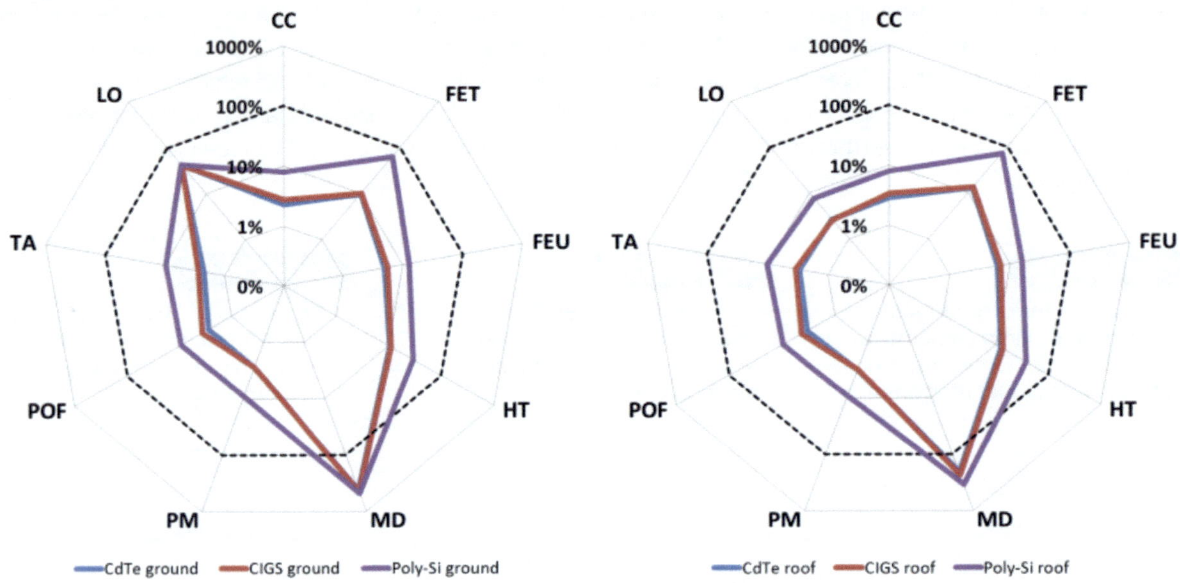

Abbreviations for the impact indicators: CC: climate change; FET: freshwater ecotoxicity; FEU: freshwater eutrophication; HT: human toxicity; MD: metal depletion; PM: particulate matter formation; POF: photochemical oxidant formation; TA: terrestrial acidification; LO: land occupation.

10.5 SCENARIOS

10.5.1 BASELINE

FIGURE 10.18

Global emissions, land use, non-renewable energy demand, material requirements, capacity increase and production broken down by technology under the Baseline scenario assumptions.

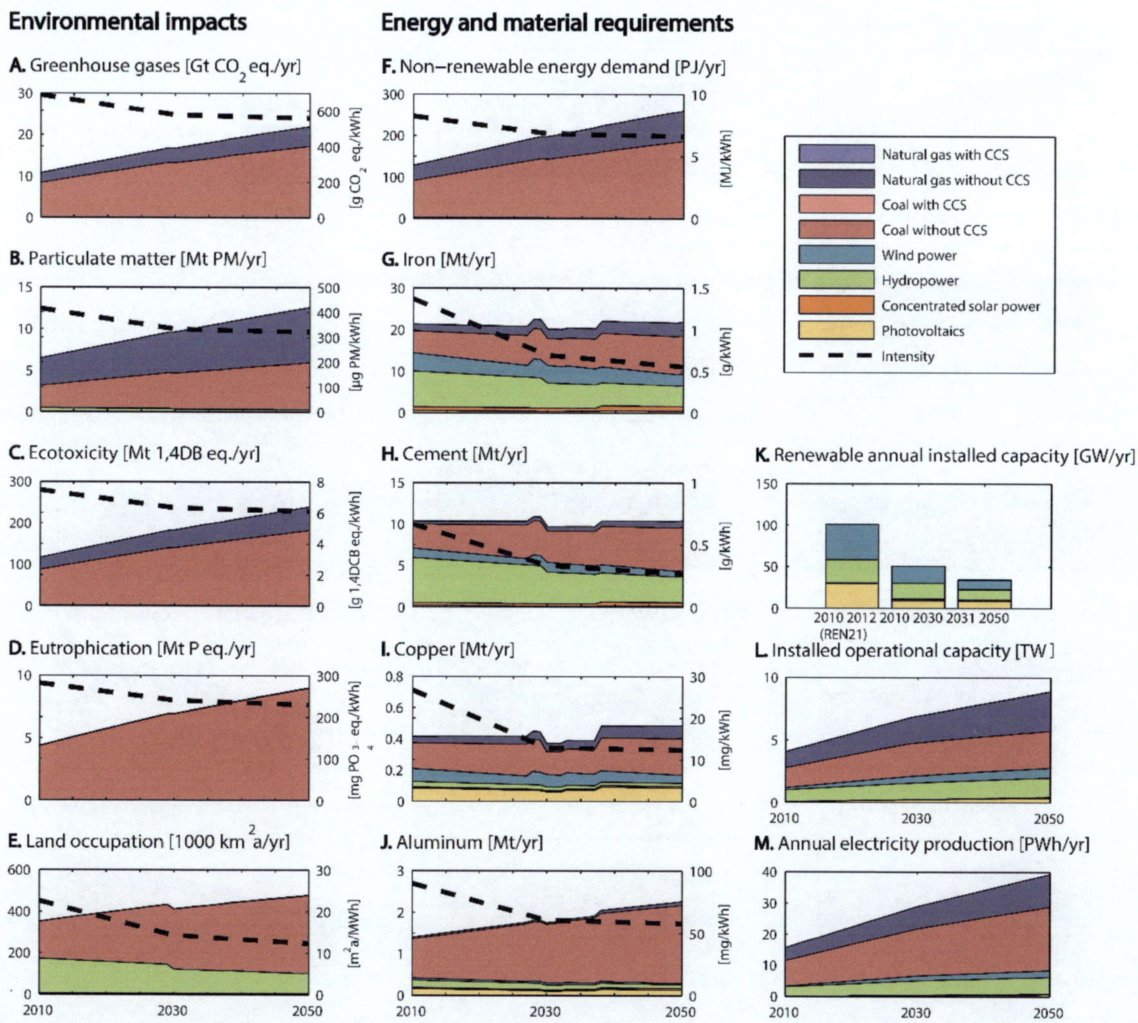

Source: Hertwich et al., 2014

Figure 10.18 shows a projection of various environmental impacts and material requirements at a global scale, broken down by technology group, under IEA Baseline (business-as-usual) assumptions. All impacts are shown to undergo an increase over the 40-year period. By 2050, GHG emissions, particulate matter emissions, freshwater ecotoxicity and eutrophication, land use and non-renewable energy demand are foreseen to grow to slightly less than double their initial value in 2010. Fossil fuels, namely coal and natural gas, are practically the sole contributors to these impacts with the exception of land use to which hydropower contributes a significant share. Material demand is expected to remain stable over the projection period with the exception of aluminium which will increase roughly by half. All emissions and resource use intensities decrease over the projection period.

10.5.2 BLUE MAP

FIGURE 10.19

Global emissions, land use, non-renewable energy demand, material requirements, capacity increase and production broken down by technology under the BLUE Map scenario assumptions

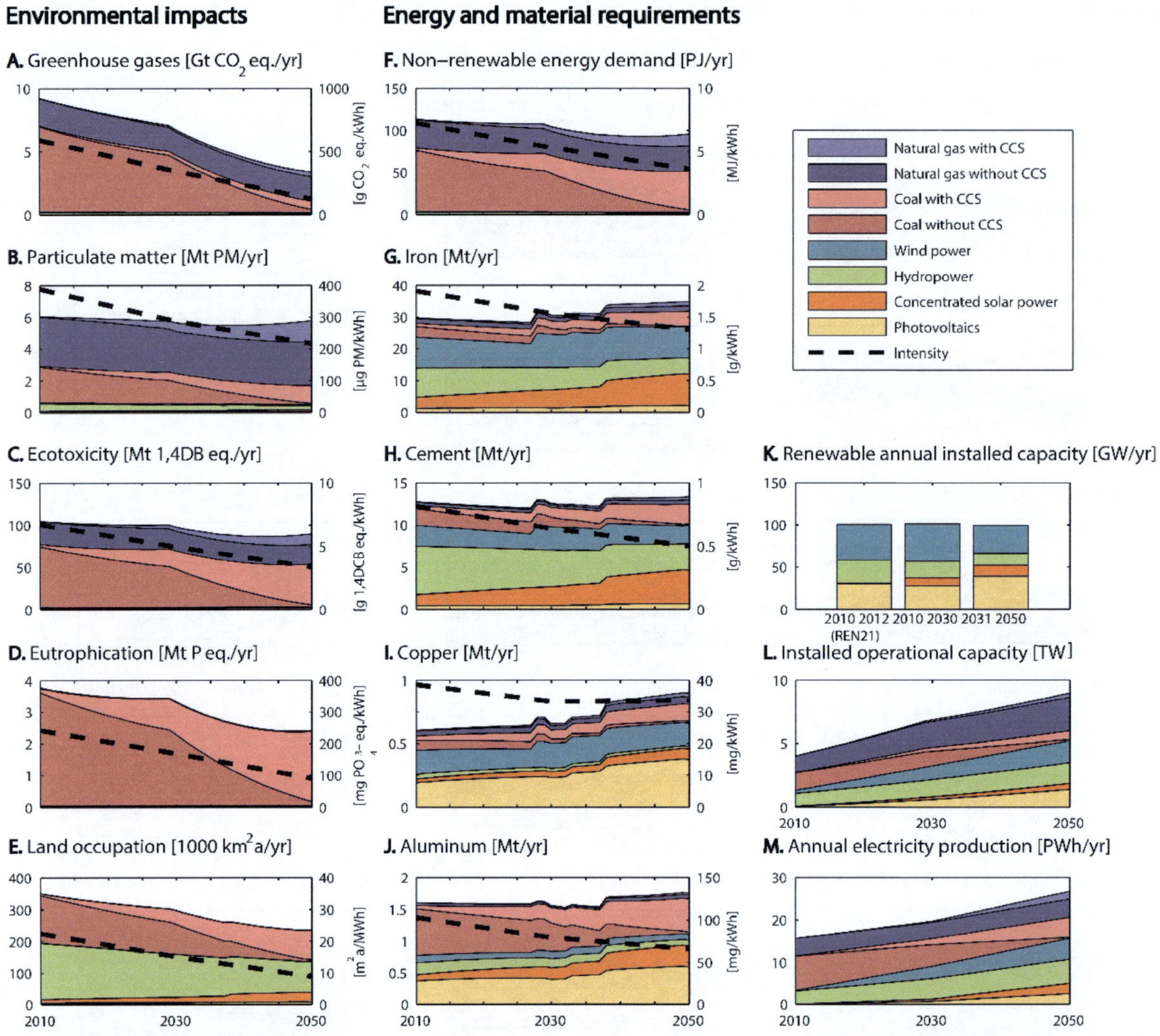

Source: Hertwich et al., 2014

As a consequence of reduced coal use, emissions associated with coal power would decrease. The increasing share of low-pollution renewable electricity supply would reduce the pollution impacts per unit of electricity generation by a factor of two or more (Figure 10.19, left-hand panels). In the face of continued growth in electricity supply, these improvements would allow us to stabilize emissions leading to particulate matter formation and freshwater ecotoxicity while reducing the emissions associated with climate change and freshwater eutrophication. This downward trend is in contrast to the Baseline scenario where the increased use of coal and gas would lead to an increase in all pollution-related environmental impacts (Figure 10.18). The mitigation scenario would lead to a reduction of the use of non-renewable energy resources and, surprisingly, land use. Interestingly, the rate of installment of new renewable power capacity in recent years is at a level that, if sustained, is consistent with the BLUE Map scenario (Figure 14, right). CSP and wind power plants would cause additional demand for cement and iron, while PV would lead to additional requirements of copper.

10.5.3 A COMPARISON OF BASELINE AND BLUE MAP SCENARIOS

FIGURE 10.20

Net difference of global emissions, land use, non-renewable energy demand, material requirements, capacity increase and production broken down by technology between the BLUE Map and the Baseline scenarios

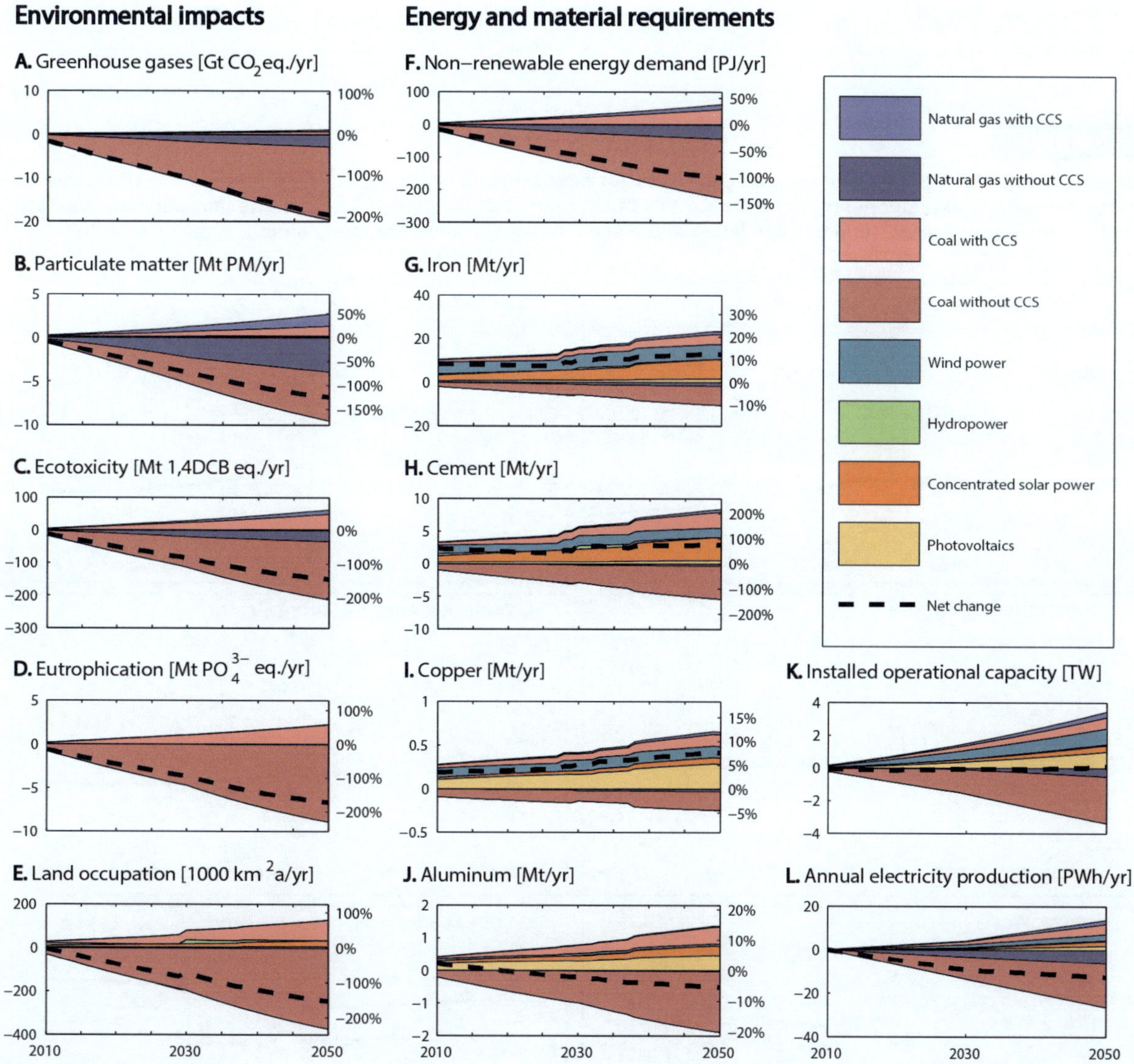

Source: Hertwich et al., 2015

The effect of pursuing a mitigation strategy instead of a business-as-usual strategy becomes apparent when comparing the scenario results for life cycle environmental impacts and material demand. Figure 10.20 displays the difference between impacts of the BLUE Map scenario and the Baseline scenario. A negative result indicates that the impacts from the BLUE Map scenario will be lower than those of the Baseline scenario. Figure 10.21 shows the results for the two scenarios and indicates the difference for every year with vertical lines. The widespread deployment of low carbon technologies does imply an increased investment in infrastructure leading to an increased demand for iron and steel, cement and copper, but potentially a small reduction in the demand for aluminium (Figure 10.20, middle panel). Compared to the Baseline scenario, the main consequences of the climate change-mitigating BLUE Map scenario are a clear reduction of all selected environmental impacts and a slight increase in material requirements, except for aluminium.

Since the implementation of mitigation scenario requires a continued, high rate of installation of renewable power plants and the higher materials of renewable power is associated with the manufacturing and construction of these power plants, the material demand and related environmental impacts occur before the power is produced. In contrast, most of the environmental impacts associated with fossil fuels occur during the use of the technology. Note that the development of low carbon technologies over the 2010-2012 period is consistent with the BLUE Map scenario in terms of total installed capacity, but differs slightly in shares for each technology type.

FIGURE 10.21

Scenario results for GHG emissions, particulate matter emissions, ecotoxicity, eutrophication, non-renewable energy demand, and material requirements for the BLUE Map and Baseline scenarios. The vertical lines indicate the difference between BLUE Map (blue lines) and Baseline (orange lines) for every year.

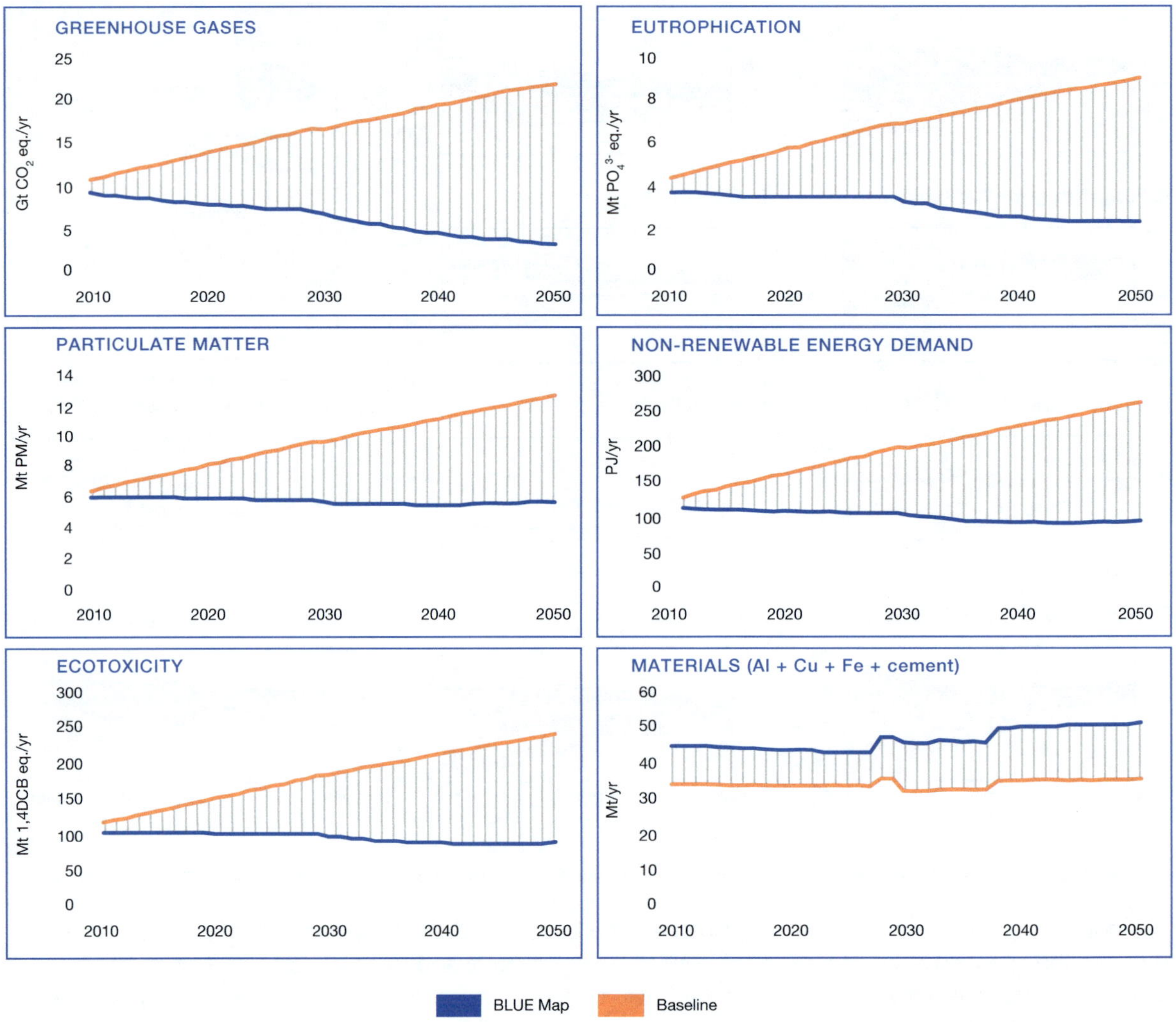

10.6 LIMITATIONS AND UNCERTAINTIES

10.6.1 REPRESENTATIVENESS

The LCI data here are a generalization of environmental pressures derived from either specific case studies (as in the case of hydropower) or generic inventories (as in the case of solar energy). This study adapts the regional and temporal scope of the databases to the greatest extent possible. The issue of representativeness is a major concern. It has been assumed that a few technologies are representative of all systems that are built over the 2010-2050 period in nine regions. This assumption is restrictive for three main reasons:

1. Technical variability: it is not feasible to model all existing technologies in 2010; a selection and averaging of available data had to be performed. It is even harder to foresee the future development and deployment of technologies.

2. Unpredictability: scenario assessments are mere models derived from a set of assumptions that obviously cannot guarantee a true representation of future technologies.

3. Limited resolution: the maximum resolution available is at the regional level. Variations occurring at finer resolutions have to be aggregated and averaged.

For all technologies, some design and performance parameters are dependent on local conditions. In addition, environmental impacts can be site-specific and therefore vary widely from location to location. Some sites have high biodiversity or provide habitat to species that are more vulnerable to a specific technology than others. The region-specific assessments presented in this study are thus not meant to be representative for all conditions. Rather, this assessment intends to address typical conditions to help orient macro-level policy. Choosing the right sites and avoiding locations with high biological diversity or low energy yield can be important for avoiding high impacts. A further understanding of the importance of site-dependent effects can be obtained through more systematic case studies.

10.6.2 IMPACT ASSESSMENT METHODS

This study adopts the midpoint and endpoint characterization factors from the ReCiPe 1.08 impact assessment method. This method is widely applied, particularly with the ecoinvent LCA data used as background data for our study. There are, however, limitations in the scope of the method, in the representation of some relevant impact mechanisms, and in the degree to which ecoinvent or our specific inventories specify environmental interventions that match the requirements of impact assessment methods. The quantitative environmental impact evaluation carried out in this study is inherently bound to the list of characterised elementary flows, consisting of emissions and resource use, and the impact categories covered by ReCiPe. An evaluation of these methods is provided in the ILCD Handbook (JRC, 2011). Typically, social impacts or specific impacts on biodiversity such as habitat degradation from wind power or hydropower cannot be quantified within the scope of this study. In particular, the occupation of water bodies is not characterised in the same fashion as urban or agricultural land occupation, since it does not affect the actual use of the land in the same way. Impacts on marine ecosystems are poorly represented. Other impact assessment methods would necessarily provide different conclusions, as they may contain a different coverage of emission flows (stressors) or a different set of characterization factors. Furthermore, there is a limitation in the specification of inventories and to what degree the emitted substances match those included in the impact assessment (Pettersen and Hertwich, 2008). This assessment can be improved with updated or new impact assessment methods as they are released. Impact assessment must also account for regional specificity, particularly in impact categories with local effects, such as human and ecotoxicity.

10.6.3 LIFE CYCLE INVENTORY MODELLING

The present assessment exercise relies upon many sources of data and information. Data are especially heterogeneous; each data provider gathered inventory information from different sources, while the model in itself is a compilation of various databases that have been adapted according to other sources. We provided instructions and common data collection sheets to different author groups for each of the technologies. Although these measures improved consistency within the data collected, the data nevertheless did not always have the same scope or resolution. Most of the data used in technology inventories are based on existing peer-reviewed literature, although in several instances we sought to fill analytical gaps through our own research efforts.

Like all LCAs, our inventories also suffer cut-off errors as a result of the omission of some inventory processes. With the hybrid approach applied here, some inputs not available in physical terms have been specified as purchases from the input-output table. We have not, however, conducted a complete hybridization of the inventory as suggested by Strømman and Solli (2008). It is difficult to assess the size of the resulting cut-off errors without conducting such a complete hybridization. For fossil-based technologies and site-specific impacts, the error is likely to be smaller, but for unevenly distributed impacts such as toxicity, it may be larger. The technologies investigated have varying levels of physical and input-output data available. Input-output data contain information that is necessarily omitted from LCI, such as services or overhead costs. As a result, uneven contributions from each data type cause discrepancies in system boundaries. These differences in system boundaries thereby make the systems difficult to compare. Data provided by our various partners did not compare in terms of scope, but efforts have been made to assemble comprehensive inventories that include connection to grid and decommissioning.

10.7 CONCLUSIONS

The present assessment is the first of its kind, where the environmental impacts of different electricity supply technologies are systematically reviewed, assessed and compared, applying an LCA modelling that is based on the same background inventory, is regionally adapted to nine different world regions, and projected to future years, namely 2030 and 2050. The assessment is supplemented with a review of site-specific ecological impacts. While there are a number of limitations resulting in uncertainties and unanswered questions, the overall picture is quite clear (Table 10.1).

TABLE 10.1

Summary of the impacts of low carbon technologies for electricity generation on climate, human health, ecosystems and resources, comparing state-of-the-art power plants at well-suited locations. The reference is the current global mix, which has high impacts compared to the levels indicated in this table.

	Climate	Human health	Ecosystem health	Resources
Wind	Low GHG (++)	Reduced particulate exposure (++) Potentially reduced human toxicity (--)	Bird and bat collisions (+=) Low ecotoxicity and eutrophication (=-)	High metal consumption (+=) Low water use and direct land use (==)
PV	Low GHG (==)	Low PM (+=) Low HT (=-)	Low eutrophication and ecotoxicity (+-)	High metal use (+=) High direct land use for ground-based systems (++)
CSP	Low GHG (==)	Low PM (=-) Low HT (=-)	Concern about heat transfer fluid (+=) Low eutrophication and ecotoxicity (+-)	High water use (++) High land use (++)
Hydropower	Low fossil GHG (++) High biogenic GHG from some dams (==)	Low air pollution impacts (=-)	Riparian habitat change (reservoir and downstream) (++)	Water use due to evaporation (+-) Land use for reservoirs (+=)
Geothermal power	Low fossil GHG (+-) Geogenic GHG for some types (=+)	Air and water pollution from geofluid flow in some sites (=-)	Aquatic habitat change/ pollution (+=)	Cooling water use (+=)
Gas CCS Coal CCS	Low GHG (++); substantial fugitive methane emissions (==) Concern about CO_2 leakage (-=)	Solvent-related emissions (==), high PM (==), high HT (++)	High eutrophication (++) and ecotoxicity (+=)	Increased fossil fuel consumption (++); limited CO_2 storage volume (++)

Key to the assessment (##). First symbol: + high agreement among studies, = moderate agreement, - low agreement Second symbol: + robust evidence (many studies), = medium evidence, - limited evidence

10.7.1 CLIMATE CHANGE MITIGATION

Climate change mitigation does not have a unique technological solution. Implementing a portfolio of varied low carbon stationary electricity production technologies has the potential to play a major role in decreasing greenhouse gas emissions from 2010 to 2050. Such a benefit is of course only possible if governments choose to initiate the implementation of these technologies. However, this rollout has alternative consequences, namely inducing other environmental impacts. These so-called "co-impacts" (IEA, 2010) have been identified in this due diligence study, and most were also quantified using impact assessment methods.

This study shows clear environmental advantages for certain low carbon technologies. Relying on renewable energy free from fossil fuels leads to a substantial decrease of the life cycle greenhouse emissions of grid electricity (or *background mix carbon content*), since direct combustion does not occur, as well as a decrease in the extraction of fossil resources. Each of the low carbon technologies analyzed in this report typically lies in the range of 10-80 g CO_2 eq./kWh for climate change impact, while fossil fuel technologies roughly range from 400-1000 g CO_2 eq./kWh. The carbon content of the low carbon technologies is likely to further diminish in a significant low carbon technology adoption scenario, a phenomenon that can be observed by analyzing the LCA results of a technology across regions and over time. This positive feedback loop is evident in Figure 10.3. Conversely, the life cycle emissions of fossil fuel technologies are not affected by a change in the background mix, since GHG emissions are almost exclusively associated with direct emissions.

Carbon content is a criterion that is used to make a clear distinction between these two groups of technologies. In addition, power plants equipped with CCS infrastructure may be considered as a third group. The main objective of CCS technologies is to prevent CO_2 from being emitted into the atmosphere. From that perspective, plants with CCS can be considered as GHG emission mitigation technologies. However, aspects such as higher toxicity, energy penalty and storage safety classify CCS as a *transition technology*, i.e., interesting in a context of climate change mitigation, while being more harmful to the environment than low-carbon renewable technologies.

10.7.2 NON-CLIMATE IMPACTS

Environmental consequences of energy production are not always climate-related. A few environmental impact categories need to be mentioned.

Metal depletion: First and foremost, the current structure of global resource use is likely to change substantially with the global energy transition that is necessary to achieve to mitigate climate change. Electricity production technologies that are mainly made of metal, or that use rare metals (with a low known reserve) have a substantial impact on metal depletion. Figure 10.8 shows that technologies with short lifetimes are even more prone to impact the use of metal resources per unit of energy delivered to final users. Recycling scenarios have not been taken into account in this study, but they are certainly an aspect that needs to be investigated if we aim at lowering the material intensity of energy. The impact assessment method used in this study accounts for metal depletion and characterizes each metal in the LCI in terms of kg of depleted iron ore. This characterization is done with respect to the current global reserves and without accounting for recycling either. The ongoing assessment of metal criticality by the International Resource Panel will hopefully give rise to more robust and appropriate methods to assess the criticality of metals used in energy technologies. The results should be analyzed with these limitations in mind.

Land occupation: or land use, is defined by the area of land occupied over a period of time for a process or service to be delivered. Wide variations in land use, and in the structure of this land use, can be observed across technologies. From a biodiversity perspective, both the original state of the land occupied and the ecological value of the land during occupation are important. These two factors vary across technologies and are site-specific. We have refrained from trying to estimate the impact of land use transformation and land occupation, as occupied land types are almost entirely case-specific. More systematic work on the biodiversity impacts of land occupation associated with energy technologies is required.

Figure 10.9 shows that ground-mounted solar technologies are visibly occupying large areas. Solar radiation can indeed be harnessed only by spreading modules over a given area; the energy delivered is proportional to the surface occupied. This is not observed for roof-mounted PV panels since the land occupation is allocated to the services delivered by the building on which they are mounted. Aside from these examples, renewable technologies generally occupy relatively little land in comparison to fossil fuel technologies, which have large land use impacts mostly attributed to mining activities.

Human health: For non-climate-related human health, fossil fuel technologies appear to have the highest impact among the investigated technologies. These impacts are largely attributed to particulate matter emissions from combustion and human toxicity impacts. Power plants with CCS have even higher health impacts despite reduced direct emissions. Both energy penalty (more coal or gas is necessary to achieve the same power output when compared to a plant without CCS) and toxic chemical use for the capture process contribute to the high health effects. Comparatively, all analysed low carbon energy technologies have a low impact on human health.

Habitat change: Renewable energy sources cause habitat change. Methods to quantify and evaluate this habitat change are in their infancy and hence not included in the comparative LCA apart from a quantification of land use. A widespread deployment of renewable energy undoubtedly represents additional pressure on ecosystems and biodiversity. Care should hence be taken with respect to project selection and design, and measures to protect vulnerable species should be considered.

10.7.3 GENERAL CONCLUSIONS

Fossil fuel-fired power plants have consistently higher impacts than the investigated renewable energy technologies in nearly all impact categories. In the endpoint assessment, the difference is often one order of magnitude, and is on occasion even greater. The result is clearer and stronger than the authors expected. Land occupation is the only indicator where solar technologies (and wind if the entire wind park area is considered) have higher impacts than fossil fuel technologies, with the exception of coal power. However, renewable power plants require larger amounts of metals and other minerals.

CCS technology, while reducing climate impacts, increases most other impact categories by 20-100 per cent compared to modern, highly regulated power plants without CCS. According to one endpoint assessment, reducing global warming impacts negates the health and ecological impacts from the emissions of other compounds. However, this result requires further study for confirmation.

10.8 REFERENCES

Arvesen, A. and E. G. Hertwich. 2011. Environmental implications of large-scale adoption of wind power: A scenario-based life cycle assessment. *Environmental Research Letters* 6(4): 045102.

Bruckner, T., I. A. Bashmakov, Y. Mulugetta, H. Chum, A. D. L. V. Navarro, J. Edmonds, A. Faaij, B. Fungtammasan, A. Garg, E. Hertwich, D. Honnery, D. Infield, M. Kainuma, S. Khennas, S. Kim, H. B. Nimir, K. Riahi, N. Strachan, R. Wiser, and X. Zhang. 2014. Energy Systems. In *Climate Change 2014: Mitigation of Climate Change. Contribution of Working Group III to the Fifth Assessment Report of the Intergovernmental Panel on Climate Change*, edited by O. Edenhofer, et al. Cambridge, United Kingdom and New York, NY, USA: Cambridge University Press.

De Baan, L., C. L. Mutel, M. Curran, S. Hellweg, and T. Koellner. 2013. Land use in life cycle assessment: Global characterization factors based on regional and global potential species extinction. *Environmental Science and Technology* 47(16): 9281-9290.

Hertwich, E. G. 2013. Addressing Biogenic Greenhouse Gas Emissions from Hydropower in LCA. *Environmental Science & Technology* 47(17): 9604-9611.

Hertwich, E. G., T. Gibon, E. A. Bouman, A. Arvesen, S. Suh, G. A. Heath, J. D. Bergesen, A. Ramirez, M. I. Vega, and L. Shi. 2015. Integrated life-cycle assessment of electricity-supply scenarios confirms global environmental benefit of low-carbon technologies. *Proceedings of the National Academy of Sciences* 120(20): 6277–6282.

IEA. 2010. *Energy Technology Perspectives 2010: Scenarios and Strategies to 2050*. Paris: OECD/IEA.

IPCC. 2011. *Special Report on Renewable Energy Sources and Climate Change Mitigation. Prepared by Working Group III of the Intergovernmental Panel on Climate Change*. Edited by O. Edenhofer, et al. Cambridge, United Kingdom, and New York, NY, USA: Cambridge University Press.

IPCC. 2014. *Climate Change 2014: Mitigation of Climate Change. Contribution of Working Group III to the Fifth Assessment Report of the Intergovernmental Panel on Climate Change*. Edited by O. Edenhofer, et al. Cambridge, United Kingdom and New York, NY, USA: Cambridge University Press.

JRC. 2011. ILCD Handbook: Recommendations for Life Cycle Impact Assessment in the European context - based on existing models and factors. Brussels: European Commission.

Koellner, T., L. De Baan, T. Beck, M. Brandão, B. Civit, M. Margni, L. M. I Canals, R. Saad, D. M. De Souza, and R. Müller-Wenk. 2013. UNEP-SETAC guideline on global land use impact assessment on biodiversity and ecosystem services in LCA. *International Journal of Life Cycle Assessment* 18(6): 1188-1202.

Kounina, A., M. Margni, J. B. Bayart, A. M. Boulay, M. Berger, C. Bulle, R. Frischknecht, A. Koehler, L. Milà I Canals, M. Motoshita, M. Núñez, G. Peters, S. Pfister, B. Ridoutt, R. Van Zelm, F. Verones, and S. Humbert. 2013. Review of methods addressing freshwater use in life cycle inventory and impact assessment. *International Journal of Life Cycle Assessment* 18(3): 707-721.

Lim, S. S. et al., 2012. A comparative risk assessment of burden of disease and injury attributable to 67 risk factors and risk factor clusters in 21 regions, 1990-2010: A systematic analysis for the Global Burden of Disease Study 2010. *The Lancet* 380(9859): 2224-2260.

Mathers, C., G. Stevens, and M. Mascarenhas. 2009. *Global health risks: Mortality and burden of disease attributable to selected major risks*. Geneva: World Health Organization.

Moomaw, W., P. Burgherr, G. Heath, M. Lenzen, J. Nyboer, and A. Verbruggen. 2011. *Annex II: Methodology. In IPCC Special Report on Renewable Energy Sources and Climate Change Mitigation*. Cambridge, UK and New York, NY, USA: Cambridge University Press.

Pettersen, J. and E. G. Hertwich. 2008. Critical Review: Life-Cycle Inventory Procedures for Long-Term Release of Metals. *Environ. Sci. Technol.* 42(13): 4639-4647.

Sathaye, J., O. Lucon, A. Rahman, J. Christensen, F. Denton, J. Fujino, G. Heath, S. Kadner, M. Mirza, H. Rudnick, A. Schlaepfer, and A. Shmakin. 2011. Renewable Energy in the Context of Sustainable Energy. In *IPCC Special Report on Renewable Energy Sources and Climate Change Mitigation*, edited by O. Edenhofer, et al. Cambridge, United Kingdom and New York, NY, USA: Cambridge University Press.

Singh, B., A. H. Strømman, and E. G. Hertwich. 2012. Environmental Damage Assessment of Carbon Capture and Storage. *Journal of Industrial Ecology* 16(3): 407-419.

Strømman, A. H. and C. Solli. 2008. Applying Leontief's Price Model to Estimate Missing Elements in Hybrid Life Cycle Inventories. *Journal of Industrial Ecology* 12(1): 26-33.

Verones, F., S. Pfister, and S. Hellweg. 2013. Quantifying area changes of internationally important wetlands due to water consumption in LCA. *Environmental Science and Technology* 47(17): 9799-9807.